Bacterial Fish Pathogens: Disease of Farmed and Wild Fish

Third (Revised) Edition

Springer
*London
Berlin
Heidelberg
New York
Barcelona
Hong Kong
Milan
Paris
Santa Clara
Singapore
Tokyo*

B. Austin and D. A. Austin

Bacterial Fish Pathogens: Disease of Farmed and Wild Fish

Third (Revised) Edition

Published in association with
Praxis Publishing
Chichester

Professor B. Austin and Dr D. A. Austin
Department of Biological Sciences
Heriot-Watt University
Riccarton
Edinburgh EH14 4AS

SPRINGER–PRAXIS SERIES IN AQUACULTURE & FISHERIES
SERIES EDITOR Dr. Lindsay Laird, M.A., Ph.D., University of Aberdeen, UK
CONSULTANT EDITOR Dr. Selina Stead, B.Sc., M.Sc., Ph.D., Scottish Agricultural College, Aberdeen, UK

ISBN 1-85233-120-8 Springer-Verlag Berlin Heidelberg New York

British Library Cataloguing in Publication Data
Austin, B. (Brian), 1951-
     Bacterial fish pathogens : disease of farmed and wild fish.
     — 3rd rev. ed. — (Springer-Praxis series in aquaculture and
     fisheries)
     1.Bacterial diseases in fishes 2.Fishes — Microbiology
     I. Title II. Austin, D. A. (Dawn A.)
     639.3
     ISBN 1852331208

Library of Congress Cataloging-in-Publication Data
Austin, B. (Brian), 1951–
     Bacterial fish pathogens: disease of farmed and wild fish/B.
     Austin and D.A. Austin. — 3rd rev. ed.
          p.   cm.
          "Published in association with Praxis Publishing, Chichester."
          Includes bibliographical references (p.   ) and indexes.
          ISBN 1-85233-120-8 (alk. paper)
       1. Bacterial diseases in fishes. I. Austin, D. A. (Dawn A.)
       II. Title.
SH177.B3A97 1999
639.3—dc21   98-52790
                                                          CIP

Cover design: Jim Wilkie
Typesetting: Heather FitzGibbon, Christchurch, Dorset, UK

Printed on paper supplied by Precision Publishing Papers Ltd, UK

To Aurelia Jean

Table of contents

Brian Austin is a Professor, in the Department of Biological Sciences, Heriot-Watt University, a position he has held since 1993. From 1975 to 1978 he was Research Associate at the University of Maryland, USA, and from 1978 to 1984 was Head of Bacteriology at the Fish Diseases Laboratory in Weymouth, UK. He joined Heriot-Watt University as a Lecturer in Aquatic Microbiology in 1984.

Professor Austin gained a B.Sc. (1972) in Microbiology, a Ph.D. (1975) also in Microbiology, both from the University of Newcastle upon Tyne, and a D.Sc. (1992) from Heriot-Watt University. He was elected FRSA and Fellow of the American Academy of Microbiology, and is a member of the American Society of Microbiology, Society of Applied Bacteriology, Society of General Microbiology, European Association of Fish Pathologists, and the UK Federation of Culture Collections; and has written previous books on bacterial taxonomy, marine microbiology, methods in aquatic bacteriology, methods for the microbiological examination of fish and shellfish, and pathogens in the environment.

Dawn Austin is a Research Associate, in the Department of Biological Sciences, at Heriot-Watt University, a position she has held since 1986. Prior to this she was Research Assistant at the University of Maryland (1977–79), Lecturer in Microbiology, University of Surrey (1983–84), and Research Fellow of the Freshwater Biological Association, The River Laboratory, Dorset (1984–85).

Dr Austin gained a B.S. (1974) from City College, The City University of New York; an M.S. (1979) and a Ph.D. (1982) both from the University of Maryland.

Preface

This third revised edition of *Bacterial Fish Pathogens* is the successor to the original version, first published by Ellis Horwood Limited in 1987, and was planned to fill the need for an up-to-date comprehensive text on the biological aspects of the bacterial taxa which cause disease in fish. We have observed some well-thumbed copies throughout the world. The first and second editions have been quoted and mis-quoted extensively. One of the common mistakes has concerned authorship/editorship—we have written NOT edited the first two editions. Notwithstanding that, between 1987 and 1993, many new pathogens were described, and these were included in the second edition. Since then, there has been a dramatic increase in the numbers of scientific articles about bacterial fish diseases, although there has been a reduction in the number of so-called new fish pathogens. It is appropriate to consider the new information in a re-structured book, i.e. this third edition. We have included information on new pathogens and new developments on old established pathogens, such as *Aeromonas salmonicida* and *Renibacterium salmoninarum*. Because of the veritable deluge of new information, we have needed to be selective. For example, we have condensed details of the pathology of the diseases, insofar as there are excellent texts already available, which cover detailed aspects of the pathological conditions. Nevertheless, this third edition will hopefully meet the needs of the readership. Again, it is emphasized that most of the information still appertains to diseases of farmed, rather than wild, fish.

The scope of the book covers all of the bacterial taxa which have at one time or another been reported as fish pathogens. Of course, it is realized that some taxa are merely secondary invaders of already damaged tissues, whereas others comprise serious, primary pathogens. Shortcomings in the literature or gaps in the overall understanding of the subject have been highlighted.

In preparing the text, we have sought both advice and material from colleagues. We are grateful to the following for the supply of information and comment: Dr A. Baya, Dr E. Benediktsdóttir, Ms A. Carnahan, Dr C. Ghittino, Dr B. K. Gudmundsdóttir, Dr P. D. Smith, Professor G. Stemke and Mr J. Wood. Despina Athanasiadou and Alex Morrison helped with the literature searches. We acknowledge publishers for permission to use copyrighted material, as follows:

Figs 8.1, 8.2, 8.3 and 8.4 (*Journal of Fish Diseases*, Blackwells)
Figs 4.1, 4.2 and 4.3 (*Systematic and Applied Microbiology*)

Finally, we express gratitude to Mrs J. Currie for excellent secretarial support.

B. & D. A. Austin
Edinburgh, 1999

Abbreviations

Aer., *Aeromonas*
AFLP, amplified fragment length polymorphism
A-layer, the additional surface layer of *Aer. salmonicida*
ATCC, American Type Culture Collection, Rockville, Maryland, USA
BHIA, brain heart infusion agar
BKD, bacterial kidney disease
BMA, basal marine agar
bp, base pair
Car., *Carnobacterium*
CBB, coomassie brilliant blue agar
CDC, Centers for Disease Control, Atlanta, USA
CE, carp erythrodermatitis
Chrys., *Chryseobacterium*
Cit., *Citrobacter*
Cl., *Clostridium*
CLB, *Cytophaga*-like bacteria
CLED, cystine lactose electrolyte deficient agar
Cor., *Corynebacterium*
Cyt., *Cytophaga*
DNA, deoxyribonucleic acid
ECP, extracellular product
EDTA, ethylene diamine tetraacetic acid
Edw., *Edwardsiella*
ELISA, enzyme linked immunosorbent assay
Ent., *Enterobacter*
En., *Enterococcus*
ERM, enteric redmouth
Esch., *Escherichia*
Eu., *Eubacterium*
FAT, fluorescent antibody test
FCA, Freunds complete adjuvant

FIA, Freunds incomplete adjuvant
Fla., Flavobacterium
Fle., Flexibacter
GCAT, glycerophospholipid:cholesterol acyltransferase
G+C, guanine plus cytosine
H., Haemophilus
Haf., Hafnia
iFAT, indirect fluorescent antibody test
i.m., intramuscular
i.p., intraperitoneal
IROMP, iron-regulated outer membrane protein
J., Janthinobacterium
HG, Hybridization Group
kb, kilobase
kDa, kiloDalton
KDM2, kidney disease medium 2
LD_{50}, lethal dose 50%, i.e. the dose needed to kill 50% of the population
LD_{100}, lethal dose 100%
Lis., Listeria
LPS, lipopolysaccharide
mDa, megaDalton
MHC, Mueller-Hinton agar supplemented with 0.1% (w/v) L-cysteine hydrochloride
MIC, minimum inhibitory concentration
Mor., Moraxella
MRVP, methyl red Voges Proskauer
MSS, marine salts solution
Myc., Mycobacterium
NCBV, non-culturable but viable
NCIMB, National Collection of Industrial and Marine Bacteria, Aberdeen, Scotland
Nec., Necromonas
Noc., Nocardia
OMP, outer membrane protein
p57, 57 kDa protein (of *Ren. salmoninarum*)
Pa., Pasteurella
PAGE, polyacrylamide gel electrophoresis
PBS, phosphate buffered saline
PCR, polymerase chain reaction
PFGE, pulsed-field gel electrophoresis
Ph., Photobacterium
PMSF, phenylmethyl-sulphonyl fluoride
Pr., Proteus
Ps., Pseudomonas
PFGE, pulsed-field gel electrophoresis
RAPD, randomly amplified polymorphic DNA
Ren., Renibacterium

RLO, *Rickettsia*-like organism
RPS, relative percent survival
rRNA, ribosomal ribonucleic acid
RTFS, rainbow trout fry syndrome
S_D, Dice coefficient
S-layer, surface layer
Sal., Salmonella
SDS, sodium dodecyl sulphate
Ser., Serratia
SKDM, selective kidney disease medium
Sta., Staphylococcus
Str., Streptococcus
TCBS, thiosulphate citrate bile salts sucrose agar
TSA, tryptone soya agar
TSB, tryptone soya broth
V., Vibrio
Vag., Vagococcus
VAM, *Vibrio anguillarum* medium
Y., Yersinia

1

Introduction

It is apparent that representatives of many bacterial taxa have, at one time or another, been associated with fish diseases. However, not all of these bacteria constitute primary pathogens. Many should be categorized as opportunistic pathogens, which colonize and cause disease in already damaged hosts. The initial weakening process may involve pollution or a natural physiological state (e.g. during the reproductive phase) in the life cycle of the fish. There remain doubts about whether other bacteria, e.g. *Aeromonas bestiarum*, should be considered as fish pathogens. In such cases, the supportive evidence is weak or non-existent. However, it is readily apparent that there is great confusion about the precise meaning of disease. A definition, from the medical literature, states that:

> a disease is the sum of the abnormal phenomena displayed by a group of living organisms in association with a specified common characteristic or set of characteristics by which they differ from the norm of their species in such a way as to place them at a biological disadvantage. (Campbell *et al.*, 1979)

This legalistic definition is confusing, and the average reader will be little wiser about its meaning. Dictionary definitions of disease are somewhat better, and include 'an unhealthy condition' and 'infection with a pathogen [= something that causes a disease]'. One conclusion is that disease is a complex phenomenon, leading to some form of measurable damage to the host. Yet, it is anticipated that there might be profound differences between scientists about just what constitutes a disease. Fortunately, infection by micro-organisms is one aspect of disease that finds ready acceptance within the general category of disease.

For his detailed treatise on diseases of marine animals, Kinne (1980) considered that disease may be caused by genetic disorders, physical injury, nutritional imbalance, pathogens and/or pollution. This list of possible causes illustrates the complexity of disease. An initial conclusion is that disease may result from biological (= biotic) factors, such as pathogens, as well as abiotic causes, e.g. the emotive issue of pollution. Disease may also be categorized in terms of epizootiology (Kinne, 1980), namely as:

— *sporadic* diseases, which occur sporadically in comparatively small numbers of a fish population;

—*epizootics*, which are large-scale outbreaks of communicable disease occurring temporarily in limited geographical areas;

— *panzootics*, which are large-scale outbreaks of communicable disease occurring over large geographical areas;

— *enzootics*, which are diseases persisting or re-occurring as low-level outbreaks in certain areas.

The study of fish diseases has concentrated on problems in fish farms (= aquaculture), where outbreaks either begin suddenly, progress rapidly often with high mortalities, and disappear with equal rapidity (= *acute* disease) or develop more slowly with less severity, but persist for greater periods (= *chronic* disease).

This text will deal with the diseases caused by bacteria. However, at this stage, it is relevant to emphasize that disease is not necessarily caused by single bacterial taxa. Instead, there may well be synergistic interactions between two or more taxa. Also, disease may be the outcome of a delicate interaction between the host (= fish), the disease-causing situation (= pathogen) and external stressor(s) (= unsuitable changes in the environment; poor hygiene; stress) (Fig. 1.1). Before the occurrence of clinical signs of disease, there may be demonstrable damage to/weakening of the host. Yet, all too often, the isolation of bacteria from an obviously diseased fish is taken as evidence of infection. Koch's Postulates may be conveniently forgotten in the rush to publication.

So, what are the bacterial fish pathogens? A comprehensive list of all the bacteria, which have been considered to represent fish pathogens, has been included in Table 1.1. Some genera, e.g. *Vibrio*, include many species that are acknowledged to be pathogens of freshwater and/or marine fish species. Some taxa (highlighted by quotation marks), namely 'Catenabacterium', 'Haemophilus piscium' and 'Myxobacterium' are of doubtful taxonomic validity. Others, such as *Proteus rettgeri* and *Staphylococcus epidermidis*, are of questionable significance in fish pathology insofar as their recovery from diseased animals has been, at most, sporadic. A heretical view would be that enteric bacteria, e.g. *Proteus*, comprise contaminants from water or from the gastro-intestinal tract. Clearly, many of the bacterial pathogens are members of the normal microflora of water and/or fish. Other pathogens have been associated only with clinical diseases or covertly infected (asymptomatic) fish. Examples of these 'obligate' pathogens include *Aeromonas salmonicida* and *Renibacterium salmoninarum*, the causal agents of furunculosis and bacterial kidney disease (BKD), respectively. In later chapters, it will be questioned whether or not bacteria should be considered as obligate pathogens of fish. It is a personal view that the inability to isolate an organism from the aquatic environment may well reflect inadequate recovery procedures. Could the organism be dormant/damaged/ senescent in the aquatic ecosystem; a concept which has been put forward for other water-borne organisms (Stevenson, 1978)?

It is undesirable that any commercially important species should suffer the problems of disease. Unfortunately, the aetiology of bacterial diseases in the wild is often improperly understood. Moreover, it seems that little if anything may be done to aid wild fish stocks, except, perhaps, by controlling pollution of the rivers and seas, assuming that

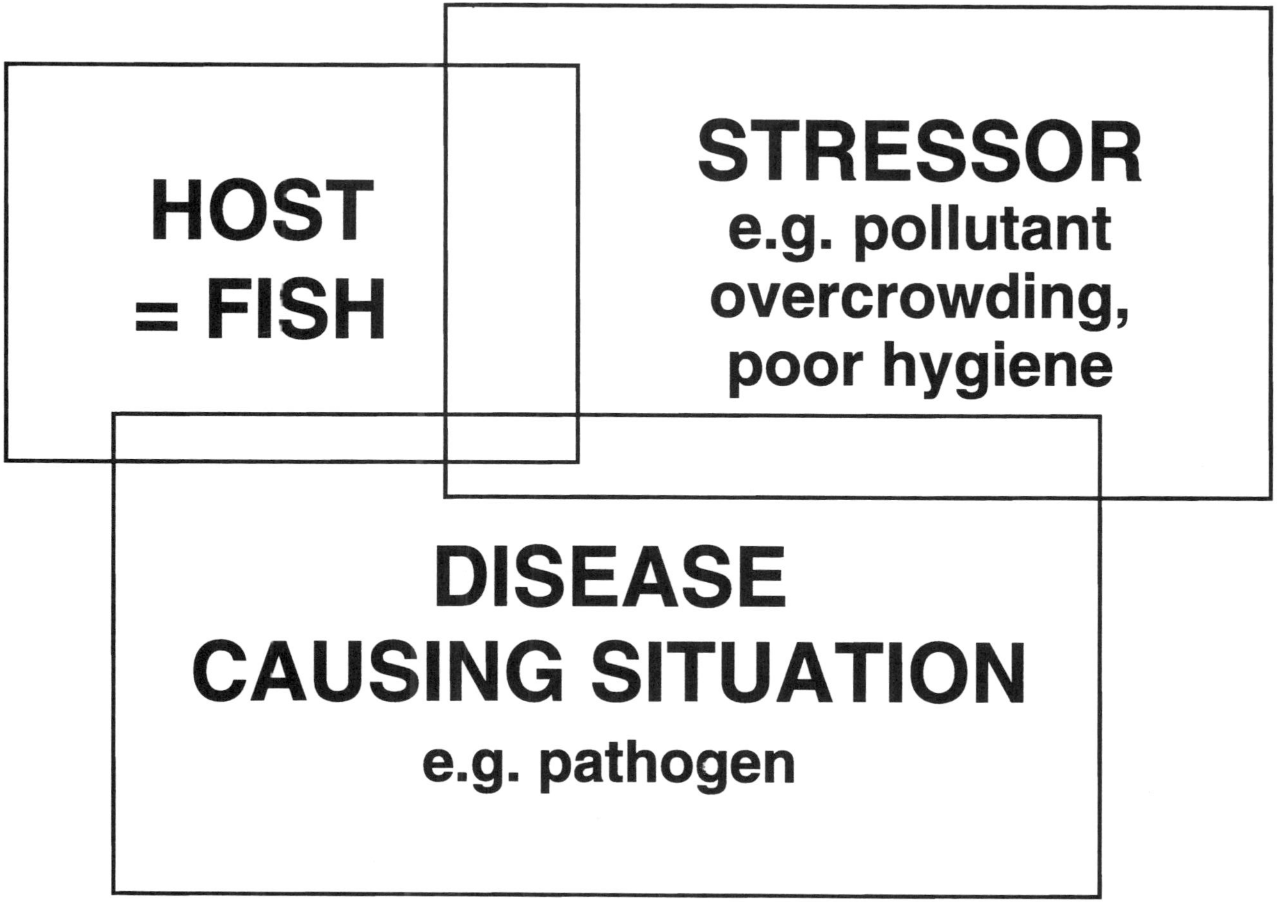

Fig. 1.1. Interactions leading to the occurrence of disease (based on Austin, 1999).

Table 1.1. Bacterial pathogens of freshwater and marine fish

Pathogen	Disease	Host range	Geographical distribution
Anaerobes			
'Catenabacterium' sp.	—	grey mullet (*Mugil auratus*), redfish (*Sebastes* sp.)	USA
Clostridium botulinum	botulism	salmonids	Denmark, England, USA
Eubacterium tarantellae	eubacterial meningitis	striped mullet (*Mugil cephalus*)	USA
Gram-positive bacteria—the 'lactic acid' bacteria			
Carnobacterium piscicola	lactobacillosis, pseudokidney disease	salmonids	North America, UK
Lactobacillus spp.	lactobacillosis, pseudokidney disease	salmonids	North America, UK
Lactococcus garvieae (= *Enterococcus seriolicida*)	streptococcicosis/streptococcosis	many fish species	Australia, Europe, Israel, Japan, Saudi Arabia, South Africa, USA
Lactococcus piscium	lactobacillosis, pseudokidney disease	rainbow trout (*Oncorhynchus mykiss*)	North America
Streptococcus difficilis (= *Str. agalactiae*)	meningo-encephalitis	carp (*Cyprinus carpio*), rainbow trout, tilapia (*Oreochromis* spp.)	Israel
Enterococcus (*Streptococcus*) *faecalis* subsp. *liquefaciens*	—	rainbow trout, catfish	Italy
Streptococcus iniae (*Str. shiloi*)	meningo-encephalitis, streptococcicosis/streptococcosis	various freshwater fish species	Australia, Europe, Israel, Japan, Saudi Arabia, South Africa, USA
Streptococcus milleri	—	koi carp (*Cyprinus carpio*)	UK
Streptococcus parauberis	streptococcicosis/streptococcosis	turbot (*Scophthalmus maximus*)	Spain
Vagococcus salmoninarum	lactobacillosis, pseudokidney disease, peritonitis, septicaemia	Atlantic salmon, brown trout, rainbow trout	Australia, France, North America

Table 1.1. (*continued*)

Pathogen	Disease	Host range	Geographical distribution
Aerobic Gram-positive rods and cocci			
Bacillus spp.	septicaemia	various freshwater fish species	Nigeria
Bacillus cereus	branchio-necrosis	carp (*Cyprinus* sp.), striped bass (*Morone saxatilis*)	USA
Bacillus mycoides	ulceration	channel catfish (*Ictalurus punctatus*)	Poland, USA
Bacillus subtilis	branchio-necrosis	carp	Poland
Corynebacterium aquaticum	exophthalmia	striped bass	USA
Coryneform bacteria	'corynebacteriosis'	salmonids	England
Micrococcus luteus	micrococcosis	rainbow trout	England
Mycobacterium spp. (Myc. abscessus, Myc. anabanti, Myc. chelonei subsp. *piscarium, Myc. fortuitum, Myc. marinum, Myc. neoaurum, 'Myc. piscium', Myc. platypoecilus', 'Myc. ranae', 'Myc. salmoniphilum', Myc. scrofulaceum, Myc. simiae)*	mycobacteriosis (fish tuberculosis)	most fish species	worldwide
Nocardia spp. (*Noc. asteroides, Noc. seriolae)*	nocardiosis	most fish species	worldwide
Planococcus sp.	—	salmonids	England
Renibacterium salmoninarum	bacterial kidney disease (BKD; Dee disease; corynebacterial kidney disease)	salmonids	Europe, Japan, North and South America

Table 1.1. (*continued*)

Pathogen	Disease	Host range	Geographical distribution
Rhodococcus sp.	ocular oedema	chinook salmon (*O. tshawytscha*)	Canada
Staphylococcus aureus	eye disease	silver carp (*Hypophthalmichthys molitrix*)	India
Staphylococcus epidermidis	—	red sea bream (*Chrysophrus major*), yellowtail (*Seriola quinqueradiata*)	Japan
Streptomyces salmonis (= *Streptoverticillum salmonis*)	streptomycosis	salmonids	USA
Aeromonadaceae representatives			
Aeromonas allosaccharophila	—	elvers	Spain
Aeromonas bestiarum	—	—	US
Aeromonas caviae	septicaemia	Atlantic salmon (*Salmo salar*)	Turkey
Aeromonas hydrophila (= *Aer. liquefaciens, Aer. punctata*)	haemorrhagic septicaemia, motile aeromonas septicaemia, redsore disease, fin rot	many freshwater fish species	worldwide
Aeromonas jandaei	—	eel (*Anguilla* sp.)	Spain
Aeromonas salmonicida (sub-species *achromogenes, masoucida, salmonicida* and *smithia*) {= *Haemophilus piscium*}	furunculosis, carp erythrodermatitis, ulcer disease	salmonids, cyprinids, and marine flatfish	worldwide
Aeromonaas sobria	—	gizzard shad (*Dorosoma cepedianum*)	USA

Table 1.1. (*continued*)

Pathogen	Disease	Host range	Geographical distribution
Enterobacteriaceae representatives			
Citrobacter freundii	—	salmonids, sunfish (*Mola mola*), carp	Europe, India, USA
Edwardsiella ictaluri	enteric septicaemia of catfish	channel catfish, danio (*Danio devario*)	USA
Edwardsiella tarda (Paracolobactrum anguillimortiferum, Edw. anguillimortifera)	redpest, edwardsiellosis, emphysematous putrefactive disease of catfish	various freshwater fish species	Japan, USA
Enterobacter agglomerans	—	dolphin fish (*Coryphaena hippurus*)	USA
Escherichia vulneris	septicaemia	various freshwater fish species	Turkey
Hafnia alvei	haemorrhagic septicaemia	cherry salmon (*O. masou*), rainbow trout	Bulgaria, England, Japan
Klebsiella pneumoniae	fin and tail disease	rainbow trout	UK
Proteus (Providencia) rettgeri	—	silver carp	Israel
Salmonella choleraesuis subsp. *arizonae (= Sal. arizonae)*	septicaemia	piarucu (*Arapaima gigas*)	Japan
Serratia liquefaciens	septicaemia	Atlantic salmon, turbot	France, Scotland
Serratia marcescens	—	white perch (*Morone americanus*)	USA
Serratia plymuthica	—	rainbow trout	Scotland, Spain
Yersinia intermedia	—	Atlantic salmon	Australia
Yersinia ruckeri	enteric redmouth (ERM), salmonid blood spot	salmonids	Australia, Europe, North America

Table 1.1. (*continued*)

Pathogen	Disease	Host range	Geographical distribution
***Cytophaga–Flavobacterium–Flexibacter* group**			
Chryseobacterium balustinum (= Flavobacterium balustinum)	flavobacteriosis	marine fish	USA
Chryseobacterium scophthalmum (= Flavobacterium scophthalmum)	gill disease; generalized septicaemia	turbot	Scotland
'Cytophaga rosea'	gill disease	salmonids	Europe, USA
Flavobacterium branchiophilum	gill disease	salmonids	Europe, Korea, Japan, USA
Flavobacterium columnare (= Flexibacter/Cytophaga columnaris)	columnaris, saddleback disease	many freshwater fish species	worldwide
Flavobacterium hydatis (= Cytophaga aquatilis)	gill disease	salmonids	Europe, USA
Flavobacterium johnsoniae (= Cytophaga johnsonae)	gill disease, skin disease bacterial gill disease	barramundi (*Lates calcarifer*) salmonids	Australia, France Europe, USA
Flavobacterium psychrophilum (= Cytophaga psychrophila)	coldwater disease, rainbow trout fry syndrome	salmonids	Australia, Europe, USA
Flexibacter maritimus	gill disease, black patch necrosis	many marine fish species	Europe, Japan
Flexibacter ovolyticus	larval and egg mortalities	halibut (*Hippoglossus hippoglossus*)	Norway
Myxococcus piscicola	gill disease	green carp (*Ctenopharyngodon idelluls*)	China
Sporocytophaga sp.	saltwater columnaris	salmonids	Scotland, USA

Table 1.1. (*continued*)

Pathogen	Disease	Host range	Geographical distribution
Pseudomonadaceae representatives			
Pseudomonas anguilliseptica	red spot (Sekiten-byo)	rainbow trout, marine fish species, and particularly eels (*Anguilla anguilla, A. japonica*)	Finland, France, Japan, Scotland, Spain
Pseudomonas chlororaphis	—	Amago trout (*Oncorhynchus rhodurus*)	Japan
Pseudomonas fluorescens	generalized septicaemia	most fish species	worldwide
Pseudomonas pseudo-alcaligenes	skin ulceration	rainbow trout	Scotland
Pseudomonas putida	haemorrhagic ascites	ayu (*Plecoglossus altivelis*)	Japan
Vibrionaceae representatives			
Moritella marina (= *Vibrio marinus*)	skin lesions	Atlantic salmon	Iceland
Photobacterium damselae	vibriosis	damsel fish (*Chromis punctipinnis*), sharks, turbot, yellowtail	Europe, USA
Photobacterium damselae subsp. *piscicida* (= *Pasteurella piscicida*)	pasteurellosis, pseudotuberculosis	gilt-head sea bream (*Sparus aurata*), striped bass (*Morone saxatilis*), white perch (*Roccus americanus*), yellowtail	Europe, Japan, USA
Pleisiomonas shigelloides	—	African catfish (*Heterobranchus bidorsalis*), eel, gourami (*Osphyronemus gourami*), rainbow trout, sturgeon (*Acipenser sturio*)	Germany, Portugal

Table 1.1. (*continued*)

Pathogen	Disease	Host range	Geographical distribution
Vibrionaceae representatives			
Shewanella putrefaciens	septicaemia	rabbit fish (*Siganus rivulatus*)	Saudi Arabia
Vibrio alginolyticus	eye disease, septicaemia	gilt-head sea bream, grouper (*Epinephelus malabanicus*), sea bream (*Sparus aurata*)	Asia, Europe, Israel
Vibrio anguillarum (= *Listonella anguillara*)	vibriosis	most marine fish species	worldwide
V. cholerae (non-O1)	septicaemia	ayu, goldfish (*Carassius aurata*)	Australia, Japan
V. fischeri	—	gilt-head sea bream, turbot	Spain
V. furnissii	—	eel	Spain
Vibrio harveyi (= *V. carchariae*)	eye disease (blindness), vasculitis	gilt-head sea bream, common snook (*Centropomus undecimalis*), milkfish, sharks (*Carcharhinus plumbeus, Negaprion brevirostris*)	Europe, Japan, USA
V. ichthyoenteri	intestinal necrosis	Japanese flounder (*Paralichthys olivaceus*)	Japan
V. logei	skin lesions	Atlantic salmon	Iceland
V. ordalii	vibriosis	most marine fish species	worldwide
V. pelagius	—	turbot	Spain
V. salmonicida	coldwater vibriosis, Hitra disease	Atlantic salmon	Canada, Norway, Scotland
V. splendidus	septicaemia	gilt-head sea bream, turbot	Spain
V. trachuri	—	horse mackerel	Japan
'*V. viscosus*'	winter ulcer disease/syndrome	Atlantic salmon	Iceland, Norway
V. vulnificus	septicaemia	eel	Europe, Japan, USA

Table 1.1. (*continued*)

Pathogen	Disease	Host range	Geographical distribution
Moraxellaceae representatives			
Acinetobacter sp.	acinetobacter disease	Atlantic salmon, channel cat-fish	Norway, USA
Moraxella sp.	—	striped bass	USA
Halomonadaceae representative			
Halomonas (= *Deleya*) *cupida*	—	black sea bream (*Acanthopagrus schlegeli*)	Japan
Small, morphologically simple bacteria			
Mycoplasma mobile	red disease	tench (*Tinca tinca*)	USA
Piscirickettsia salmonis	coho salmon syndrome, salmonid rickettsial septicaemia	salmon	Canada, Chile, Norway
Miscellaneous pathogens			
Aquaspirillum sp.	epizootic ulcerative syndrome	snakeheads (*Ophicephalus striatus*) and catfish (*Clarias batrachus*)	Thailand
Janthinobacterium lividum	anaemia	rainbow trout	Scotland
Streptobacillus	—	Atlantic salmon	Ireland

Names in quotation marks are not included in the Approved Lists of Bacterial Names (Skerman *et al.*, 1980) or their supplements.

when environmental quality deteriorates this influences disease cycles. In contrast, much effort has been devoted to controlling diseases of farmed fish.

- The list of fish pathogens has extended substantially since 1980, and encompasses a larger number of *Aeromonas* and *Vibrio* species.
- A question mark hangs over the significance of some organisms, e.g. *Aer. bestiarum*, to fish pathology—are they truly pathogens or chance contaminants?
- There has been considerable improvement in the taxonomy of some groups, notably *Cytophaga–Flavobacterium–Flexibacter* and the causal agents of streptococcicosis/streptococcosis.

2

Characteristics of the diseases

A. ANAEROBES

Clostridium botulinum

The first report of botulism as a disease of fish came from Denmark, with the detective work of Huss and Eskilden (1974). It was shown that botulism was a chronic disease, termed 'bankruptcy disease', of farmed trout. The causal agent was recognized as *Cl. botulinum* type E. Subsequently, the disease was found on one farm of rainbow trout in Great Britain (Cann and Taylor, 1982; 1984) and has been similarly identified among farmed coho salmon in the USA (Eklund *et al.*, 1982). Characteristic disease symptoms appear to be very vague. It has been observed that fish exhibited sluggish, erratic swimming, appeared to be listless, and may alternately float and sink, before showing temporary rejuvenation. This pattern was repeated until death eventually ensued (Cann and Taylor, 1982).

Eubacterium tarantellae

The term 'eubacterial meningitis' was coined for this disease (Winton *et al.*, 1983). Essentially, it is a neurological condition, with which the infected fish display twirling until death results (Udey *et al.*, 1976). Cells of the organism may be readily observed in sections of brain tissue. There was little, if any, external pathology observed. It is interesting to note from Udey's work that some fish were also infected with other organisms, namely, trematodes (*Bucephalus* sp.) and *Vibrio* spp. which were found in some fish; whereas approximately 20% possessed low numbers of *Myxosoma cephalus* spores in the brain cavity. Therefore, it is relevant to enquire whether this anaerobe represented a primary or secondary pathogen during the outbreak of disease in Biscayne Bay.

B. GRAM-POSITIVE BACTERIA—THE 'LACTIC ACID' BACTERIA

Carnobacterium piscicola **and the lactobacilli**

According to Ross and Toch (1974), the abdomen of moribund fish were distended because of the presence of ascitic fluid. However, it was readily admitted that mortalities could not be directly attributed to the 'pathogen'. The subsequent report of Cone (1982) indicated that the condition was stress mediated, insofar as it was recognized mostly in post-spawning fish. In these specimens, there was an accumulation of ascitic fluid in the peritoneal cavity, and extensive damage in the liver, kidney and spleen. Fin rot and other external signs of disease were absent. However, petechial haemorrhages in the muscle and hyperaemic air bladder were observed in some fish. The heart and gills appeared normal. Disease signs, similar to those described by Cone (1982), were observed in diseased rainbow trout in England (B. Austin, unpublished data). According to Hui *et al.* (1984), the disease occurred in fish more than one year old, which may have undergone stress, namely handling and spawning. Disease symptoms were varied, including septicaemia, the abdomen distended with ascitic fluid, muscle abscesses, blood blisters just beneath the skin, and internal haemorrhaging.

Enterococcus (Streptococcus) faecalis subsp. *liquefaciens*
Although a question mark remains over the accuracy of the identification of this pathogen, its inclusion is justified for the sake of completion. Also, this is the first indication of streptococci as fish pathogens in Croatia. Farmed brown bullhead (*Amiurus nebulosus*) developed deep ulcers predominantly between the dorsal and caudal fins. Internal organs displayed unspecified changes, and fluid was present in the digestive tract. There was haemorrhaging at the anus. Gram-positive cocci were observed in the kidney and liver (Teskeredzic *et al.*, 1993).

Lactococcus garvieae
There appears to be pronounced variation in disease signs, including exophthalmia and distended abdomen (these are also common features with BKD), haemorrhaging in the eye (this is a characteristic of salmonid blood spot, which is synonymous with ERM), in the opercula and at the base of the fins, and on the surface (this could be confused with vibriosis) (Bullock, 1981; Kusuda *et al.*, 1991), and darkening of the skin (Barham *et al.*, 1979). Moribund fish swim erratically just below the surface of the water.

Internal signs of disease were absent in golden shiners, although raised lesions were apparent on the body surface (Robinson and Meyer, 1966). Yellowtails were damaged in the liver, kidney, spleen and intestine, and there was a concomitant accumulation of ascitic fluid in the peritoneal cavity (Kusuda *et al.*, 1976a; 1991; Ugajin, 1981). Marine fish showed pronounced enteritis, pale livers and blood in the peritoneal cavity, although the kidneys were apparently unaffected (Plumb *et al.*, 1974).

Lactococcus piscium
The disease condition encompasses lactobacillosis or pseudokidney disease of rainbow trout.

Streptococcus difficilis (= Str. agalactiae)
Str. difficilis was named as a result of an outbreak of disease in St. Peter's fish (*Tilapia*) and rainbow trout within Israel during 1986. The disease spread rapidly, and caused severe economic losses in the farmed fish (Eldar *et al.*, 1994). Diseased tilapia were lethargic, swam erratically, and showed signs of dorsal rigidity. In rainbow trout, the disease signs were consistent with a septicaemia, with brain damage (Eldar *et al.*, 1994).

Streptococcus iniae (= Str. shiloi)

Str. shiloi was named as a result of an outbreak of disease in St. Peter's fish and rainbow trout within Israel during 1986. The disease spread rapidly, and caused severe economic losses in the farmed fish (Eldar *et al.*, 1994). Disease tilapia were lethargic, swam erratically, and showed signs of dorsal rigidity. In rainbow trout, the disease signs were consistent with a septicaemia, with brain damage (Eldar *et al.*, 1994). In a separate development, an organism, subsequently identified as *Str. iniae*, was diagnosed as causing disease in two tanks of hybrid striped bass in a commercial farm, using recirculating freshwater, in the USA (Stoffregen *et al.*, 1996).

Streptococcus milleri

During 1992, Gram-positive chaining cocci and atypical *Aer. salmonicida* were recovered from newly imported koi carp, which displayed pronounced surface ulcers of 4–20 mm in diameter on the flank or tail. Internal damage was not recorded (Austin and Robertson, 1993).

Streptococcus parauberis

This form of streptococcosis was originally recognized in farmed turbot (weight: 0.8–2 kg) from five sites in northern Spain during 1993 and 1994 (Doménech *et al.*, 1996). Overall, the farms, which used the same fish food, reported losses of 0.1–5%. Disease signs, which were more severe during summer, included weight loss, haemorrhaging on the anal and pectoral fins, petechial haemorrhages on the abdomen, bilateral exophthalmia, haemorrhaging and pus in the eyes, pale liver, congested liver and spleen, ascites, and mucohaemorrhagic enteritis (Doménech *et al.*, 1996).

Vagococcus salmoninarum

In France, the organism was attributed to significant losses, i.e. up to 50% in a year, in rainbow trout farmed at low water temperature (Michel *et al.*, 1997). Disease signs included listless behaviour, impaired swimming, unilateral exophthalmia, external haemorrhages, petechial haemorrhages on the gills, and enlarged liver and spleen (Michel *et al.*, 1997) and peritonitis (Schmidtke and Carson, 1994).

C. AEROBIC GRAM-POSITIVE RODS AND COCCI

Bacillus spp.

The initial outbreak of disease led to mortalities of 10–15% of the fish in earthen ponds (Oladosu *et al.*, 1994). Diseased fish were characterized by weakness, lethargy, emaciation and generalized necrotizing dermatitis, with death occurring in a few days. Blood-tinged fluid was present in the peritoneal cavity. Petechia and focal necrosis was evident in the liver and kidney. The spleen was enlarged, soft and friable. The myocardium was described as soft and flabby. The stomach was hyperaemic (Oladosu *et al.*, 1994). Gram-positive rods of 1–4 μm in length were observed. It should be emphasized that skin lesions revealed the presence of *Aeromonas* and *Fla. columnare*.

Bacillus cereus

The organism has been associated with branchionecrosis in common carp (Pychynski *et al.*, 1981) and striped bass (Baya *et al.*, 1992a).

Bacillus mycoides

An epizootic occurred in channel catfish in Alabama during 1992. The fish were darker in colour, inappetent, displayed pale areas or ulcers on the dorsal surface, focal necrosis of the epaxial muscle, and opaque muscle (Goodwin *et al.*, 1994). Histopathological examination revealed the presence of chains of Gram-positive rods.

Bacillus subtilis

The organism has been associated with branchionecrosis in common carp (Pychynski *et al.*, 1982).

Corynebacterium aquaticum

The organism was associated with three-year-old striped bass in an experimental aquaculture facility in Maryland, USA, during December, 1990. Fish displayed pronounced bilateral exophthalmia, and contained the organism in brain tissue (Baya *et al.*, 1992b). Fish stopped feeding, swam more slowly, and died (at this point the eyes were ruptured). Internally, the only disease sign was that the brain was haemorrhagic, and the cranium was full of blood.

Micrococcus luteus

Disease signs were consistent with the notion of RTFS. Thus, moribund fish, in the size range of 0.5–5.0 g, displayed exophthalmia, pale gills, enhanced skin pigmentation and swollen abdomen. Internally, the kidney was swollen, the spleen was pale and elongated, and some ascitic fluid was present in the peritoneal cavity (Austin and Stobie, 1992a).

Mycobacterium spp.

Mycobacteriosis (fish tuberculosis) is a chronic progressive disease, with various external signs, including emaciation, inflammation of the skin, exophthalmia, open lesions and ulceration (e.g. Lansdell *et al.*, 1993). Internally, greyish white nodules (granulomas) develop on various organs, particularly the liver, kidney, heart and spleen (Dulin, 1979; Van Duijn, 1981). The disease may take several years to progress from the asymptomatic state to clinical illness. Initially, the pigment may fade, and the fish appear sluggish with loss of appetite. If the skin is affected, blood spots develop with the ultimate formation of ulcers. In addition, fin and tail rot and the loss of scales may be seen.

Myc. abscessus was associated with 2–27-month-old Japanese Meduka (*Oryzias latipes*), which had been cultured in the USA for aquatic toxicology testing (Teska *et al.*, 1997). During a routine examination, granulomas, notably in the buccal cavity and vent, and a few acid-fast bacteria were noted in <1% of the otherwise healthy fish. On clinically diseased fish, the disease signs would include listlessness, inappetence, swollen abdomen and visible granulomas (Teska *et al.*, 1997).

Myc. marinum was first recognized from the liver, spleen and kidney of tropical coral fish kept in the Philadelphia Aquarium (Aronson, 1926). As the name implies, the organism was considered to be only pathogenic to marine fish. But, it is now recognized to infect both marine and freshwater fish and also human beings (Van Duijn, 1981). Currently, *Myc. marinum* is a major constraint on the farming of sea bass in Israel, leading to stunting and therefore loss of market value of the infected fish (Knibb *et al.*, 1993). It is relevant to note that the first case of 'tuberculosis' reported in wild stocks, was in cod (*Gadus morhua*) landed at Fleetwood (UK), although isolation of the pathogen was not achieved (Alexander, 1913).

Myc. neoaurum has been associated with ocular lesions (oedema; exophthalmia) in Atlantic salmon (Bachman *et al.*, 1990). Nodules may form in the muscle, where they are visible on the outside of the fish. These nodules may burst, releasing bacteria into the aquatic environment. Internally, nodules may develop on the organs, leading to emaciation, or oedema or peritonitis may ensue. Infection may spread to the skeleton, in which case deformities become apparent. Death will ultimately occur (Van Duijn, 1981).

Nocardia spp.

It is appreciated that nocardiosis may be problematical in fresh water (Valdez and Conroy, 1963; Conroy, 1964; Snieszko *et al.*, 1964b; Heuschmann-Brunner, 1965a; Campbell and MacKelvie, 1968; Ghittino and Penna, 1968) and marine fish (Wood and Ordal, 1958). Symptoms similar to mycobacteriosis develop in affected fish. All age groups may be infected, with lesions, manifested as small white spots, present in the dermis, muscle, gills and internal organs.

Planococcus sp.

The organism has been associated with small off-white (2–4 mm diameter) round recessed spots on the heads of Atlantic salmon (Austin *et al.*, 1988). In large rainbow trout (average weight = 500 g), the only disease sign was the presence of watery kidney and small amounts of ascitic fluid in the peritoneal cavity (B. Austin, unpublished data). In addition during 1990, the organism was associated with two populations of rainbow trout fry deemed to have RTFS. These animals were anaemic, with pale gills, swollen kidney, pale liver and elongated spleen (Austin and Stobie, 1992a).

Renibacterium salmoninarum

In 1935, the disease was reported in the USA, where it occurred in hatchery reared brown trout (*S. trutta*), rainbow trout (*Oncorhynchus mykiss*) and brook trout (*Salvelinus fontinalis*) (Belding and Merrill, 1935). Additional evidence indicated the presence of BKD in chinook salmon (*Oncorhynchus tshawytscha*), coho salmon (*O. kisutch*) and sockeye salmon (*O. nerka*) (Rucker *et al.*, 1951). Since then, the disease has been reported to occur in 13 species of salmonids in Canada, Chile (especially during transfer of chinook salmon from fresh to seawater; Uribe *et al.*, 1995), England, France, Germany, Iceland, Italy, Spain, USA and Yugoslavia (see Fryer and Sanders, 1981; Hoffman *et al.*, 1984; Uribe *et al.*, 1995). During 1997, BKD was demonstrated for the first time in Denmark (Lorenzen *et al.*, 1997). Workers have highlighted the presence of the disease in farmed salmonid stocks but only occasionally has it been found in wild fish populations

(Rucker *et al.*, 1951, 1953; Smith, 1964; Pippy, 1969; Evelyn *et al.*, 1973; Wood, 1974; Ellis *et al.*, 1978; Paterson *et al.*, 1978, 1981; Mitchum and Sherman, 1981; Banner *et al.*, 1986). External signs include exophthalmia, lesions in the eyes, swollen abdomen (full of ascitic fluid), blood-filled blisters on the flank, and the presence of ulcers/abscesses (Fryer and Sanders, 1981; Hoffman *et al.*, 1984). Internally, lesions may develop in the kidney (this may become swollen), brain (= meningo-encephalitis; Speare, 1977), liver, heart and spleen. The lesions contain a fluidy mass of leucocytes, bacteria and cellular debris (Fryer and Sanders, 1981). A false membrane, covering some internal organs, has been described (Snieszko and Griffin, 1955; Bell, 1961). This membrane, the presence of which may be influenced by water temperature (Smith, 1964), consists of layers of fibroblasts and histiocytes, degenerating leucocytes with macrophages, and fibrin (Wolke, 1975). In Atlantic salmon, petechial haemorrhages have been described on the muscle lining the peritoneum (Smith, 1964). However, *Ren. salmoninarum* may be found in salmonids, e.g. char and grayling in Alaska, without any evidence of clinical disease (Meyers *et al.*, 1993).

Rhodococcus sp.

In chinook salmon, there was evidence of melanosis and ocular oedema, leading to rupture of the cornea. There was no evidence of involvement of any internal organs (Bachman *et al.*, 1990). With Atlantic salmon, the presence of granulomas in the kidney was apparent (Claveau, 1991). Progressive low-level mortalities were recorded.

Staphylococcus aureus

Dead animals displayed pronounced eye disease, with the cornea becoming reddish, due to vascularization, and then opaque. Thereafter, there was degeneration of the eye tissues, leaving a hollow cup. The brain and optic nerves were affected. In addition, diseased fish became lethargic and darker in colour (melanosis). The internal organs did not appear to be affected (Shah and Tyagi, 1986).

Staphylococcus epidermidis

The initial description of the disease was not exhaustive, but typical signs included exophthalmia, congestion, and ulceration on the tail (Kusuda and Sugiyama, 1981). Later, *Sta. epidermidis* was recovered from moribund cultured grass carp in Taiwan (Wang *et al.*, 1996). Here, the fish displayed haemorrhages on the opercula and pelvic fins. Internally, petechial haemorrhages and bloody ascites were observed. Although tapeworms were present in the lumen of the intestine, smears revealed the presence of oval bacteria, which were isolated, and considered to represent *Sta. epidermidis* (Wang *et al.*, 1996).

Streptomyces salmonis

The exact disease signs association with this organism are unclear from the published information.

D. AEROMONADACEAE REPRESENTATIVES

Aeromonas allosaccharophila

A description of the exact pathology present in the diseased elvers was not provided (Martinez-Murcia *et al.*, 1992).

Aeromonas caviae

In 1991, a septicaemic condition was diagnosed on four Atlantic salmon farms located on the Black Sea in Turkey (Candan *et al.*, 1995). Diseased fish displayed signs of haemorrhagic septicaemia, namely haemorrhages on the body, intestine filled with bloody exudate, enlarged liver and spleen, and liquefying kidney. Subsequently, the organism, has been associated with eye disease and haemorrhagic septicaemia in farmed rainbow trout from Kenya (Ogara *et al.*, 1998).

Aeromonas hydrophila

Since its initial recognition as the causal agent of haemorrhagic septicaemia (Sanarelli, 1891; Schäperclaus, 1930; Haley *et al.*, 1967), *Aer. hydrophila* has been recovered as a pathogen from a wide variety of freshwater fish species, including ornamental fish (Hettiarachchi and Cheong, 1994) and occasionally from marine fish, e.g. ulcer disease of cod (Larsen and Jensen, 1977). The aetiological agent may be considered to have worldwide distribution. However, some doubt has been expressed over its precise role as a fish pathogen (Heuschmann-Brunner, 1965b; Eurell *et al.*, 1978; Michel, 1981) with some workers contending that it may be merely a secondary invader of already compromised hosts. Conversely, other groups have insisted that *Aer. hydrophila* constitutes a primary pathogen.

Aer. hydrophila has been credited with causing several distinct pathological conditions, including tail/fin rot and haemorrhagic septicaemias (e.g. Hettiarachchi and Cheong, 1994). The organism may be found commonly in association with other pathogens, such as *Aer. salmonicida*, although there is no evidence for the presence of synergistic interactions leading to exacerbated disease conditions. Haemorrhagic septicaemia (also referred to as motile aeromonas septicaemia) is characterized by the presence of small surface lesions (which lead to the sloughing-off of scales), local haemorrhages particularly in the gills and vent, ulcers, abscesses, exophthalmia and abdominal distension. Llobrera and Gacutan (1987) described the presence of necrotic ulcers in a variety of fish from the Philippines. Internally, there may be accumulation of ascitic fluid, anaemia and damage to the organs, notably kidney and liver (Huizinga *et al.*, 1979; Miyazaki and Kaige, 1985). Also, redsore disease in bass has been attributed to *Aer. hydrophila* (Hazen *et al.*, 1978). This condition, which may reach epizootic proportion, is characterized by erosion of the scales and pin-prick haemorrhages, which may cover up to 75% of the body surface. There is often a high mortality rate. Hettiarachchi and Cheong (1994) described *Aer. hydrophila* as the cause of disease in freshwater ornamental fish in Sri Lanka, with disease signs including the presence of eroded fins, haemorrhages on the skin and at the base of the caudal fin, sloughing scales and haemorrhaging in the intestinal wall.

Aeromonas salmonicida

Traditionally, *Aer. salmonicida* was thought to have a predilection for salmonids. Over the years, however, the apparent host range of the pathogen has steadily expanded. Thus infections are known to occur among representatives of several major families of Osteicthys, including Cyprinidae, Serranidae and Anoplopomatidae, in addition to the Salmonidae (Herman, 1968), and also the Agnatha (family Petromyzontidae) (Hall, 1963). Non-salmonids which have been documented as suffering from diseases of *Aer. salmonicida* aetiology include minnow and goldfish (Williamson, 1929), carp (Bootsma *et al.*, 1977), perch (Bucke, 1979) and bream, roach, dace, chub, tench, pike, bullheads, sculpins and catfish (McCarthy, 1975a), wrasse (Treasurer and Cox, 1991) and sea bream (Real *et al.*, 1994). In some of the fish, particularly the non-salmonids, the disease may manifest itself in a different form to the classical furunculosis, and, the causal agent is often regarded as 'atypical'.

Traditionally, *Aer. salmonicida* has been known as the causative agent of furunculosis. However, it has become apparent that the pathogen manifests itself with other conditions, notably ulcerative dermatitis (Brocklebank, 1998)/ulcerations, especially in non-salmonids. Furunculosis, named because of the sub-acute or chronic form of the disease, is recognized by the presence of lesions resembling boils, i.e. furuncles, in the musculature. In fact, the term furunculosis is a misnomer, because the lesions do not resemble those found in a similarly named condition of human beings (McCarthy, 1975a). The name has, however, become established in the fisheries literature, so that it has been retained for convenience and to avoid the confusion which could result from a new name.

The sub-acute or chronic form of furunculosis, which is more common in older fish, is characterized by lethargy, slight exophthalmia, blood-shot fins, bloody discharge from the nares and vent, and multiple haemorrhages in the muscle and other tissues. Internally, haemorrhaging in the liver, swelling of the spleen, and kidney necrosis may occur (Snieszko, 1958a; McCarthy and Roberts, 1980). This form of the disease usually causes low rates of mortality, and fish may survive, although survivors have scar tissue in the vicinity of the furuncles (McCarthy, 1975a). Oddly enough, the chronic form of the disease is not the most frequently occurring, nor is the presence of furuncles the most typical symptom of the disease (Snieszko, 1958a).

The acute form of furunculosis, which is most common particularly in growing fish and adults, is manifested by a general septicaemia accompanied by melanosis, inappetence, lethargy, and small haemorrhages at the base of the fins. This form of the disease is of short duration, insofar as the fish usually die in 2–3 days, and causes high mortalities. The bacteria occur in the blood, disseminated throughout the tissues, and in the lesions. Internally, haemorrhaging occurs over the abdominal walls, viscera and heart. The spleen may appear enlarged. The acute disease is of sudden onset, with few, if any, external signs (McCarthy, 1975a).

McCarthy and Roberts (1980) discussed a third clinical form of furunculosis, termed peracute furunculosis, which is confined to fingerling fish. The infected animals darken in colour, and may quickly die with only slight external symptoms, such as mild exophthalmia. Haemorrhages may occur at the base of the pectoral fin, if the fish manage to survive for long enough periods. Losses in farmed stock may be extremely high (Davis, 1946).

Yet another form of furunculosis was discussed by Amlacher (1961), i.e. intestinal furunculosis. The symptoms were described as inflammation of the intestine, and anal inversion. This description is similar to a report by Herman (1968) of chronic furunculosis, i.e. low, relatively constant rate of mortality with intestinal and variable haemorrhages.

In addition to furunculosis, *Aer. salmonicida* has been implicated in several conditions, especially of non-salmonids. The best known of these conditions is undoubtedly carp erythrodermatitis (CE). Fijan (1972), who is credited with the name of the disease, demonstrated that CE was caused by a transmissible, antibiotic-sensitive organism, which manifested itself as predominantly a skin infection. Bootsma *et al.* (1977) isolated a small, Gram-negative, rod-shaped organism from skin lesions in mirror carp in Yugoslavia. This organism was subsequently identified as an 'atypical' strain of *Aer. salmonicida*. CE was described as a sub-acute to chronic contagious skin disease, which varied in its morbidity and mortality (Bootsma *et al.*, 1977). It appears that the infection often starts at the site of injury to the epidermis. A haemorrhagic inflammatory process then develops between the epidermis and dermis. This red inflammatory zone gradually extents as the infection spreads. The breakdown of tissue leads to the formation of a central ulcer, which may occur in any location on the body surface, although it is most frequently located on the flanks. Infected fish exhibit inappetence, and appear darker in colour. Secondary invasion of the ulcer by fungi or other bacteria is common. If the fish recovers, the healed ulcer is recognizable as a grey-black scar. Frequently, contraction of the collagen of the scar tissue can result in serious deformity, which reduces the commercial value of the fish (Fijan, 1972). In some instances, CE may also result in generalized septicaemia and death. Unlike furunculosis, which usually occurs only when water temperatures exceed 16°C, CE may occur at all water temperatures.

More recently, *Aer. salmonicida* has been reported to cause a cutaneous ulcerative disease in goldfish (*Carassius auratus* L.), although the disease itself had been known for a long time previously. Ulcer disease of cyprinids, in general, has occurred in widely separated geographical locations, including the USA, Japan and England (Shotts *et al.*, 1980). Mawdesley-Thomas (1969) studied, in detail, an outbreak of an ulcerative disease of goldfish, and recovered *Aer. salmonicida*. Symptoms included lethargy, loss of orientation, and abnormal swimming behaviour. The ulcers were of various sizes and depths, and some fish died shortly after infection. Secondary invasion of the ulcers by *Saprolegnia* frequently was observed. More recently, we have recovered an extremely fastidious form of the pathogen from ulcerated carp, goldfish and roach in England. Here, there was evidence of secondary invasion of the ulcers by *Aer. hydrophila*. According to McCarthy and Roberts (1980), ulcer disease differed from CE in the following ways:

(a)   the ulcers were deeper and more extensive,
(b)   renal and splenic changes, such as those found in sub-acute furunculosis of salmonids, were present at an earlier stage in the course of the disease than in CE (where internal lesions were restricted to later stages in cases where septicaemia developed).

Another variation of *Aer. salmonicida* infection, termed 'head ulcer disease', has been described in Japanese eels (Hikada *et al.*, 1983; Ohtsuka *et al.*, 1984; Kitao *et al.*, 1984).

An atypical strain of *Aer. salmonicida* was implicated as the aetiological agent. The progression of this disease is worthy of note because the pathogen is apparently capable of causing mortalities solely as a result of localized proliferation, with no evidence for the development of a generalized septicaemia. Results of natural and laboratory-based infections of eels with head ulcer disease revealed that *Aer. salmonicida* was not recovered in significant numbers from internal organs, i.e. brain, kidney or spleen or, indeed, blood (Ohtsuka *et al.*, 1984; Nakai *et al.*, 1989a). However, the pathogen proliferated substantially in the muscle of eels (Nakai *et al.*, 1989a). Similar, localized ulcerative infections caused by atypical *Aer. salmonicida* have been recognized in goldfish (Elliott and Shotts, 1980) and carp (Csaba *et al.*, 1981a).

Aeromonas sobria
Evidence has been presented for *Aeromonas sobria* as a fish pathogen. Yet, fish pathogenic isolates have often been mis-identified, whereas *bona fide Aer. sobria* has not been harmful to salmonids (Austin *et al.*, 1989). However, organisms, considered as *Aer. sobria*, have been isolated from wild spawning gizzard shad (*Dorosoma cepedianum*) in Maryland, USA, during 1987 (Toranzo *et al.*, 1989). Yet, moribund fish did not display any external or internal signs of disease.

E. ENTEROBACTERIACEAE REPRESENTATIVES

Citrobacter freundii
Diseased sunfish displayed erratic swimming, inappetence, eroded skin, surface haemorrhages, enteritis, deep red spleen, pale liver and tumorous masses (granulomas) on the kidney (Sato *et al.*, 1982). Similar signs have been observed in salmonids within the UK (B. Austin, unpublished data). In the initial disease outbreak, 25/29 animals (= 86% of the total) died (Sato *et al.*, 1982).

Edwardsiella ictaluri
There is a great variability in the clinical signs associated with the disease. Just prior to death, the fish may hang listlessly in an almost vertical position at the water surface, spin rapidly in circles, or exhibit spiral swimming. External signs, often absent in fish over 15 cm in length, include the presence or petechial (pin-prick) haemorrhages on the skin in the vicinity of the throat and mouth, pale gills, exophthalmia and open lesions on the head, particularly on the frontal bone of the skull between the eyes on the lateral body surface. Internally, there may be swelling of the kidney and spleen, haemorrhaging and necrotic areas in the liver, blood-filled ascitic fluid in the peritoneum, and petechial haemorrhages throughout the internal muscle walls (Hawke, 1979). In short, the disease is a typical bacterial septicaemia.

Edwardsiella tarda
Excellent descriptions of the clinical manifestations of *Edw. tarda* infections in channel catfish were published by Meyer and Bullock (1973). These authors reported that with mild infections, the only external signs of disease are the presence of small cutaneous lesions of approximately 3–5 mm in diameter, which are located in the postero-lateral

region of the body. With progression of the disease, abscesses develop in the muscle of the body and tail. These abscesses may enlarge, and develop into gas-filled hollow areas. From the surface, these are seen as poorly pigmented, convex swollen areas which, if punctured, emit a foul odour. This condition has given rise to the name emphysematous putrefactive disease of catfish. Although catastrophic losses of catfish have not been recorded, the disease has a severe economic effect on infected sites. When infected fish enter processing plants, the noxious odours effectively stop production by necessitating disinfection and deodorization. Thus, heavy financial losses to the processors may result from the presence of only a small number of infected animals (Meyer and Bullock, 1973).

Examination of naturally diseased tilapia has indicated a range of symptoms including loss of pigmentation, the presence of a swollen abdomen filled with ascitic fluid, protruded haemorrhaged anus and opaqueness to the eyes. Internally, small white nodules may be observed in the gills, kidney, liver and spleen, and occasionally the intestine. These nodules are packed with bacteria (Kubota *et al.*, 1981).

Enterobacter agglomerans
Pronounced haemorrhaging was noted in the eyes of dead and moribund animals. Haemorrhages were also recorded in the musculature. Otherwise, there was an absence of disease signs in the internal organs (Hansen *et al.*, 1990).

Escherichia vulneris
Clinical signs included haemorrhagic lesions on the skin, pale gills, digestive tract full of bloody exudate, haemorrhaging in the gonads, and yellow liver with hyperaemic areas (Aydin *et al.*, 1997).

Hafnia alvei
The disease was described as a haemorrhagic septicaemia. Fish became darker, moved slowly, developed exophthalmia, haemorrhaging in the eye and petechial haemorrhages on the body surface. Internally, petechial haemorrhages were apparent on the spleen (enlarged) and kidney (hyperaemic). The disease appeared in fish after transportation or during cultivation in inappropriate conditions (Gelev and Gelev, 1988).

A second outbreak occurred in cherry salmon (*O. masou*) on Japanese farms. Here, the disease was characterized by melanosis, swollen abdomen, and slow swimming. There were grey/white furuncles on the kidney (Teshima *et al.*, 1992).

Klebsiella pneumoniae
The organism was recovered from the diseased tails and fins of twelve rainbow trout in Scotland. No other disease signs were noted (Daskalov *et al.*, 1998).

Proteus rettgeri
The fish displayed large red ulcers on the abdomen, base of the pectoral fin, and around the head. At 20–23°C, mortalities occurred in 3 days, whereas at 12–20°C, the deaths ensued in 8 days (Bejerano *et al.*, 1979).

Salmonella choleraesuis subsp. *arizonae*

Pirarucu was deemed to have died of a septicaemic condition. Externally, there was minimal evidence of disease, with the eyes (corneas) displaying opacity. A bloody exudate was found in the body cavity. Lesions and congestions were recorded in the mucous membranes of the stomach and intestine (Kodama *et al.*, 1987).

Serratia liquefaciens

Moribund fish did not display any external signs of disease. Yet, internally the kidney was swollen, and nodules were present on both the kidney and spleen; the liver appeared speckled, and some ascitic fluid was present in the peritoneal cavity. Initially, it was considered likely that the animals were infected with *Ren. salmoninarum*, the causal agent of BKD, but the presence of Gram-negative bacteria and complete absence of any Gram-positive micro-organisms precluded this possibility (McIntosh and Austin, 1990b). In a separate development during 1990, the organism was attributed to low level mortalities in turbot, farmed in floating cages in France. With this outbreak, the disease signs included swelling and liquefaction of the kidney and spleen, which were also characterized by the presence of yellowish nodules (Vigneulle and Baudin-Laurençin, 1995).

Serratia marcescens

It should be emphasized that the organisms were recovered only from apparently healthy fish, which were devoid of overt signs of disease (Baya *et al.*, 1992c).

Serratia plymuthica

Nieto *et al.* (1990) noted that diseased fish did not display any external or internal clinical signs. However in Scotland, the diseased fish possessed extensive skin lesions over the entire flank, from the operculum to the tail (Austin and Stobie, 1992b).

Yersinia intermedia

Affected Atlantic salmon were of 40–50 g in weight, and were held at a water temperature of 5°C. Disease signs included lazy movement with the fish congregating at the surface of the water, darkening of the body pigment, tail erosion, haemorrhaging on the flank and abdominal inflammation (Carson and Schmidtke, 1993).

Yersinia ruckeri

The disease is mostly restricted to salmonids (Bullock and Snieszko, 1975) within the geographical locations of North and South America (Ross *et al.*, 1966; Stevenson and Daly, 1982), Denmark (Dalsgaard *et al.*, 1984), Great Britain (Austin, 1982; Roberts, 1983), France (Lesel *et al.*, 1983), Germany (Fuhrmann *et al.*, 1983), Italy (Busch, 1978), Ireland (McCormick and McLoughlin, 1993), Norway (Richards and Roberts, 1979) and Australia (Bullock *et al.*, 1978a; Llewellyn, 1980). With the recent upsurge in the number of cases of the disease, particularly in Europe, it would appear that the disease is still spreading. So far, it has been a severe problem mostly in rainbow trout, although outbreaks of disease have been reported among populations of brown trout, brook trout, chinook salmon and coho salmon (Dulin *et al.*, 1976) and Atlantic salmon and Pacific salmon (Bullock *et al.*, 1978a). A few non-salmonid fish species have also been reported to harbour the pathogen, and these include emerald dace (*Notropis atherinoides*; Mitcham, 1981), goldfish (McArdle and Dooley-Martin, 1985), carp (B. Austin, unpublished observation), lake herring or cisco (*Coregonus artedii*; Bullock and Anderson, 1984) and minnows (*Pimephales promelas*) (Michel *et al.*, 1986a). Also, sea bass and turbot appear to be susceptible (Vigneulle, 1990). Much excellent work has been accomplished, but there are many basic facets of the biology of the pathogen which remain unknown or unclear.

The name of the disease, i.e. enteric redmouth (ERM), is fairly descriptive insofar as one of the most common symptoms is reddening of the mouth and throat, which is caused by subcutaneous haemorrhaging (Busch, 1973). Other external signs include inflammation and erosion of the jaws and palate, melanosis, haemorrhaging around the base of the fins, bilateral exophthalmia, and a tendency for sluggishness (Fuhrmann *et al.*, 1983; Bullock and Anderson, 1984). Internally, there may be haemorrhaging in the muscle, body fat and in the intestine, which may also contain a yellow fluid. A generalized bacteraemia occurs in the principal organs, with slight enlargement of the kidney and spleen. The disease has been held responsible for greater financial loss of the trout farming industry in western areas of the USA than any other disease (Hester, 1973). Onset of an epizootic is often gradual with resulting heavy losses (Newman and Majnarich, 1982). It is relevant to note that a second disease, known as salmonid blood spot, in Australia (Llewellyn, 1980) is attributable to the same aetiological agent (Green and Austin, 1982).

F. *CYTOPHAGA–FLAVOBACTERIUM–FLEXIBACTER* GROUP

Many of these organisms cause gill disease, which may be characterized histologically as hyperplasia (swelling) of the gill epithelia. Frequently, the condition in juvenile fish involves fusion at the distal tips of adjoining gill lamellae. The involvement of hyperplasia-inducing agents has also been implicated in some cases by *Flavobacterium* spp. (Kudo and Kimura, 1983a,b).

Chryseobacterium scophthalmum
This organism caused lethargy, gill hyperplasia, haemorrhaging in the gills, distended abdomen, surface haemorrhaging, and extensive internal haemorrhaging (Mudarris and Austin, 1989; 1992).

Flavobacterium columnare
Columnaris has been recognized to have worldwide distribution in a wide range of freshwater fish, including Arctic char, bass, black bullheads, carp, channel catfish, chub, eel, goldfish, killifish, loach, perch, rainbow trout, roach, Atlantic salmon, chinook salmon, sheatfish, squawfish, tilapia, white crappie, whitefish and white-suckers (Nigrelli, 1943; Nigrelli and Hutner, 1945; Wakabayashi and Egusa, 1966; Ajmal and Hobbs, 1967; Fijan, 1969; Bowser, 1973; Wobeser and Atton, 1973; Chun, 1975; Bootsma and Clerx, 1976; Ferguson, 1977; Farkas and Olàh, 1980; Kou *et al.*, 1981; Morrison *et al.*, 1981; Chen *et al.*, 1982; Koski *et al.*, 1993). The disease was considered to be sufficiently

serious to warrant inclusion in the list of notifiable fish diseases, as defined by the (British) Diseases of Fish Act, 1937. However, columnaris was omitted from the 1983 Act.

Infection with *Fla. columnare* may result in several discrete disease conditions. In young fish, there is often negligible pathology before death ensues. The gill is usually the major site of damage. Typically, congestion (blockage) of the blood vessels supplying the gills occurs, with dissociation of the surface epithelium of the lamellae from the capillary bed. There may be scattered areas of haemorrhaging (Pacha and Ordal, 1967). With adult fish, the lesions may occur on the gills, skin and/or in the musculature. Systemic infections may develop (Wolke, 1975). Gill lesions consist normally of yellow-orange areas of necrosis. These start usually at the periphery of the gill, and extend towards the base of the gill arch. Eventually, extensive erosion may completely destroy the gill filament (Pacha and Ordal, 1970). On the body, small lesions start as areas of pale discoloration at the base of the dorsal fin or occasionally at the base of the pelvic fin, and lead to deterioration of the fins. These areas increase in size and may become as large as 3–4 cm in diameter, covering as much as 20–25% of the total surface area of the fish. This may have the characteristic appearance of a saddle, and hence the descriptive term, 'saddleback'. Frequently, the skin becomes completely eroded away, exposing the underlying muscle. Large numbers of bacteria are present at the advancing edge of the lesion. It is not uncommon for the fish to die within 48 h of the appearance of the skin discoloration (Pacha and Ordal, 1970; Becker and Fujihara, 1978; Morrison *et al.*, 1981).

Flavobacterium johnsoniae

In the initial outbreak in Queensland, Australia, 2–5% mortalities occurred in farmed barramundi, *Lates calcarifer*, over a two-week period when the water temperature was 27–28°C. The fish were listless, anorexic, and displayed elevated scales and superficial erosion of the skin (particularly on the posterior flank). Some fish presented eroded pectoral fins and lower jaws (Carson *et al.*, 1993).

Flavobacterium psychrophilum

This organism causes saddle-like lesions, containing myriads of organisms, near the dorsal fin. The fish may darken and, in advanced cases, develop bacteraemia with the pathogen ramifying throughout the animal (Wood and Yasutake, 1956; Winton *et al.*, 1983; Lehmann *et al.*, 1991).

Flexibacter maritimus

Principal signs of disease caused by *Fle. maritimus* include mouth erosions, gill erosion and tail rot (Handlinger *et al.*, 1997), especially in juvenile fish. In older animals, lesions develop initially as grey-white cutaneous areas on the fins, head and trunk. These lesions degenerate into ulcers (Hikida *et al.*, 1979; Wakabayzshi *et al.*, 1984). Black patch necrosis of Dover sole in Scotland has been attributed to this pathogen (Bernadet *et al.*, 1990).

Flexibacter ovolyticus
Fle. ovolyticus led to mortalities among halibut eggs and larvae. The chorion became dissolved, and the underlying zona radiata was damaged by exotoxins resulting in puncturing of the egg, leakage of cell constituents and larval death (Hansen *et al.*, 1992).

G. PSEUDOMONADACEAE REPRESENTATIVES

Pseudomonas anguilliseptica
Typically, the disease manifests itself by the presence of petechial (pinprick) haemorrhages in the skin of the mouth regions, opercula and ventral side of the body. Haemorrhaging in the eye has been seen in infected Baltic herring (*Clupea harengus membras*) (Lönnström *et al.*, 1994). Reddening of the fins (as with vibriosis or *Aer. hydrophila* infections) does not usually occur. Small petechial haemorrhages may develop in the peritoneum, and the liver may be pale and haemorrhaged. The kidney may be soft and liquefying. Alternatively, in some cases of disease, there may be a dearth of internal signs of distress (Wakabayashi and Egusa, 1972).

Pseudomonas chlororaphis
In moribund fry, it was observed that symptoms included the presence of distended abdomen with ascitic fluid, and haemorrhages on the body surface (Hatai *et al.*, 1975).

Pseudomonas fluorescens
It has been reported to cause disease in a wide range of fish species, including silver carp (*Hypophthalmichthys molitrix*) and big head (*Aristichthys nobilis*) (Csaba *et al.*, 1981; Markovic *et al.*, 1996), goldfish (*Carassius auratus*) (Bullock, 1965), tench (*Tinca tinca*) (Ahne *et al.*, 1982), grass carp (*Ctenopharyngodon idella*) and black carp (*Myopharyngodon piceus*) (Bauer *et al.*, 1973), unnamed species of carp (Schäperclaus, 1959; Schäperclaus and Brauer, 1964; Heuschmann-Brunner, 1978) and rainbow trout (Li and Fleming, 1967; Li and Traxler, 1971, Sakai *et al.*, 1989a). Generally, *Ps. fluorescens* is associated with fin or tail rot in which the infected area is eroded away (Schäperclaus, 1979). In tench fry, high mortalities (up to 90% of the population) have been reported, in which visual signs of disease included haemorrhagic lesions on the skin and at the base of the fins. Ascitic fluid accumulated in the peritoneal cavity, and petechial haemorrhages were evident in the gills, kidney, liver and in the lumen and submucosa of the gut, i.e. a typical generalized bacterial septicaemia (Ahne *et al.*, 1982). Similar symptoms were apparent in silver carp and big head (Csaba *et al.*, 1981). Stress, including a lowered water temperature, may trigger outbreaks of disease (Markovic *et al.*, 1996). With rainbow trout the presence of ulcers at haemorrhages on the gills and fins were reported (Sakai *et al.*, 1989a).

Pseudomonas pseudoalcaligenes
The fish displayed extensive skin lesions, which extended over the entire flank from the operculum to the tail. The skin and underlying muscle to a depth of approximately 1 mm were totally eroded (Austin and Stobie, 1992b).

H. VIBRIONACEAE REPRESENTATIVES

Moritella marina
The organism was associated with shallow skin lesions of Atlantic salmon farmed in Iceland at low temperatures, i.e. ~10°C (Benediktsdóttir *et al.*, 1998).

Photobacterium damselae
The organism was associated initially with ulcerative lesions along the flank of blacksmith (*Chromis punctipinnis*), one of the damselfish. These ulcers were noted in summer and autumn among fish populations in the coastal waters of southern California. Surveys of wild fish populations led to a conclusion that the ulcers were restricted to species of damselfish. Additional information pointed to a role in human pathogenicity, insofar as the organism has been isolated from human wounds (Love *et al.*, 1981). Subsequent work demonstrated this organism in sharks (Grimes *et al.*, 1984a; Fujioka *et al.*, 1988), turbot (Fouz *et al.*, 1991; 1992) and yellowtail (Sakata *et al.*, 1989).

Characteristic skin lesions, i.e. ulcers, are formed, particularly in the region of the pectoral fin and caudal peduncle. These ulcers may reach a size of 5–20 mm in diameter. Typically, muscle lysis occurs. The results of histopathological examination suggests the presence of granulomatous ulcerative dermatitis.

Photobacterium damselae subsp. *piscicida*
This disease has been responsible for heavy losses among menhaden and striped mullet in Galveston Bay, Texas (Lewis *et al.*, 1970). However, it is in Japan that the disease has become of considerable economic importance, causing significant losses in farmed yellowtail (Egusa, 1983). Since its initial recognition in yellowtails, pasteurellosis appears to have spread to other fish species, including gilt-head sea bream (Balebona *et al.*, 1998), red sea bream (Yasunaga *et al.*, 1983) and black sea bream (Muroga *et al.*, 1977a; Ohnishi *et al.*, 1982). More recently, the disease seems to have spread to farmed and wild fish stocks in the Mediterranean area, notably France, Italy and Spain (Magariños *et al.*, 1992).

Essentially, pasteurellosis is a septicaemia in which acute cases exhibit only a few pathological signs. Internally, granulomatous-like deposits, which have led to the coining of the descriptive name of pseudotuberculosis, may develop on the kidney and spleen. These deposits comprise many greyish-white bacterial colonies of 0.5–1.0 mm^2 in size (Kusuda and Yamaoka, 1972). Purulent material may accumulate in the abdominal cavity (Lewis *et al.*, 1970).

Plesiomonas shigelloides
Symptoms included emaciation, reddening of the anus with yellow exudation, petechial haemorrhages in the internal muscle wall, and sometimes the accumulation of ascitic fluid in the peritoneal cavity (Cruz *et al.*, 1986). Inappetence was noted by Klein *et al.*, (1993).

Shewanella putrefaciens

During spring of 1985, a disease occurred which resulted in high mortalities in rabbitfish, *Siganus rivulatus*, farmed in sea cages in the Red Sea. From diseased animals, a Gram-negative bacterium was recovered, which was capable of re-infecting healthy fish (Saeed *et al.*, 1987). To date, the disease has not been described in any other fish species, or, for that matter, elsewhere.

Disease signs included lethargy, discoloration, exophthalmia, haemorrhaging and necroses on the body and mouth, and fin damage. Internal damage was not reported (Saeed *et al.*, 1987).

Vibrio alginolyticus

From the description of Colorni *et al.* (1981) and Austin *et al.* (1993), the disease may be classified as a typical bacterial septicaemia. Infected fish were observed to become sluggish, the skin darkened, scales loosened and sloughed off, and ulcers developed. The liver, capillaries in the intestinal wall, air bladder, and peritoneum become congested. Simultaneously, the intestine and gall bladder became distended with clear fluid and bile, respectively. Anaemia and gill rot were also reported. Austin *et al.* (1993) attributed the organism with gill disease leading to progressive low-level mortalities in turbot, which were maintained at supra-optimal temperatures in a recirculating aquarium. The later study of Lee (1995) attributed the organism with causing exophthalmia and corneal opaqueness in grouper; signs that are reminiscent of the pathology caused by *V. harveyi*. Woo *et al.* (1995) and Ye *et al.* (1997) considered that *V. alginolyticus* was responsible for heavy mortalities in silver sea bream (*Sparus sarba*) in Hong Kong. The organism has also been associated with disease in gilt-head sea bream in Spain (Balebona *et al.*, 1998).

Vibrio anguillarum

'Red-pest' (referred to historically as *pestis rubra anguillarum* and *erysipelosis anguillarum*) caused catastrophic losses among eels held in sea water sites within Italy during the eighteenth and nineteenth centuries. The excellent description of an outbreak of 'red-pest' in eels during 1718 is undoubtedly the first reference to a bacterial fish disease in the European literature (Bonaveri, 1761). For a detailed account of the early narratives, reference is made to the splendid review of Drouin de Bouville (1907). However, confusion may result from the multiplicity of names used to describe the disease. Thus, references may be found to 'salt-water furunculosis' (Rucker, 1963), 'boil-disease' (Kubota and Takakuwa, 1963), and ulcer-disease' (Bagge and Bagge, 1956), as well as to the universally accepted name of 'vibriosis'.

Apt but gory descriptions have been made about the nature of vibriosis in fish. To microbiologists, the disease may be regarded as yet another haemorrhagic septicaemia. Typically, infected fish show skin discoloration, the presence of red necrotic lesions in the abdominal muscle, and erythema (bloody blotches) at the base of the fins, around the vent, and within the mouth (in this respect there is a resemblance to ERM, caused by *Yersinia ruckeri*). The gut and rectum may be distended, and filled with clear viscous fluid. Exophthalmia may be evident (Anderson and Conroy, 1970).

In Pacific salmon fingerlings, a bacteraemia occurs in the initial stages of disease. From histological examination, it may be concluded that there are pathological changes in the blood, connective tissue, gills, kidney, liver (an anaemia) and posterior gastro-intestinal tract, and swelling in the spleen. The bacterial cells appear to be uniformly distributed throughout the affected tissues, although the greatest concentration is in the blood (Tajima *et al.*, 1981; Ransom *et al.*, 1984). Usually infected fish become inactive, cease feeding (this may cause problems for chemotherapy), and suffer heavy mortalities.

Vibrio cholerae (non-O1)

Petechial haemorrhages developed on the body surface. Internally, there was congestion of the organs (Muroga *et al.*, 1979; Kiiyukia *et al.*, 1992). Reddacliff *et al.* (1993) reported that septicaemia developed in infected goldfish.

Vibrio fischeri

Most diseased fish possessed whitish nodules on the skin (dorsal surface), haemorrhagic ulceration, and tumours involving the pancreas and bile duct. Within a year, 39% losses occurred in the fish population (Lamas *et al.*, 1990).

Vibrio harveyi (including *V. carchariae*)

V. carchariae was originally isolated from a dead sandbar shark (*Carcharhinus plumbeus*) which died at the National Aquarium in Baltimore, USA, in 1982 (Grimes *et al.*, 1984a). Subsequently, a similar organism was recovered from lemon sharks (*Negraprion brevirostris*) (Colwell and Grimes, 1984). Grimes *et al.* (1984b) and Colwell and Grimes (1984) described the disease as a 'vasculitis'. Infected animals became lethargic, stopped feeding, appeared disorientated, and developed necrotic subdermal cysts. On post-mortem examination, encephalitis, meningitis, kidney necrosis, vasculitis, and unspecified liver and spleen damage were noted. Evidence has been forthcoming that the pathogen is more serious in compromised than healthy hosts (Grimes *et al.*, 1985). A similar organism has been isolated from a chronic skin ulcer on a shark (Bertone *et al.*, 1996). *V. carchariae* was associated with gastro-enteritis leading to heavy mortalities among cultured groupers (*Epinephelus coioides*) during 1993 in Taiwan (Yii *et al.*, 1997).

A parallel development involved gamefish, namely common snook (*Centropomus undecimalis*). These were found to suffer with opaque white corneas within 24 h of capture in Florida, USA (Kraxberger-Beatty *et al.*, 1990). From such damaged specimens, *V. harveyi* was recovered. Common snook developed opaque white corneas within 24 h of capture. In the absence of treatment, blindness resulted. With a second species, jack crevalle (*Caranx hippos*), deep dermal lesions were noted in wild specimens, which were also captured in Florida. Internal abnormalities were not recorded (Kraxberger-Beatty *et al.*, 1990). In another example from an aquarium in Barcelona, eye lesions in the short sunfish (*Mola mola*)—due to biting by other fish—were colonized by *V. harveyi* (Hispano *et al.*, 1997).

Vibrio ichthyoenteri

Since 1971, opaque intestines and intestinal necrosis accompanied by high mortalities have been reported in Japanese hatcheries rearing Japanese flounder (Ishimaru *et al.*, 1996).

Vibrio logei

An organism, with similarities to *V. logei*, was associated with shallow skin lesions of Atlantic salmon farmed in Iceland at low temperatures, i.e. ~10°C (Benediktsdóttir *et al.*, 1998).

Vibrio ordalii

Essentially, the disease may be categorized as a haemorrhagic septicaemia. However, there are subtle differences in the pathologies of the diseases caused by *V. anguillarum* and *V. ordalii*. In the case of *V. ordalii* in Pacific salmon, there is a tendency for the formation of micro-colonies in the skeletal and heart muscle, gill tissue, and in both the anterior and posterior regions of the gastro-intestinal tract (Ransom, 1978; Ransom *et al.*, 1984). Moreover, bacteraemia developed much later in the disease cycle than with *V. anguillarum*. Perhaps, this accounted for the lower numbers of bacterial cells in the blood. A further difference concerned the marked decrease in the numbers of leucocytes in the blood, i.e. leucopenia (Ransom, 1978; Harbell *et al.*, 1979; Ransom *et al.*, 1984).

Vibrio pelagius

An epizootic of juvenile farmed turbot in northwest Spain occurred during January and February 1991 when the water temperature was 12–15°C, with fish displaying eroded dorsal fins and tail, haemorrhages at the base of the fins, haemorrhages on the internal organs, and intestines full of mucous liquid (Angulo *et al.*, 1992). The total losses amounted to 3% of the turbot population.

Vibrio salmonicida

With the tremendous increases in production of Atlantic salmon in Norway, it was perhaps inevitable that at some time a new or emerging disease would cause havoc to the industry. Thus in 1979, such a 'new' disease appeared in salmon farms located around the island of Hitra, south of Trondheim in Norway. In 1983, the disease appeared in Stavanger and, in particular, the large number of fish farms in the Bergen region. The disease, coined coldwater vibriosis or Hitra disease (Egidius *et al.*, 1981) occurs mainly during the period of late autumn to early spring. The disease is now widespread throughout Norway, and there are some reports of outbreaks in Scotland, Shetland (Bruno *et al.*, 1985) and Canada. The disease resembles a generalized haemorrhagic septicaemia. Externally, haemorrhaging may be evident around the abdomen (Holm *et al.*, 1985). Internally, there is often evidence of anaemia, haemorrhaging on the organs, swim bladder and abdominal wall and posterior gastro-intestinal tract (Poppe *et al.*, 1984; Holm *et al.*, 1985; Egidius *et al.*, 1986).

Vibrio splendidus

During 1987, a disease occurred in cultured turbot in northwest Spain. During the outbreak, there was a continuous low-level mortality amounting to 4% of the total stock. Infected fish contained a virus, deemed to be a reovirus, and a bacterium, which was considered to resemble *V. splendidus* (Lupiani *et al.*, 1989). Interestingly, a similar organism has been recovered from diseased Atlantic salmon in Scotland (B. Austin, unpublished data), turbot and sea bass in Norway (Myhr *et al.*, 1991) and gilt-head sea bream in Spain (Balebona *et al.*, 1998). Diseased turbot displayed swollen abdomen and haemorrhaging in the mouth, at the anus and base of the fins. The swimming behaviour was not unusual. Internally, the stomach and intestine were swollen, and filled with a mucoid liquid. Haemorrhaging was apparent on the walls of the peritoneal cavity, which also contained a reddish liquid. The liver was pale (Lupiani *et al.*, 1989; Angulo *et al.*, 1994).

Vibrio trachuri

A disease, resembling vibriosis, has been long associated with Japanese horse mackerel (*Trachurus japonicus*) especially during summer when the seawater temperature exceeds 25°C (Iwamoto *et al.*, 1995). Infected fish displayed erratic swimming, darkened in colour, developed pronounced bilateral exophthalmia, and developed haemorrhages on the internal organs.

Vibrio vulnificus

Between 1975 and 1977 in Japan, there were serious outbreaks of disease among cultured eels from six separate localities (Muroga *et al.*, 1976a,b; Nishibuchi and Muroga, 1977; 1980; Nishibuchi *et al.*, 1979; 1980). The disease has certainly spread to Europe, with incidences in Spain (Biosca *et al.*, 1981; Amaro *et al.*, 1992), The Netherlands and England (B. Austin, unpublished data). This is a haemorrhagic condition characterized by redness on the body, notably flank and/or tail. In advanced cases, pathological changes may be observed in the gastro-intestinal tract, gills, heart, liver and spleen (Miyazaki *et al.*, 1977). Superficially, the disease resembles classical vibriosis.

I. MORAXELLACEAE REPRESENTATIVES

Acinetobacter sp.

During autumn 1978 when the water temperature was between 8°C and 11°C, an outbreak of disease occurred in a group of 60 sexually mature Atlantic salmon. The fish, each of 5–12 kg in weight, were wild stock from the River Surma, Norway, and were held in brackish water during the occurrence of disease. During the 5-week period of disease, the cumulative mortalities amounted to 92% of the population. However, only about 40% of the animals displayed clinical signs of disease, namely hyperaemia of the dermal blood vessels, and haemorrhaging in the scale pockets, with severe oedema extending into the lower epidermis in the vicinity of the base of the fins. Ulceration developed. Lesions appeared in the kidney, liver and spleen, and small haemorrhages occurred in the air bladder and on the visceral peritoneal surfaces (Roald and Hastein, 1980).

Moraxella sp.

During winter 1987, mortalities were recorded among juvenile striped bass, *Morone saxatilis*, in the Potomac River, Maryland, USA. Gills of diseased fish were affected with the parasites *Trichodina* and *Ergasilus*. In addition, a reo-like virus and a bacterium was recovered from   some individuals. Large haemorrhagic lesions and missing scales occurred on the dorsal surface of the eleven affected fish. Haemorrhages were apparent in the swim bladder. The liver was enlarged, pale and mottled in appearance. Membranous material appeared to connect the liver with the body wall (Baya *et al.*, 1990b).

J. HALOMONADACEAE REPRESENTATIVE

Halomonas (=Deleya) cupida

During April to June 1984, heavy mortalities occurred in black sea bream, *Acanthopagrus schlegeli*, fry (up to 14 days after hatching) in hatcheries in western Japan. Generally, the fish were too small to discern disease signs (Kusuda *et al.*, 1986).

K. SMALL, MORPHOLOGICALLY SIMPLE BACTERIA

Mycoplasma mobile

The mycoplasma was associated with 'red disease', a condition in the gills of tench (*Tinca tinca*) (Kirchhoff *et al.*, 1987).

Piscirickettsia salmonis

Infected fish gathered at the surface of cages, became sluggish and were inappetent. External signs included melanosis, epidermal indurations, and paleness of the gills, which was indicative of anaemia. The haemacrits fell to $\leq 27\%$. Internally, haemorrhages were evident on the abdominal walls, visceral fat and on the air bladder. A mottled liver and swollen spleen was apparent in heavily infected animals. The kidney was inflamed and swollen. The intestine was full of yellowish mucoid material (Schäfer *et al.*, 1990; Branson and Diaz-Munoz, 1991). Initially, the disease was not considered to occur during the freshwater stage of fish culture. Instead, mortalities have been noted to begin 6–12 weeks after transfer of fish from fresh to seawater (Fryer *et al.*, 1992). However, in a later development, the pathogen was recovered from the freshwater stage of coho salmon and rainbow trout (Bravo, 1994; Gaggero *et al.*, 1995).

Rickettsia-like organism (RLO)

Chern and Chao (1994) reported that the RLO caused mass mortalities in tilapia, with disease signs including the presence of white nodules and microscopic granulomas on all the organs, and an enlarged spleen. Subsequently, Comps *et al.* (1996) described a small coccoid organism from the brain of juvenile sea bass obtained from the South of France. These were derived from a population which suffered 20% mortalities.

L. MISCELLANEOUS PATHOGENS

Aquaspirillum sp.
There has been a report of putative *Aquaspirillum* sp., along with *Aer. hydrophila*, *Pseudomonas* sp. and *Streptococcus* sp., being associated with a disease, termed epizootic ulcerative syndrome, in snakeheads and catfish obtained from two fish farms in Thailand (Lio-Po *et al.*, 1998). However, the evidence for the involvement of *Aquaspirillum* is not convincing.

Janthinobacterium lividum
During 1991, purple-pigmented Gram-negative rod-shaped bacteria were associated with mortalities at two fish farms. At one site in Scotland, moribund rainbow trout (size range = 0.5–1.0 g) were diagnosed with RTFS. The second site in Northern Ireland also experienced high mortalities (~35% of the stock) in rainbow trout fry of 0.2–0.5 g in size, two to three weeks after the introduction of feeding. At this site, the rise in mortalities coincided with a change from the use of spring to river water (Austin *et al.*, 1992b). In addition during January 1992, we found similar purple-pigmented bacteria to be associated with skin lesions, on larger rainbow trout (100–200 g in weight), which were otherwise debilitated with ERM. It is relevant to note that this fish population had received prolonged and varied chemotherapy.

Small fish, considered to be displaying RTFS, became lethargic, displayed exophthalmia, pale gills, enhanced skin pigmentation, swollen abdomen and (sometimes) skin lesions. Internally, the kidney was swollen, the spleen was pale and elongated, and some ascitic fluid was present in the peritoneal cavity (Austin *et al.*, 1992b). On the larger rainbow trout, the organism was associated with surface lesions. In particular, the skin was sloughed off along the entire flank of the animals, from operculum to tail, exposing the underlying (necrotic) muscle.

Unidentified Gram-negative rod
The human and animal pathology literature abounds with references to hard-to-identify or unidentified pathogens. An example from fish pathology concerns a hitherto unknown intracellular bacterial pathogen of farmed Atlantic salmon in Ireland (Palmer *et al.*, 1994). During 1992–93 when the water temperature was between 8–9°C and 15–16°C, fish became lethargic, swam close to the water surface, and displayed loss of balance. Apart from an infestation of salmon lice, the fish revealed the presence of petechia or haemorrhagic areas on the abdominal walls, petechia on the pyloric caeca and swim bladder, congestion of the kidney and spleen, splenomegaly, and kidney swelling. The fore and hind guts contained white mucus. Some fish had pale friable livers, pale spleens, visceral adhesions, and false membranes in the peritoneum (Palmer *et al.*, 1994).

Winter ulcer disease
Ulcers, of indeterminate cause, have been appearing on the flanks of Atlantic salmon in seawater during winter, principally in Iceland and Norway (Salte *et al.*, 1994; Lunder *et al.*, 1995; Benediktsdóttir *et al.*, 1998), and more recently in Scotland. Since its first recognition, a view has emerged that a new *Vibrio*, coined *V. viscosus*, may be responsible.

3

Characteristics of the pathogens: Gram-positive bacteria

Taxonomists rarely consider the ecological relevance of their findings. Similarly, ecologists ignore taxonomy, which is perceived to be old-fashioned and/or boring.

Within the realms of fish diseases, it is apparent that the names of bacterial species are often used with little supporting evidence to justify the use of the names. Also, many studies are based on the examination of single isolates the relevance of which to fish pathology or science in general is doubtful.

A. ANAEROBES

Although only two species of anaerobic bacteria, namely *Clostridium botulinum* and *Eubacterium tarantellae*, have been implicated as fish pathogens, it is considered likely that detailed biological investigations may reveal that anaerobes cause more widespread problems than has been hitherto realized. In the first place, diagnostic laboratories do not normally use anaerobic methods. Therefore, it is unlikely that isolation of a offending anaerobic pathogen would ever be achieved. Consequently, the cause of disease may not be recognized, or maybe attributed to an aerobic secondary invader. Although there is no evidence that anaerobic pathogens have been missed, there are puzzling cases of mortalities among fish populations for which the aetiological agent has never been isolated. Of course, this could reflect the use of inappropriate methods. It is recognized, however, that anaerobes occur in aquatic sediments (Davies, 1969; Rouhbakhsh-Khaleghdoust, 1975) and in the gastro-intestinal tract of fish (Sakata *et al.*, 1978; 1980; Trust *et al.*, 1979), where they would be readily available for initiation of a disease cycle. Trust and colleagues recovered *Actinomyces*, *Bacteroides*, *Fusobacterium* and *Peptostreptococcus* from grass carp, gold fish and rainbow trout. Sakata *et al.* (1980) described two groups of anaerobes from the intestines of freshwater fish, i.e. ayu, gold fish and tilapia, in Japan. These Gram-negative, non-motile, asporogenous rods were considered to be representative of the family Bacterioidaceae. It is questionable whether or not these bacteria will be recognized as fish pathogens in the future.

Clostridium botulinum

Descriptions of fish pathogenic clostridia have tended towards extreme brevity. However, it is clear that the outbreaks of botulism in fish have been caused by predominantly *Cl. botulinum* type E.

Clostridium botulinum

Cultures comprise anaerobic, chemo-organotrophic, Gram-positive, non-acid-fast rods, of 3.4–7.5 × 0.3–0.7 μm in size, which are motile by means of peritrichous flagella. It is important to note that care should be taken in interpreting Gram-stained smears, because cells may appear Gram-negative with age. Oval, sub-terminally positioned endospores are formed, which have characteristic appendages and exosporia. The cell wall contains diaminopimelic acid. Surface colonies are 1–3 mm in diameter, slightly irregular with lobate margins and raised centres, and translucent to semi-opaque with a matt appearance. Poor to moderate growth occurs in cooked meat broth, but abundant growth occurs in broth containing fermentable carbohydrates at the optimum temperature of 25–30°C. Lecithinase, lipase, neurotoxins and haemolysins, but not catalase, caseinase, H_2S, indole or urease, are produced. Gelatinase is not usually produced by *Cl. botulinum* type E. Nitrates are not reduced, nor is the Voges Proskauer reaction positive. Some carbohydrates, such as fructose and glucose, but not aesculin, cellobiose, dulcitol, glycogen, inulin, mannitol, melezitose, melibiose, raffinose, rhamnose, salicin, sorbose, starch, sucrose and xylose, are fermented to acetic and butyric acids. The G + C ratio of the DNA is in the range of 26 to 28 moles % (Cato *et al.*, 1986).

Eubacterium tarantellae

It was reported initially by Udey *et al.* (1976) that a novel anaerobic organism was capable of producing a neurological disease in striped mullet (*Mugil cephalus*). This conclusion resulted from an investigation into major fish mortalities in Biscayne Bay, Florida, USA. The anaerobe was subsequently elevated to species status, as *Eubacterium tarantellus* (Udey *et al.*, 1977), and then corrected to *Eu. tarantellae* (Trüper and de'Clari, 1997). It seems likely that *Catenabacterium*, previously described as an anaerobic patho-gen of fish (Henley and Lewis, 1976), should probably be equated with *Eu. tarantellae* (Udey *et al.*, 1977).

Eubacterium tarantellae

On BHIA, the organism produces flat, translucent colonies, approximately 2–5 mm in diameter, which are colourless, rhizoidal and slightly mucoid. These contain long, unbranched, filamentous, Gram-positive, asporogenous rods, which fragment into smaller bacilli of 1.3–1.5 × 1.0–17.0 μm. Good growth occurs at 25–37°C. All isolates degrade blood (β-haemolysis) and lecithin, but not aesculin, gelatin or starch. Catalase, H$_2$S and indole are not produced; nitrates are not reduced, and carbohydrates are generally not fermented. However, there is some evidence for the production of acid from fructose, glucose and lactose, but not from aesculin, amygdalin, arabinose, cellobiose, maltose, mannitol, mannose, melezitose, raffinose, rhamnose, salicin, starch, sucrose, trehalose or xylose.

Unfortunately, the G + C ratio of the DNA was not determined. From these characteristics, Udey *et al.* (1977) proposed that the organisms should be classified in a new species, as *Eubacterium tarantellus*. Clearly, the fish isolates possess the general characteristics of *Eubacterium*, i.e. Gram-positive anaerobic, asporogenous, chemo-organotrophic, non-motile, catalase-negative rods, which grow well at 37°C (Moore and Holdeman-Moore, 1986). A comparison between the descriptions of *Eu. limosum* and *Eu. tarantellae* reveals that, among comparative tests, there are similarities (Table 3.1). In fact, the only major differences concern hydrolysis of aesculin and acid production from lactose and mannitol.

Table 3.1. Comparison of *Eubacterium limosum* with *Eu. tarantellae*

Character	*Eu. limosum*[a]	*Eu. tarantellae*[b]
Catalase production	–	–
Growth at 37°C	+	+
Degradation of:		
Aesculin	+	–
Blood	NS	+
DNA	NS	+
Gelatin	V	–
Lecithin	NS	+
Starch	–	–
Production of acid from:		
Adonitol	+	NS
Aesculin	–	–
Amygdalin	–	–
Arabinose	–	–
Cellobiose	–	–
Dulcitol	–	NS
Fructose	+	+
Glucose	+	+
Lactose	–	+
Maltose	–	–
Mannitol	+	–

Table 3.1. Comparison of *Eubacterium limosum* with *Eu. tarantellae*

Character	*Eu. limosum*[a]	*Eu. tarantellae*[b]
Production of acid from:		
Mannose	–	–
Melezitose	–	–
Raffinose	–	–
Rhamnose	–	–
Salicin	–	–
Starch	–	–
Sucrose	–	–
Trehalose	–	–
Xylose	–	–

[a]   From Moore and Holdeman-Moore (1986).
[b]   From Udey *et al.* (1977).
NS = not stated.
V = variable result.

B. GRAM-POSITIVE BACTERIA—THE 'LACTIC ACID' BACTERIA

Carnobacterium piscicola and the lactobacilli

There have been only a few reports citing *Lactobacillus* spp. as fish pathogens. The initial work was by Ross and Toth (1974), who described mortalities among 3-year-old rainbow trout in a hatchery in California. The term pseudokidney disease was coined by these workers to distinguish the condition from BKD. However, it is not clear what pathological significance, if any, could be attributed to the lactobacilli, because pure cultures were incapable of reproducing infection. Nevertheless, lactobacilli were again implicated in an infection of female 2–3-year-old rainbow trout from a hatchery in Newfoundland, Canada (Cone, 1982). However, this was a mixed bacterial infection involving primarily lactobacilli but also *Aer. hydrophila*, *Ps. fluorescens* and Enterobacteriaceae representatives. There is no report about any pathogenicity experiments using the Canadian *Lactobacillus* strain; therefore, its precise role as a fish pathogen is open to question. From these two initial reports, it is doubtful whether *Lactobacillus* should be regarded as a true fish pathogen. However, additional lactic-acid producing organisms have been isolated from the kidneys of moribund rainbow trout (B. Austin, unpublished data). Moreover, these isolates produced clinical disease upon intraperitoneal injection into salmonids. Similar isolates have been studied by Professor Stemke (personal communication; Table 3.2). Hui *et al.* (1984) proposed a new species, namely *Lactobacillus piscicola*, to accommodate a group of 17 isolates recovered from diseased chinook salmon, cutthroat trout and rainbow trout. Similar isolates were recovered from juvenile salmonids and carp in Belgium and France (Michel *et al.*, 1986b) and from striped bass and catfish in the USA (Baya *et al.*, 1991). *Lactobacillus piscicola* became subsequently reclassified as *Carnobacterium piscicola* (Collins *et al.*, 1987). In addition, fish-pathogenic lactobacilli,

Table 3.2. Characteristics of fish pathogenic lactobacilli

Character	Isolates of Starliper *et al.* (1992)	Isolates of Austin and G. Stemke[a]	*Car. piscicola*
Production of:			
Arginine dihydrolase	−	+	+
Catalase	+/−	−	−
β-Galactosidase	−	−	−
H$_2$S	−	−	−
Indole	−	−	ND
Lysine decarboxylase	−	−	ND
Ornithine decarboxylase	−	−	ND
Oxidase	−	−	−
Phenylalanine deaminase	−	−	ND
Phosphatase	−	+	ND
Gluconate oxidation	−	−	−
Methyl red test	+	+	ND
Nitrate reduction	−	−	−
Degradation of:			
Aesculin	+	+	+
Blood	+ (α)	−	ND
Casein	ND	+	ND
Cellulose, chitin, DNA	ND	−	ND
Elastin, gelatin	ND	−	ND
Hypoxanthine, lecithin	ND	−	ND
RNA, sodium hippurate	ND	−	ND
Starch	−	−	ND
Tween 20	ND	−	ND
Tween 40	ND	+	ND
Tween 60, Tween 80	ND	−	ND
Tyrosine, xanthine	ND	−	ND
Growth at 37°C	−	−	+
Production of acid from:			
Arabinose	+	−	−
Fructose, glucose	+	+	+
Galactose	−	−	+
Glycerol	−	+	+
Lactose, xylose	−	−	−
Maltose, mannitol,	+	+	+
Sucrose	+	+	+
G + C ratio of the DNA	−	35.2–36.5 moles %	34–36 moles%

[a] Unpublished data.

ND = not determined.

with similarities to *Lactobacillus alimentarius* and *Lactobacillus homohiochi*, have been recovered from post-spawning rainbow trout in the USA (Starliper *et al.*, 1992; Table 3.2).

Carnobacterium piscicola
On TSA, *Car. piscicola* produces small round, entire, shiny, opaque colonies that develop within 48 h. These contain non-motile, non-acid-fast, Gram-positive fermentative cocco-bacilli or rods of approximately $1.1–1.4 \times 0.5–0.6\ \mu$m in size. Other phenotypic traits of the lactobacilli include the inability to produce catalase (catalase-positive isolates were described by Starliper *et al.*, 1992), H_2S, indole, gelatinase, urease, the Voges Proskauer reaction, arginine hydrolysis, or lysine or ornithine decarboxylases or reduce nitrates. However, the isolates produce acid from fructose, galactose, glucose, glycerol, inulin, lactose, maltose, mannitol, melibiose, salicin, starch, sucrose and trehalose, but not raffinose, sorbitol or xylose (Schmidtke and Carson, 1994). There are a few differences between the Californian and Canadian isolates, namely hydrolysis of aesculin, and acid production from lactose. Growth occurs at 10°C but not 40°C, in 6.5% (w/v) sodium chloride, and at pH 9.6.

However in most respects, all the isolates were markedly similar. By a comparison of these phenotypic traits with conventional identification schemes, it is apparent that an identification of *Lactobacillus* or *Streptococcus* could result. Furthermore, the photomicrographs of cells, published by Ross and Toth (1974), could be interpreted as chains of cocci rather than short rods, which would be more in keeping with an identification of *Streptococcus*. The isolates recovered from England and Canada, by Austin and Stemke, respectively, possessed more of the characteristics attributable to *Lactobacillus* or the related genera. Cultures comprised Gram-positive, non-motile, fermentative rods of approximately $3.0 \times 1.0\ \mu$m in size, forming round, raised entire white colonies on tryptone soya agar after incubation for 48 h at 20°C. Growth occurred at 4–26°C but not 37°C, and in 2.5% but not 7.5% (w/v) sodium chloride. A resemblance to *Car. piscicola* is readily apparent. Indeed, the only differences reflect growth at 37°C and acid production from lactose. Both the Canadian isolates of Stemke and the USA isolates of Hui *et al.* (1984) and Starliper *et al.* (1992) produced lactic acid from the fermentation reactions. Clearly, it seems that all these lactobacilli may belong in the same taxon, i.e. *Car. piscicola*. However, it will be necessary for further study to determine the true relationship of the English and Canadian isolates to those of Hui *et al.* (1984).

The cell wall peptidoglycan of *Car. piscicola* was found to comprise diaminopimelic acid, alanine and glutamic acid, but no lysine. DNA:DNA hybridization revealed negligible, i.e. 10%, homology with the reference cultures of *Lactobacillus acidophilus*, *Lactobacillus crispatus*, *Lactobacillus jensenii*, *Lactobacillus salivarius* and *Lactobacillus yamanashiensis*, compared to 70% re-association between isolates of *Car. piscicola* (Hui *et al.*, 1984). Thus in their publication, it was concluded that the fish pathogens were most closely related to *Lactobacillus yamanashiensis* subsp. *yamanashiensis* in terms of

G + C ratio of the DNA and fermentation profile. Subsequently, Collins *et al.* (1987) demonstrated closer relationships with other carnobacteria.

Gram-positive cocci in chains (general comments)
Since the first publication in 1958, there has been considerable confusion about the number of and the nature of the bacterial species involved in streptococcicosis/ streptococcosis. Thus at various times, the fish pathogenic streptococci have been linked with *Str. agalactiae, Str. dysgalactiae, Str. equi, Str. equisimilis, Str. (= En.) faecium, Str. pyogenes* and *Str. zooepidemicus*. In addition, we have found that *Enterococcus faecalis* NCTC 775^T, *En. faecium* NCTC 7171^T, *Lactococcus lactis* NCFB 604 and *Str. mutans* NCFB 2062 will cause similar diseases in Atlantic salmon and rainbow trout. Certain traits of the causal agent(s) have been repeatedly emphasized as having supposedly taxonomic significance. In particular, the ability to attack blood has been highlighted. Thus fish pathogenic strains have been described, at one time or another, as either α- (Kusuda *et al.*, 1976a; Al-Harbi, 1994) or β-haemolytic (Robinson and Meyer, 1966; Minami *et al.*, 1979; Kitao *et al.*, 1981; Ugajin, 1981; Iida *et al.*, 1986) or as non-haemolytic (Plumb *et al.*, 1974; Cook and Lofton, 1975; Iida *et al.*, 1986). Superficially, this information could infer heterogeneity among the pathogens, although some well-established taxa, e.g. *Str. agalactiae*, contain both α- and β-haemolytic strains; Table 3.3). Nevertheless, many characteristics are shared by the majority of the fish pathogens. Yet, there is also some variance in the overall descriptions reported by different groups of workers (Table 3.3). For example, Boomker *et al.* (1979) reported that isolates, recovered from the Transvaal in South Africa, grew on MacConkey agar and at 45°C, hydrolysed sodium hippurate, and produced acid from a range of carbohydrates, including galactose, glucose, lactose, maltose, salicin, starch and trehalose, but not from arabinose, glycerol, inulin, mannitol, raffinose, sorbitol, sucrose or xylose. In contrast, Japanese isolates did not grow at 45°C or hydrolyse sodium hippurate (Minami *et al.*, 1979; Kitao *et al.*, 1981; Ugajin, 1981). Of course, such differences may reflect the lack of standardization in the testing regimes or, indeed, point to heterogeneity in the species composition of the organisms (Fig. 3.1).

A comparison of the characteristics of fish pathogenic streptococci and lactobacilli with the results of the comprehensive taxonomy study by Bridge and Sneath (1983) revealed that the isolates, described by Hoshina *et al.* (1958), Robinson and Meyer (1966), Boomker *et al.* (1979), Minami *et al.* (1979), Cone (1982) and Kitao (1982a), approximated to *En. faeculis, Str. equinus, Str. lactis, Str. casseliflavus*, pediococci and the 'aerococcal' group, respectively. Subsequently, some of these taxa have been reclassified in the genus *Enterococcus*. However, the organisms recovered by Kusuda *et al.* (1976a), Ohnishi and Shiro (1978), Minami (1979), Kitao *et al.* (1981) and Ugajin (1981) did not match the descriptions of any of the 28 phena defined by Bridge and Sneath (1983).

With such information, it could readily be assumed from the early literature that streptococcicosis is a syndrome caused by more than one species. To some extent, geographical differences have been implied. For example, the South African isolates of Boomker *et al.* (1979) have been described as comprising the unidentified Lancefield Group D *Streptococcus*; Japanese isolates linked with, but not identical to, *En. faecalis* and *En. faecium* [note: the type strain of *En. faecium* has been determined to be

Table 3.3. Characteristics of fish pathogenic lactobacilli and streptococci

Character	1	2	3	4	5	6	7	8	9	10	11	12	13
Production of arginine dihydrolase	–	NS	NS	NS	+	NS	NS	+	–	+	+	NS	+
Methyl red test	+	NS	NS	NS	NS	NS	NS	NS	NS	+	+	NS	+
Voges Proskauer reaction	–	–	NS	NS	+	NS	NS	NS	NS	–	–	+	+
Degradation of:													
Arginine	NS	–	NS	NS	+	+	NS	+	+	+	+	+	NS
Blood (haemolysis)	α	α/–	–	–	α	β	β	α	β	β	β	NS	–
Sodium hippurate	NS	NS	–	+	–	–	+	–	–	–	–	–	NS
Starch	–	NS	–	–	–	+	NS	–	+	+	+	NS	–
Growth at/in:													
10°C	NS	–	–	–	+	–	–	+	–	–	–	+	–
45°C	–	–	+	–	+	–	+	+/–	–	–	–	+	–
pH 9.6	NS	NS	+	NS	+	–	NS	+	–	–	–	+	+
6.5% (w/v) sodium chloride	+	–	NS	–	+	–	NS	+	–	–	–	+	–
10% bile salts	NS	NS	NS	NS	NS	NS	NS	NS	+	+	NS	+	NS
40% bile salts	NS	–	NS	–	+	–	NS	+	–	–	–	+	NS
0.1% methylene blue	NS	NS	–	–	+	–	NS	+	–	+	–	NSS	NS
Acid production from:													
Arabinose	+	+	+	–	–	–	–	–	–	–	–	–	–
Galactose	–	NS	NS	NS	NS	NS	+	NS	NS	NS	+	NS	–
Glycerol	–	NS	+	–	–	–	–	–	+	–	–	–	NS
Lactose	–	+/–	+	–	–	–	+	+	+	–	–	–	NS
Mannitol	+	+	NS	NS	+	NS	–	+	+	+	–	+	–
Salicin	+	+	+	–	+	+	+	+	+	+	+	+	–
Sorbitol	–	+	+	–	+	–	–	–	–	–	–	–	–
Starch	NS	NS	NS	NS	NS	NS	+	NS	NS	NS	+	NS	NS
Sucrose	+	+	+	+	–	+	–	–	+	+	+	–	+
Trehalose	+	NS	+	–	+	+	+	+	+	+	+	+	–

1 = *Lactobacillus* (Starliper *et al.*, 1992), 2 = *Lactobacillus* (Cone, 1982), 3 = *Streptococcus* (Hoshina *et al.*, 1958), 4 = *Streptococcus* (Robinson and Meyer, 1966), 5 = *Streptococcus* (Kusuda *et al.*, 1976), 6 = *Streptococcus* (Ohnishi and Shiro, 1978), 7 = *Streptococcus* (Boomker *et al.*, 1979), 8 = *Streptococcus* (Minami, 1979), 9 = *Streptococcus* (Minami *et al.*, 1979), 10 = *Streptococcus* (Kitao *et al.*, 1981), 11 = *Streptococcus* (Ugajin, 1981), 12 = *Streptococcus* (Kitao, 1982), 13 = *Streptococcus* (Baya *et al.*, 1990).

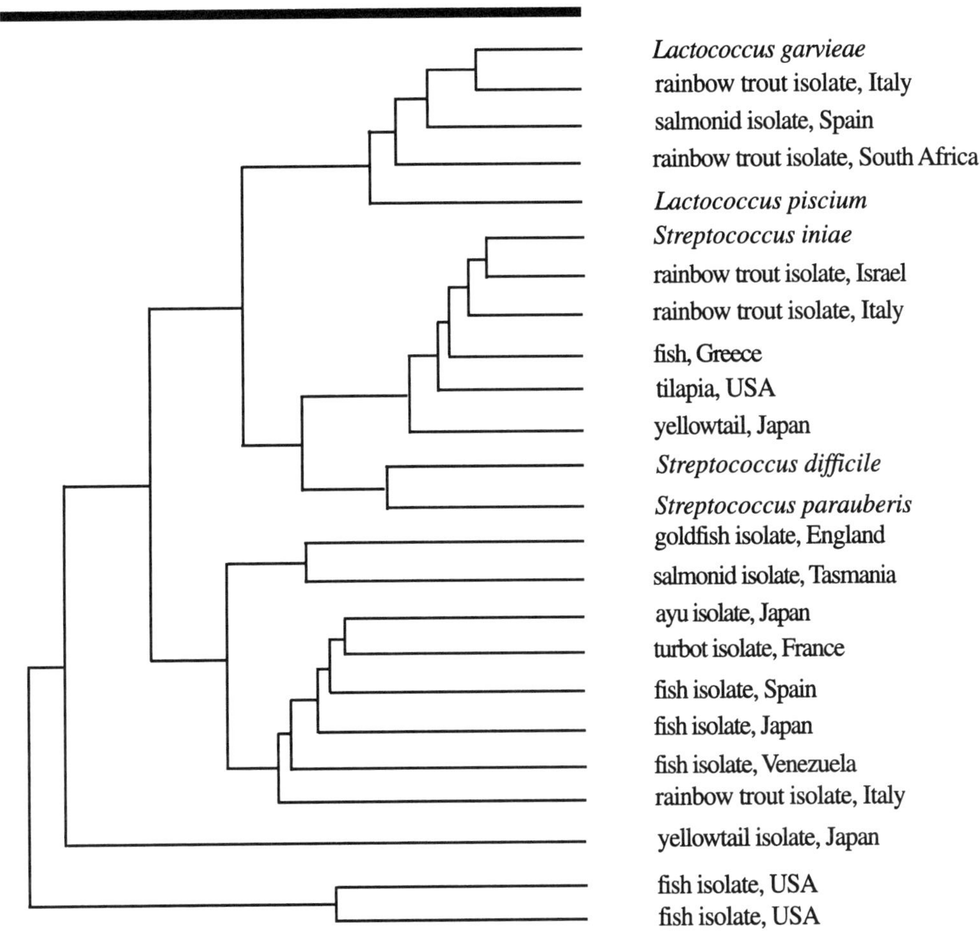

Fig. 3.1. Dendrogram of the relationship between fish-pathogenic Gram-positive cocci in chains, based on a comparison of >100 phenotypic traits. The analyses reveal separate clusters for *Lactococcus garvieae, Lactococcus piscium, Streptococcus iniae, Streptococcus difficilis* and *Streptococcus parauberis*. However, many fish isolates have not been recovered with these taxa, indicating that other groups of fish-pathogens remain to be described.

pathogenic to salmonids in laboratory-based infectivity experiments; Austin, unpublished data]; whereas American strains approximate to the description of *Str. agalactiae* (Kusuda and Komatsu, 1978). It is interesting that isolates from cases of streptococcicosis in rainbow trout farmed in Italy were originally linked with *En. faecalis* and *En. faecium* (Ghittino and Prearo, 1982; 1993) before taxonomic re-appraisal, as indicated below.

Serology, although indicating a multiplicity of serotypes, has confirmed that the fish pathogens are indeed *bona fide* representatives of *Streptococcus/Enterococcus*. Thereafter, serological techniques have not improved the understanding of the precise taxonomic status of the strains. The organisms described by Cook and Lofton (1975) and considered as identical to those of Plumb *et al.* (1974) were identified as Group B type 1_b *Streptococcus* by the Centers for Disease Control, Atlanta. Also, Baya *et al.* (1991) identified their isolates as Group B. However, Boomker *et al.* (1979) regarded the isolates as Group D. To further complicate the issue, Kitao *et al.* (1981) reported a new serotype in Japan, which did not react with specific antisera to Lancefield groups A, B, C, D, E, F, G, H, K, L, N, O and M; this conclusion was also reached by Kitao (1982a) and Kusuda *et al.* (1982).

Enterococcus (Streptococcus) faecalis subsp. *liquefaciens*

We are not satisfied with the identification of these isolates. However, in the absence of an alternative, details are included here (Teskeredzic *et al.*, 1993).

Enterococcus faecalis subsp. *liquefaciens*

Yellow colonies of 1–2 mm in diameter were obtained from kidney and liver on TSA. Colonies comprised Gram-positive cocci which did not produce catalase, H_2S or indole and were negative for the methyl red test and Voges Proskauer reaction. Nitrates were reduced. DNA and gelatin were attacked. Citrate was utilized.

Although there are insufficient data for a meaningful comparison with other taxa, it is interesting to note that streptococci are normally associated with the production of white colonies.

Lactococcus garvieae (= Enterococcus seriolicida)

The first attempt at clarifying the taxonomic status of the causal agents of streptococcicosis/streptococcosis was the landmark publication of Kusuda *et al.* (1991), who described a new species, i.e. *Enterococcus seriolicida*, to accommodate twelve isolates recovered from eels and yellowtail in Japan.

Enterococcus seriolicida

Cultures were described as comprising non-motile facultatively anaerobic Gram-positive cocci in short chains, which do not produce catalase, H_2S, indole or oxidase. Blood is degraded (α-haemolysis). Aesculin and arginine are hydrolysed, but not so casein, gelatin or sodium hippurate. Acid is produced from a wide range of carbohydrates, namely aesculin, cellobiose, D-fructose, galactose, D-glucose, maltose, mannitol, D-mannose, salicin, sorbitol and trehalose, but not from adonitol, D-arabinose, glycerol, glycogen, inositol, lactose, melezitose, melibiose, raffinose, L-rhamnose, starch, sucrose or D-xylose. The methyl red and tetrazolium reduction tests and the Voges-Proskauer reaction are positive, but not nitrate reduction. Growth occurs at 10–45°C but not 50°C, in 0–6.5% (w/v) sodium chloride, and at pH 4.5–9.6. The G + C ratio of the DNA is 44 moles %. The organisms do not belong to Lancefield groups, A, B, C, D, E, F, G, H, K, L, M, N or O.

Interestingly in the original publication describing *En. seriolicida*, only low DNA homology values were obtained with reference species of *Enterococcus*. Indeed, the greatest DNA homology, i.e. 24%, was with *En. hirae* (Kusuda *et al.*, 1991). Perhaps, it was inevitable that the association of these fish pathogens with *Enterococcus* would be challenged. Thus, it was determined that *En. seriolicida* was really identical (77% DNA:DNA homology) with a previously described lactococcus, namely *Lactococcus garvieae* (Teixeira *et al.*, 1996). This view has been reinforced by others. For example, Pot *et al.* (1996) detailed research by SDS-PAGE of whole cell proteins, concluding that *En. seriolicida* was closely related to *Lactococcus garvieae*. Also, Eldar *et al.* (1996) reached the same conclusion after studying the type strains phenotypically and by DNA:DNA hybridization. However, *Lactococcus garvieae* appears to be similar, in terms of phenetic data, to *Lactococcus lactis* (Zlotkin *et al.*, 1998). This similarity could result in mis-identification of fresh isolates.

Lactococcus piscium

The taxonomy of the Group N streptococci has undergone extensive revision. On the basis of 23S rRNA-DNA hybridization and superoxide dismutase studies, the genus *Lactococcus* was defined to accommodate these organisms. A group of fish pathogenic lactococci/Group N streptococci have been studied, and named as a new species, i.e. *Lactococcus piscium* (Williams *et al.*, 1990).

Lactococcus piscium

From the available information, it is apparent that cultures comprise Gram-positive non-motile facultatively anaerobic short (ovoid) rods, which are catalase negative, grow at 5–30°C, produce acid from amygdalin, L-arabinose, arbutin, *N*-acetylglucosamine, cellobiose, D-fructose, galactose, β-gentiobiose, gluconate, glucose, lactose, maltose, D-mannose, mannitol, melibiose, melezitose, D-raffinose, ribose, salicin, sucrose, trehalose, D-turanose and D-xylose, but not adonitol, D-arabinose, D- or L-arabitol, dulcitol, erythritol, D- or L-fucose, glycogen, glycerol, inositol, inulin, 2 and 5-ketogluconate, D-lyxose, α-methyl-D-glucoside, α-methyl-D-mannoside, β-methyl-xyloside, rhamnose, L-sorbose, sorbitol, D-tagatose, xylitol and L-xylose. Aesculin and starch (slow, weak reaction) but not arginine are degraded. H_2S is not produced. The long-chain cellular fatty acids are considered to be of the straight-chain saturated, mono-unsaturated and cyclopropane-ring types. The major acids correspond to hexadecanoic acid, Δ 11-octadecanoic acid and Δ 11-methylenoctadecanoic acid. The G + C ratio of the DNA is calculated as 38.5 moles %.

Streptococcus difficilis (= Str. agalactiae)

Str. difficilis was described to accommodate what was perceived to be a new species of fish pathogen causing meningo-encephalitis in cultured fish, which was first recognized in Israel during 1984 (Eldar *et al.*, 1994; 1995b). The initial work with diseased fish resulted in the recognition of two groups of streptococci; the separation being achieved by use of API 50 CH and API 20 STREP, and by growth and haemolysis characteristics (Eldar *et al.*, 1994). A fairly unreactive non-haemolytic mannitol negative group was labelled as *Str. difficile* (Eldar *et al.*, 1994), and the specific epithet corrected to *difficilis*, i.e. *Str. difficilis* (Euzéby, 1998), whereas a second more reactive α-haemolytic, mannitol positive group became known as *Str. shiloi*.

Streptococcus difficilis
Colonies on BHIA are 1 mm in diameter, and non-pigmented after incubation aerobically for 24 h at 30°C. Cultures comprise fermentative, catalase-negative Gram-positive cocci of varying diameters in small chains, which do not grow at 10, 37 or 45°C, or in 40% bile or 6.5% (w/v) sodium chloride, but do grow at pH 9.6. The isolates attack (produce acid from) *N*-acetyl-glucosamine, D-fructose, D-glucose, maltose, D-mannose, ribose and saccharose, but not adonitol, aesculin, amygdalin, L- or D-arabinose, L- or D-arabitol, arbutin, cellobiose, dulcitol, erythritol, L- or D-fucose, galactose, gentiobiose, *M*-D-glucoside, glycerol, glycogen, inositol, inulin, lactose, melibiose, D-raffinose, rhamnose, salicin, sorbitol, L-sorbose, starch, turanose, xylitol, L- or D-xylose or *M*-xyloside. Alkaline phosphatase, arginine dihydrolase and leucine arylamidase are produced, but not α- or β-galactosidase, pyrrolidonylarylamidase or β-glucuronidase. The Voges Proskauer reaction is positive (Eldar *et al.*, 1994). Bovine blood is not attacked.

These isolates were considered to belong to a separate and distinct DNA homology group, with DNA relatedness between members of 89–100% (Eldar *et al.*, 1994). The level of DNA relatedness with *Str. shiloi (= Str. iniae)* was 17% (Eldar *et al.*, 1994). Then, whole-cell protein electrophoresis revealed that the type strain of *Str. difficilis* was indistinguishable from *Str. agalactiae* (Vandamme *et al.*, 1997). Moreover, it was determined that *Str. difficilis*, which was originally regarded as serologically untypable (Eldar *et al.*, 1994), cross-reacted with group B *Streptococcus*, namely capsular polysaccharide antigen type Ib (Vandamme *et al.*, 1997). Certainly, there were biochemical differences between *Str. difficilis* and *Str. agalactiae*. Yet, in terms of biochemistry, *Str. difficilis* was similar to other group B, type Ib streptococci.

Streptococcus iniae

Streptococcus iniae was initially recovered from an Amazon freshwater dolphin, *Inia geoffrensis* (Pier and Madin, 1976). The association with fish diseases came when it was described as a cause of mortality in tilapia hybrids (*Tilapia nilotica* × *T. aurea*) (Perera *et al.*, 1994) and later in dusky spinefoot (*Siganus fuscenscens*) (Sugita, 1996) and hybrid striped bass (Stoffregen *et al.*, 1996). Then, it was realized that on the basis of DNA:DNA hybridization, i.e. 77–100% DNA homology, *Str. iniae* was synonymous with *Str. shiloi*, which had been previously named as the causal agent of a septicaemic condition in cultured fish, which occurred in Israel in 1984 (Eldar *et al.*, 1994; 1995a,b). This change in the taxonomy was confirmed by others (e.g. Teixeira *et al.*, 1996).

Streptococcus iniae (= Str. shiloi)
Colonies on BHIA are 1 mm in diameter, and non-pigmented after incubation aerobically for 24 h at 30°C. Cultures comprise fermentative, catalase-negative Gram-positive cocci in pairs and chains (some degree of pleomorphism has been observed), which grow at 37°C but not at 10 or 45°C, or in 40% bile or 6.5% (w/v) sodium chloride, but do grow at pH 9.6. Isolates attack (produce acid from) N-acetyl-glucosamine, aesculin, arbutin, cellobiose, D-fructose, gentiobiose, D-glucose, glycogen, maltose, mannitol, D-mannose, melezitose, ribose, salicin, starch and trehalose, but not adonitol, amygdalin, L- or D-arabinose, L- or D-arabitol, dulcitol, erythritol, L- or D-fucose, galactose, gluconate, glycerol, inositol, inulin, lactose, melibiose, D-raffinose, rhamnose, sorbitol, L-sorbose, tagatose, turanose, xylitol, or L- or D-xylose. Alkaline phosphatase, arginine dihydrolase, β-glucuronidase, leucine arylamidase and pyrrolidonylarylamidase are produced, but not α- or β-galactosidase. The Voges Proskauer reaction is negative. α-haemolysis is recorded for bovine blood (Eldar *et al.*, 1994).

By serology using streptococcal specific antisera, the original isolates equated with *Str. shiloi* were untypable. Moreover, these isolates were considered to belong to a separate and distinct DNA homology group, with DNA relatedness between members of 89–100% (Eldar *et al.*, 1994).

Streptococcus milleri

Two cultures were obtained from kidney samples in ulcerated koi carp (Austin and Robertson, 1993). The following characteristics were displayed:

Streptococcus milleri
Cultures contain catalase and oxidase-negative fermentative cocci in chains, that produce acid and alkaline phosphatase, arginine dihydrolase, chemotrypsin, esterase (caprylate and lipase), β-galactosidase, leucine and valine arylamidase and pyrrolidonylarylamidase but not cystine arylamidase, α-fucosidase, α-galactosidase, α- and β-glucosidase, β-glucuronidase, H_2S, indole, α-mannosidase, nitrate reductase, trypsin or tryptophan deaminase. The methyl red and bile aesculin tests are negative. Aesculin, arginine, casein and horse blood (weakly β-haemolytic) are degraded, but not DNA, gelatin, sodium hippurate, starch or urea. Growth occurs in 0–1.5% but not 8% (w/v) sodium chloride and on MacConkey agar. Citrate is utilized. Acid is produced from N-acetyl-glucosamine, amygdalin, D-fructose, D-glucose, inositol, maltose, D-mannose, α-methyl-D-glucoside, ribose, saccharose, sucrose, D-trehalose, xylitol and xylose, but not from L-arabinose, glycogen, inulin, lactose, D-mannitol, D-melibiose, raffinose, rhamnose or sorbitol.

An identification to *Str. milleri* (probability of a correct identification = 98%) resulted from use of the Bacterial Identifier Program (Bryant and Smith, 1991). By means of the API 20 STREP system, an identification of *Lactococcus lactis* resulted (probability of a correct identification = 96.7%). Yet from the published description, there was a closer fit with *Str. milleri*. The only discrepancies concerned utilization of citrate, and acid production from inositol, xylitol and xylose. The isolates from koi carp were positive for these tests.

Streptococcus parauberis

During 1993 and 1994, low-level (0.1–5%) mortalities occurred in farmed turbot of 0.8 to 2 kg in size in northern Spain (Doménech *et al.*, 1996). Isolates were obtained and identified by phenotypic (Rapid ID32 and API 50CH systems) and genotypic data (16S rRNA sequencing) as *Str. parauberis*; an organism known previously as *Str. uberis* genotype II.

Streptococcus parauberis

After overnight incubation, pure cultures produce whitish slightly α-haemolytic colonies of 1.5 to 2 mm in diameter. These contain non-motile Gram-positive short rods/cocco-bacilli in pairs or short chains, which produce alkaline phosphatase, α-galactosidase, β-glucuronidase and pyrrolidonylarylamidase, but not catalase, indole or catalase, grow at 10–37°C but not at 4 or 45°C, in 4.5 but not 6.5% (w/v) sodium chloride or at pH 9.6 or on MacConkey agar, and degrade arginine and hippurate (some strains).

There was a 100% sequence homology between the fish isolates and *Str. parauberis*. The only reliable difference between the turbot isolates and the type strain concerned the action on D-raffinose which was negative for the latter (Doménech *et al.*, 1996).

Vagococcus salmoninarum

In 1968, a so-called lactobacillus was recovered from diseased adult rainbow trout in Oregon, USA and later subjected to detailed taxonomic examination. This isolate, designated OS1-68[T], has become the type strain of *Vagococcus salmoninarum* (Wallbanks *et al.*, 1990). Further isolates have been studied by Schmidtke and Carson (1994).

Vagococcus salmoninarum

Cultures comprise short or oval non-motile, facultatively anaerobic Gram-positive rods, which produce H_2S but not arginine dihydrolase or catalase. Aesculin and blood (α-haemolysis), but not sodium hippurate or urea, are degraded. Nitrates are not produced. The Voges Proskauer reaction is negative. Acid is produced from amygdalin, arbutin, *N*-acetylglucosamine, cellobiose, fructose, β-gentiobiose, glucose (gas is not produced), maltose, mannose, α-methyl-D-glucoside, ribose, salicin, starch, sucrose, D-tagatose and trehalose, but not from D- or L-arabinose, D- or L-arabitol, adonitol, dulcitol, erythritol, D- or L-fucose, galactose, gluconate, glycogen, glycerol, 2- or 5-keto-gluconate, inulin, inositol, lactose, D-lyxose, melibiose, melezitose, methyl-xyloside, methyl-D-mannoside, mannitol, rhamnose,

> raffinose, sorbose, sorbitol, D-turanose, D- or L-xylose or xylitol. Growth occurs at
> 5–30°C but not 40°C, and at pH 9.6. The major cellular fatty acids are of the
> straight-chain saturated and mono-unsaturated types. The G + C ratio of the DNA
> has been calculated as 36.0–36.5 moles %.

Based on an examination of only one culture of *Vag. salmoninarum*, it was established
that 96.3% homology with *Vag. fluvialis* was recorded in the 1340-nucleotide region of
the 16S rRNA. Slightly lower homology values of 94.5%, 94.1%, 94.0%, 93.8% and
93.7% were obtained with *En. durans*, *Car. divergens*, *En. avium*, *Car. piscicola* and
Car. mobile, respectively. Despite the very high similarity to *Vag. fluvialis*, strain
$0/S1–68^T$ was described in a new species, as *Vagococcus salmoninarum*. It remains for
the examination of further isolates to determine the level of genetic variability within the
taxon.

C. AEROBIC GRAM-POSITIVE RODS AND COCCI

Aerobic heterotrophic Gram-positive rods and cocci have received great attention from
fisheries microbiologists, largely because of the severity of the diseases caused by pathogenic representatives. There are ten genus groupings which will be considered in this
chapter, namely *Bacillus, Corynebacterium, Micrococcus, Mycobacterium, Nocardia,
Planococcus, Rhodococcus, Renibacterium, Staphylococcus* and *Streptomyces*.

Bacillus spp.
During 1989 to 1991, low-level mortalities (10–15%) were noted in farmed populations
of *Clarias carpis, Clarias gariepinus, Clarias nigrodigitatus,* 'Heteroclarias' and
Heterobanchus bidorsalis in Nigeria (Oladosu *et al.*, 1994).

> *Bacillus* sp.
> Using nutrient agar plates with an incubation temperature of 37°C, cream, rough
> opaque colonies may be obtained. These colonies comprise non-motile, fermenta-
> tive Gram-positive rods of 1–4 μm in length, which contain central and oval
> endospores. The cells grow at 45°C but not 50°C, and are not haemolytic.

From these data, a link with *Bacillus* was made. However, there is insufficient information to achieve a proper identification.

Bacillus cereus
There has been occasional mention of *B. cereus* as a fish pathogen (Pychynski *et al.*,
1981; Baya *et al.*, 1992a). However, the supporting evidence is weak.

Bacillus mycoides
Cultures were considered to possess the key characteristics of *B. mycoides*, as follows:

> *Bacillus mycoides*
> Cultures are rhizoidal, and contain non-motile Gram-positive rods with oval endospores. Parasporal crystals are not observed. Indole is not produced. The Voges Proskauer reaction is positive. Blood (haemolysis), casein, gelatin, lecithin and tyrosine are degraded. Acid is produced from D-glucose. Resistance is recorded to penicillin. Growth does not occur at 45°C (Goodwin *et al.*, 1994).

Bacillus subtilis

There has been one reference to *B. subtilis* as a fish pathogen (Pychynski *et al.*, 1981). However, the supporting evidence is weak.

Corynebacterium aquaticum

So far, *Corynebacterium aquaticum* has been associated with fish diseases on only one occasion, namely bilateral exophthalmia and cerebral haemorrhaging in striped bass (Baya *et al.*, 1992b). Characteristics of the organism were as follows.

> *Corynebacterium aquaticum*
> Colonies are 1–3 mm in diameter and exhibit a yellow non-diffusible pigment after incubation at 25°C for 48 h. Cultures comprise motile, non-spore-forming, non-acid-fast, slightly pleomorphic (club shapes and angular arrangements) Gram-positive rods, which are neither fermentative nor oxidative, and produce alkaline phosphatase, catalase, β-galactosidase, α-glucosidase, pyrazinamidase, pyrrolido-nylarylamidase, but not N-acetyl-β-glucosaminidase, arginine dihydrolase, β-glucuronidase, H_2S, indole, lysine or ornithine decarboxylase, oxidase or phospholipase. Aesculin, blood (β-haemolysis; only at 37°C), casein and gelatin are degraded, but not urea. Nitrates are not reduced. Citrate is not utilized, nor is acid produced from any of the carbohydrates examined. The Voges Proskauer reaction is positive. Growth occurs at 4–42°C and in 0–5% but not 8% (w/v) sodium chloride.

Identification was achieved using the API-Coryne system, and comparison to the type culture of *Cor. aquaticum* ATCC 14665. The fish isolate and reference culture agglutinated with antisera prepared against both strains. Discrepancies with the named reference culture included growth at 4°C and in 5% (w/v) sodium chloride, degradation of casein and gelatin, nitrate reduction, pyrrolidonylarylamidase, and N-acetyl-β-glucosaminidase (Baya *et al.*, 1992b). Also, the fish isolate differed from the reference culture in the precise composition of the membrane proteins, as determined by Western blotting. However, both cultures shared a 68 kDa major antigenic protein (Bayer *et al.*, 1992b).

Coryneform bacteria

Occasional mention has been made of the role of coryneforms as fish pathogens. Ajmal and Hobbs (1967) referred to *Corynebacterium* infections in rudd, salmon and trout. However, there may have been confusion with BKD, the aetiological agent of which used to be regarded as *Corynebacterium* but is now classified as *Ren. salmoninarum*. Nevertheless during a routine examination of apparently healthy rainbow trout, Austin *et al.* (1985) recovered an organism with some of the salient features of *Ren. salmoninarum*. Cultures were subsequently assigned to the coryneform group of bacteria.

Coryneforms

Growth occurs on BHIA and plate count agar, weakly on CLED, but not on MacConkey agar or TCBS. Characteristically, isolates are non-motile, non-acid-fast rods of $0.75 \times 1.5–3.0\ \mu m$ in size, which contain darkly stained intracellular granules. Growth occurs at 15°C and 30°C but not at 4°C or 37°C, and in 0% and weakly in 2% (w/v) sodium chloride but not at all in 4% (w/v) sodium chloride. Catalase is produced, but not arginine dihydrolase, β-galactosidase, H_2S, indole, lysine or ornithine decarboxylase, oxidase, phenylalanine deaminase or phosphatase. The methyl red test and Voges Proskauer reaction are negative. Nitrates are reduced to nitrites, weakly. Aesculin is degraded, but not blood, DNA, gelatin, lecithin or urea. Sodium citrate is utilized slowly. Acid is not produced from glucose.

However, it must be emphasized that the status of this organism to fish pathology is uncertain.

Micrococcus luteus

Conroy (1966) described a single outbreak of disease, termed micrococcosis, in farmed rainbow trout from Argentina. However, the identification of that aetiological agent is uncertain. Nevertheless, during 1990, a 'micrococcus' was associated with diseased rainbow trout fry in the UK (Austin and Stobie, 1992a). On a rainbow trout farm, deemed to harbour RTFS, large ($\sim 2\ \mu m$ diameter) Gram-positive cocci, displaying a characteristic tetrad arrangement, were recovered from moribund fish.

Micrococcus luteus

The cultures (eight in total were examined) comprise yellow-pigmented non-motile, oxidative Gram-positive cocci, which display a characteristic tetrad arrangement. Acid and alkaline phosphatase, catalase, esterase, leucine arylamidase, lipase, oxidase and phosphoamidase are produced, but not so α- or β-galactosidase, H_2S, indole, lysine or ornithine decarboxylases or tryptophan deaminase. Nitrates are not reduced, nor is the Voges Proskauer reaction positive. Casein and gelatin are degraded.

From these characteristics, it was apparent that the organisms matched the description of *Micrococcus luteus* (Kocur, 1986).

Mycobacterium spp.

The first report of acid-fast bacteria in freshwater fish (carp) was published by Bataillon *et al.* (1897). This was followed over a decade later by an observation in marine fish (von Betegh, 1910). Interest in infections caused by acid-fast bacteria continued with the isolation of *Myc. fortuitum* from diseased neon fish (*Paracheirodon innesi*) in 1953, although its identification was not reported until six years later (Ross and Brancato, 1959). To date, mycobacteriosis (a term suggested by Parisot and Wood, 1960) has been observed in more than 150 species of marine and freshwater fish (Nigrelli and Vogel, 1963). The aetiological agents have been classified, at various times, into a wide assortment of species including *Myc. anabanti, Myc. chelonei, Myc. chelonei* subsp. *piscarium, Myc. fortuitum, Myc. marinum, Myc. piscium, Myc. platypoecilus, Myc. ranae, Myc. salmoniphilum, Myc. scrofulaceum* and *Myc. simiae*. In addition, *Myc. neoarum* has been recovered as mixed culture growth from Atlantic salmon with ocular lesions (Bachman *et al.*, 1990). Unspeciated *Mycobacterium* have also been reported by Kusuda *et al.* (1987). Descriptions of fish pathogenic mycobacteria are generally poor (e.g. Gomez *et al.*, 1993; Hatai *et al.*, 1993). Many publications have been based on purely morphological descriptions, which have resulted from the examination of histological sections. It has been established that the pathogens are Gram-positive, acid-fast, non-motile, pleomorphic rods of approximately 1.5–2.0×0.25–0.35 μm in size (Dulin, 1979). They produce pale-cream to yellow/orange colonies on solid media. The optimum temperature for growth is 25°C, although some isolates grow well at 37°C. From these descriptions and lack of molecular genetic data, it is difficult to determine whether the isolates belong in *Mycobacterium* or *Nocardia*. Conceivably the problem arises from the inherent difficulty in isolating the organisms, and a lack of interest among workers. Of the nomenspecies listed above, it is relevant to note that the taxonomic validity of *Myc. piscium* is in doubt (Van Duijn, 1981), *Myc. anabanti* and *Myc. platypoecilus* are regarded as synonyms of *Myc. marinum* (Van Duijn, 1981), and the slow-growing *Myc. salmoniphilum* is synonymous with *Myc. fortuitum* (Gordon and Mihm, 1959). Therefore, from the early literature, it would appear that *Myc. fortuitum* and *Myc. marinum* were the only *bona fide* species of fish pathogenic mycobacteria, which could be differentiated, as follows:

	Myc. fortuitum	*Myc. marinum*[a]
Nitrate reduction	+	–
Production of nicotinamidase	–	+
Production of pyrazinamidase	–	+

[a] Data from Runyon *et al.* (1974).

A limited range of phenotypic tests were used to study the pathogen equated with *Myc. abscessus* (Teska *et al.*, 1997).

Mycobacterium abscessus
Described as homogeneous, the pathogen produces arylsulphatase, catalase and
pyrazinamidase, degrades Tween 80 and urea, grows in 7 d and in 6.5% (w/v)
sodium chloride, 2% (w/v) thiophenecarboxylic acid and on MacConkey agar
(without crystal violet), but does not reduce nitrate, accumulates niacin and is
negative for iron uptake (Teska *et al.*, 1997).

Six isolates, recovered between 1964 and 1982, were recognized as a new subspecies,
i.e. *Myc. chelonei* subsp. *piscarium*, by Arakawa and Fryer (1984). Essentially, the testing regime was quite extensive, but the value of using only six isolates is questionable.
Colonies on Ogawa medium were off-white with a smooth texture.

Mycobacterium chelonei subsp. *piscarium*
These comprise pleomorphic, acid-fast, weakly Gram-positive rods 1–4 × 0.3–
0.6 μm in size. Neither branching nor aerial hyphae have been observed. Growth
occurs at 10°C and 30°C, but not at all at 37°C, weakly in 3% (w/v) sodium
chloride but not at all in 5% (w/v) sodium chloride, and on MacConkey agar and
potassium tellurite agar, and in 250 μg/ml of azoguanine, 5 μg/ml of ethambutol,
250 μg/ml of hydroxylamine, 0.1% (w/v) malachite green, 0.01% (w/v) methyl
violet, 0.2% (w/v) picric acid, 0.01% (w/v) pyronin B, 1% (w/v) sodium deoxycholate and 0.1% (w/v) sodium nitrate, and variably in 0.01% (w/v) chlorophenol
red and 20 μg/ml of sodium azide. Acid phosphatase, aryl sulphatase and catalase
are produced, but not acetamidase, benzamidase, isonicotinamidase, nicotinamidase, pyrazinamidase or succinimidase. *p*-Aminobenzoate, *p*-aminosalicylate,
sodium salicylate and urea are attacked, but not allantoin, sodium hippurate or
Tweens. Acid is produced from glucose and mannose, but not arabinose, dulcitol,
fructose, galactose, inositol, mannitol, rhamnose, sorbitol, sucrose, trehalose or
xylose. Neither sodium benzoate, sodium malonate nor sodium oxalate are utilized
as the sole source of carbon, but variable responses may be recorded with sodium
citrate, sodium fumarate and sodium succinate. L-serine and sodium L-glutamate
but not acetamide, benzamide, glucosamine hydrochloride or nicotinamide are
utilized as the sole source of carbon and nitrogen. Mycolic acids are present. The
G + C ratio of the DNA is in the range of 61–65 moles %; with an average value of
63 ± 1.7 moles % (Arakawa and Fryer, 1984).

The fish isolates were considered to be related to *Myc. chelonei*, although somewhat
distinct from the current subspecies, i.e. *chelonei* and *abscessus*. For example, the fish
isolates were unable to grow at 37°C, produce nicotinamidase or pyrazinamidase and
degrade sodium hippurate or produce acid from trehalose, in contrast to the two validly
published subspecies. However, there was an overall phenotypic similarity in excess of
85%. Thus, from this similarity together with the results of mycolic acid determination
(the fish isolates were identical with *Myc. chelonei* subsp. *chelonei* and *Myc. chelonei*

subsp. *abscessus* when examined by two-dimensional thin-layer chromatography of acid methanolysates), it was proposed to establish a new subspecies, i.e. *Myc. chelonei* subsp. *piscarium*. Nevertheless to date, this subspecies has not been formally proposed in the referred scientific literature.

Myc. chelonae is slowly increasing in significance in farmed fish. For example, *Myc. chelonae* isolates, not allocated to subspecies, were recovered from diseased Atlantic salmon in two farms in Shetland (Bruno *et al.*, 1998).

Myc. neoaurum was described by Bachman *et al.*, 1990) as follows.

Mycobacterium neoaurum
Contains yellow-pigmented acid-fast rods, which virtually fails to be stained by the Gram's method. Growth occurs on blood agar at room temperature (but not at 37°C) in 5–7 days, but not in 5% (w/v) sodium chloride. Aryl sulphatase is produced. Resistance is recorded to penicillin. The cell wall chemotype is IVA. Glycolated muramic acids, mycolic acids and MK-9, as the predominant isoprenoid quinone, are present (Bachman *et al.*, 1990).

Nocardia spp.

There has been some confusion between distinguishing infections caused by *Mycobacterium* and *Nocardia*, the latter leading to nocardiosis. Thus it is often difficult to determine from largely histological reports the genus to which an acid-fast pathogen belongs. Conroy and Valdez (1962) isolated tubercle bacilli from neon fish (*Paracheirodon innesi*), which were also pathogenic to paradise fish (*Macropodus opercularis*) and three-spot gouramis (*Trichogaster trichopterus*), but not to goldfish. These organisms were subsequently identified by Dr R. E. Gordon as *Nocardia asteroides*. A second species, *Noc. kampachi*, was described as a causal agent of nocardiosis, in yellowtail farmed in Japan, by Kariya *et al.* (1968) and Kubota *et al.* (1968). However, the name was not validated. Instead, *Noc. seriolae* was formally proposed, as a causal agent of fish nocardiosis (Kudo *et al.*, 1988).

There is limited information available about the characteristics of fish pathogenic strains of *Noc. asteroides*.

Nocardia asteroides
Most of the description appertains to morphological characters. For example, it has been observed that cultures undergo a complete life cycle, including germination from resting microcysts, simple and complex fission, and branching. Thus, nocardias may appear in coccal to oval forms, and as long, slender, multiseptate rods. All these stages have been seen in infected fish (Van Duijn, 1981). From Valdez and Conroy (1963), it would appear that the organisms reduce nitrates and degrade starch but not gelatin or urea. Neither H_2S nor indole is produced. The methyl red test and Voges Proskauer reaction are negative. Acid is produced from glucose, but not from any other carbohydrate tested.

A better description exists for *Noc. kampachi* (Kariya *et al.*, 1968; Kubota *et al.*, 1968; Kusuda *et al.*, 1974). However, in retrospect it would appear that Kariya and co-workers exercised considerable taxonomic intuition in elevating the isolates into a new species, i.e. *Noc. kampachi*. Unfortunately, the results of G + C determinations were not reported. Moreover, it is perhaps surprising that the original authors did not provide detailed reasons explaining why *Noc. kampachi* should be regarded as distinct from other species of *Nocardia*. This is especially relevant as there is some resemblance between the description of *Noc. kampachi* and *Noc. caviae* (Table 3.4). Nevertheless, a more recent publication has formally proposed other nomenspecies, i.e. *Noc. seriolae*, which effectively replaces *Noc. kampachi* (Kudo *et al.*, 1988).

Nocardia seriolae

The isolates of *Noc. seriolae* contain *meso*-diaminopimelic acid, arabinose and galactose, suggesting chemotype IVA. The major components of the cellular fatty acids are n-$C_{16:0}$, n-$C_{16:1}$ and n-$C_{18:1}$; 10-methyl-$C_{19:0}$ is also present as a major component in four of the five isolates examined. Iso- and anteiso- branched acids have not been detected. The total number of carbon atoms in the mycolic acids is from 44 to 58. The predominant isoprenoid quinone is tetrahydrogenated menaquinone with eight isoprene units. The G + C ratio of the DNA is 66.8–67.4 moles %.

Table 3.4. Characteristics of nocardias

Character	*Nocardia asteroides*[a]	*Noc. caviae*[a]	*Noc. kampachi*[b]	*Noc. seriolae*[c]
Gram-positive rods and cocci	+	+	+	+
Weakly acid-fast staining	+	+	+	ND
Aerial hyphae	+	+	+	+
Motility	−	−	−	−
Growth at 10°C	v	v	−	−
Production of:				
Catalase	v	+	+	ND
H_2S	ND	ND	+	ND
Indole	ND	ND	−	ND
Oxidase	−	−	−	ND
Nitrate reduction	+	+	+	ND
Degradation of:				
Casein	−	−	−	−
Gelatin	−	−	−	ND
Hypoxathine	−	+ (weak)	+ (weak)	−
Starch	ND	ND	+	ND
Tyrosine	−	−	+ (weak)	−

Table 3.4. (*continued*)

Character	*Nocardia asteroides*[a]	*Noc. caviae*[a]	*Noc. kampachi*[b]	*Noc. seriolae*[c]
Urea	+ (weak)	+ (weak)	−	−
Xanthine	−	+	−	−
Acid production from:				
Fructose	+	+	+	ND
Glucose, glycerol	+	+	+	+
Utilization of:				
Adonitol, arabinose	−	−	−	ND
Cellobiose	−	v	−	ND
Dextrin, maltose	v	v	−	ND
Dulcitol, glycogen	−	−	−	ND
Fructose, glucose	+	+	+	ND
Glycerol	+	−	v	ND
Inositol	−	+	−	ND
Inulin	−	−	v	ND
Lactose, salicin	−	−	−	ND
Mannitol	v	+	−	ND
Mannose	+	v	v	ND
Rhamnose	v	−	−	ND
Sodium acetate	+	+	+	ND
Sodium benzoate	−	−	−	−
Sodium citrate	−	v	+	+
Sodium lactate	+	+	+	ND
Sodium malate	+	+	+	ND
Sodium malonate	−	−	v	ND
Sodium propionate	+	+	+	ND
Sodium pyruvate	+	+	+	ND
Sodium tartrate	−	−	v	−
Sorbitol, xylose	−	−	−	ND
Starch, trehalose	v	v	−	ND

[a]   From Goodfellow (1971).
[b]   From Kusuda *et al.* (1974).
[c]   From Kudo *et al.* (1988).
ND = not determined.
v = variable response.

Planococcus sp.

Since 1988, there has been a steady increase in the incidence of motile Gram-positive cocci, tentative *Planococcus*, associated with diseases of Atlantic salmon and rainbow trout in the UK. In some cases, it appeared that the Gram-positive cocci were inhabiting fish which had previously received extensive chemotherapy to control diseases caused by Gram-negative bacteria.

Planococcus sp.
Cultures comprise motile (single polar flagellum) often paired Gram-positive cocci of 1–2 μm in diameter, which possess a strictly aerobic metabolism for glucose, and produce catalase, β-galactosidase and oxidase, but not gelatinase, H$_2$S, indole or lysine decarboxylase. Growth occurs at 37°C and in 0–15% (w/v) sodium chloride.

From these traits and despite a freshwater rather than a marine origin, it was considered that the organisms belonged in the genus *Planococcus*, possibly related to *P. citreus* (Hao and Komagata, 1985).

Renibacterium salmoninarum

Bacterial kidney disease (BKD, Dee disease, corynebacterial kidney disease, salmonid kidney disease) was described initially in 1930 for a condition in Atlantic salmon (*Salmo salar*) found in the Rivers Dee and Spey in Scotland (Mackie *et al.*, 1933; Smith, 1964). Histological examination of fixed material revealed the presence of large numbers of Gram-positive rods in lesions. At various times, the causal agent of BKD has been linked with *Corynebacterium* (Ordal and Earp, 1956; Smith, 1964; Sanders and Fryer, 1978; Austin and Rodgers, 1980), *Brevibacterium* (Smith, 1964), *Listeria* (Bullock *et al.*, 1975), *Lactobacillus* (Vladik *et al.*, 1974) and *Rickettsia* (Snieszko and Griffin, 1955). Subsequently, it was appreciated that the organisms were sufficiently unique to warrant separate species status, so, *Corynebacterium salmoninus* was described (Sanders and Fryer, 1978). With further information, these authors realized that the pathogen belonged in a new, as yet undescribed genus and, therefore, proposed *Renibacterium*. Thus, the causal agent of BKD became classified as *Renibacterium salmoninarum* (Sanders and Fryer, 1980).

The initial difficulties experienced in culturing the pathogen contributed significantly to the uncertainty over its precise taxonomic status. Early work emphasized a few morphological features, namely the presence of small (0.3–1.5 × 0.1–1.0 μm), Gram-positive, asporogenous, non-motile, non-acid-fast rods, which frequently occurred in pairs. Evidence of pleomorphism, metachromatic granules and a 'coryneform' appearance (Ordal and Earp, 1956; Smith, 1964) led to the initial, tenuous association with the coryneform group of bacteria, namely *Corynebacterium*. It is interesting to note that the later investigation of Young and Chapman (1978) did not substantiate the 'coryneform' morphology. However, transmission electron microscopy of negatively stained cells, obtained from 28-day-old cultures on growth medium, i.e. KDM2 (see Chapter 5), revealed the presence of pleomorphism and intracellular vacuoles/granules (B. Austin, unpublished data). Therefore, morphological studies should be treated cautiously.

Earp (1950) and Ordal and Earp (1956) demonstrated catalase and proteolytic activity, and realized that there was a growth requirement for cysteine. Additional attributes of the organism were slowly realized; in particular, Smith (1964) indicated the temperature range of growth, i.e. most rapid at 15°C, slow at 5°C and 22°C, and not at all at 37°C, and determined an inability to degrade gelatin. During the period of the late 1970s to early 1980s, a wealth of knowledge was accumulated on *Renibacterium*. A low genetic diversity among North American isolates has been indicated from multilocus enzyme electrophoresis using 44 enzymes (Starliper, 1996). Thus, from 40 isolates, 21 electrophoretic types were recognized.

Renibacterium salmoninarum

Characteristically, *Ren. salmoninarum* produces cream (non-pigmented), shiny, smooth, round, raised, entire, 2-mm diameter colonies on KDM2 after incubation at 15°C for 20 days. Old cultures, i.e. 12 weeks, may become extremely granular and crystalline in appearance. Indeed, a transverse section through such colonies will reveal the presence of a few Gram-positive rods embedded in a crystalline matrix. Subculturing at this stage often leads to the development of more crystalline 'colonies'. It is thought that the material is principally cystine, which has been precipitated from the medium. For some strains, a uniformly turbid growth occurs in broth, but for others, a sediment may develop. The cell wall peptidoglycan of renibacteria contains D-alanine, D-glutamic acid, glycine and lysine as the diamino acids (Fiedler and Draxl, 1986). The principal cell wall sugar is glucose, but arabinose, mannose and rhamnose are also present (Sanders and Fryer, 1980). Here, there is a discrepancy with the more recent work of Kusser and Fiedler (1983). These authors reported that the principal cell wall sugar is galactose, with less amounts of *N*-acetyl-glucosamine, rhamnose and *N*-acetyl-fucosamine. This is a curious anomaly insofar as the same strain, i.e. the type strain (ATCC 33209), is common to both studies. Mycolic acids are absent. Methyl-branched fatty acids form over 92% of the total fatty acid component of the cells, with 12-methyltetradecanoic (anteiso-C_{15}), 13-methyldecanoic (iso-C_{15}) and 14-methylhexadecanoic (anteiso-C_{17}) as the major components. Straight chain fatty acids generally account for 1% of the total fatty acids, and unsaturated fatty acids are not detected at all. Over 81% of the total fatty acids are composed of the lower melting point anteiso acids, which may contribute to membrane fluidity at low temperatures. Unsaturated menaquinones with nine isoprene units are present. All strains contain diphosphatydylglycerol, two major and six or seven minor glycolipids and two unidentified minor phospholipids (Embley *et al.*, 1983). Although renibacteria were considered to be serologically homogeneous (Bullock *et al.*, 1974; Getchell *et al.*, 1985), two antigenic groups have been described recently (Bandín *et al.*, 1992). These groups have been defined after analyses of membrane proteins, which determine the presence of 57 kDa and 30 kDa molecules in the respective groups. The G + C ratio of the DNA has been calculated as 53.0 ± 0.46 moles % by Sanders and Fryer (1980) and as 55.5 moles % by Banner *et al.* (1991). Additional characteristics of *Ren. salmoninarum* have been included in Table 3.5.

Table 3.5. Characteristics of *Renibacterium salmoninarum*[a]

Character	Response
Production of:	
Acid and alkaline phosphatase	+
Butyrate esterase	−
Caprylate esterase	+
Catalase	+
Chymotrypsin	−
Cystine arylamidase	−
α-Fucosidase	−
α- and β-Galactosidase	−
β-Glucosaminidase	−
α-Glucosidase	+
β-Glucosidase	−
β-Glucuronidase	−
Leucine arylamidase	+
α-Mannosidase	+
Myristate esterase	−
Oxidase	−
Trypsin	+
Valine arylamidase	−
Nitrate reduction	−
Degradation of:	
Adenine, aesculin, arbutin, chitin, chondroitin, DNA	−
Casein, Tributyrin, Tween 40 and 60	+
Elastin, gelatin, guanine, hyaluronic acid, hypoxanthine	−
Lecithin, RNA, starch, testosterone Tween 80, tyrosine	−
Xanthine	−
Acid production from sugars	−
Growth on/at:	
pH 7.8	+
0.025% (w/v) bile salts, 0.001% (w/v) methylene blue	−
0.0001% (w/v) crystal violet, 0.00001% (w/v) nile blue	+
0.005% (w/v) phenol, 1% (w/v) potassium thiocyanate	−
1% (w/v) sodium chloride	+ (poor)
0.01% (w/v) sodium selenite, 0.001% (w/v) thallous acetate	−
Utilization of:	
4-Umbelliferyl-acetate, 4-umbelliferyl-butyrate	+
4-Umbelliferyl-β-D-cellobiopyranoside monohydrate	−
4-Umbelliferyl-elaidate, 4-umbelliferyl-α-L-arabinopyranoside	−
4-Umbelliferyl-2-acetamido-2-deoxy-β-D-galactopyranoside	−
4-Umbelliferyl-β-L-fucopyranoside	−

Table 3.5. (*Continued*)

Character	Response
Utilization of:	
4-Umbelliferyl-heptanoate, 4-umbelliferyl-laurate	+
4-Umbelliferyl-nonanoate, 4-umbelliferyl-oleate	+
4-Umbelliferyl-palmitate	−
4-Umbelliferyl-propionate	+

[a] From Embley (1983) and Goodfellow *et al.* (1985).

The exact taxonomic position of *Ren. salmoninarum* is uncertain. However, the numerical phenetic study of Goodfellow *et al.* (1985) confirmed the homogeneity of the taxon, and demonstrated its dissimilarity to *Lactobacillus* and *Listeria* (*Lis. denitrificans*) (Fig. 3.2). The results of the chemotaxonomy study discussed above, also indicated the unique position of renibacteria. On the basis of the fatty acid data, *Renibacterium* is distinguishable from *Corynebacterium sensu stricto* and other representatives of mycolic-acid-containing taxa, which have predominantly straight chain and mono-unsaturated fatty acids. In short, the data indicate that *Renibacterium* is distinct from other Gram-positive organisms (Embley, 1983; Embley *et al.*, 1983; Goodfellow *et al.*, 1985), although its relationship to *Cor. (Actinomyces) pyogenes* needs clarification. On the basis of 16S rRNA cataloguing, *Ren. salmoninarum* was considered to be a member of the actinomycete subdivision, being related to *Arthrobacter*, *Brevibacterium*, *Cellulomonas*, *Jonesia*, *Micrococcus*, *Promicromonospora*, *Stomatococcus* and *Terrabacter* (Stackebrandt *et al.*, 1988; Gutenberger *et al.*, 1991).

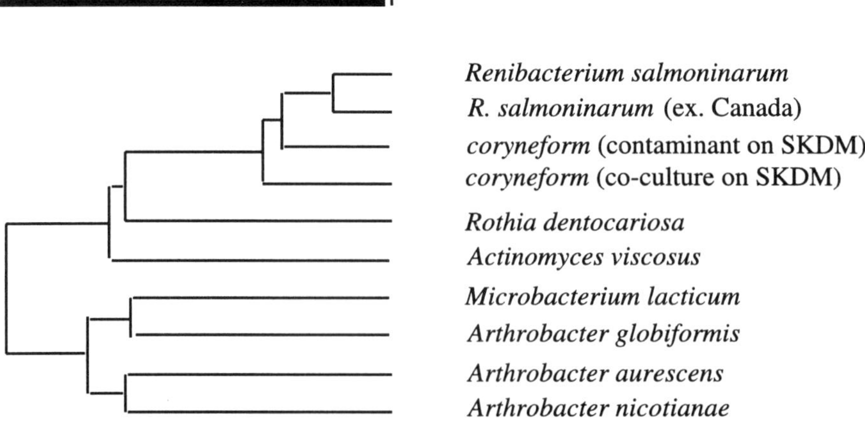

Fig. 3.2. Dendrogram, based on the examination of phenotypic data, showing the relationship of *Ren. salmoninarum* to other Gram-positive bacteria (modified from Goodfellow *et al.*, 1985). The relative uniqueness of *Ren. salmoninarum* is apparent. However, some coryneforms, which were recovered from diseased salmonids on SKDM, showed some similarity with *Ren. salmoninarum*. Yet, there was no evidence that these coryneforms posed any problems to salmonids in laboratory-based infectivity experiments.

Rhodococcus sp.

In Canada, farmed Chinook salmon, *Oncorhynchus tshawytscha*, have been found with exophthalmia, from which Gram-positive bacteria were recovered. These organisms have been cultured, and identified as *Rhodococcus* (Bachman *et al.*, 1990). Subsequent examination of similar isolates from Atlantic salmon has also suggested an affinity with *Rhodococcus* (Claveau, 1991).

One colony type was equated with *Mycobacterium* whereas the second was regarded as *Rhodococcus*.

Rhodococcus sp.

Isolates comprise non-acid-fast facultatively anaerobic Gram-positive rods (2–3×0.6 μm in size) [slightly club-shaped], which grow aerobically at room temperature (but not at 37°C) on blood agar in 3–4 days. The cell wall components include *meso*-diaminopimelic acid, arabinose, and galactose, which equates with chemotype IVA, *N*-glycolated muramic acid, mycolic acids, and MK-8 as the predominant isoprenoid menaquinone. Neither catalase nor oxidase is produced. Urea is degraded. Xylose is fermented, but not so glucose, lactose, maltose or sucrose. Acid is produced from *meso*-inositol but not from dulcitol, mannitol or sorbitol. Growth occurs in 5% (w/v) sodium chloride.

Staphylococcus aureus

During 1982 and 1983, mortalities occurred among silver carp, *Hypophthalmichthys molitrix*, at a farm in India. These mortalities were associated with eye disease, from which Gram-positive cocci were recovered, and identified as *Staphylococcus aureus* (Shah and Tyagi, 1986).

Staphylococcus aureus

The cultures were described as comprising Gram-positive cocci, which produce coagulase and phosphatase, degrade blood (β-haemolysis) and DNA, and ferment glucose and mannitol. Zones of opalescence develop around (black) colonies on Baird Parker agar.

From this description, an identification of *Staphylococcus aureus* was achieved (Shah and Tyagi, 1986). However, it is apparent that there are insufficient data to differentiate between *Sta. aureus* and *Sta. intermedius* (Kloos and Schleifer, 1986).

Staphylococcus epidermidis

The only reports of fish pathogenic strains of *Staphylococcus epidermidis* have emanated from Japan where, from July 1976 to September 1977, severe epizootics occurred in farmed yellowtail (*Seriola quinqueradiata*) and red sea bream (*Chrysophrus major*) (Kusuda and Sugiyama, 1981; Sugiyama and Kusuda, 1981a,b). From these outbreaks, six isolates, identified as *Sta. epidermidis*, were recovered.

Staphylococcus epidermidis
All cultures comprise non-motile Gram-positive, fermentative spherical cells of approximately 0.6–1.8 μm in diameter, which form white to white/yellow colonies on BHIA. The cells occur singly, in pairs, and in irregular clusters. Catalase, β-galactosidase and phosphatase are produced, but not arginine dihydrolase, coagulase, H_2S, indole, lysine or ornithine decarboxylase or oxidase. Nitrates are reduced. The methyl red test and Voges Proskauer reaction are positive. Blood (β-haemolysis), gelatin, sodium hippurate, tributyrin and urea are degraded, but not starch, Tween 80, tyrosine or xanthine. Growth occurs at 45°C and in 0–15% (w/v) sodium chloride. Neither citrate, mucic acid nor D-tartrate is utilized (Sugiyama and Kusuda, 1981a).

These isolates matched closely the species description of *Sta. epidermidis* (Kloos and Schleifer, 1986), and approximated to biotypes II, V and VI of Baird-Parker (1963; 1965). It is curious that a diverse range of serotypes, i.e. five serotypes, were recognized (Sugiyama and Kusuda, 1981a). The G + C content of the DNA was not assessed.

Streptomyces salmonis (= Streptoverticillum salmonis)
Streptomycosis in fish was described initially by Rucker (1949), who classified the aetiological agent in *Streptomyces*, as *Streptomyces salmonicida*. This was amended initially to *Verticillomyces salmonicida* (Shinobu, 1965), then to *Streptoverticillium salmonicida* (Baldacci *et al.*, 1966), *Streptoverticillium salmonis* (Locci *et al.*, 1969) and more recently to *Streptomyces salmonis* (Witt and Stackebrandt, 1990). In general, a dearth of information exists about this fish pathogen. Essentially, since the work of Rucker (1949), the disease has received limited attention. Therefore, it is difficult to decide whether or not streptomycosis represents a genuine problem.

Streptomyces salmonis
The organism produces brick-red to orange pigmented substrate mycelia with white (with pink and yellow shades) aerial mycelia. On primary isolation, the colonies are small and, initially, smooth, but later they develop aerial mycelia that appear velvety. Characteristically, the aerial mycelia produce whorls (verticils) at frequent intervals, giving an appearance of barbed wire. The Gram-positive mycelia contain LL-diaminopimelic acid (DAP) and glycine but not *meso*-DAP, arabinose or galactose in the cell wall (i.e. Type I). Melanin, but generally not H_2S, is produced. Casein, DNA, gelatin, hypoxanthine, starch, Tween 20 (some isolates) and tyrosine are degraded. Growth occurs at 12°C and 37°C but not at 45°C, in 5% (w/v) sodium chloride and 0.01% (w/v) potassium tellurite, but not in 0.01% (w/v) crystal violet or 0.01% (w/v) malachite green. Acid is produced from glucose, inositol, ribose and trehalose, and slightly from sucrose, but not from arabinose, cellulose, fructose, mannitol, raffinose, rhamnose or xylose. L-proline is utilized (Williams *et al.*, 1985). The G + C ratio of the DNA has not been reported.

4

Characteristics of the pathogens: Gram-negative bacteria

Some Gram-negative bacteria have been described as fish pathogens with little if any supportive evidence. Consequently a question mark hangs over whether or not such organisms should be regarded as *bona fide* pathogens. In the following text, it will become apparent that varying amounts of information are available for the Gram-negative bacterial fish pathogens. For some, e.g. *Aer. salmonicida*, a wealth of information is available. Unfortunately for others, such as *Aer. bestiarum*, barely enough information is available to enable scientists to recognize fresh isolates.

A. AEROMONADACEAE REPRESENTATIVES

Aeromonas is classified in the family Aeromonadaceae (Colwell *et al.*, 1986). In addition to the aeromonads detailed below, *Aer. veronii* has been implicated as a potential fish pathogen but only in laboratory-based experiments, where intramuscular injection of 10^7 cells/ml resulted in muscle necrosis in Atlantic salmon (McIntosh and Austin, 1990b).

Aeromonas allosaccharophila

During an examination of 16S rRNA sequences of motile aeromonads, two isolates, which were originally recovered from diseased elvers in Spain during 1988, were considered as sufficiently distinct from existing genospecies to warrant description as a new species, *Aer. allosaccharophila* (Martinez-Murcia *et al.*, 1992).

Aeromonas allosaccharophila
Cultures comprise Gram-negative, motile fermentative rods, which produce catalase, β-galactosidase, indole, lysine decarboxylase and oxidase but not H_2S, reduce nitrates, degrade casein, DNA, egg yolk, gelatin, starch and Tween 80 but not elastin, sodium dodecyl sulphate or urea, and grow in 0–3% (w/v) sodium chloride,

at 4–42°C and at pH 9.0. The Voges Proskauer reaction is negative. Acid is produced from D-cellobiose, D-galactose, glucose (plus gas), glycerol, maltose, D-mannitol, D-mannose and D-trehalose, but not from adonitol, arbutin, dulcitol, *m*-erythritol, *m*-inositol, lactose, salicin, D-sorbitol or D-xylose. A wide range of compounds are utilized as sole sources of carbon for energy and growth, including L-arabinose, L-arginine, D-cellobiose, fumarate, D-galactose, D-gluconate, L-glutamate, glycerol, L-histidine, maltose, D-mannitol, D-mannose, L-proline, succinate, sucrose and D-trehalose, but not L-alanine, γ-aminobutyrate, L-citrulline, dulcitol, *m*-erythritol, ethanol, D-glucuronate, L-glutamine, glycine, DL-3-hydroxybutyrate, *m*-inositol, α-ketoglutarate, lactose, L-leucine, propionate, putrescine, salicin or L-serine. Susceptibility has been recorded to chloramphenicol, erythromycin, fosfomicin, gentamicin, kanamycin, nalidixic acid, nitrofurantoin, oxolinic acid, polymyxin B and rifampicin, but not to ampicillin, streptomycin, sulfadimethoxine or trimethoprim. The G + C content of the DNA is 59.6 moles % (Martinez-Murcia *et al.*, 1992).

The basis of allocating the isolates to a new species stemmed from the examination of 16S rRNA sequences, where homology values of >97.7% were exhibited to other validly described *Aeromonas* species (Martinez-Murcia *et al.*, 1992). Three isolates were found to be highly related, i.e. 70–100%, by DNA:DNA hybridization (Esteve *et al.*, 1995a). On the basis of AFLP fingerprinting, *Aer. allosaccharophila* has been determined to be genetically related to *Aeromonas* HG 8/10 (Huys *et al.*, 1996).

Aeromonas bestiarum
Aer. bestiarum appears to be an example of a taxon which emerged from the taxonomic chaos surrounding the understanding of *Aer. hydrophila* (Ali *et al.*, 1996). Originally, classified in DNA HG 2 (*Aer. hydrophila*), isolates have apparently been recovered from diseased fish (Huys *et al.*, 1996). However, negligible information is available about the characteristics of the supposed fish-pathogenic cultures.

Aeromonas caviae
Cultures were identified by recourse to the scheme in Popoff (1984).

Aeromonas caviae
Cultures produce arginine dihydrolase, β-galactosidase and indole, but not H_2S, lysine or ornithine decarboxylase or tryptophan deaminase, degrade aesculin, blood and gelatin but not urea, ferment amygdalin, arabinose, glucose, mannitol, sorbitol and sucrose, but not inositol, melibiose or rhamnose, utilize potassium cyanide but not citrate, grow in 0% (w/v) sodium chloride, and do not reduce nitrates (Candan *et al.*, 1995).

Aeromonas hydrophila
The significance of *Aer. hydrophila* and its synonyms *Aer. formicans* and *Aer.*

liquefaciens, has been overshadowed by *Aer. salmonicida*. Nevertheless, since its initial recognition as the causal agent of haemorrhagic septicaemia (Sanarelli, 1891; Schäperclaus, 1930; Haley *et al.*, 1967), *Aer. hydrophila* has been recovered as a pathogen from a wide variety of freshwater fish species (e.g. Pathiratne *et al.*, 1994) and occasionally from marine fish, e.g. ulcer disease of cod (Larsen and Jensen, 1977). The aetiological agent may be considered to have worldwide distribution. However, some doubt has been expressed over its precise role as a fish pathogen (Heuschmann-Brunner, 1965; Eurell *et al.*, 1978, Michel, 1981) with some workers contending that it may be merely a secondary invader of already compromised hosts. Conversely, other groups have insisted that *Aer. hydrophila* constitutes a primary pathogen. Yet with improvements in the taxonomy of the 'motile' aeromonads (see Carnahan and Altwegg, 1996), it is a matter of speculation whether or not the fish isolates belong as *Aer. hydrophila* or in any of the other *Aeromonas* Hybridization Groups. To some extent, the improvements in aeromonad taxonomy may reflect the sudden emergence of other taxa as fish pathogens. Certainly, there is marked phenotypic, serological and genotypic heterogeneity within the descriptions of fish pathogenic *Aer. hydrophila* (MacInnes *et al.*, 1979; Leblanc *et al.*, 1981; Allen *et al.*, 1983a).

Aeromonas hydrophila
The fish pathogens comprise Gram-negative fermentative rods of approximately 0.8–1.0×1.0–$3.5\ \mu$m in size, which are motile by single polar flagella. Arginine dihydrolase, catalase, β-galactosidase, H_2S, indole, lysine decarboxylase (a variable response may occur), oxidase and phosphatase are produced, but not ornithine decarboxylase or phenylalanine deaminase. Nitrates are reduced. The Voges Proskauer reaction is positive, but not so the methyl red test. Growth occurs in 0–4% (w/v) but not 5% (w/v) sodium chloride, at 5–37°C and in potassium cyanide. Aesculin, blood (β-haemolysis; by some isolates), casein, DNA, gelatin, lecithin, RNA, starch and Tween 80 are degraded, but not pectin or urea. Sodium citrate is utilized by some isolates. Acid is produced from cellobiose (a variable response), fructose, galactose, glucose, glycerol (a variable response), lactose (a variable response), maltose, mannitol, salicin, sucrose and trehalose, but not from adonitol, dulcitol, erythritol, inositol, raffinose, rhamnose, sorbitol or xylose (Paterson, 1974; Larsen and Jensen, 1977; Allen *et al.*, 1983a; Popoff, 1984) The G + C ratio of the DNA falls in the range of 58–61.6 moles % (Larsen and Jensen, 1977; MacInnes *et al.*, 1979).

In contrast to the usual characteristics of *Aer. salmonicida*, the majority of isolates of *Aer. hydrophila* are capable of growth at 37°C and are, indeed, motile. Some isolates have also been determined to produce diffusible brown pigments, as does *Aer. salmonicida*, which could superficially confuse diagnosticians (Ross, 1962; Paterson, 1974; Allen *et al.*, 1983a,b; Austin *et al.*, 1989). Santos *et al.* (1991) serotyped 62 motile *Aeromonas* spp. from rainbow trout. Of these, 55 isolates (89% of the total) were distributed between 17 serogroups, of which O3, O6, O11 and O19 were dominant. Moreover, 40 (63% of the total) of these isolates were pathogenic to fish. Nevertheless, antigenic cross-reactivity with *Aer. salmonicida* and *Aer. sobria* has been noted (Leblanc *et al.*, 1981). However, Shaw and Hodder (1978) reported that the core region of the LPS of *Aer. hydrophila* was distinct from that of *Aer. caviae* and *Aer. sobria*.

Aeromonas jandaei

Aer. jandaei has been reported as pathogenic to eel in Spain (Esteve *et al.*, 1993, 1994; Esteve, 1995). Initially, eight isolates were recovered using unstated procedures during 1987 and 1988. Whereas initially, the method of identification was not stated (Esteve *et al.*, 1994), a subsequent numerical taxonomy study equated isolates with *Aer. jandaei* (Esteve, 1995).

Aeromonas jandaei
Cultures are motile Gram-negative rods, that produce arginine dihydrolase, indole and lysine decarboxylase but not ornithine decarboxylase, degrade casein, chitin, DNA, gelatin, starch and Tween 80, grow in 0% but not 6% (w/v) sodium chloride and at 4–42°C, and produce acid from galactose, glycogen, mannose, sucrose and trehalose, but not arabinose, lactose, melibiose, raffinose, rhamnose or salicin.

Aeromonas salmonicida

Aer. salmonicida, which comprises the so-called non-motile taxon, is one of the most important fish pathogens because of its widespread distribution, diverse host range and economically devastating impact on cultivated fish, particularly the valuable salmonids. More has been written about *Aer. salmonicida* than any other bacterial fish pathogen. Comprehensive reviews, i.e. McCraw (1952), Herman (1968), McCarthy and Roberts (1980) and an excellent textbook (Bernoth *et al.*, 1997) have adequately summarized the available knowledge on the pathogen in its context as a fish pathogen. The following narrative will emphasize taxonomic aspects, which have not been considered adequately by others.

Aer. salmonicida is one of the oldest described fish pathogens. It is generally accepted that the first authentic report of the organism was by Emmerich and Weibel (1894), who isolated the pathogen from diseased brown trout, obtained from a hatchery in Germany, and named it 'Bacillus der Forellenseuche' or bacillus of trout contagious disease. Because of the importance of the organism, some authors have examined the early literature seeking evidence for the occurrence of furunculosis prior to 1894 (Williamson, 1928; McCarthy and Roberts, 1980). However, although several reports exist which suggest that furunculosis occurred earlier, it has proved impossible because of poor descriptions, to be certain that the bacterial isolates in question were *Aer. salmonicida* (Forel, 1868; Fabre-Domerque, 1890; Fischel and Enoch, 1892). In North America, the first report of furunculosis was made by Marsh (1902) who described an organism, which was named as *Bacillus truttae*. This caused an epizootic in hatchery fish in Michigan, USA.

The placement of *Aer. salmonicida* in the bacterial taxonomic hierarchy should be put into context. Like so many of the bacterial taxa described in the nineteenth century, the organism has undergone a series of changes in its classification. Shortly after Emmerich and Weibel (1894) reported it, the pathogen was placed in the genus *Bacterium*, as *Bacterium salmonicida*, by Lehmann and Neumann (1896). This was probably its best-known epithet before its eventual transfer to the genus *Aeromonas*. To confuse the issue, however, *Bacillus devorans* (Zimmermann, 1890), *Bacterium salmonica* (Chester, 1897), *Bacterium truttae* (Marsh, 1902) and *Bacillus salmonicida* were also names assigned to the pathogen in the past. With the publication of the seventh edition of *Bergey's Manual of Determinative Bacteriology* in 1957, the pathogen was transferred to the genus *Aeromonas*, then placed in the family Pseudomonadacaeae. Later, the genus *Aeromonas* was moved again, to the family Vibrionaceae and more recently to its own family, i.e. the Aeromonadaceae (Colwell *et al.*, 1986). The initial re-classification was based primarily on the work of Griffin *et al.* (1953a). This team undertook the first detailed characterization of the organism, providing the information necessary to formulate the description which resulted in its re-classification. However, the division of *Aer. salmonicida* strains into subspecies remains an ongoing bone of taxonomic contention.

Two aspects of the status of *Aer. salmonicida* in the bacterial taxonomic hierarchy require discussion. One centres on the intraspecific relationships of *Aer. salmonicida* strains; the other is involved with questions that have been raised regarding the retention of the species in the genus *Aeromonas*.

Aer. salmonicida has been known by its present name since the 1950s when, on the basis of work by Griffin *et al.* (1953a), Snieszko (1957) in his contribution on the genus *Aeromonas* in the seventh edition of *Bergey's Manual of Determinative Bacteriology* assigned the pathogen to this genus, where it has remained. It is curious, in view of its seriousness as a pathogen, that it was over 50 years from the initial discovery of the organism to the characterization and description of *Aer. salmonicida*. Griffin *et al.* (1953a) provided the first detailed description of the organism, and were of the opinion that attempts to recognize and identify isolates were being hampered by a lack of a complete description. This resulted in confusion due to disagreement concerning the physiological and biochemical characterization of the organism. From the results of a study of ten isolates, it was concluded that *Bacterium salmonicida* was extremely consistent in its general cultural and biochemical traits, and that problems in the past had arisen primarily from the use of media which varied in composition among laboratories. Although Griffin *et al.* (1953a) ended their report by recommending the re-classification of *Bacterium salmonicida* to the newly created genus *Aeromonas*, as *Aer. salmonicida*, no definite reasons were given for this move. However, with time, additional data have accumulated, and the homogeneity and authenticity of the taxon has generally been supported.

Subsequent investigators have re-examined the homogeneity of the taxon, using conventional and numerical phenetic methods (Eddy, 1960; Ewing *et al.*, 1961; Schubert, 1961; Smith, 1963; Eddy and Carpenter, 1964; Popoff, 1969). Thus, the traditional description of *Aer. salmonicida* is of non-motile, fermentative, Gram-negative rods, which produce a brown water-soluble pigment on tryptone-containing agar, which do not grow at 37°C, and which produce catalase and oxidase (Table 4.1). The circular chromosome is 4658 ± 30 kb (Umelo and Trust, 1998). Cells are found in, and are pathogenic to, salmonids and increasingly other fish species. Traditionally, the lack of motility has been accepted as one of the reliable diagnostic traits used for the division of the aeromonads.

Table 4.1. Characteristics of *Aeromonas salmonicidaa*

Character	*achromogenes*	*Aer. salmonicida* subsp. *mascoucida*	*salmonicida*	*smithia*
Production of:				
Brown, diffusible pigment	–	–	+	–
Arginine dihydrolase	+	+	v	–
Catalase	+	+	+	+
β-Galactosidase	+	+	+	+
H_2S	–	+	v	+
Indole	–	–	–	–
Lysine decarboxylase	–	+	v	–
Ornithine decarboxylase	–	–	–	–
Oxidase	+	+	+	+
Phenylalanine deaminase	.	.	–	–
Phosphatase	.	.	–	+
Fermentative metabolism	+	+	+	+
Gluconate oxidation	–	+	–	–
Methyl red test	–	+	v	–
Motility	–	–	–	–
Nitrate reduction	+	+	+	.
Voges Proskauer reaction	–	+	–	–
Degradation of:				
Aesculin	–	+	v	–
Blood (β-haemolysis)	–	+	+	–
Casein	+	–	+	+
Chitin	–	–	–	–
DNA	+	+	+	+
Elastin	.	.	+	–
Gelatin	–	+	+	+
Lecithin	.	.	+	–
RNA	+	+	+	+
Starch	.	.	+	+
Tweens	+	+	+	–
Tyrosine	.	.	+	–
Urea	–	–	–	.
Xanthine	–	–	–	.
Growth at/on				
4–5°C	v	v	v	+
30°C	+	+	+	–
37°C	–	–	–	–
Cystine lactose electrolyte deficient agar	–	–	–	.

Table 4.1. (*Continued*)

Character	*Aer. salmonicida* subsp.			
	achromogenes	*mascoucida*	*salmonicida*	*smithia*
MacConkey agar	+	+	+	−
Potassium cyanide	−	−	−	·
Thiosulphate citrate bile salt sucrose agar	−	−	−	·
0–2% (w/v) NaCl	+	+	+	+
3% (w/v) NaCl	v	v	v	−
4% (w/v) NaCl	−	−	−	−
Utilization of sodium citrate	−	−	−	−
Production of acid from:				
Adonitol	−	−	−	·
Amygdalin	−	−	−	·
Arabinose	−	+	+ (slow)	·
Cellobiose	−	−	−	−
Dulcitol	−	−	−	·
Erythritol	·	·	−	·
Fructose	·	·	+	·
Galactose	+	+	+	−
Glucose	+	+	+	v
Glycerol	·	·	−	−
Glycogen	·	·	+	·
Inulin	·	·	−	·
Lactose	−	−	−	−
Maltose	+	+	+	−
Mannitol	−	+	+	−
Mannose	·	+	+	·
Melezitose	·	·	−	·
Melibiose	·	·	−	·
Raffinose	−	−	−	−
Rhamnose	−	−	−	−
Salicin	v	v	+ (slow)	·
Sorbitol	−	−	−	−
Sucrose	+	+	−	v
Trehalose	+	+	+	−
Xylose	−	−	−	·
G + C ratio of the DNA (moles %)	·	·	57–59	56

[a] Based on Griffin *et al.* (1953a), Schubert (1967a,b; 1974), McCarthy (1977a; 1980) and Austin *et al.* (1989).
v = variable result;
· = not done.

However, this criterion has been challenged by the report of McIntosh and Austin (1991a) of motility (by polar flagella) in a strain of *Aer. salmonicida* subsp. *salmonicida* grown at elevated temperatures, i.e. 30–37°C. The appearance of motility was also accompanied by variation in sugar fermentation patterns, the loss of ability to degrade complex molecules and an increase in antibiotic resistance. Further evidence for a motile mode of existence of *Aer. salmonicida* was provided by the recovery of eight atypical isolates from ulcers (but not from kidney tissue) on goldfish, carp and roach. The ulcerated fish were obtained from aquaria, garden ponds and rivers in England (Austin, 1992). Interestingly, these isolates did not dissociate into different colony types, but grew at 37°C. Of course, there is always the concern that motile contaminants may have been present in cultures which were predominantly *Aer. salmonicida*. However, the isolation of flagella genes, *flaA* and *flaB* which coded for unsheathed polar flagella at low frequency, has clinched the argument that *Aer. salmonicida* can be motile under certain circumstances (Umelo and Trust, 1997).

Certain traits, such as pigment production, captured the attention of fisheries scientists particularly because they were readily observable. In an examination of pigment production, Griffin *et al.* (1953b) showed that its development was dependent upon medium composition, insofar as tyrosine or phenylalanine was deemed to be essential. This was confirmed by O'Leary *et al.* (1956). However, it was initially assumed that this pigment was related to melanin, although subsequent investigation has refuted this possibility. Thus, Donlon *et al.* (1983) discovered that biosynthesis of the pigment from tyrosine differed substantially from melanogenesis and not 3,4-dihydroxyphenylalanine as would have been expected of melanin synthesis. Although production of the brown, water-soluble pigment constitutes a major diagnostic feature of *Aer. salmonicida* (Popoff, 1984), caution is advised against relying too heavily on the presence of pigment, insofar as there are variations among pigmented strains in the quantity of compound produced and of the time needed for its appearance (Horne, 1928; Mackie and Menzies, 1938). In addition, non-pigmented variants may arise (Wiklund *et al.*, 1993), particularly upon subculture (Duff and Stewart, 1933; Evelyn, 1971a). It has also been observed that other *Aeromonas* species, namely *Aer. hydrophila* and *Aer. media*, may produce such pigments when grown on media containing tryptone (see Paterson, 1974; Allen *et al.*, 1983a). Obviously, this questions the reliability of using pigment production as a differential characteristic. To further complicate the issue, the existence of achromogenic or slowly pigmenting strains of *Aer. salmonicida* have been described.

Another intriguing trait of *Aer. salmonicida* is the ability to dissociate into different colony types, i.e. rough, smooth and G-phase (intermediate) colonies. This phenomenon was extensively studied by Duff (1937), and will be discussed further in connection with its relevance to pathogenicity (see Chapter 9). Electron microscopy demonstrated that 'rough' and 'smooth' forms were attributed to the presence or absence of an extracellular layer (= the A-layer), respectively.

The notion of homogeneity could be dispelled by the results of PFGE of 44 isolates of *Aer. salmonicida* subsp. *salmonicida* which generated 30 different profiles and 40 distinct types (Chomarat *et al.*, 1998). However, numerical analysis using the S_D coefficient revealed that all the isolates were genomically related (Chomarat *et al.*, 1998). So, the *status quo* is maintained.

Atypical isolates of *Aeromonas salmonicida*
When compared to the so-called motile aeromonads, the description of *Aer. salmonicida* suggests a very homogeneous group of organisms. Alas, as so often happens in biology, a multitude of exceptions have disturbed the apparent idyllic situation. 'Atypical' strains deviate from the classical description of the taxon over a number of biochemical and physiological properties, making typing difficult (Hirvelä-Koski *et al.*, 1994; Austin *et al.*, 1998; Wiklund and Dalsgaard, 1998; Dalsgaard *et al.*, 1998). The most common reasons for describing isolates as 'atypical' are:

- lack of, weak or slow pigment production (Nakatsugawa, 1994)
- oxidase-negativity (Wiklund and Bylund, 1993; Wiklund *et al.*, 1994; Wiklund and Dalsgaard, 1995; Pedersen *et al.*, 1994; 1996a)
- nutritional fastidiousness, i.e. for blood or blood products (Austin, 1993)
- slow growth (Austin, 1993)
- different hosts from salmonids, i.e. cyprinids (e.g. Austin, 1993) and marine fish, including shotted halibut (*Eopsetta grigorjewi*; Nakatsugawa, 1994), dab, plaice and flounder (*Platichthys flesus*) (Wiklund *et al.*, 1994; Wiklund and Dalsgaard, 1995), common wolf-fish (*Anarhichas lupus*) (Hellberg *et al.*, 1996), turbot (*Scophthalmus maximus*) (Pedersen *et al.*, 1994), greenling (*Hexagrammos otakii*), Japanese flounder (*Paralichthys olivaceus*) and Schlegel's black rockfish (*Sebastes schlegeli*) (Iida *et al.*, 1997), where the disease is often ulceration

With the last-mentioned example, the justification for describing the isolates as atypical was based on the host rather than the characteristics of the cultures.

One of the earliest indications that aberrant strains occurred was provided by Smith (1963), who examined six isolates of non-pigmented cultures which were clustered as Group I from a numerical taxonomy study. These organisms were related at the 75.6% similarity level to typical pigment-producing isolates. Smith (1963) proposed a separate new species for Group I, i.e. with the specific epithet *achromogenes*, although the recommendation was not adopted. A second non-pigmented group was recognized by Kimura (1969a) as *Aer. salmonicida* subsp. *masoucida*. This subspecies differed from typical strains on account of indole production, Voges Proskauer reaction, H_2S and lysine decarboxylase production, and fermentation of sucrose (Table 7.1). In the eighth edition of *Bergey's Manual of Determinative Bacteriology*, Schubert (1974) regarded these non-pigmented isolates as *Aer. salmonicida* subsp. *achromogenes* and *Aer. salmonicida* subsp. *masoucida*, respectively. Typical strains were classified by him as *Aer. salmonicida* subsp. *salmonicida*. This classification, with three subspecies, has been retained in the recent literature (Popoff, 1984). This is interesting, because in an earlier publication (Popoff, 1970), it was contended that subspecies *achromogenes* and *masoucida* were more closely related to *Aer. hydrophila* than to *Aer. salmonicida*. In fact, Paterson *et al.* (1980) suggested that '*masoucida*' bridged the gap between typical *Aer. salmonicida* cultures and *Aer. hydrophila*, insofar as the subspecies possessed similar physiological and growth characteristics to the latter. However, *Aer. salmonicida* subsp. *masoucida* is non-motile, sensitive to *Aer. salmonicida* bacteriophages, possesses an antigenic component specific to *Aer. salmonicida*, and has a DNA homology of 103% with *Aer.*

salmonicida (MacInnes *et al.*, 1979). The relationship between the subspecies is certainly not sacrosanct, insofar they could be combined or kept separate according to which methods happen to be in vogue (Austin *et al.*, 1998). For example by PCR, there would be good reason to consider combining subspecies *achromogenes* and *masoucida*, a view which is not substantiated by ribotyping and RAPD analyses (Austin *et al.*, 1998). Phenetic data suggested that there would be a case for combining subspecies *masoucida* with *salmonicida*, and subspecies *achromogenes* with *Haemophilus piscium* (Fig. 4.1). Yet, all methods point to the comparative uniqueness of subspecies *smithia* (Austin *et al.*, 1998).

Certainly, a species subdivided into four subspecies should not be considered as unworkable; however, the classification is complicated by other factors. Thus, the existence of aberrant strains from a wide range of fish hosts and geographical locations is well established, and new reports are continually being made. According to McCarthy (1980), the existence of such strains is no doubt more common than even their documentation in the published literature suggests. As Mawdesley-Thomas (1969) points out, there is no sound reason why a pathogen which affects one family of freshwater fishes should not infect others. He contended that emphasis had been placed on food and game fish, and that only the absence of detailed investigations of fish diseases generally had given the false impression that each fish has its own specific set of diseases. McCarthy (1980) and McCarthy and Roberts (1980) made the valid point that the original description of atypical strains, as reported by Schubert (1974), was based upon data for only a few isolates. These authors submitted that revision is now both possible and necessary. According to a comparative phenetic and genotypic analysis of 29 atypical isolates, in addition to 144 other *Aeromonas* spp., McCarthy (1977a) delineated four phenetic groups. Of these, one cluster comprised typical isolates of *Aer. salmonicida*, a second group was composed of atypical isolates of *Aer. salmonicida* derived from salmonids including representatives of subspecies *achromogenes* and *masoucida*, a third group contained atypical isolates (*Aer. salmonicida*) from non-salmonids, and the fourth group was equated with *Aer. hydrophila*. Not surprisingly, the typical isolates of *Aer. salmonicida* formed an extremely compact group, which could not be readily differentiated from the atypical strains. However, the results of numerical phenetic analyses were not unequivocally confirmed by guanine plus cytosine (G + C) ratio determinations or DNA:DNA homology studies. McCarthy and Roberts (1980) proposed that, from the results of their studies, there should be three subspecies of *Aer. salmonicida*, as follows:

Group 1   *Aer. salmonicida* subsp. *salmonicida*
Group 2   *Aer. salmonicida* subsp. *achromogenes* (incorporating subsp. *masoucida*)
Group 3   *Aer. salmonicida* subsp. *nova*

However, the 'Approved Lists of Bacterial Names' (Skerman *et al.*, 1980) retained the classification of Schubert (1974), namely of subspecies *achromogenes*, *masoucida* and *salmonicida*. Thus, a growing consensus of opinion suggested that it was timely to elevate certain of the better characterized atypical isolates to subspecies status; the problem concerns the number and composition of such groups (Austin *et al.*, 1998). On

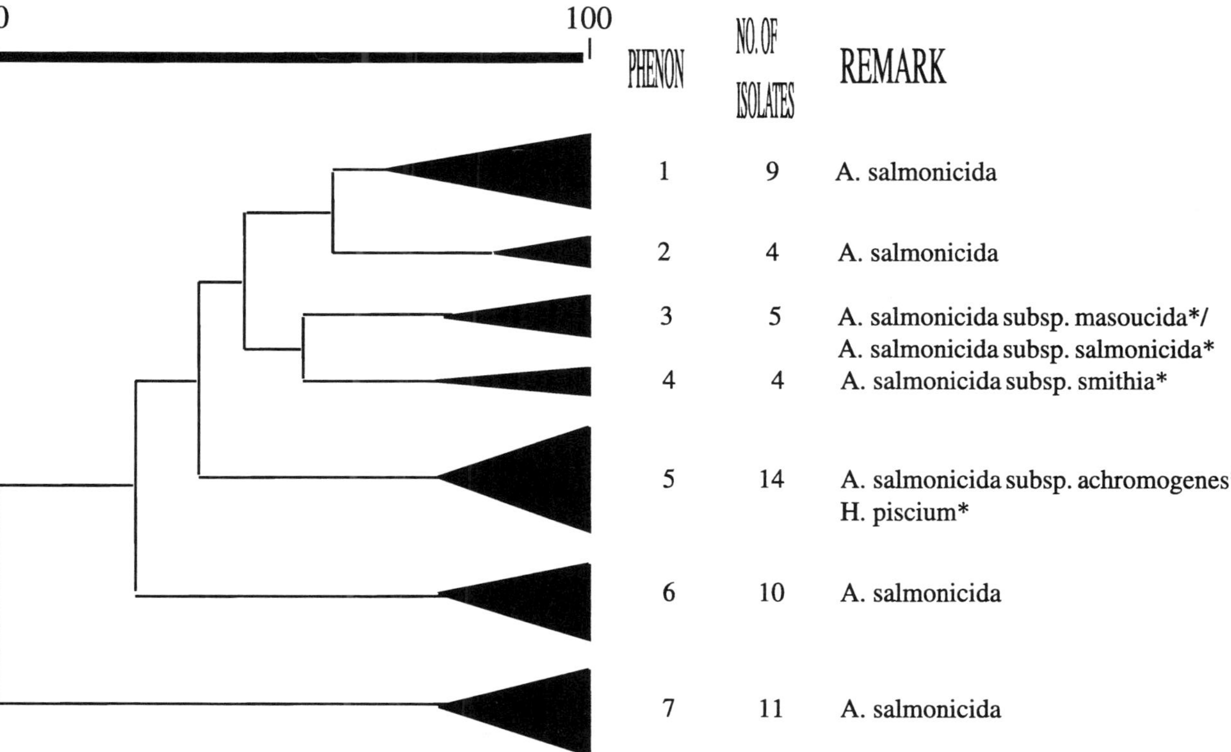

Fig. 4.1. A dendrogram showing the relations between atypical isolates and *Aer. salmonicida* subsp. *achromogenes*, *masoucida*, *salmonicida* and *smithia*, and *Haemophilus piscium*, based on a comparison of phenotypic data (from Austin *et al.*, 1998). The distinctiveness of atypical isolates is readily apparent.

the basis of DNA:DNA reassociation studies, Belland and Trust (1988), also supported the rationale of McCarthy and Roberts (1980) to create a new subspecies, i.e. *Aer. salmonicida* subsp. *nova*, to accommodate atypical isolates from non-salmonids. Yet, the subspecies was not formally proposed. Moreover, they also agreed with the suggestion of McCarthy and Roberts (1980) to combine the subspecies *achromogenes* and *masoucida*. In another (numerical) taxonomy and DNA:DNA hybridization study, Austin *et al.* (1989) elevated a group of 18 non- or slowly pigmenting 'atypical' isolates into a new subspecies, as *Aer. salmonicida* subsp. *smithia*. In addition to the phenotypic studies, there is an increasing trend to employ molecular genetic techniques to elucidate interspecific and intraspecific relationships within the genus. Thus, DNA:DNA and RNA:DNA hybridization, 16S RNA cataloguing, and 5S and 16S rRNA sequencing techniques have been used. Work on DNA homologies by MacInnes *et al.* (1979) revealed that all isolates of *Aer. salmonicida* (including *Aer. salmonicida* subsp. *masoucida*) possessed very high homologies, i.e. 96–106%, when hybridized against a representative strain of *Aer. salmonicida* subsp. *salmonicida*. Indeed, these authors concluded that the non-motile aeromonads comprise a genetically homogeneous taxon. In this study, the culture of *Aer. salmonicida* subsp. *masoucida* hybridized at 103% with *Aer. salmonicida* subsp. *salmonicida*. MacInnes and co-workers determined that this homology level was also achieved with a strain of *Aer. salmonicida* subsp. *salmonicida*, which was an ultraviolet induced mutant of NCIMB (= National Collection of Industrial and Marine Bacteria, Aberdeen, Scotland) 74. An interpretation was reached, therefore, that *Aer. salmonicida* subsp. *masoucida* and, perhaps, some of the biochemically atypical isolates do not warrant separate subspecies status, insofar as they may be merely mutants of other well-recognized groups. However, we emphasize that dramatic conclusions should not be made from an examination of only eleven isolates. It is interesting to note, however, that McCarthy (1980) also reported, as a result of the genotypic part of his analyses, that typical and atypical isolates were very closely related, with minimal divergence. It is a pity that the exact homology values were not presented. In a study of 26 typical and atypical isolates by DNA:DNA re-association methods, Belland and Trust (1988) found that typical isolates were recovered in a homogeneous group, whereas the atypical representatives were more diverse. Of these, one biotype consisted of isolates obtained from goldfish (obtained from a wide geographical range), whereas the second group accommodated isolates derived from carp in Europe. From the results of a numerical taxonomic and DNA:DNA hybridization study, Austin *et al.* (1989) made similar conclusions regarding the homogeneity of typical isolates of *Aer. salmonicida*. However, using 16S rRNA sequencing techniques, Martínez-Murcia *et al.* (1992) reported that subspecies *achromogenes* and *masoucida* were indistinguishable, and only differed from subspecies *salmonicida* by two bases.

There is overwhelming evidence that the so-called 'atypical' isolates are distinct from typical *Aer. salmonicida* (e.g. Hänninen and Hirvelä-Koski, 1997; Austin *et al.*, 1998; Wiklund and Dalsgaard, 1998; Umelo and Trust, 1998; Fig. 4.1). Yet, it has so far proved to be impossible to include the atypical isolates into a meaningful classification (Fig. 4.1). Moreover, there has been incongruence reported between the results of molecular (PCR, RAPD and ribotyping) and phenotypic methods, in terms of group membership (Austin *et al.*, 1998). The problems of inter-laboratory differences and lack of standardization in test methods has been highlighted by Dalsgaard *et al.* (1998). Some studies have indicated homogeneity among atypical isolates—a sentiment which is not endorsed by others. For example, Kwon *et al.* (1997) carried out a RAPD analyses of 29 atypical isolates from eight species of fish in Japan, and concluded that the profiling was identical, thereby indicating genetic homogeneity. However, heterogeneity was apparent between these atypical isolates and reference cultures from the validly described subspecies (Kwon *et al.*, 1997); a notion which has been confirmed by others (e.g. Austin *et al.*, 1998; Fig. 4.1). After studying 51 isolates from Finland by ribotyping, plasmid profiling and phenotyping, it was concluded that pigment-producing strains could be separated from achromogenic cultures. Also, oxidase-negative isolates were distinct from oxidase-positive atypical isolates in terms of ribotypes and phenotypes (Hänninen and Hirvelä-Koski, 1997).

There have been several reports of acid production in sucrose fermentation tests among typical isolates (see Fryer *et al.*, 1988; Wiklund *et al.*, 1992). This is relevant because hitherto this was one of the tests used to differentiate typical from atypical isolates of the pathogen (Popoff, 1984).

Plasmid profiles of *Aeromonas salmonicida*
Plasmids carried by typical (14 strains) and atypical (11 strains) forms of *Aer. salmonicida* have also provided additional genetic evidence for the classification of typical and atypical isolates into separate taxa (Belland and Trust, 1989). These workers found that typical isolates possessed a very homologous plasmid content comprised of a single large (70–145 kb) plasmid and three low molecular weight plasmids. Livesley *et al.* (1997) reported that five plasmids were most common among the 18 isolates examined; Giles *et al.* (1995) found four or six plasmids with four smaller plasmids of 4.3–8.1 kb being often observed in isolates from the Atlantic coast of Canada, but six plasmids of 4.2–8.9 kb among cultures from the Pacific coast of Canada. A total of 23 plasmids and 40 different plasmid profiles were recognized among 124 isolates from Denmark, Norway, Scotland and North America (Nielsen *et al.*, 1993). An earlier theme was repeated insofar as all isolates had one large plasmid of 60–150 kb, and two low molecular weight plasmids of 5.2 and 5.4 kb. In addition, two plasmids of 5.6 and 6.4 kb were frequently present (Nielsen *et al.*, 1993). A larger investigation of 383 isolates over a 6-year period concluded that 1–4 plasmids of 52–105 mDa were inevitably present, casting doubt on the relevance of plasmid typing for epizootiology (Sørum *et al.*, 1993b). Again, oxytetracycline and streptomycin resistant isolates from the Atlantic and Pacific coasts contained four or six plasmids, with four smaller plasmids of 4.3 to 8.1 kb being often observed. Some slight variation in plasmid content was noted between source of the isolates (Giles *et al.*, 1995). Atypical isolates possessed two to four different plasmid types (Belland and Trust, 1989). Moreover, there was a correlation between plasmid composition and source of the atypical isolates. This observation may prove useful in epizootiological studies, where plasmid content of atypical isolates could serve as useful markers (Belland and Trust, 1989).

The taxonomic dilemma
Using molecular techniques, a consistent view about the genetic relatedness of the long-

established subspecies of *Aer. salmonicida* emerges. The outstanding dilemma concerns the poor correlation between phenetic and genotypic data (Austin *et al.*, 1989). This problem needs to be addressed before the definitive classification of *Aeromonas* results. Nevertheless, it may be concluded that unlike the atypical isolates, *Aer. salmonicida* subsp. *salmonicida* is extremely homogeneous; a conclusion which is supported by phenotypic and molecular data (Austin *et al.*, 1989; Dalsgaard *et al.*, 1994; Hänninen *et al.*, 1995; Miyata *et al.*, 1995; Umelo and Trust, 1998).

Despite the existence of typical and atypical representatives of *Aer. salmonicida*, the current theme in operation for speciation has proved tenable. Nevertheless, new 'atypical' isolates, which do not fit into existing classifications of *Aer. salmonicida*, are regularly reported. Debate has centred around the relationship of *Aer. salmonicida* within the genus *Aeromonas*. In the study by Eddy (1960), attention was focused on the inability of *Aer. salmonicida* to produce 2,3-butanediol from glucose, and the absence of motility, both characters of which contravened the genus description of Kluyver and van Niel (1936). However, Eddy did not dispute the retention of *Aer. salmonicida* within the genus. Instead, this placement was challenged by Smith (1963), who expressed doubt as to whether *Aer. salmonicida* belonged in the genus *Aeromonas*. Her recommendation was the establishment of a new genus, i.e. *Necromonas*, with two species, namely *Nec. salmonicida* for the typical isolates and *Nec. achromogenes* for the non-pigmented strains. The evidence appertained to the morphological, biochemical and metabolic traits of 42 isolates of '*Bacterium salmonicida*', six non-pigmented pathogens, and 42 other bacterial cultures. Thus, there were pronounced differences between *Bacterium salmonicida* and other *Aeromonas* cultures. It emphasized, for instance, that the production of gas from glucose was an important genus characteristic. Although many previous reports had stated that the pathogen produced gas from glucose (Griffin *et al.*, 1953a; Eddy, 1960, 1962; Ewing *et al.*, 1961; Schubert, 1961), the hundreds of isolates examined at the Marine Laboratory, Aberdeen, between 1953 and 1962 produced either very little or no gas from glucose. Instead, they produced gas from mannitol (Smith, 1963). As for the production of 2,3-butanediol from glucose, Smith (1963) contended that previously this test required a tedious procedure, and consequently was often not applied to presumptive aeromonads. However, in her laboratory, *Bacterium salmonicida* isolates did not so produce the compound. Because of such discrepancies with the genus description, it was proposed that the species should be removed from the genus *Aeromonas* and placed in a new genus, i.e. *Necromonas*. Although Smith's proposal was not formally adopted, it should be mentioned that Cowan (1974) followed her suggested classification, by including *Nec. salmonicida* in the diagnostic tables. However, it is our opinion that the deviations of *Aer. salmonicida* from the initial genus description of *Aeromonas* should, for several reasons, be viewed less stringently than may be to the approval of some taxonomist purists. For example, it is often a difficult decision in bacterial systematics as to how much variation to allow within the definition of a species or a genus, before the line is drawn and relationships, or lack of, declared. If examples are taken from the characteristics of motile aeromonads, it is certainly the case that discrepancies occur for some members of these species as regards agreement with the genus description (Holder-Franklin *et al.*, 1981; Allen *et al.*, 1983b). Schubert (1974) included the production of 2,3-butanediol (a generic trait) as occurring in *some* species. Non-motility is also taken into account, and carbohydrates are cited as being broken down to acid or to acid and gas. These modifications to the genus description thus eliminate the major objections of Smith (1963) regarding the retention of *Aer. salmonicida* within the genus *Aeromonas*. In addition, subsequent serological and bacteriophage sensitivity data have provided strong evidence for a relationship between *Aer. salmonicida* and the motile aeromonads. The existence of a common antigen between *Aer. hydrophila* and *Aer. salmonicida* subsp. *masoucida* and some other strains of *Aer. salmonicida* was demonstrated by Kimura (1969b) and Paterson *et al.* (1980). In an examination of the specificities of aeromonad extracellular antigens, Liu (1961) observed that cross-reactions occurred between *Aer. salmonicida* and the motile aeromonads. Liu suggested that the absence of serological cross-reactions with other Gram-negative bacteria indicated that organisms belonging to the genus *Aeromonas* comprise a distinct group. Studies, using bacteriophage, demonstrated the sensitivity of some *Aer. hydrophila* cultures to *Aer. salmonicida* bacteriophages, whereas organisms not belonging to *Aeromonas* showed complete resistance to the virus (Popoff, 1971a,b). In particular, studies employing molecular genetic techniques support the retention of *Aer. salmonicida* in the genus *Aeromonas*. MacInnes *et al.* (1979) determined that percentage DNA homologies of motile aeromonads with *Aer. salmonicida* subsp. *salmonicida* ranged between 31% and 80%, whereas those hybridized against *Aer. hydrophila* were between 31% and 100%. The *Aer. salmonicida* strains exhibited a relatively high degree of homology when hybridized against *Aer. hydrophila*. Thus, the values for ten out of eleven isolates were in the range of 51% to 69%. From these results, MacInnes *et al.* (1979) concluded that non-motile aeromonads demonstrated a legitimate genetic relationship to the motile species of *Aeromonas*. In fact, some motile strains shared a higher level of sequence homology with *Aer. salmonicida* than with the reference motile aeromonads which according to MacInnes could be attributed to the smaller genome size of *Aer. salmonicida*. In another investigation, 56–65% binding between *Aer. salmonicida* and *Aer. hydrophila* DNA strands was recorded (McCarthy, 1978). These homology values indicate strong genetic relationships between the principal *Aeromonas* species (Paterson *et al.*, 1980; Belland and Trust, 1988).

It is well established that *Aer. salmonicida*, as a species, is phenotypically distinct from its motile counterparts. The increasing evidence from genetic and other molecular biology studies pertaining to intrageneric relationships between *Aer. salmonicida* and other aeromonad species, however, appears to support the retention of the pathogen within the genus. Certainly, the situation is not aided by ongoing manoeuvres in the classification of the motile aeromonads. Therefore, since the arrangement of the genus is in transition, and yet to be definitively resolved, we believe that nothing would be gained by the re-classification of *Aer. salmonicida*.

To reiterate, it appears to be the consensus of opinion that the area in need of further work is the intraspecific relationships between typical and atypical isolates. It is anticipated that such work would improve the classification of the genus *Aeromonas*.

Serology

Additional approaches to ascertaining the intra- and interspecific relationships of *Aer. salmonicida* have been adopted. These include serological techniques and bacteriophage typing. Certainly, the antigenicity of *Aer. salmonicida* has been the focus of much attention, primarily with a view to its significance in vaccine development. Unfortunately, the early investigations of Williamson (1929) were halted by the persistent agglutination of the strains in saline. This problem was circumvented by Blake and Anderson (1930), who employed complement fixation for the examination of 82 isolates of *Aer. salmonicida*. All of these isolates gave a positive response. Ewing *et al.* (1961) examined agglutinin absorption with reference to 'O' and 'H' antigens. They concluded that the 21 cultures reacted to *ca.* 25% of the titre of the O-antiserum prepared with *Aer. hydrophila*). Other researchers have also found serological homogeneity among strains of *Aer. salmonicida*, but some degree of cross-reactivity with *Aer. hydrophila*. For example, common antigens among two *Aer. salmonicida* strains, as determined by gel-diffusion, and cross-reactions between *Aer. salmonicida* antiserum and three out of four isolates of *Aer. hydrophila*, but not of *Aer. salmonicida* strains and *Aer. hydrophila* antiserum, was reported by Liu (1961). Karlsson (1962), using the antigenic properties of a haemolysin from *Aer. salmonicida*, found no serological differences among the six strains tested. Although these six strains did not cross-react with other aeromonads recovered from humans, it was later established that there were indeed common thermolabile antigens between *Aer. salmonicida* and other *Aeromonas* species as assessed using precipitin, agglutination and double diffusion precipitin tests. Bullock (1966) also found evidence of cross-reactions between soluble antigens of *Aer. salmonicida* and *Aer. hydrophila*. Serological cross-reactions between casein-precipitating enzymes of these two *Aeromonas* species were reported by Sandvick and Hagan (1968). Within the *Aer. salmonicida* group, Popoff (1969) found no serological differences among large numbers of typical pigment-producing isolates. Indeed, Popoff (1984), citing the work of Karlsson (1964), Spence *et al.* (1965) and his own previous studies, concluded that *Aer. salmonicida* is a serologically homogeneous species. Several workers have, in contrast, reported antigenic differences within the species (Duff, 1939; Liu, 1961; Klontz and Anderson, 1968; Kimura, 1969b). Paterson *et al.* (1980) considered that these conflicting findings reflected the choice of cultures, because comparable methodology had been used throughout. This group also reported results similar to those of Kimura (1969a,b), suggesting that *Aer. salmonicida* may be separated serologically into two groups based upon antigenicity of a given strain. In their studies, *Aer. salmonicida* NCIMB 1110 and 1102 and *Aer. salmonicida* subsp. *masoucida* contained an extra antigenic component, termed the 'c' component. Kimura (1969b) demonstrated the heat-sensitivity of an additional antigenic component (shared with *Aer. hydrophila*) in the subspecies *masoucida*. Klontz and Anderson (1968) observed smears prepared from 24 cultures of *Aer. salmonicida* with three antisera, by means of an indirect FAT. They postulated the existence of at least seven different serotypes, based upon non-reactivity of certain strains with one or more of the antisera. However, McCarthy and Roberts (1980) questioned the suitability of this technique for serological analysis of laboratory cultures, due to the potential for technical difficulties.

McCarthy and Rawle (1975) carried out an extensive serological study of both thermolabile and thermostable somatic antigens of *Aer. salmonicida* and their relationship to other bacteria. They employed whole-cell agglutination and double cross-absorption of smooth strains, and passive haemagglutination and double cross-absorption of rough colony types. It was determined that cross-reaction titres for both antigen types were, in general, high, and cross-reactions between the *Aer. hydrophila* isolate and three out of six thermostable *Aer. salmonicida* antisera were very weak. In contrast, a strain of *V. anguillarum* and *Ps. fluorescens* gave no reaction with the *Aer. salmonicida* antisera. The passive haemagglutination method was more sensitive than whole-cell agglutination, as titres obtained for positive reactions were ten-fold higher in the former. When a double-diffusion method was employed to study cell-free extracts, prepared from the bacteria used for the somatic antigen study, strong cross-sections among *Aer. salmonicida, Aer. hydrophila* and *V. anguillarum*, and, to a lesser extent, with *Ps. fluorescens* occurred. McCarthy and Rawle (1975) concluded that no qualitative differences in serological composition among *Aer. salmonicida* strains had been demonstrated, but noted that laboratory maintenance of some *Aer. salmonicida* cultures resulted in progressive loss of serological reactivity, giving negative responses with their antisera. Therefore, they suggested that, when embarking on serological (or vaccination) studies, it is important to include only fresh isolates. Hahnel *et al.* (1983) used micro-agglutination and double-diffusion precipitin tests to study serological relatedness among virulent and avirulent forms of eight isolates of *Aer. salmonicida* subsp. *salmonicida*. No serological differences were detected in the virulent isolates, but antigenic differences were observed between the virulent and avirulent form of each culture. Thus, in double-diffusion precipitin tests, the antigens of virulent sonicated cells formed an additional precipitin line when compared with the homogeneous avirulent form.

Bacteriophage typing
Bacteriophages have been used to study taxonomic relatedness between strains. The first isolation of bacteriophage specific for *Aer. salmonicida* was made by Todd (1933), although the usefulness for typing purposes was not demonstrated until the work of Popoff (1971a,b). It is considered that phage typing has value in epizoological studies (Popoff, 1984; Bast *et al.*, 1988; Belland and Trust, 1989). Essentially, the bacteriophages may be divided into three morphological groups and ten serological types (Popoff, 1984). Thus, using a set of eight phages, Popoff (1971b) recognized 14 phage types. Also, Paterson *et al.* (1980) studied phage sensitivity as a means of determining relationships between typical and atypical cultures. Pigmented and achromogenic as well as aggregating and non-aggregating strains showed a high sensitivity to two out of the three bactcriophages. In a further investigation by Rodgers *et al.* (1981), 27 groups of *Aer. salmonicida* were defined on the basis of sensitivity patterns to 18 bacteriophage isolates. Significantly, the morphological characteristics of the host bacterium, i.e. whether a rough, smooth or G-phase form, influenced attachment of the bacteriophage. This was apparently attributed to the varying quantities of LPS in the cell wall of the different morphological types.

Haemophilus piscium
What about *Haemophilus piscium*, the causal agent of ulcer disease? The name was coined by Snieszko *et al.* (1950). However, the detailed taxonomic study of Kilian (1976) showed that the organism did not belong in the genus *Haemophilus*. In particular, the strains did not exhibit requirements for haemin or NAD, which contrasted with the genus

description. *H. piscium* differed from the type species, *H. influenzae*, in the inability to reduce nitrate or alkaline phosphatase and to grow at 37°C, together with a relatively high G + C ratio of the DNA. Unfortunately, Kilian did not establish the most appropriate taxonomic position of the pathogen. The validity of the taxonomic position was similarly questioned by Broom and Sneath (1981), as a result of a detailed numerical taxonomic study. The low similarity of *H. piscium* with other *Haemophilus* spp., i.e. only 65%, suggested that the organism should be excluded from the genus. From examination of DNA, biochemical, serological and bacteriophage sensitivity data, it is apparent that *H. piscium* represents an atypical, achromogenic variant of *Aer. salmonicida* (Paterson *et al.*, 1980). Evidence for this conclusion consists of the G + C ratio of the DNA (55.1 moles %), which is well within the range reported for *Aer. salmonicida* by McCarthy (1978). Moreover, Paterson *et al.* (1980) regarded *H. piscium* to be serologically indistinguishable from *Aer. salmonicida*. Also, the pathogen was sensitive to several *Aer. salmonicida* bacteriophages, and exhibited biochemical reactions similar to those expected for some achromogenic variants of *Aer. salmonicida*. Trust *et al.* (1980a) also concluded, on the basis of bacteriophage sensitivity, that *H. piscium* is, in fact, atypical *Aer. salmonicida*. Thus, a virus that produced lysogeny in *Aer. salmonicida* but displayed no such activity in *Aer. hydrophila*, caused plaque formation in several isolates of *H. piscium*. On the basis of one strain, Austin *et al.* (1998) concurred with the view that *H. piscium* should probably be classified with *Aer. salmonicida* (Fig. 4.1). However its precise relationship to the four subspecies reflected the nature of the phenotypic and molecular methods used.

Aeromons sobria

Evidence has been presented for *Aeromonas sobria* as a fish pathogen. Yet, fish pathogenic isolates have often been mis-identified, whereas *bona fide Aer. sobria* has not been harmful to salmonids (Austin *et al.*, 1989). However, organisms, considered as *Aer. sobria*, have been isolated from wild spawning gizzard shad (*Dorosoma cepedianum*) in Maryland, USA, during 1987 (Toranzo *et al.*, 1989).

Aeromonas sobria

Cultures comprise motile (single polar flagellum) fermentative Gram-negative rods, which produce arginine dihydrolase, catalase, β-galactosidase, indole, lysine decarboxylase and oxidase, but not H_2S or ornithine decarboxylase. Blood, casein, gelatin, starch and Tween 80 are degraded, but not so aesculin, elastin or urea. Acid is produced from glucose (with gas), mannitol and sucrose, but not from arabinose, inositol, rhamnose, salicin or sorbitol. Nitrates are reduced, and the Voges Proskauer reaction is positive. Growth occurs in 0–5% but not 10% (w/v) sodium chloride and on TCBS. Resistance has been recorded to ampicillin, novobiocin and the vibriostatic agent, 0/129. The G + C ratio of the DNA is 60.4 moles % (Toranzo *et al.*, 1989).

Overall, there was good agreement with the description of *Aer. sobria* (Popoff, 1984). In particular, the important differentiating traits included production of acid from glucose (with gas), mannitol and sucrose but not salicin, and the inability to degrade aesculin (Popoff, 1984). The only difference to the species description centred on the production of H_2S.

B. ENTEROBACTERIACEAE REPRESENTATIVES

In addition to the organisms detailed below, there has been fleeting mention of a fish pathogenic role for *Erwinia* sp. (Starr and Chatterjee, 1972) and *Salmonella typhimurium* (Morse *et al.*, 1978).

Citrobacter freundii

Historically, there have been brief references to *Cit. freundii* as a fish pathogen (see Conroy, 1986). Yet, definitive evidence was not forthcoming until a report of the organism as a pathogen of sunfish, *Mola mola* in a Japanese aquarium (Sato *et al.*, 1982). Subsequently, *Cit. freundii* has been implicated with disease in Atlantic salmon and rainbow trout in Spain and the USA (Baya *et al.*, 1990a; Sanz, 1991) and with carp in India (Karunasagar *et al.*, 1992). In addition, the organism has been recovered in mixed culture from diseased salmonids in the UK (B. Austin, unpublished data). Consequently, it appears that *Cit. freundii* is an emerging fish pathogen.

Citrobacter freundii
Cultures contain Gram-negative motile, fermentative rods, which produce catalase, β-galactosidase and H_2S but not arginine dihydrolase, indole, lysine or ornithine decarboxylase, oxidase or tryptophan deaminase. Neither gelatin nor urea is attacked, nor is citrate utilized. The Voges Proskauer reaction is negative. Acid is produced from arabinose, glucose, mannitol, melibiose, rhamnose, sorbitol and sucrose but not from amygdalin or inositol.

From these traits, mostly obtained with the API 20E rapid identification system, an identification as *Cit. freundii* was obtained (Sato *et al.*, 1982; Baya *et al.*, 1990; Sanz, 1991). Certainly, these characteristics match the description of *Cit. freundii*, except in the inability to utilize citrate (Sakazaki, 1984).

Edwardsiella ictaluri

Enteric septicaemia of catfish was initially recognized in 1976 among populations of pond-reared fingerlings and yearling fish in Alabama and Georgia, USA (Hawke, 1979). Later, reports described its presence in Mississippi, Arkansas, Idaho, Colorado, Indiana and Maryland. All the published reports point to the presence of a very homogeneous group of bacteria phenotypically, including use of isozyme analysis (Starliper *et al.*, 1988; Bader *et al.*, 1998), although subgroups may be recognized by molecular methods. Thus, Bader *et al.* (1998) recognized four subgroupings among 20 isolates by use of arbitrary primed PCR.

Edwardsiella ictaluri
Cultures comprise Gram-negative rod-shaped fermentative organisms, which are motile by peritrichous flagella. Catalase, β-galactosidase (a variable response has been recorded between various laboratories) and lysine and ornithine decarboxylase are produced but not H_2S, indole, oxidase or phenylalanine deaminase. The methyl red test is positive, but not so the Voges Proskauer reaction. Nitrates are reduced. Growth occurs in 1.5% but not 2% (w/v) sodium chloride. The optimum growth temperature is between 20°C and 30°C, which coincides with the water temperature during severe outbreaks of the disease. Blood is degraded, but not casein, DNA, elastin, gelatin, Tween 20, 40, 60 and 80 or urea. Acid is produced from fructose, galactose, glycerol, maltose, mannose and ribose but not from adonitol, aesculin, amygdalin, arabinose, arbutin, cellobiose, dulcitol, erythritol, inositol, inulin, lactose, melezitose, raffinose, rhamnose, salicin, sodium malonate, sorbitol, sorbose, starch or sucrose (Waltman *et al.*, 1986). The G + C ratio of the DNA is 53 moles % (Hawke *et al.*, 1981).

Cryptic plasmids, i.e. pCL1 and pCL2 of 5.7 kb and 4.9 kb respectively, have been found in isolates of *Edw. ictaluri* (Lobb *et al.*, 1993). However, an isolate from green knife fish differed insofar as four plasmids, of 3.1, 4.1, 5.7 and 6.0 kb, were present. Of these, the 4.1 and 5.7 kb plasmids hybridized strongly (Lobb *et al.*, 1993).

Similarities have been observed with *Edw. tarda*, except that isolates of *Edw. ictaluri* did not produce H_2S or indole, or ferment glucose with the production of gas at 37°C. Moreover, the isolates did not agglutinate with antiserum to *Edw. tarda* (Hawke, 1979). However, in terms of DNA relatedness, the causal agent of enteric septicaemia of catfish was determined to be most similar to *Edw. tarda*, i.e. 56–62% DNA homology, but sufficiently distinct to warrant separate species recognition (Hawke *et al.*, 1981). It remains for further work to elucidate the relationship, of any, with *Edwardsiella* phenon 12 of Johnson *et al.* (1975).

Edwardsiella tarda (= Paracolobactrum anguillimortiferum, Edw. anguillimortifera)
Whereas *Edw. ictaluri* has remained confined within the broadly defined geographical limits of the USA, *Edw. tarda* has emerged with some frequency in the Far East, particularly in Japan and Taiwan (Wakabayashi and Egusa, 1973; Miyazaki and Egusa, 1976; Kou, 1981). The history of edwardsiellosis, as the disease is called, may be traced to two separate parallel developments, i.e. those of Wakabayashi and Egusa (1973) in Japan, and Meyer and Bullock (1973) in the USA. The situation is further complicated by an earlier report of Hoshina (1962), who described *Paracolobactrum anguillimortiferum* as a pathogen of pond-cultured eels. This organism is probably synonymous with *Edw. tarda*.

The organisms from fish match closely the description of *Edw. tarda* (Ewing *et al.*, 1965; Cowan, 1974; Farmer and McWhorter, 1984), insofar as they are fairly reactive (Table 4.2) Gram-negative rods, motile by means of peritrichous flagella. The original isolates of Meyer and Bullock (1973) were identified definitively as *Edw. tarda* by the CDC. Thereafter, each successive report emphasized the similarity to the species description of Ewing *et al.* (1965), highlighting the homogeneity of the fish isolates. In fact, it is

Table 4.2. Characteristics of *Edwardsiella tarda*[a] and *Paracolobactrum anguillimortiferum*[b]

Character	*Edw. tarda*	*Paracolobactrum anguillimortiferum*
Fermentative metabolism	+	+
Production of:		
Arginine dihydrolase	−	NS
Catalase	+	+
β-galactosidase	−	NS
H_2S	+	+
Indole	+	+
Lysine decarboxylase	+	NS
Ornithine decarboxylase	−	NS
Oxidase	−	NS
Phenylalanine deaminase	−	NS
Methyl red test	+	NS
Nitrate reduction	+	+
Voges Proskauer reaction	−	−
Degradation of:		
Aesculin, casein, gelatin	−	−
Tributyrin, urea	−	NS
DNA, elastin, lecithin	−	−
Tween 20, 40, 60 and 80	−	−
Chitin	−	−
Blood	+	+
Utilization of:		
Sodium citrate, sodium malonate	−	NS
Growth at 42°C	+	+
Growth in 3% (w/v) sodium chloride	+	+
Growth in 8% (w/v) sodium chloride	−	−
Acid production from:		
Adonitol, aesculin, erythritol	−	NS
Arabinose, cellobiose, dulcitol	−	−
Fructose, galactose, glycerol	+	NS
Glucose, maltose	+	+
Inositol	−	NS
Lactose, mannitol, raffinose	−	−
Mannose	+	NS
Rhamnose, salicin, sorbitol	−	−
Sucrose, trehalose, xylose	−	−

[a] From Sakazaki and Tamura (1975), Wakabayashi and Egusa (1973), Amandi *et al.* (1982), Nakatsugawa (1983) and Waltman *et al.* (1986).
[b] From Hoshina (1962).
NS = not stated.

astonishing how little variation has been recorded in the phenotypic characters; despite recovery from as far afield as Japan (Wakabayashi and Egusa, 1973; Nakatsugawa, 1983) and the USA (Amandi *et al.*, 1982). The only difference reflects serology. Thus, four serotypes (A, B, C and D) have been recognized from O-agglutination analyses of 270/445 isolates recovered from eel ponds in Japan during 1980 and 1982 (Park *et al.*, 1983). Among the cultures, serotype B predominated (22–35% of the total), followed by serotype A (13–17% of the isolates), serotype C (4–13% of the isolates) and serotype D (2–4% of the total). Most of the isolates in serotype A were derived from kidney samples. Moreover, this group was by far the most virulent as assessed from experimental infections of eels, loach and tilapia (Park *et al.*, 1983) and natural outbreaks in Japanese flounder (Rashid *et al.*, 1994). The G + C ratio of the DNA was calculated as 59 moles % (Amandi *et al.*, 1982).

Confusion has enveloped the taxonomic status of this pathogen, insofar as the 'Approved Lists of Bacterial Names' (Skerman *et al.*, 1980) include two specific epithets for what is apparently the same organism, *Edw. tarda* and *Edw. anguillimortifera. Edw. tarda* was proposed initially by Ewing *et al.* (1965), and it would appear that there was a lack of familiarity with the earlier work of Hoshina (1962), who described *Paracolobactrum anguillimortiferm*. The characteristics of these two nomenspecies are identical (Table 4.2). Recognizing the efforts of Hoshina (1972), Sakazaki and Tamura (1975) proposed that *Paracolobactrum anguillimortiferum* should be reclassified in the genus *Edwardsiella*, as *Edw. anguillimortifera*. Furthermore, Sakazaki and Tamura (1975) emphasized that priority should be attached to the specific epithet of *anguillimortiferum* rather than *tarda*. Thus two names have resulted for what is surely the same organism. Clearly, this is an unacceptable situation and, despite the pioneering work of Hoshina (1962), it is our contention that, to avoid further confusion in medical and veterinary microbiology, the name of *tarda* should be retained.

Enterobacter agglomerans

There has been some debate over the precise taxonomic relationship of *Enterobacter agglomerans* to *Erwinia herbicola* (see Richard, 1984). The situation has still not been resolved. Nevertheless, an organism corresponding to the description of *Ent. agglomerans* was recovered from dead mahi-mahi (100 g) (dolphin fish; *Coryphaena hippurus*). Juvenile mahi mahi (average weight = 1 g) were transported from Florida, USA, to Bermuda for ongrowing in sea cages. During January 1986, mortalities were noted with dead specimens displaying marked haemorrhaging in the eyes. From these animals, *Ent. agglomerans* was recovered (Hansen *et al.*, 1990).

There was excellent agreement between the characteristics of the fish isolate and the description of biogroup 1 of anaerogenic *Ent. agglomerans* (Richard, 1984).

Enterobacter agglomerans
Cultures comprise pale yellow pigmented, motile, fermentative Gram-negative rods, which produce catalase and β-galactosidase but not arginine dihydrolase, H_2S, indole, lysine or ornithine decarboxylase, or oxidase. The methyl red test, Voges

> Proskauer reaction, and nitrate reduction test are positive. Gelatin, but not blood, chitin or starch is degraded. Acid is produced from glucose, D-mannitol and sucrose. Citrate is utilized. Growth occurs at 4–37°C but not at 40°C, and in 0–6% (w/v) but not 8% (w/v) NaCl (Hansen *et al.*, 1990).

Escherichia vulneris

An organism, subsequently identified as *Esch. vulneris*, was first isolated in 1994 from naturally infected balloon moly (*Poecilia* sp.), silver moly (*Poecilia* sp.) and Caucasian carp (*Carassius carassius*) from Turkey (Aydin *et al.*, 1997).

> *Escherichia vulneris*
> Cultures comprise Gram-negative, motile, fermentative rods, which produce arginine dihydrolase, catalase and lysine decarboxylase but not H$_2$S, indole, oxidase or ornithine decarboxylase, degrade aesculin and blood, but not casein, DNA, gelatin, starch or urea, reduce nitrates, utilize malonate, and produce acid from arabinose, glucose, lactose, maltose, mannitol, melibiose, rhamnose, salicin, trehalose and xylose, but not adonitol, cellobiose, dulcitol, glycerol, inositol, sorbitol or sucrose. The methyl red test is positive, but the Voges Proskauer reaction is negative. Growth occurs at 37°C but not at 5°C or 42°C.

The organisms coincided with the description of *Esch. vulneris* (Brenner *et al.*, 1982).

Hafnia alvei

In 1988, an apparently new *Brucella*-like bacterium was described as a pathogen of rainbow trout in Bulgaria (Gelev and Gelev, 1988). Although antigenic relationships to *Brucella abortus* were reported, the organism displayed marked similarities to the Enterobacteriaceae. Subsequent investigation led to a realization that the organism was, in fact, *Hafnia alvei* (Gelev *et al.*, 1990). Subsequently, *Haf. alvei* was recognized as the cause of mortalities among cherry salmon (*O. masou*) in Japanese farms (Teshima *et al.*, 1992). Isolates matched the species description, with the exception of lack of motility and utilization of D-tartrate.

> *Hafnia alvei*
> The isolates comprise small coccoid or slightly elongated motile Gram-negative rods which produce catalase, β-galactosidase and lysine and ornithine decarboxylase, but not arginine dihydrolase, H$_2$S, indole or oxidase. The methyl red, nitrate reduction and citrate utilization tests and Voges Proskauer reaction (weakly positive) are positive, but not malonate utilization or phenylalanine deaminase. Neither aesculin, DNA, gelatin, lipids nor urea is degraded. Acid is produced from arabinose, cellobiose, glucose (but not gas), glycerol, maltose, mannose, mannitol, rhamnose, trehalose and xylose, but not from adonitol, dulcitol, erythritol, lactose, melibiose, raffinose, salicin, sorbitol or sucrose. Growth occurs at 41°C.

It was deemed that there was a common antigenic determinant in the LPS with *Brucella abortus* and *Y. ruckeri* (Gelev and Gelev, 1988). However, the presence of motility and the negative reaction in the oxidase test precludes a relationship with *bona fide Brucella* taxa. There was 82–100% re-association between the fish isolates and *Haf. alvei* (Gelev *et al.*, 1990).

Klebsiella pneumoniae
Twelve pure cultures were obtained. The characteristics were, as follows:

Klebsiella pneumoniae
Cultures comprise fermentative Gram-negative encapsulated rods that produce catalase but not arginine dihydrolase, H_2S, indole, lysine or ornithine decarboxylase, oxidase nor phenylalanine deaminase, produce acid from arabinose, cellobiose, glucose, maltose and raffinose, reduce nitrates to nitrite, and do not degrade gelatin (Daskalov *et al.*, 1998).

Proteus (Providencia) rettgeri
During 1976, there was a mass mortality among farmed silver carp (*Hypophthalmichthys molitrix*) in Israel. From moribund animals, *Pr. rettgeri* was isolated. To date, this has been the only report implicating this organism as a fish pathogen (Bejerano *et al.*, 1979).

Proteus rettgeri
Cultures comprise fermentative Gram-negative rods, which are motile by means of peritrichous flagella. Catalase, indole, phenylalanine deaminase and tryptophan deaminase are produced, but not β-galactosidase, H_2S or oxidase. DNA and urea are degraded, but not casein, gelatin, starch or the Tweens. The methyl red test is positive. Nitrates are reduced. Acid is produced from adonitol, aesculin, erythritol, galactose, inositol, mannitol, mannose, melezitose, rhamnose and salicin, but not from amygdalin, arabinose, cellobiose, dulcitol, glycerol, glycogen, inulin, lactose, maltose, melibiose, raffinose, ribose, sorbitol, sucrose, trehalose or xylose. The G + C ratio of the DNA is 39.2 moles % (Bejerano *et al.*, 1979).

Generally, the isolates matched the description of *Pr. rettgeri* (Cowan, 1974; Johnson *et al.*, 1975; McKell and Jones, 1976; Penner, 1984).

Salmonella choleraesuis **subsp.** *arizonae (= Salmonella arizonae)*
In a previous study by Austin *et al.* (1982), a possible taxonomic relationship was discussed between *Y. ruckeri* and *Sal. choleraesuis* subsp. *arizonae*. Therefore, it is ironic that the latter organism has now been involved in fish pathogenicity. Thus, *Sal. choleraesuis* subsp. *arizonae* was recovered during 1986 from a single dead pirarucu, *Arapaima gigas*, which had been exhibited in an aquarium at Sapporo, Japan. To date,

this is the only report implicating *Sal. choleraesuis* subsp. *arizonae* as a fish pathogen (Kodama *et al.*, 1987).

The culture was identified biochemically as *Sal. choleraesuis* subsp. *arizonae*. Yet, agglutination was not recorded with commercial *Salmonella* antisera, although not specifically a product to *Sal. choleraesuis* subsp. *arizonae* (Kodama *et al.*, 1987).

Salmonella choleraesuis subsp. *arizonae*
Cultures contain Gram-negative fermentative, motile rods, which produce catalase, β-galactosidase, H_2S and lysine decarboxylase but not indole or oxidase. Neither aesculin, blood, casein, gelatin, starch nor urea are degraded. Citrate is utilized, nitrate is reduced, and the methyl red test is positive. The Voges Proskauer reaction is negative. Acid is produced from glucose (plus gas), lactose, maltose, mannitol, raffinose, sorbitol, sucrose, trehalose and xylose, but not arabinose or salicin. Growth occurs at 15–41°C in 0–6% (w/v) NaCl and on MacConkey agar (Kodama *et al.*, 1987).

With the exception of acid production from arabinose, raffinose and sucrose, there is good agreement with the description of *Sal. choleraesuis* subsp. *arizonae* (Le Minor, 1984).

Serratia liquefaciens

During spring and early summer of 1988, heavy mortalities, i.e. up to 30% of the stock, were noted among three separate populations of Atlantic salmon (*Salmo salar* L.), two of which (average weight = 450 g) were in Scottish marine cage sites, and the third population (average weight = 25 g) was in freshwater. Microscopic examination of formalin-fixed kidney sections indicated the presence of Gram-negative bacteria which exhibited bipolar staining. These organisms were recovered as dense pure culture growth from diseased tissues, and were examined bacteriologically (McIntosh and Austin, 1990b). Since the initial occurrence, the disease has been recognized on another two marine sites in Scotland during 1990. Also in 1990, the organism was recognized as the cause of low continuous mortalities among farmed turbot in France (Vigneulle and Baudin-Laurençin, 1995).

Serratia liquefaciens
Cultures comprise Gram-negative fermentative catalase positive rods of 2–3 μm in length which, in contrast to the normal characteristics of the Enterobacteriaceae (Brenner, 1984), are motile by single polar flagella. Intracellular bipolar staining properties are exhibited. Catalase, β-galactosidase and lysine and ornithine decarboxylase are produced, but not H_2S, indole or tryptophan deaminase. The results of the oxidase test is variable, i.e. from negative to weakly positive depending to some extent on the age of the culture. The methyl red test and Voges Proskauer reaction is negative. Nitrates are not reduced. Blood (β-haemolysis), casein, DNA, lecithin (weak), tributyrin (slow), Tween 20, 40, 60 and 80, and tyrosine are degraded, but

not chitin, elastin, starch or urea. Growth occurs at 4–37°C, in 0–5% (w/v) but not 7% (w/v) sodium chloride, and on deoxycholate citrate agar, eosin methylene blue agar and MacConkey agar, but not in 40% (v/v) bile salts. Sodium citrate but not sodium malonate is utilized. Acid is produced from amygdalin, arabinose, inositol, mannose, melibiose and saccharose, but not rhamnose. Acid and gas are produced from fructose, galactose, glucose, maltose, mannitol, melibiose, raffinose, sorbitol, sucrose, trehalose and xylose. The G + C ratio of the DNA is 55 moles % (McIntosh and Austin, 1990b).

Initially, it was considered that the organisms bore similarities to *Aer. veronii* (Hickman-Brenner *et al.*, 1987) and *Ser. liquefaciens* (Grimont and Grimont, 1984). Similarities with the former included the presence of Gram-negative fermentative, catalase (and oxidase) positive rods, which were motile by single polar flagella, and with a G + C ratio of the DNA of 55 moles %. Indeed, the only discrepancies with the definition of *Aer. veronii* concerned some sugar fermentation reactions. Yet, similarities were also apparent with *Ser. liquefaciens*. However, the use of the API 20E rapid identification system (profile = 5305363 or 5305367) and whole cell agglutination using a polyclonal antiserum raised against the type strain, confirmed an identification as *Ser. liquefaciens*.

Serratia marcescens

The initial association of *Ser. marcescens* with fish diseases stemmed from a publication by Clausen and Duran-Reynals (1937). More detailed work waited for over half a century when in July 1990 during a survey of white perch (*Morone americanicus*) from the Black river, a tributary of Chesapeake Bay (USA), a red-pigmented organism, i.e. *Ser. marcescens*, was recovered, and, on subsequent examination, deemed to be potentially pathogenic for fish. It was considered likely that the presence of the organisms reflected the polluted nature of the river (Baya *et al.*, 1992c).

Serratia marcescens
Cultures comprise red (prodigiosin) pigmented fermentative motile Gram-negative rods, which produce catalase, β-galactosidase and lysine and ornithine decarboxylase but not arginine dihydrolase, H_2S, indole, oxidase or tryptophan deaminase. Nitrate is reduced; but the methyl red test and the Voges Proskauer reaction are negative. Growth occurs at 4–45°C, and in 0–8% (w/v) sodium chloride. Blood (sheep), casein, gelatin, starch and Tween 80 are degraded, but not so urea. A wide range of carbohydrates are attacked, including amygdalin, fructose, galactose, glycerol, inositol, maltose, mannitol, mannose, salicin, sorbitol, sucrose and trehalose, but not so arabinose, cellobiose, dulcitol, lactose, melibiose, raffinose, rhamnose or xylose (Baya *et al.*, 1992c).

From these characteristics, it is apparent that there is reasonable agreement with the description of *Ser. marcescens* (Grimont and Grimont, 1984).

Serratia plymuthica

During September to December 1987, a new pathogen was associated with diseased rainbow trout fingerlings (average weight = 7 g) in a hatchery in northwestern Spain. Progressive low-level mortalities (cumulative total = 35% of the stock) were reported, with which there appeared to be a correlation with rainfall (Nieto *et al.*, 1990). Then in 1992, a similar organism was associated with skin lesions in farmed rainbow trout in Scotland (Austin and Stobie, 1992b). In this disease outbreak, there was an association with pollution by domestic sewage, i.e. leakage from a septic tank.

Serratia plymuthica

Nieto *et al.* (1990) reported that cultures comprise red (prodigiosin) pigmented fermentative non-motile Gram-negative rods, producing catalase and β-galactosidase but not arginine dihydrolase, H_2S, indole, lysine or ornithine decarboxylase, or oxidase. The nitrate reduction, citrate utilization and Voges Proskauer reaction tests are positive. A negative response is recorded for the methyl red test. Gelatin is degraded, but not so blood, casein or urea. Acid is produced from L-arabinose, D-fructose, D-galactose, D-glucose (but not gas), inositol, D-maltose, D-mannitol, D-mannose, melibiose and sucrose, but not from D-adonitol, lactose, L-rhamnose or D-sorbitol.

From these results, it was considered that the cultures resembled *Ser. plymuthica* and *Ser. rubidaea* (Brenner, 1984). Yet as a result of acid production from inositol but not adonitol, it was considered that the pathogen should be assigned to *Ser. plymuthica*. Similar traits were reported by Austin and Stobie (1992b).

Yersinia intermedia

Characteristics matched the species description (Farmer and Kelly, 1991) at 36°C, except that the fish-isolate utilized sodium citrate and was weakly positive in the Voges Proskauer reaction (Carson and Schmidtke, 1993).

Yersinia intermedia

The culture comprised fermentative, motile (at 25°C but not 36°C) cells, that produced β-galactosidase and indole, but not arginine dihydrolase, H_2S, lysine or ornithine decarboxylase or oxidase. Aesculin and urea were degraded. The methyl red test and Voges Proskauer reaction (at 36°C but not 25°C) were positive. Nitrates were reduced. Acid was produced from glycerol, inositol, mannitol, melibiose (at 36°C but not 25°C), rhamnose, sorbitol, sucrose, trehalose and xylose, but not adonitol or lactose. Sodium citrate was utilized at 36°C but not 25°C (Carson and Schmidtke, 1993).

Yersinia ruckeri

Enteric redmouth (ERM, Hagerman redmouth disease, redmouth, salmonid blood spot) was initially diagnosed as a systemic infection among farmed rainbow trout in the Hagerman Valley of Idaho, USA, during the early 1950s, and subsequently described in detail by Ross *et al.* (1966).

Yersinia ruckeri
The organisms comprise a homogeneous group of fermentative Gram-negative, slightly curved rods of 1.0×2.0–3.0 μm in size, which are motile usually by means of seven or eight peritrichously arranged flagella. Catalase, β-galactosidase, and lysine and ornithine decarboxylase are produced, but not H_2S, indole, oxidase, phenylalanine deaminase or phosphatase. The methyl red test is positive, but not the Voges Proskauer reaction. Nitrates are reduced. Gelatin, and Tween 20, 40 and 60 are degraded, but not aesculin, chitin, DNA, elastin, pectin, tributyrin, Tween 80 or urea. Growth occurs in 0–3% (w/v) sodium chloride. Sodium citrate is utilized. Acid is produced from fructose, glucose, maltose, mannitol and trehalose, but not from inositol, lactose, raffinose, salicin, sorbitol or sucrose. The G + C content of the DNA is 47.5–48.5 moles % (Ewing *et al.*, 1978).

The precise taxonomic position of the causal agent of ERM has intrigued bacteriologists since the initial isolation of the organism. Ross *et al.* (1966) realized that heated O-antigens prepared from 14 cultures of the ERM organism agglutinated strongly (titre = 1:320 or 1:640) with the corresponding antigens of *Sal. choleraesuis (= Arizona)* subsp. *arizonae* O group 26, and weakly (titre = 1:20) with O group 29. Conversely, there was no reaction with O-antigens prepared from *Salmonella, Ent. liquefaciens, Citrobacter* or *Serratia*. In addition, this team pointed to the biochemical similarities with *Ent. liquefaciens, Ser. marcescens* subsp. *kiliensis* and *Sal. choleraesuis* subsp. *arizonae*. Furthermore, Stevenson and Daly (1982) indicated serological cross-reactions with *Haf. alvei*. However, after an examination of the phenotypic and molecular traits of 33 isolates, Ewing *et al.* (1978) elevated the pathogen to species status, as *Y. ruckeri*. It is interesting to note that the pathogen was included in the genus *Yersinia* because of only a 30–31% DNA homology to *Y. enterocolitica* and *Y. pseudotuberculosis*. This compares to DNA homologies of 24–28% and 31% with *Ser. marcescens* and *Ser. liquefaciens*, respectively (Steigerwalt *et al.*, 1976). Therefore from the DNA hybridization experiments, it is difficult to determine the reasons for including the ERM organism with *Yersinia* rather than *Serratia*. However, it would be relevant to extend the existing databases by comparing the homology with *Sal. choleraesuis* subsp. *arizonae*. This is especially relevant because Green and Austin (1982) have shown a greater phenotypic relationship of *Y. ruckeri* with *Sal. choleraesuis* subsp. *arizonae* than with *Y. enterocolitica* or *Y. pseudotuberculosis*. In fact, the causal agent of ERM may belong in a new genus of the Enterobacteriaceae, an idea which has been mooted by Bercovier and Mollaret (1984). The distinctiveness was confirmed by an examination of phylogeny based on 16S rRNA sequences, when *Y. ruckeri* formed a distinct node, albeit associated with other yersinias (Ibrahim *et al.*, 1993).

Whereas the taxon is phenotypically homogeneous, it is serologically diverse insofar as five serotypes have been recognized (O'Leary, 1977; Bullock *et al.*, 1978a; Stevenson and Airdrie, 1984a). Three serotypes have been referred to colloquially as the 'Hagerman strain' (the most common and the most virulent), the 'Big Creek strain' (relatively avirulent) and the 'Australian strain' (appears to be avirulent) (Busch, 1981). These serotypes have been designated as Type 1 (Hagerman), Type 2 (O'Leary) and Type 3 (Australian) (Bullock and Anderson, 1984) and the newly described serovars IV and V (Stevenson and Airdrie, 1984a). Serotype 2 may be distinguished from serotype 1 by its ability to ferment sorbitol (O'Leary, 1977). However, caution needs to be advocated since this may well be a plasmid-mediated trait. Nevertheless, serotypes/serovars 2, 3, 4 and 5 appear to be highly related in terms of DNA homology (De Grandis *et al.*, 1988).

C. *CYTOPHAGA–FLAVOBACTERIUM–FLEXIBACTER* GROUP

Numerous reports have centred around the role of Gram-negative chromogens as agents of fish disease. Genera, which have been mentioned frequently, include *Cytophaga, Flavobacterium, Flexibacter, Myxobacterium, Myxococcus* and *Sporocytophaga*. The common factor is that these genera comprise difficult-to-identify species, a taxonomic re-evaluation of which has at long last occurred. Moreover from the early literature it is often uncertain into which of these genera unknown isolates should have been placed. In particular, the distinction between *Cytophaga* and *Flexibacter* was confusing (Christensen, 1977; Allen *et al.*, 1983). The authenticity of *Flavobacterium* has been questioned repeatedly, insofar as it became a recipient of problematical pigmented bacteria. Fortunately, more recent work has improved the taxonomy of *Flavobacterium* (Holmes *et al.*, 1984; Bernardet *et al.*, 1996). *Myxobacterium*, considered to be a causal agent of gill disease (Ghittino, 1967; Wood, 1968; Bullock and McLaughlin, 1970; Ashburner, 1978), is not included in the 'Approved Lists of Bacterial Names' (Skerman *et al.*, 1980) or their supplements. Therefore, this genus lacks taxonomic meaning. It is possible that the organisms, identified as *Myxobacterium* (and referred to as myxobacters), belong in either *Cytophaga, Flavobacterium* or *Flexibacter*.

Historically, interest in chromogenic Gram-negative bacteria started with a publication by Davis (1922), who reported serious mortalities (columnaris) among warm-water fish, namely small mouth bass and common perch, from the Mississippi River. These fish were held at the US Biological Station at Fairport, Iowa, when the disease occurred. Davis recognized two important features of the disease, namely that occurrence was primarily in injured (damaged/stressed) fish, and that the water temperature was high, i.e. >21.1°C. Unfortunately, Davis did not succeed in isolating the pathogen. In fact, this was not achieved for two decades, until Ordal and Rucker (1944) succeeded in 1943 during an outbreak in hatchery-reared sockeye salmon.

Fish pathogenic flexibacters have been recognized (see Masumura and Wakabayashi, 1977; Hikida *et al.*, 1979; Pyle and Shotts, 1980, 1981; Wakabayashi *et al.*, 1984). For example, during 1976 and 1977, a bacterial disease developed in juvenile (usually ≤60 mm in length) red sea bream and black sea bream maintained at marine sites in Hiroshima Prefecture, Japan. The outbreak occurred 1–2 weeks after the fish were transported from the hatchery to sea cages. An organism was isolated by Masumura and Wakabayashi (1977), and later considered to be a new species of *Flexibacter*, for which the name of *Fle. marinus* was coined (Hikida *et al.*, 1979). This organism was subsequently re-named as *Fle. maritimus* (Wakabayashi *et al.*, 1986). However, it has been suggested that this taxon is synonymous with *Cyt. marina* (Holmes, 1992). *Fle. ovolyticus* has been named as a new pathogen of halibut eggs and larvae (Hansen *et al.*, 1992). Additional reports of fish pathogenic *Flexibacter* spp. have been noted. In particular, the genus has been found in cases of fin rot (Bullock and Snieszko, 1970; Ziskowski and Murchelano, 1975; Schneider and Nicholson, 1980) and general 'myxobacterial' diseases (e.g. Pyle and Shotts, 1980, 1981).

Flavobacterium became the second genus of yellow-pigmented Gram-negative fish pathogens to be recognized, with the description of *Fla. balustinum* by Harrison and Sadler (1929). This taxon was re-classified into a newly established genus, as *Chryseobacterium balustinum* (Vandamme *et al.*, 1994). However, we believe that this organism was really a fish spoilage agent rather than a pathogen, insofar as the only work describing 'infectivity' referred to dead not living fish. Historically, the organism was first described as being a problem on freshly landed halibut, on which it produced a yellowish slime (Harrison and Sadler, 1929). It would seem likely that the first *bona fide* report of fish pathogenic flavobacteria was by Bein (1954), who described *Fla. piscicida* as the causal agent of mass mortalities (referred to as 'red tide') in marine fish from Florida, US. Incidence of the disease appeared to be associated with the proliferation of phytoplankton and may, consequently, be considered as influenced by water quality. The name of *Fla. piscicida* does not appear in the 'Approved Lists of Bacterial Names' (Skerman *et al.*, 1980) or their supplements, and is, therefore, of dubious taxonomic standing.

A filamentous Gram-negative organism, isolated from gill lesions in rainbow trout and Yamame salmon, formed the basis of an article by Kimura *et al.* (1978a). These isolates from Japan, together with similar strains recovered from Oregon, USA, were included in a detailed investigation by Wakabayashi *et al.* (1980). The outcome was recognition of a novel group of *Flavobacterium*, i.e. *Fla. branchiophila* (Wakabayshi *et al.*, 1989). This name has now been corrected to *Fla. branchiophilum* (von Gravenitz, 1990), and its authenticity verified (Bernardet *et al.*, 1996). *Fla. branchiophilum* appears to be spreading as a fish pathogen in the Far East, e.g. in Korea (Ko and Hwo, 1997).

There have been other reports of 'flavobacteriosis' caused by unknown species of *Flavobacterium* (e.g. Brisou *et al.*, 1964; Roberts, 1978; Acuigrup, 1980a; Farkas, 1985). Whereas there has been reticence to equate these organisms with existing nomenspecies (or to name new species), the descriptions of the aetiological agents have been quite reasonable. For example, Acuigrup (1980a) discussed flavobacteriosis among coho salmon held in Spanish seawater sites. This disease, which resembled a generalized septicaemia, caused 20–25% mortality in the fish population during the summer of 1978. The subsequent description of the organism was superior to the initial work with *Fla. balustinum* or *Fla. piscicida*.

Cytophagas became implicated as fish pathogens with the work on coldwater (low-temperature) disease. The causal agent was initially isolated and described by Borg (1948), and subsequently named as *Cyt. psychrophila* (Borg, 1960), and then to *Fla. psychrophilum* (Bernadet *et al.*, 1996). Coldwater disease affects predominantly juvenile salmonid fish, notably coho salmon in the northwest US, and is most prevalent in winter and spring when the water temperature is <10°C. Cranial and vertebral lesions may occur

(Kent *et al.*, 1989). More recently, the organism has been associated with systemic disease in eels and cyprinids in Europe (Lehmann *et al.*, 1991) and with an anaemic condition of juvenile rainbow trout in Chile (Bustos *et al.*, 1995) and Europe (e.g. Lorenzen *et al.*, 1991), referred to as rainbow trout fry syndrome [RTFS] (Baudin-Laurençin *et al.*, 1989; Lorenzen *et al.*, 1991).

A causal mention was given initially to the role of *Cyt. johnsonae* and *Cyt. rosea* as fish pathogens (Christensen, 1977). *Cyt. johnsonae* has emerged as a problem in Australia. However, only scant information is available about *Cyt. rosea*; therefore, this organism will not be considered further.

An organism was recovered initially from the gills of diseased hatchery-reared salmon, trout and suckers in Michigan, USA (Strohl and Tait, 1978). Thirteen isolates were recovered, and although similarities were noted to organisms previously described by Borg (1960), Pacha and Porter (1968) and Anderson and Conroy (1969), it was decided to elevate them into a new species, as *Cyt. aquatilis* (Strohl and Tait, 1978), and thence to *Fla. hydatis* (Bernardet *et al.*, 1996). It must be emphasized that Strohl and Tait (1978) did not prove that the organisms were capable of causing disease. Nevertheless, we have recovered similar organisms from outbreaks of gill disease in farmed rainbow trout from England.

Cytophaga sp. has been associated with skin and muscle lesions on Atlantic salmon in the USA (Kent *et al.*, 1988), and a previously undescribed *Cytophaga*-like bacterium (CLB) has been associated with a gill and systemic disease in turbot (Mudarris and Austin, 1989). This organism was described as a new species, as *Fla. scophthalmum* (Mudarris *et al.*, 1994), which was re-classified to *Chrys. scophthalmum* (Vandamme *et al.*, 1994).

Little is known about the role of *Sporocytophaga* as a fish pathogen. Mixed infections attributed to *Sporocytophaga* and *V. anguillarum* occurred as surface lesions, termed saltwater columnaris, on salmon and trout held in marine conditions (Wood, 1968). However, apart from discussing the presence of microcysts which began to form at 2–7 days, there is little information about this suspected pathogen (Pacha and Ordal, 1970).

There has been one report of gill disease (in 1972) caused by a supposed new species of the fruiting organism *Myxococcus*, for which the name of *Myxococcus piscicola* was suggested (Xu, 1975). This organism was associated with an epizootic in grass carp (*Ctenopharyngodon idelluls*) fingerlings held in ponds in Wuhan, China.

Chryseobacterium balustinum (= Flavobacterium balustinum)

The original description was extremely brief, referring to the growth of colonies at very low temperatures, i.e. 1–3°C, rod-shaped micromorphology, motility, and the ability to degrade gelatin (Harrison and Sadler, 1929). The revised description of *Flavobacterium* excludes motile organisms (Holmes *et al.*, 1984); therefore the validity of *Fla. balustinum* was doubtful. The organisms were re-classified to *Chryseobacterium*, as *Chrys. balustinum* on the basis of rRNA clustering (Vandamme *et al.*, 1994). Interestingly, the extensive list of characters in *Bergey's Manual of Systematic Bacteriology* (Holmes *et al.*, 1984) precluded reference to motility or, for that matter, gliding movement among representative strains of any species. Other characteristics of the taxon included:

> *Chryseobacterium balustinum*
> The presence of yellow-pigmented, mucoid colonies comprising non-motile, oxidative rods of 1.0–1.8 × 0.5 μm in size. Catalase, indole, oxidase and phosphatase, but not β-galactosidase, arginine dihydrolase, H_2S or phenylalanine deaminase, are produced. Nitrates are reduced. Aesculin, casein, DNA, gelatin, tributyrin, Tween 20 and 80, but not starch, tyrosine or urea, are degraded. Growth occurs at 37°C, but not at 5°C or 42°C. This is in contrast to the original work of Harrison and Sadler (1929), who reported growth at almost the freezing point of water. Acid is produced from ethanol and glucose, but not from arabinose, cellobiose, lactose, mannitol, raffinose, salicin, sucrose or xylose. The G + C ratio of the DNA is 33.1 moles %.

Chryseobacterium scophthalmum (= Flavobacterium scophthalmum)

> *Chryseobacterium scophthalmum*
> Colonies (2–3 mm in diameter after 48 h at 25°C) contain orange-pigmented (flexirubin pigment) uniformly shaped, short, fermentative Gram-negative rods of approximately 2.0 × 0.8 μm in size. Gliding movement is exhibited. Catalase and oxidase are produced, but not H_2S, indole, or lysine or ornithine decarboxylase. Nitrates are not reduced. The Voges Proskauer reaction is negative. Aesculin, casein, DNA, gelatin, tributyrin and tyrosine are degraded, but not chitin or starch. Acid is produced from cellobiose and glucose, but not arabinose, mannitol, raffinose, sucrose or xylose. Thin sections reveal the presence of a thick (43.5 nm) cell envelop, which could be mistaken for an extracellular layer. The G +C ratio of the DNA is 31.7–34.2 moles % (Mudarris and Austin, 1989; Mudarris *et al.*, 1994).

From these characteristics, the 50 isolates from turbot were equated with a new species of *Flavobacterium*, as *Fla. scophthalmum* (Mudarris *et al.*, 1994). With subsequent improvements in taxonomy, the species was transferred to the newly established genus *Chryseobacterium*, as *Chrys. scophthalmum* (Vandamme *et al.*, 1994).

Flavobacterium sp.

The organisms, discussed by Acuigrup (1980a), were characterized, as follows:

> *Flavobacterium* sp.
> The isolates produce yellow-orange colonies after 24 h (the temperature of incubation was not stated). Cultures comprise motile, pleomorphic, Gram-negative rods, which produce arginine dihydrolase and oxidase but not β-galactosidase, H_2S, indole or lysine or ornithine decarboxylase, and are neither oxidative nor fermentative. The Voges Proskauer reaction is positive. Gelatin is degraded. Acid is produced from amygdalin, arabinose, inositol, mannitol, melibiose and rhamnose, but not from glucose or lactose. Sodium citrate is not utilized (Acuigrup, 1980a).

Most of the data originate from use of an unstated API product, possibly the API 20E
rapid identification system. The presence of motility contrasts with the revised definition
of *Flavobacterium* (Holmes *et al.*, 1984). Therefore, it appears that the organisms are not
bona fide Flavobacterium, but probably represent a closely allied taxon.

Flavobacterium branchiophilum

Flavobacterium branchiophilum
This pathogen forms yellow, translucent, smooth colonies of 0.5–1.0 mm in
diameter after incubation at 18°C for 5 days. Isolates (ten isolates from Oregon,
five from Japan and one from Hungary; Farkas, 1985) comprise slender, strictly
aerobic, non-motile, Gram-negative rods of $5–8 \times 0.5$ μm in size, which are
surrounded by pili (Heo *et al.*, 1990). Usually, they occur as short chains, each
comprising two to three cells. Catalase and oxidase are produced, but not H_2S or
indole. Nitrates are not reduced. Casein, gelatin, lecithin, starch, tributyrin, Tweens
and tyrosine are degraded, but not cellulose or chitin (Bernardet *et al.*, 1996).
Growth occurs at 5°C and 30°C but not at 37°C, and in 0 and 0.1% (w/v) sodium
chloride. It is noteworthy that Japanese isolates grow at 30°C, whereas American
strains do not. Acid is produced from a wide range of compounds, including cellobiose, fructose, glucose, inulin, maltose, melibiose, raffinose, sucrose and trehalose,
but not from adonitol, arabinose, dulcitol, galactose, inositol, lactose, mannitol,
salicin, sorbitol or xylose (Ko and Heo, 1997). The $G+C$ ratio of the DNA is
29–34 moles %. Another geographical difference concerns serology, insofar as the
Japanese isolates are antigenically distinct from the American and Hungarian
cultures (Huh and Wakabayashi, 1989).

Flavobacterium columnare (= Flexibacter/Cytophaga columnaris)

Flavobacterium columnare
Colonies appear to be flat, spreading, yellowish green (flexirubin-type pigment),
with rhizoidal edges, and contain slender, strictly aerobic, Gram-negative rods of
$4–8 \times 0.5–0.7$ μm in size, which move by gliding. Catalase, H_2S and oxidase are
produced, but not indole, or lysine or ornithine decarboxylase. Nitrates are reduced.
The methyl red test and Voges Proskauer reaction are negative. Casein, gelatin and
tributyrin are degraded, but not aesculin, agar, cellulose, chitin, starch and tyrosine.
Growth occurs at 4–30°C (some isolates grow also at 37°C), and in 0–0.5% (w/v),
but not 1% (w/v), sodium chloride (Bernardet, 1989; Bernardet and Grimont,
1989a; Bernardet *et al.*, 1996). Dead bacterial cells, notably *Escherichia coli* are
lysed. Carbohydrates are generally not utilized. Moreover, acid is not produced
from arabinose, cellobiose, glucose, lactose, mannitol, raffinose, salicin, sucrose or
xylose. The $G+C$ ratio of the DNA is 32.9–35.9 moles % (Wakabayashi and
Egusa, 1966; Pacha and Porter, 1968; Pacha and Ordal, 1970; Chun, 1975; Bootsma

and Clerx, 1976; Morrison *et al.*, 1981). Four major serological groups and one miscellaneous group have been recognized among the 325 strains examined by Anacker and Ordal (1959). Moreover, separate and distinct DNA homology groups have been described (Song *et al.*, 1988).

The taxonomic status of the pathogen has undergone radical change since the pioneering work of Davis (1922). Davis coined the name *Bacillus columnaris* because, in wet preparations of infected material, the bacteria congregated in column-like masses. According to Ordal and Rucker (1944), pure cultures of the organism exhibited gliding, and therefore should be associated with the myxobacteria. The taxonomy became further complicated by the observation in cultures of oval and spherical structures, which were thought to be microcysts (this notion was subsequently rejected). This led to a re-classification, as *Chondrococcus columnaris*. Shortly afterwards, Garnjobst (1945) suggested that the pathogen was actually a *Cytophaga*, and thus, the transition was made to *Cytophaga columnaris*. With further deliberation, the organism was transferred to *Flexibacter*, as *Flexibacter columnaris* (Leadbetter, 1974), and from phylogenetic data to *Fla. columnare* (Bernardet *et al.*, 1996).

Flavobacterium hydatis (= Cytophaga aquatilis)

Flavobacterium hydatis (= Cytophaga aquatilis)
Following incubation at 20°C for 14 days, yellow-orange colonies develop, which contain Gram-negative facultatively anaerobic rods of 8.0×0.5 μm in size. Copious quantities of extracellular slime are produced. Cells demonstrate gliding movement. Microcysts and fruiting bodies are absent. Catalase is produced, but not H_2S, indole, lysine or ornithine decarboxylase, oxidase or phenylalanine deaminase, and the methyl red test and Voges Proskauer reaction are negative. A wide range of complex molecules are degraded, including aesculin, blood, casein, DNA, gelatin, pectin, starch, tributyrin, Tween 40, 60 and 80, and tyrosine (slowly), but not cellulose or urea. Some strains attack chitin. Growth occurs at 5–35°C but not 42°C, in 0–2% (w/v) sodium chloride, which is an indication that the organism is unlikely to be present in full strength seawater. Growth occurs also at pH 5.5–11.0. Nitrates are reduced to ammonia. Acid is produced from arabinose, cellobiose, glucose, lactose, mannitol, raffinose, sucrose and xylose. The G + C content of the DNA is 33.7 moles %. Chemical analyses have established that the major pigments are similar to the 'flexirubin' of *Fle. elegans* (Strohl and Tait, 1978).

On the basis of the micromorphology, gliding movement, G + C ratio, and the ability to degrade complex molecules, this group was considered to represent a previously undescribed species of *Cytophaga*, for which the name of *Cyt. aquatilis* was proposed (Strohl and Tait, 1978). However, on the basis of phylogenetics, the species was transferred to *Flavobacterium*, as *Fla. hydatis* (Bernardet *et al.*, 1996).

Flavobacterium johnsoniae (= Cytophaga johnsonae)

Fla. johnsoniae was identified by an examination of phenotypic tests, and associated with mortalities in salmonids (Rintamäki-Kinnunen *et al.*, 1997). The characteristics for the fish isolates were as follows:

Flavobacterium johnsoniae

Colonies are yellow (flexirubin pigment) and contain Gram-negative gliding, filaments. Catalase, β-galactosidase and oxidase are produced, but not H_2S or indole. Nitrates are reduced. Acid is produced from glucose. Growth occurs at 10–30°C but not at 37°C, and in 0.5–1.0% (w/v) sodium chloride. Casein, chitin, DNA, gelatin, starch, tributyrin, Tween 20, Tween 40 and tyrosine (with pigment production) are degraded, but not Tween 80. Ammonium, asparagine, glutamate, potassium nitrate and urea are utilized as sole sources of nitrogen (Carson *et al.*, 1993).

Flavobacterium psychrophilum (= Cytophaga psychrophila = Flexibacter psychrophilum)

The best of the descriptions of this organism stems from the work of Pacha (1968), Bernardet and Kerouault (1989), Bernardet and Grimont (1989b) and Schmidtke and Carson (1995).

Flavobacterium psychrophilum

Cultures produce non-diffusible yellow-pigmented (flexirubin pigment) colonies with thin spreading margins. Cells are strictly aerobic, Gram-negative, slender, flexible rods of 1.5–7.5 × 0.75 μm in size. With increasing age the cells appear to be shorter. As with *Fla. hydatis*, gliding movement is exhibited, and fruiting bodies and microcysts are absent. Catalase is produced, oxidase may appear to be positive, but H_2S, indole, and lysine and ornithine decarboxylase are not produced. Nitrates are not reduced, and the Voges Proskauer reaction is negative. Casein, gelatin and tributyrin are degraded; tyrosine is attacked by some isolates, and aesculin, chitin, starch and xanthine not at all. Generally, growth occurs at 4–23°C, but not at 30°C, and in 0.8% but not 2% (w/v) sodium chloride. This demonstrates clearly that the organisms are suited to low-temperature, freshwater environments. The organisms are capable of degrading autoclaved cells of *Esch. coli* (Pacha, 1968). The G +C ratio of the DNA is 32.5–34 moles %. The original ten isolates of Pacha (1968) were deemed to be serologically homogeneous. Yet, later work identified two major serogroups, with O-1 accommodating isolates from Japan and the USA and O-2 comprising only Japanese cultures (Wakabayashi *et al.*, 1994). This number increased to three, with European isolates included in one major serogroup (Lorenzen and Olesen, 1997). Several ribotypes have been recognized among 85 isolates, with most harbouring one or more plasmids [eleven plasmid profiles defined] (Chakroun *et al.*, 1998).

On the basis of these characteristics, the conclusion was reached that the causal agent of coldwater disease could be equated with the genus *Cytophaga*, as defined by Stanier (1942), but was sufficiently distinct from existing species to warrant description as a new species. Hence, *Cyt. psychrophila* was named (Borg, 1960). However, Lewin and Lounsberry (cited in Leadbetter, 1974) disagreed with this notion, and proposed that the organisms would be better classified in the genus *Flexibacter*, as *Fle. aurantiacus*. Bernardet and Grimont (1989b) agreed with the opinion that the organisms belong in the genus *Flexibacter*, but proposed *Fle. psychrophilum*. Then with general improvements in the understanding of the yellow-pigmented bacteria resulting from phylogenetic data, the taxon was transferred to *Flavobacterium*, as *Fla. psychrophilum* (Bernardet *et al.*, 1996). It is agreed that the taxon comprises a homogeneous group of bacteria (Lorenzen *et al.*, 1997).

Flavobacterium piscicida
Only a poor description of *Fla. piscicida* exists in the fisheries literature. Therefore, further discussion will not be attempted.

Flexibacter **spp.**
The fish pathology literature abounds with references to unspeciated *Flexibacter*. Pyle and Shotts (1980; 1981) studied 17 strains for phenotypic traits, as determined by use of the API 20E rapid identification system, and by DNA homology. The conclusion reached was that the isolates from warm-water fish were distinct to those recovered from cold-water fish. At least three separate groups were recognized. However, whether or not these organisms belong in existing species or represent new taxa must await further study.

Flexibacter maritimus (= Cytophaga marina)

Flexibacter maritimus

A very homogeneous group (Bernadet *et al.*, 1994), with cultures forming pale yellow flat, thin colonies, containing strictly aerobic Gram-negative rods of $2–30 \times 0.5$ μm in size, which occasionally form filaments of up to 100 μm in length. In older cultures, cells become spherical (~0.5 μm in diameter). Microcysts do not occur. Gliding movement is a characteristic feature of all isolates. Catalase and oxidase are produced, but not H_2S or indole. Nitrates are not reduced. The methyl red test is negative. Casein, gelatin, tributyrin and tyrosine are degraded, but not aesculin, agar, cellulose, chitin starch or urea. Growth occurs from 14.6°C to 34.3°C, and at pH 6–9. There are requirements for potassium chloride and sodium chloride. Acid is not produced from arabinose, cellobiose, glucose, lactose, mannitol, raffinose, salicin, sucrose or xylose. Isolates are capable of lysing dead cells of *Aer. hydrophila*, *Edw. tarda*, *Esch. coli* and *V. anguillarum*, but not *Bacillus subtilis*. The G + C ratio of the DNA is 31.3–32.5 moles % (Hikida *et al.*, 1979; Wakabayashi *et al.*, 1986). All isolates share a common antigen (Wakabayashi *et al.*, 1984).

From these traits, Hikida *et al.* (1979) proposed that the organism belonged in an as yet undescribed species of *Flexibacter*, for which the name of *Fle. marinus* was mooted. Subsequently, Wakabayshi *et al.* (1986) formally proposed the name of *Fle. maritimus* to accommodate the pathogen. Independently, Reichenbach (1989) proposed the name of *Cyt. marina* for the same organism. We concur with the view of Holmes (1992) that the organism should remain in the genus *Flexibacter*, as *Fle. maritimus* rather than be transferred to *Cytophaga*, as *Cyt. marina*.

Flexibacter ovolyticus
The 35 isolates were described as comprising:

Flexibacter ovolyticus
Pale yellow-pigmented long, slender (0.4×2–$20\ \mu$m) Gram-negative rods, which demonstrate gliding movement. Catalase and oxidase are produced, but not so arginine dihydrolase, β-galactosidase, H_2S, indole or lysine or ornithine decarboxylase. Nitrates are reduced, DNA, gelatin, Tween 80 and tyrosine but not agar, cellulose, chitin, starch or urea, are degraded. Acid is not produced aerobically from arabinose, cellobiose, glucose, lactose, mannitol, raffinose, salicin, sucrose or xylose. Growth occurs at 4–30°C but not 35°C, and in >3% (w/v) sodium chloride. The G + C ratio of the DNA was reported as 31.0 moles % (Hansen *et al.*, 1992).

On the basis of DNA:DNA hybridization, there is 26–42% homology with *Fle. maritimus*. Certainly, the strains are distinct from *Fle. maritimus*, but it would be relevant to enquire about possible relationships with other flexibacteria.

Myxococcus piscicola
Conical fruiting bodies were produced on agar. These fruiting bodies were surrounded by a thin membrane, without peduncle or branches (Xu, 1975).

Sporocytophaga **spp.**
Negligible information is available about these Gram-negative microcyst-forming organisms (see Pacha and Ordal, 1970).

D. PSEUDOMONADACEAE REPRESENTATIVES

Pseudomonas anguilliseptica
The description of the pathogen is not extensive.

Pseudomonas anguilliseptica
A homogeneous group of Gram-negative asporogenous rods, which are motile by means of single polar flagella. Electron microscopy of 18-h-old cultures on TSA reveal the presence of long, slightly curved rods with rounded ends. The size of these cells has been estimated as 5–$10 \times 0.8\ \mu$m. In addition, many bizarre forms

have been observed. Fluorescent pigment is not produced. There is no reaction in the oxidative-fermentative test. Catalase and oxidase are produced, but not arginine dihydrolase, β-galactosidase, H_2S or indole. Nitrates are not reduced. Gelatin and Tween 80 are degraded, but not blood, DNA, starch or urea. Acid is not produced from arabinose, fructose, galactose, glucose, glycerol, inositol, lactose, maltose, mannitol, mannose, raffinose, rhamnose, salicin, sucrose or xylose. Growth occurs at 5–30°C but not 37°C, in 0–4% (w/v) sodium chloride, and at pH 5.3–9.7. The G + C ratio of the DNA is 56.5–57.4 moles % (Wakabayashi and Egusa, 1972; Muroga *et al.*, 1977b; Nakai and Muroga, 1982; Stewart *et al.*, 1983).

A comprehensive examination of 96 isolates indicated the presence of two antigenic groups. Type I was not agglutinated in unheated antisera (this was prepared against heat-killed cells), although clumping (agglutination) of the cells subsequently occurred after the antiserum was heated to 100°C for 2 h (or 121°C for 30 min). Type II lacked this inhibition. It was speculated that this thermolabile agglutination-inhibition antigen corresponds to the so-called K-antigens of coliforms (Nakai *et al.*, 1981, 1982a,b).

From the phenotypic traits, Wakabayashi and Egusa (1972) concluded that the causal agent of Sekiten-byo corresponded to a new centre of variation within 'Group III' or 'Group IV' of the genus *Pseudomonas*. This opinion was reached because the pathogen was Gram-negative, rod-shaped, motile by polarly located flagella, insensitive to the vibriostatic agent (0/129), and produced catalase and oxidase, but not acid from glucose or, for that matter, diffusible (fluorescent) pigment. Because the strains were dissimilar to other fish pathogenic pseudomonads, namely, *Ps. fluorescens*, a new taxon was proposed, i.e. *Ps. anguilliseptica*. We are sceptical about the validity of this proposal because the description could equally fit *Alcaligenes* or *Deleya* as well as *Pseudomonas* (see Cowan, 1974; Kersters and De Ley, 1984; Palleroni, 1984). In some respects, the G +C ratio and the inability to produce acid in peptone water sugars is more conducive to the concept of *Alcaligenes* or *Deleya*, although the pathogen is clearly distinct from existing nomenspecies (Kersters and De Ley, 1984). Moreover, it may not be ruled out that the causal agent of Sekiten-byo should be classified in a newly described genus. Certainly, the distinctive micro-morphology adds weight to this supposition. Maybe this explains the pronounced dissimilarity of *Ps. anguilliseptica* to other species of *Pseudomonas* as revealed by analyses of fatty acids and outer-membrane proteins (Nakajima *et al.*, 1983).

Pseudomonas chlororaphis
To date, there has been only one report of *Pseudomonas chlororaphis* as a fish pathogen. This involved a heavy mortality among farmed Amago trout (*Oncorhynchus rhodurus*) in Japan (Hatai *et al.*, 1975). For the present, it is uncertain whether *Ps. chlororaphis* represents an emerging problem, or a secondary (opportunistic) invader of already diseased hosts.

The isolates matched the description of *Ps. chlororaphis*, insofar as cultures comprised Gram-negative motile rods, which produced distinctive colonies. These produced green pigment, which crystallized as needles in the colonies (Stanier *et al.*, 1966; Palleroni,

1984). Other phenotypic traits were not reported, although the authors inferred that further tests had been carried out, and that these agreed with the definition of *Ps. chloro-raphis*.

Pseudomonas fluorescens

Ps. fluorescens is a dominant component of the freshwater ecosystem (Allen *et al.*, 1983). At various times, *Ps. fluorescens* has been considered as a fish spoilage organism (Shewan *et al.*, 1960), a contaminant or secondary invader of damaged fish tissues (Otte, 1963), as well as a primary, but poor pathogen (Roberts and Horne, 1978). All the published descriptions of the organism (e.g. Bullock, 1965; Csaba *et al.*, 1981; Ahne *et al.*, 1982) agree closely with the definition of *Ps. fluorescens* (Stanier *et al.*, 1966; Palleroni, 1984).

Pseudomonas fluorescens

Cultures comprise Gram-negative, oxidative, arginine dihydrolase, catalase and oxidase producing rods, which are motile by polar flagella. Growth occurs at 4°C but not at 42°C. Fluorescent pigment (fluorescein) and gelatinase, but not β-galactosidase, H_2S, indole, amylase or urease, are produced. The Voges Proskauer reaction is negative. Citrate is utilized, and acid is produced from arabinose, inositol, maltose, mannitol, sorbitol, sucrose, trehalose and xylose, but not from adonitol or salicin.

It seems likely that other fish pathogenic pseudomonads, as discussed by Li and Fleming (1967) and Li and Traxler (1971), correspond to *Ps. fluorescens.*

Pseudomonas pseudoalcaligenes

During 1992 at a site in the UK, rainbow trout (average weight = 100 g) were observed with extensive skin lesions and signs of enteric redmouth (from which *Yersinia ruckeri* was recovered). From these animals, an organism with the key characteristics of *Ps. pseudoalcaligenes* was recovered (Austin and Stobie, 1992b).

Pseudomonas pseudoalcaligenes

The cream-coloured colonies (with a 'gummy' consistency) comprise motile oxidative (alkali is produced) short Gram-negative rods, which produce arginine dihydrolase, catalase, ornithine decarboxylase and oxidase, but not β-galactosidase, H_2S, indole, lysine decarboxylase or tryptophan deaminase, degrade gelatin, tyrosine (with the production of melanin) and Tween 80, but not DNA, starch or urea, and grow at 15°C and 25°C, but not at 4°C or 40°C. Acid is produced from arabinose and glucose, but not from amygdalin, inositol, mannose, melibiose, rhamnose or sorbitol. Citrate is utilized. The Voges Proskauer reaction is negative.

The organism matched the description of *Pseudomonas* (Palleroni, 1984), and approximated *Ps. pseudoalcaligenes* as determined from the probability matrix of Holmes *et al.*

(1986). The only discrepancies concerned the degradation of gelatin and starch, and the production of ornithine decarboxylase.

Pseudomonas putida

There has been a casual mention of *Ps. putida* as a fish pathogen in Japan (Muroga, 1990). The organism was recovered from diseased (the disease was described as bacterial haemorrhagic ascites) ayu in Japan, and equated with *Pseudomonas* (Wakabayashi *et al.*, 1996). Similarities were noted to *Ps. putida*, but there was an absence of fluorescent pigment and a lack of agglutination with antiserum prepared to the type strain (of *Ps. putida*). The profile, obtained with the API 20NE rapid identification system, was '1140452' (Wakabayashi *et al.*, 1996).

E. VIBRIONACEAE REPRESENTATIVES

Vibrios have emerged as the scourge of marine fish and shellfish. Renewed interest has resulted in the description of new species and a better understanding of the biology of long-recognized taxa. To date, many species have been described as fish pathogens. In addition, hard-to-speciate *Vibrio* have been regularly recovered (e.g. Yasunobu *et al.*, 1988; Masumura *et al.*, 1989; Muroga *et al.*, 1990). There is controversy over the role of *V. parahaemolyticus* as a fish pathogen, and we are not entirely satisfied with the evidence; therefore for the purpose of this chapter, it has been concluded that the organism does not constitute a *bona fide* fish pathogen. One article has mentioned challenging tilapia with *V. parahaemolyticus*, but there was insufficient information about the authenticity of the isolates (Balfry *et al.*, 1997). However, organisms with inter-mediate characteristics between *V. alginolyticus* and *V. parahaemolyticus* have been recovered from diseased milkfish in the Philippines (Muroga *et al.*, 1984a). A recent publication has suggested that *V. campbellii*, *V. nereis* and *V. tubiashi* may be associated with disease in gilt-head sea bream in Spain (Balebona *et al.*, 1998). However, confirmation is desirable before these taxa become recognized as *bona fide* fish pathogens.

In the taxonomic study of Austin *et al.* (1997), a single isolate from Atlantic salmon in Tasmania was equated with *V. aestuarianus* using the diagnostic scheme of Alsina and Blanch (1994a, b). By phenotypic means, more isolates were associated with disease in gilt-head sea bream in Spain (Balebona *et al.*, 1998). Also, a single isolate from a case of winter ulcer syndrome/disease in Atlantic salmon, was recovered as a supposed new species, for which the name *V. viscosus* has been mooted, but not formally proposed.

Moritella marina (= V. marinus)

Nineteen Icelandic and one Norwegian isolate from shallow skin lesions on Atlantic salmon, and the type strain of *V. marinus* NCIMB 1144 were identified as *V. marinus* after an examination of phenotypic data and analyses by numerical taxonomy (Benediktsdóttir *et al.*, 1998). On the basis of 16S rRNA sequencing the taxon was transferred to a newly established genus, as *Moritella marina* (Urakawa *et al.*, 1998). However, apart from emphasizing a relationship to *Shewanella*, the authors did not make any comment about family membership. Until the hierarchy is clarified, *Moritella* will be left with the other representatives in the family Vibrionaceae.

> *Moritella marina*
> Cultures produce lysine decarboxylase and oxidase but not arginine dihydrolase or ornithine decarboxylase, reduce nitrates, are positive for the methyl red test but not the Voges Proskauer test, degrade blood (β-haemolysis), chitin, DNA, gelatin, lipids and starch, do not produce acid from carbohydrates except *N*-acetyl glucosamine, maltose, mannitol, mannose and ribose, are resistant to the vibriostatic agent. O/129, and grow at 4–20°C but not at 25°C.

Photobacterium damselae

The validity and distinctiveness of *Ph. damselae* has been confirmed, with isolates homogeneous by BIOLOG-GN fingerprints, API 20E profiles and LPS profiles, but heterogeneous by ribotyping and serology (four serogroups were defined) (Fig. 4.2; Austin *et al.*, 1997).

> *Photobacterium damselae*
> Cultures comprise facultatively anaerobic Gram-negative, weakly motile (by one or more unsheathed polar flagella) rods. Arginine dihydrolase, catalase and oxidase are produced, but not β-galactosidase, H_2S, indole, lysine or ornithine decarboxylase or phenylalanine deaminase. Chitin, DNA, starch and urea, but not corn oil (lipids) or gelatin, are degraded. The methyl red test and Voges Proskauer reaction are positive. Nitrates are reduced. Growth occurs in 1–6% (w/v) but not 0% or 7% (w/v) sodium chloride. Acid is produced from D-glucose, maltose and mannose, but not D-adonitol, arabinose, cellobiose, dulcitol, erythritol, inositol, lactose, mannitol, melibiose, raffinose, L-rhamnose, salicin, D-sorbitol, sucrose, trehalose or D-xylose. Acetate, citrate and malonate are not utilized. Sensitivity is recorded to the vibriostatic agent, O/129. The G + C ratio (for one strain) is 43 moles %.

Although Love *et al.* (1981) did not publish detailed reasons for the dissimilarity of 'V. damsela' to other species of *Vibrio*, they did mention that DNA:DNA hybridization studies have been completed. Unfortunately, the results were not published. This situation was corrected by Grimes *et al.* (1984b), who demonstrated low DNA homology values with other vibrios. Therefore, it is not surprising that the pathogen was re-classified out from *Vibrio*, initially to *Listonella* (MacDonell and Colwell, 1985), then to *Photobacterium*, as *Photobacterium damsela* (Smith *et al.*, 1991) and finally corrected to *Ph. damselae* (Trüper and de'Clari, 1997).

Photobacterium damselae subsp. *piscicida* (= *Pasteurella piscicida*)

During the summer of 1963, an epizootic was reported in white perch and striped bass in the upper region of the Chesapeake Bay, USA. From the diseased fish, 30 cultures of an organism were recovered which possessed some of the salient features of *Pasteurella* (Snieszko *et al.*, 1964a). Hence the condition was termed 'pasteurellosis'. However, the literature became confusing, insofar as the disease is also referred to in Japan as 'pseudotuberculosis' because of the distinctive pathology. There are also some reports indicating the presence of fish pathogenic *Pasteurella* in Great Britain (Ajmal and Hobbs, 1967) and Norway (Hastein and Bullock, 1976). However, it is possible that these organisms should have been identified as atypical *Aer. salmonicida* (see Paterson *et al.*, 1980). The results of many investigations led to the conclusion that the pathogen consists of a phenotypically and serologically homogeneous taxon (e.g. Magariños *et al.*, 1992). By ribotyping of 29 isolates, two major ribotypes were recognized which effectively separated European and Japanese isolates. A third ribotype accommodated an unique strain (Magariños *et al.*, 1997b).

Photobacterium damselae subsp. *piscicida*
Cultures comprise fairly unreactive, Gram-negative, non-motile, fermentative rods of $0.5 \times 1.5 \ \mu m$ in size, with pronounced bipolar staining. Pleomorphism may be evident, especially in older cultures. Catalase and oxidase are produced, but not alanine deaminase, β-galactosidase, H_2S, indole, lysine or ornithine decarboxylase or phenylalanine deaminase. Nitrates are not reduced. The methyl red test is strongly positive, whereas the Voges Proskauer reaction is weakly positive. Arginine and Tween 80 are degraded, but not blood, casein, chitin, gelatin, starch or urea. Growth occurs at 25–30°C but not 10 or 37°C, in 0.5–3.0% (w/v) sodium chloride and at pH 5.5–8.0, but not on MacConkey agar or in potassium cyanide broth. Uniform turbidity is recorded in broth cultures. Sodium citrate is not utilized. Acid is produced from fructose, galactose, glucose (weak) and mannose, but not amygdalin, arabinose, dulcitol, inositol, lactose, maltose, mannitol, melibiose, rhamnose, salicin, sorbitol, sucrose or trehalose. Unfortunately, the G +C ratio of the DNA has not been determined for any *bona fide* strains. So far, only one serotype has been recognized. In general, the organism possesses one heat-stable and four heat-labile somatic antigens, and three heat-labile extracellular antigens (presumably enzymes) (Kusuda *et al.*, 1978a). The LPS has been found to comprise <1% protein, 18–24% sugar and 34–36% fatty acids. The sugar component includes hexose, heptose, pentose, 6-deoxyhexose, 2-keto-3-deoxyoctonate and hexosamine. The fatty acids include lauric acid, 3-hydroxy lauric acid, myristic acid and palmitic acid (Salati *et al.*, 1989a,b).

The morphology and physiology of this pathogen led Snieszko *et al.* (1964a) to suspect a similarity to the genus *Pasteurella*. This view was reinforced by cross-precipitin reactions with *Pasteurella (Yersinia) pestis*. From this deduction, Janssen and Surgalla (1968) realized that, from an examination of 27 isolates, the organism was very homogeneous and different from existing species of *Pasteurella*. Therefore, the name of *Pa. piscicida* was coined. Independently, Kusuda gave it an alternative name, i.e. *Pa. seriola*, but quickly realized its synonymy with *Pa. piscicida*, which was accorded preference. However, *Pa. piscicida* was not included in the 'Approved Lists of Bacterial Names' (Skerman *et al.*, 1980) or their supplements. Consequently, the name of *Pa. piscicida* lacked taxonomic validity.

A detailed taxonomic evaluation based on small-subunit rRNA sequencing and DNA:DNA hybridization revealed that the organism was highly related to *Ph. damsela* (there was >80% relatedness of the DNA), and it was proposed that the organism be accommodated in a new subspecies, as *Ph. damsela* subsp. *piscicida* (Gauthier *et al.*, 1995), the epiphet of which was corrected to *damselae* (Trüper and de'Clari, 1997), as *Ph. damselae* subsp. *piscicida*.

To complicate matters, there is controversy over interpretation of the Gram-staining reaction. The majority opinion is that the organism is Gram-negative. However, Simidu and Egusa (1972) considered that cells displayed Gram-variability when young, i.e. in 12–18 h cultures incubated at 20–25°C. In addition, they presented photographic evidence which showed that cells shortened with age. In fact, the suggestion was made that the pathogen is related to *Arthrobacter*. It is ironic that a similar phenomenon, concerning the interpretation of Gram-stained smears, was reported by Killian (1976) and Broom and Sneath (1981) for *Haemophilus piscium*, the causal agent of ulcer disease.

Plesiomonas shigelloides

Although there have been unauthenticated verbal reports that *Plesiomonas shigelloides* may be pathogenic to fish, it was not until the summer of 1984 that the organism was definitely recovered from diseased rainbow trout in northern Portugal (Cruz *et al.*, 1986). Subsequently, the organism was recovered in Germany from catfish, which had originated in Nigeria, sturgeon fingerlings, which had been sent from Russia, gourami imported from Thailand and eels (Klein *et al.*, 1993). Many of the characteristics of the pathogen were derived from use of the API 20E rapid identification system (Cruz *et al.*, 1986) and API 20NE system (Klein *et al.*, 1993).

Plesiomonas shigelloides
Colonies contain motile, fermentative Gram-negative rods, which produce arginine dihydrolase, catalase, β-galactosidase, indole, lysine and ornithine decarboxylase and oxidase, but not H_2S, phenylalanine deaminase, tryptophan deaminase or urease. Nitrates are reduced to nitrite, but negative reactions are recorded for gelatin liquefaction, the methyl red test, utilization of citrate or malonate, and the Voges Proskauer reaction. Acid is produced from glucose, inositol and trehalose, but not from adonitol, aesculin, amygdalin, L-arabinose, dulcitol, lactose, mannitol, melibiose, raffinose, rhamnose, salicin, sorbitol, sucrose or xylose. Growth occurs at 22°C and 37°C but not at 40°C, and in 0% (w/v) sodium chloride (Cruz *et al.*, 1986; Klein *et al.*, 1993).

It should be emphasized that phylogenetic studies have indicated that the taxon should really belong in the genus *Proteus* as *Proteus shigelloides* (MacDonell and Colwell, 1985), although the true taxonomic position remains to be resolved.

Shewanella putrefaciens

During spring of 1985, a disease occurred which resulted in high mortalities in rabbitfish, *Siganus rivulatus*, farmed in sea cages in the Red Sea. From diseased animals, a

Gram-negative bacterium was recovered, which was capable of re-infecting healthy fish (Saeed *et al.*, 1987). To date, the disease has not been described in any other fish species, or, for that matter, elsewhere.

Shewanella putrefaciens
Cultures comprise motile Gram-negative rods, which are neither fermentative nor oxidative for glucose, but which grow at 15–42°C, in 0.85–9.0% (w/v) sodium chloride, at pH 6.2–9.6, and on MacConkey agar. Catalase, H_2S, ornithine decarboxylase and oxidase are produced, but not arginine dihydrolase, β-galactosidase, indole, lysine decarboxylase, or phenylalanine or tryptophan deaminase. Gelatin but not urea is attacked. Nitrates are reduced, and citrate is utilized. The Voges Proskauer reaction is negative. Acid is not produced from amygdalin, arabinose, glucose, inositol, mannitol, melibiose, rhamnose, sorbitol or sucrose. Growth occurs at 37°C, and in 7% but not 0% (w/v) sodium chloride (Saeed *et al.*, 1987).

From the results of the API 20E rapid identification system, it was considered that the pathogen was *Ps. (= Alteromonas) putrefaciens*, a taxon which has been subsequently re-classified as *Shewanella putrefaciens* (MacDonell and Colwell, 1985).

Vibrio alginolyticus
The most extensive study of the role of *V. alginolyticus* as a fish pathogen concerned the observations of mortalities in farmed sea bream (*Sparus aurata*) in Israel. Mortalities were recorded after extensive handling of the fish (Colorni *et al.*, 1981). However, these workers were unable to reproduce the infection under laboratory conditions, which casts some doubt on the importance of this organism in fish pathology. Nevertheless, there is additional evidence linking this organism with a pathogenic mode. Lee (1995) recovered one isolate, which was identified phenotypically, from diseased grouper (*Epinephelus malabaricus*). Also, occasional isolations have been made from ulcers (Akazawa, 1968). Isolates have been recovered from diseased gilt-head sea bream in Spain (Balebona *et al.*, 1998). Additionally, *V. alginolyticus* has also been implicated with mortalities in cultured black sea bream fry (Kusuda *et al.*, 1986). Therefore, the implication is that *V. alginolyticus* constitutes an opportunistic invader of already damaged tissues, or a weak pathogen of stressed fish.

Vibrio alginolyticus
Typically, swarming growth develops on the surface of solid media. Cultures comprise motile, fermentative Gram-negative rods, that produce catalase, H_2S, indole, lysine and ornithine decarboxylase, and oxidase, but not arginine dihydrolase or β-galactosidase. Blood (haemolysis), chitin, gelatin, lipids, starch and urea are degraded, but not aesculin. Nitrates are reduced. The methyl red test and Voges Proskauer reaction are positive. Acid is produced from glycerol, maltose, mannitol, mannose, salicin and sucrose, but not arabinose, inositol or lactose. Growth occurs

at 37°C, and in 7% but not 0% or 10% (w/v) sodium chloride. Sensitivity is recorded to the vibriostatic agent, O/129. The G + C ratio of the DNA is 45–47 moles %.

Generally, fresh isolates matched the species description of *V. alginolyticus* (Baumann *et al.*, 1984).

Vibrio anguillarum (= Listonella anguillara/anguillarum)

The causal agent of 'red-pest' in eels was first isolated by Canestrini (1893), who designated the organism as *Bacterium anguillarum*. A further case among eels in Sweden during 1907 was investigated by Bergman (1909), and it was directly attributable to this scientist that the name of *V. anguillarum* was coined.

The taxonomy of the pathogen has had a chequered history, which culminated in the description of a second species, i.e. *V. ordalii*. This accommodated strains previously regarded as biotype II of *V. anguillarum* (Schiewe, 1981). However, there is cross-reactivity with the LPS of *V. anguillarum* serogroup O2 and *V. ordalii* (Muthiara *et al.*, 1993). The complexity in the taxonomic understanding of the pathogen began with the recognition by Nybelin (1935) of two biotypes. These were differentiated on the basis of a few biochemical reactions. A further group, i.e. biotype C, was recognized by Smith (1961). These biotypes were distinguished, as follows:

— type A, known as *V. anguillarum forma typica*, produced indole and acid from mannitol and saccharose.
— type B, referred to as *V. anguillarum forma anguillicida*, did not produce indole or acid from mannitol or saccharose.
— type C, coined as *V. anguillarum forma opthalmica*, produced acid from mannitol and saccharose, but did not produce indole.

Two further biotypes, i.e. D and E, were later described, both of which produced indole, but not acid from mannitol. Biotype D, but not E, produced acid from saccharose. In view of modern thought and approaches on bacterial taxonomy, these descriptions are inadequate. Nevertheless, they heralded an appreciation of heterogeneity within the species. Type C deserves special mention because this was proposed for Japanese strains derived from rainbow trout, and labelled as *V. piscium* var. *japonicus* (David, 1927; Hoshina, 1956). Interestingly, they were originally recognized as dissimilar to *V. anguillarum*. Alternatively, it must be conceded that the separate name may have reflected ignorance of the existence of *V. anguillarum* as described by Bergman (1909). The relationship of these so-called biotypes with *V. ichthyodermis*, as described by Wells and ZoBell (1934), and the organism tentatively assigned as *V. anguillicida* (Nishibuchi and Muroga, 1977) needs clarification. Indeed, the latter was considered to resemble both *V. anguillarum* and *V. fischeri*.

The multiplicity of studies of Harrell *et al.* (1976), Ohnishi and Muroga (1976), Hastein and Smith (1977), Schiewe *et al.* (1977), Baumann *et al.* (1978), Kusuda *et al.* (1979), Ezura *et al.* (1980), Lee *et al.* (1981), Kaper *et al.* (1983) and West *et al.* (1983) demonstrated very clearly the heterogeneity within *V. anguillarum*. Hastein and Smith

(1977) distinguished two subgroups after a principal components analysis on data collected for 163 isolates and 28 tests. A similar conclusion, i.e. two subgroupings, was voiced by Schiewe *et al.* (1977), Baumann *et al.* (1978), Ezura *et al.* (1980) and Lee *et al.* (1981). Of the 50 isolates studies by numerical taxonomy, Kusuda *et al.* (1979) defined three groups. These were equated with *V. anguillarum* (divided into three subgroups), a group closely related to *V. parahaemolyticus*, and a cluster considered to have affinity with *V. ichthyodermis*. Kaper *et al.* (1983) recognized four homogeneous phena among isolates received as *V. anguillarum*. This view was reinforced in a later study by West *et al.* (1983). Pazos *et al.* (1993) studied 46 isolates of *V. anguillarum*-like organisms from diseased fish and shellfish and the environment by numerical phenetic methods, and recognized four phena. The apparent heterogeneity was reinforced by the results of ribotyping, with 44 ribotypes recognized among isolates of *V. anguillarum* (Olesen and Larsen, 1993). Yet, Austin *et al.* (1995a, 1997) recognized a single taxon, homogeneous by ribotyping—isolates were recovered in a single ribotype—and outer membrane protein patterns, but heterogeneous in terms of LPS profiles, plasmid composition, serogrouping, and BIOLOG-GN fingerprints and API 20E profiles (Figs 4.2 and 4.3).

The taxon was re-classified to *Listonella*, as *Listonella anguillara* (MacDonell *et al.*, 1986). However, the name change was not widely accepted. Consequently, the organism is still regarded as *V. anguillarum*.

Vibrio anguillarum
Cultures comprise cream-coloured, round, raised, entire, shiny colonies [dissociation into two or three colony types may occur; Austin *et al.*, 1996] comprising short $(0.5 \times 1.5~\mu m)$, fermentative, Gram-negative rods, which are motile by single polar flagella. Arginine dihydrolase, catalase, β-galactosidase, indole and oxidase, but not H_2S, lysine or ornithine decarboxylase, phenylalanine deaminase or urease, are produced. Along with most other vibrios, sensitivity is displayed to the vibriostatic agent, O/129. A positive result is usually recorded for the Voges Proskauer reaction, but not for the methyl red test. Chitin, gelatin, DNA, lipids and starch, but not aesculin, are degraded. Nitrates are reduced. Growth occurs at 15–37°C, and in 0.3–3.0% (w/v) but not 0% and 7% (w/v) sodium chloride. Citrate, malonate and tartrate are utilized. The organisms produce acid from amygdalin, arabinose, cellobiose, galactose, glycerol, maltose, mannitol, sorbitol, sucrose and trehalose, but not from adonitol, dulcitol, erythritol, inositol, lactose, melibiose, raffinose, rhamnose, salicin or xylose. Cultures comprise a single dominant ribotype. The G +C ratio of the DNA is 45.6–46.3 moles % (Smith, 1961; Kiehn and Pacha, 1969; Evelyn, 1971b; Muroga *et al.*, 1976a,b; Schiewe *et al.*, 1981; Baumann *et al.*, 1984; Austin *et al.*, 1995a).

It has been observed that some so-called *bona fide* isolates of *V. anguillarum* possess fascinating micromorphologies, insofar as broth cultures appear to contain two types of cells. Apart from the typical short, motile rods, we have observed very small, highly motile cells, some of which are capable of passing through 0.22 μm porosity filters. From transmission electron microscopy, we believe that these cells comprise extremely small,

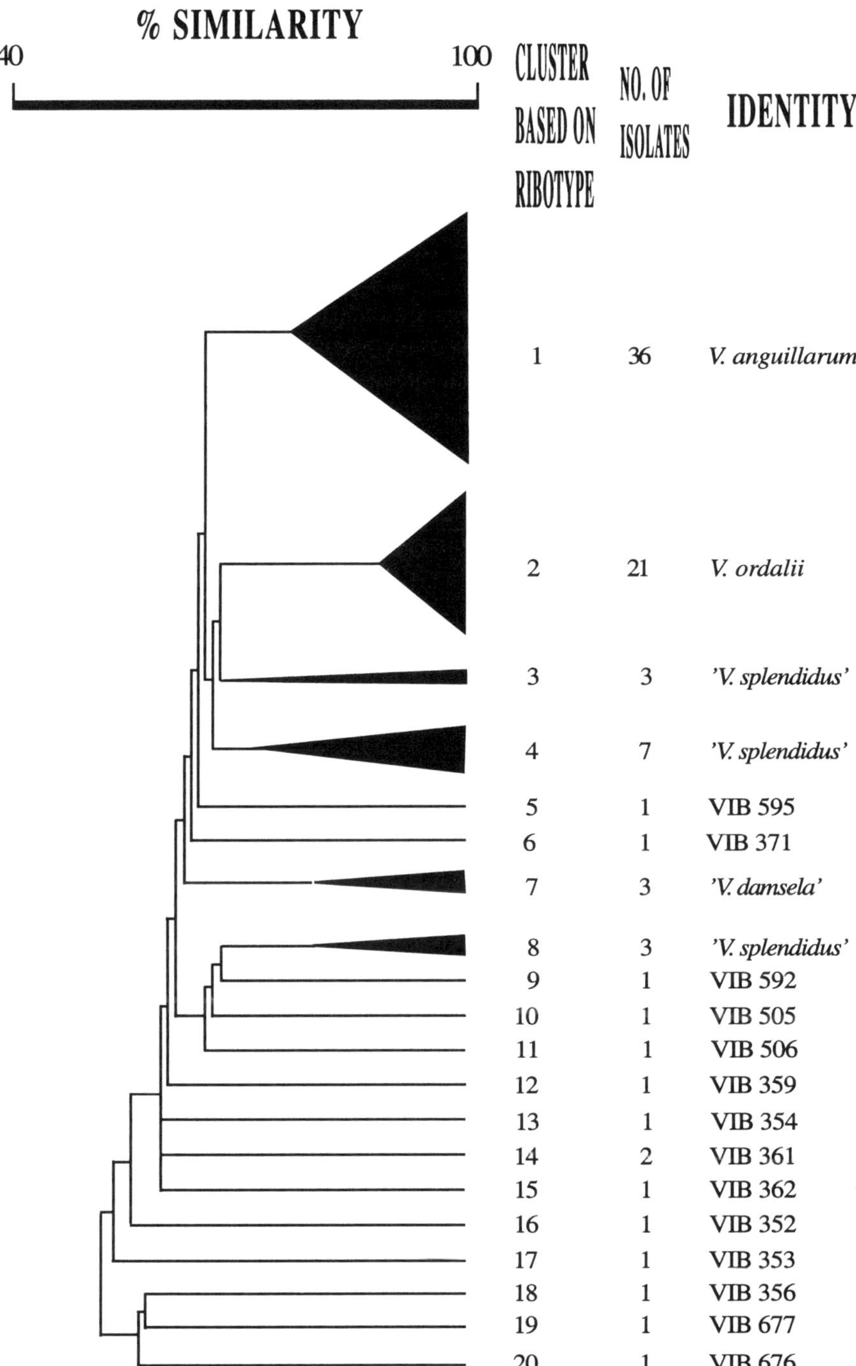

Fig. 4.2. A dendrogram showing the relationships between fish-pathogenic vibrios based on a comparison of phenotypic data (from Austin *et al.*, 1997). Heterogeneity was apparent in cultures received as *V. anguillarum*, *V. ordalii*, *V. salmonicida* and especially *V. splendidus*. However, the isolates of *Photobacterium damselae* (= *V. damsela*) were markedly homogenous.

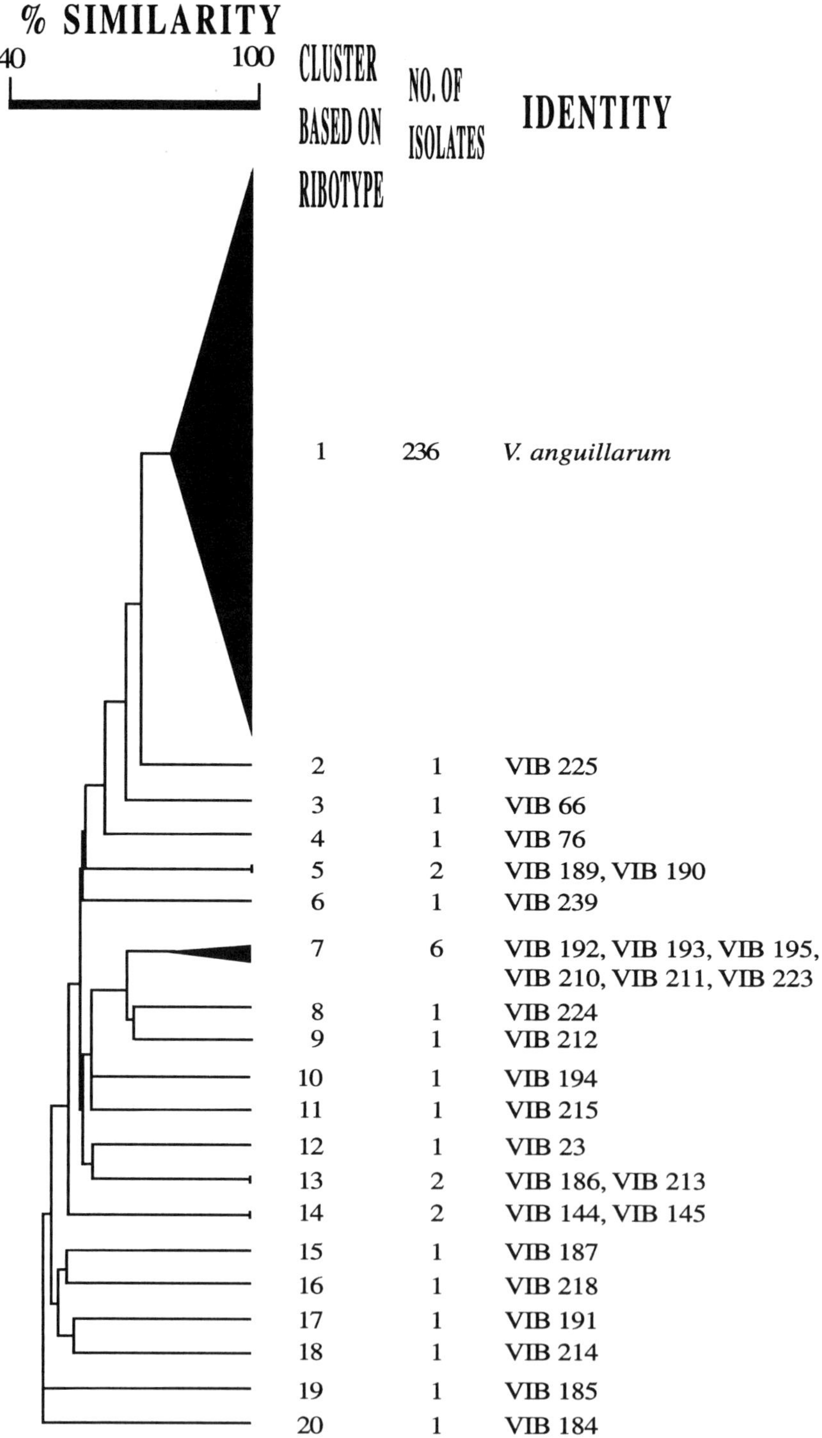

Fig. 4.3. A dendrogram, based on the examination of phenotypic traits, showing the homogeneity among 236 *bona fide* isolates of *V. anguillarum* (from Austin *et al.*, 1995a).

spherical bodies, each attached to a single polar flagellum. The significance of these small cells is unclear, but they may represent a stage in a life cycle, a laboratory artefact, or some form of survival mechanism.

The results of serology further complicated the understanding of *V. anguillarum* (Bolinches *et al.*, 1990). The establishment of serotypes has traversed species (or phenetic) boundaries. Initially, three serotypes were recognized for isolates from north-western (USA) salmonids, Europe, and Pacific-northwest (USA) (Pacha and Kiehn, 1969). This was supported by the work of Japanese scientists (Aoki *et al.*, 1981; Muroga *et al.*, 1984b). With further study, the number of serotypes increased to six (Kitao *et al.*, 1983). Thus in a mammoth study of 267 isolates from ayu, eel and rainbow trout, Kitao and co-workers defined serotypes A, B, C, D, E, and F as a result of cross-agglutination and cross-absorption tests with thermo-stable 'O' (somatic) antigens. The majority (243) of these Japanese isolates were recovered in serotype A. It is noteworthy, however, that avirulent isolates were not recovered in any of these serotypes (Muroga *et al.*, 1984b). This reflects the nature of the LPS in the cell wall, which accounts for both the nature of the serotype (Johnson, 1977; Aoki *et al.*, 1981) and the immunogenicity. Later, Sørensen and Larsen (1986) reported the presence of ten O-antigen serotypes, based upon examination of 495 isolates, representatives of which shared a common 40 kDa protein (Simón *et al.*, 1996). A common 47 kb plasmid was reported by Giles *et al.* (1995). Serovars O1 and O2 contained plasmids, but there was no apparent correlation between the presence of such extrachromosomal DNA and biochemical properties (Larsen and Olsen, 1991), although serogroup O1 was regarded as biochemically homogeneous (Pedersen and Larsen, 1995). The same ten serogroups together with six non-typable groups were described by Olsen and Larsen (1993). The number of serogroups was now increased to 16, i.e. O1 to O16 (Grisez and Ollevier, 1995).

A consensus view would be that serogroup O1 has dominated both in the number of isolates available for study and the relative importance to fish pathology (Austin *et al.*, 1995a; Pedersen *et al.*, 1996a,c). The homogeneity of serogroup O1 has been established (Austin *et al.*, 1995a), yet data have pointed to variability among isolates. For example, after studying 75 isolates of serogroup O1, eight plasmid profiles—with one predominating—and six ribotypes were recognized (Skov *et al.*, 1995). An even larger examination of 103 isolates of serogroup O1 recognized 15 plasmid profiles (Pedersen and Larsen, 1995). PFGE had high discriminatory power, recognizing 35 profiles.

It is a personal view that isolates of serogroup O2 have seemed to be more aggressive than serogroup O1. Serogroup O2, which has been subdivided into serogroup O2a and O2b, has revealed heterogeneity in LPS profiles—six different profiles have been recognized among 129 isolates (Tiainen *et al.*, 1997). By Western blotting and slide agglutination, four different patterns have emerged. By comparing LPS profiling, Western blotting and slide agglutination, nine different groupings were formed (Tiainen *et al.*, 1997). A view was expressed that additional sub-groups within serogroup O2 remain to be described (Tiainen *et al.*, 1997).

Vibrio cholerae (non-O1)

During the summer of 1977, an epizootic occurred in a wild population of ayu in the River Amano, Japan. From diseased animals, an organism conforming to the description of *V. cholerae* was isolated (Muroga *et al.*, 1979; Kiiyukia *et al.*, 1992). Subsequently, *V. cholerae* was associated with a disease of goldfish in Australia (Reddacliff *et al.*, 1993).

Vibrio cholerae

Cultures comprise small (1.5–3.0 × 0.7–1.0 μm in size) Gram-negative, fermentative rods, which are motile by single polar flagella. Growth occurs in 0–6% (w/v) sodium chloride, at 10–42°C, and at pH 7–10. Catalase, β-galactosidase, indole, lysine decarboxylase and oxidase are produced, but not arginine dihydrolase, H_2S, ornithine decarboxylase or phenylalanine deaminase. Aesculin, blood (haemolysis) chitin, gelatin, lipids and starch, but not urea, are degraded. The methyl red test and Voges Proskauer reaction are positive. Nitrates are reduced. Citrate, fructose, galactose, glucose, maltose, sucrose and tartrate are utilized, but not adonitol, arabinose, cellobiose, dulcitol, inositol, inulin, malonate, mannose, melezitose, melibiose, raffinose, rhamnose, salicin, sorbitol or xylose. Acid is produced from sucrose, but not lactose. Growth occurs at 37°C and in 0% but not 7% (w/v) sodium chloride. Sensitivity is recorded to the vibriostatic agent, O/129. The G + C ratio of the DNA is 47–49 moles % (Muroga *et al.*, 1979; Yamanoi *et al.*, 1980; Kiiyukia *et al.*, 1992).

The phenotypic traits published by Muroga *et al.* (1979) and Kiiyukia *et al.* (1992) were in close agreement with the description of *V. cholerae* (Baumann *et al.*, 1984). Indeed, there were only two discrepancies with the characteristics of the 'el tor' biotype, i.e. production of ornithine decarboxylase and utilization of mannose (these were reported as negative for the fish pathogenic isolates). However, the fish isolates did not react, by slide agglutination, with antisera to *V. cholerae*. Therefore, the conclusion was that the fish isolates comprised *V. cholerae* non-O1. However, the fish isolates demonstrated 86% DNA:DNA homology with a reference strain of *V. cholerae*. This coincided well with the intraspecific homology values of *V. cholerae* (Citarella and Colwell, 1970). Again, there was reasonably close agreement between *V. cholerae* and the isolates studied by Reddacliff *et al.* (1993). Yet, a detailed taxonomic study by Austin *et al.* (1997) reported heterogeneity among six isolates received as *V. cholerae*, insofar as they were recovered as single member ribotype clusters, and displayed diverse BIOLOG-GN fingerprints and API 20E profiles (Fig. 4.2). However, five cultures corresponded with *V. cholerae* serogroups, namely O8, O9, O23, O32 and O63.

Vibrio fischeri

During Autumn 1988, visceral tumours (neoplasia) and skin papillomas were observed in juvenile turbot, farmed in northwest Spain. Although viral involvement was suspected, bacteria were evident in the majority of affected fish. The bacteria approximated the description of *V. fischeri*, albeit with similarities to *V. harveyi* (Lamas *et al.*, 1990). However, detailed characteristics of the cultures were not presented. Interestingly, other isolates have been recovered from gilt-head sea bream, also in Spain (Balebona *et al.*, 1998).

Vibrio harveyi (= V. carchariae)

A comparative newcomer to the growing list of vibrio fish pathogens, *V. carchariae* was originally isolated from a dead sandbar shark (*Carcharhinus plumbeus*) which died at the National Aquarium in Baltimore, USA, in 1982 (Grimes *et al.*, 1984a). Subsequently, a similar organism was recovered from lemon sharks (*Negraprion brevirostris*) (Colwell and Grimes, 1984). Yii *et al.* (1997) reported *V. carchariae* as the cause of gastro-enteritis and mortalities in grouper. One isolate, EmI82KL, was noted to be motile, did not auto-agglutinate, but was haemolytic to grouper, rabbit, sheep and tilapia blood. In a separate development, gamefish, namely common snook (*Centropomus undecimalis*), were found to suffer with opaque white corneas within 24 h of capture in Florida, USA (Kraxberger-Beatty *et al.*, 1990). From such damaged specimens, *V. harveyi* was recovered. Twelve cultures were recovered from corneas, and were identified as *V. harveyi* as a result of biochemical tests and DNA:DNA hybridization (90–94% DNA homology with the type strain of *V. harveyi*). Certainly, the characteristics of the isolates were in good accord with the species description of *V. harveyi* (Baumann *et al.*, 1984). Subsequently, there has been a gradual awareness of the increasing significance of *V. harveyi* as a killer of marine fish and penaeids. Saeed (1995) blamed *V. harveyi* with causing mortalities in cultured brown spotted grouper (*Epinephelus tauvina*) and silvery black porgy (*Acanthopagrus cuvieri*) in Kuwait. Then, as a result of DNA:DNA hybridization, Ishimru and Muroga (1997) determined that pathogenic vibrios recovered from milkfish in Japan were indeed *V. harveyi*. The organism has also been recovered from ocular lesions in the short sunfish (*Mola mola*) in Spain (Hispano *et al.*, 1997).

The synonymy of *V. harveyi* and *V. carchariae* was realized by Pedersen *et al.* (1998) as a result of phenotypic and genotypic studies. In terms of taxonomic standing, *harveyi* has precedence, and therefore this name will be used in preference to *carchariae*.

Vibrio harveyi

V. harveyi shows similarities to *V. alginolyticus*, principally because of the presence of swarming on agar medium. Essentially, cultures comprise pleomorphic fermentative Gram-negative rods (1.0–1.6 × 0.5–0.7 μm in size), which are motile by polar and/or lateral flagella. Catalase, indole, lysine and ornithine decarboxylase and oxidase are produced, but not arginine dihydrolase. The Voges Proskauer reaction is negative. Nitrates are reduced. Alginate, blood, DNA, gelatin and lecithin are degraded, but not aesculin, casein, cellulose, pectin or starch. Many compounds are utilized, including arabinose, aminobutyrate, cellobiose, ethanol, glucose, glycine, α-ketoglutarate, propanol, sucrose and trehalose but not from inositol or lactose. Growth occurs in 3–8% but not 0% or 10% (w/v) sodium chloride, and at 11–40°C. Sensitivity is demonstrated to the vibriostatic agent, O/129. The G +C ratio of the DNA is 45–47 moles %. The only differences with the description of *V. carchariae* centre on the degradation of starch (*V. carchariae* = +), and growth in 7% (w/v) sodium chloride (*V. carchariae* = –). The G +C ratio of the DNA of *V. carchariae* is 47 moles %.

Results of DNA:DNA homology experiments showed a high degree of homology with *Ph. damselae*, at 88% re-association.

Vibrio ichthyoenteri

V. ichthyoenteri was described as a result of an examination of seven isolates from flounder larvae (Ishimaru *et al.*, 1996).

Vibrio ichthyoenteri

Cultures are non-pigmented on marine 2216E agar, but produce yellow colonies on TCBS, which contain Gram-negative fermentative rods, which are motile by single polar flagella. Catalase and oxidase are produced, but not arginine dihydrolase, β-galactosidase, H_2S, indole or ornithine decarboxylase. Nitrates are reduced. Neither agar, chitin, gelatin, lipids nor starch are degraded. Polyhydroxybutyrate is not accumulated intracellularly. Growth occurs at 15–30°C but not at 4°C or 35°C, and in 1–6% but not 0 or 8% (w/v) sodium chloride. Acid is produced from fructose, D-glucose, maltose, D-mannose, sucrose and trehalose, but not from adonitol, L-arabinose, D-cellobiose, dulcitol, erythritol, D-galactose, glycerol, inulin, inositol, lactose, D-mannitol, melibiose, raffinose, L-rhamnose, salicin or D-sucrose. Neither D-cellobiose, citrate, D-gluconate, L-leucine nor D-xylose are utilized. Sensitivity is recorded to the vibriostatic agent, O/129. The G + C ratio of the DNA is 43–45 moles %.

Vibrio logei

Fifteen Icelandic isolates and one Norwegian isolate from shallow skin lesions in Atlantic salmon were considered to be similar to *V. logei* (Benediktsdóttir *et al.*, 1998).

Vibrio logei

Cultures do not produce arginine dihydrolase, indole or lysine or ornithine decarboxylase. Blood (haemolysis) and chitin are degraded, but not starch. Acid is produced from *N*-acetyl glucosamine, glycerol, maltose, mannose, ribose, sucrose and trehalose.

Vibrio ordalii

The establishment of a new species to accommodate strains previously classified as *V. anguillarum* biotype II, i.e. *V. ordalii* (Schiewe, 1981; Schiewe *et al.*, 1981), generated an awareness that vibriosis could be caused by more than one bacterial taxon. Disease caused by *V. ordalii* has been documented in Japan (e.g. Muroga *et al.*, 1986) and the Pacific Northwest, USA. *V. ordalii* was homogeneous by plasmid profiling, ribotyping and serogrouping, accommodated two LPS groups, and was heterogeneous by BIOLOG-GN fingerprints and API 20E profiles (Fig. 4.2; Austin *et al.*, 1997).

Vibrio ordalii

Cultures comprise fermentative Gram-negative curved rods of $2.5–3.0 \times 1.0$ μm in size, motile by means of single polar flagella. Growth occurs quite slowly, insofar as 4–6 days incubation at 22°C are required for the production of off-white, circular, convex colonies of 1–2 mm in diameter on seawater agar. *V. ordalii* is, however, not especially active. Catalase and oxidase are produced, but not arginine dihydrolase, β-galactosidase, H_2S, indole, lysine or ornithine decarboxylase, or phenylalanine deaminase. DNA, chitin (by some isolates) and gelatin are degraded, but not aesculin, lipids, pectate, starch or urea. Nitrates are reduced by some isolates. The methyl red test and Voges Proskauer reaction are negative. Tartrate is utilized. Only a few carbohydrates, e.g. galactose (variable result), maltose, mannitol and sucrose are attacked with the production of acid. Negative responses are recorded for adonitol, arabinose, cellobiose, dulcitol, erythritol, glycerol, inositol, lactose, melibiose, raffinose, rhamnose, salicin, sorbitol, trehalose and xylose. Growth occurs at 15–22°C but not at 37°C and in 0.5–3.0% but not 0% or 7% (w/v) sodium chloride. Sensitivity is recorded to the vibriostatic agent, O/129. The G +C ratio of the DNA is 43–44 moles % (Schiewe, 1981; Schiewe *et al.*, 1981; Austin *et al.*, 1997).

Results of DNA:DNA hybridization studies have confirmed the homogeneity and validity of *V. ordalii*, with intraspecific homologies of approximately 80%. There is only a 58–59% association with *V. anguillarum* (Schiewe *et al.*, 1981). *V. ordalii* contains plasmids (Tiainen *et al.*, 1995), but the profile is inevitably different to *V. anguillarum*. In one study, Schiewe and Crosa (1981) determined that eleven isolates of *V. ordalii* contained a common plasmid type (pMJ101) with a molecular weight of 20 mDa. Indeed, this plasmid (= ~32 kb) is common to all *V. ordalii* isolates (Pedersen *et al.*, 1996b).

There is serological (antigenic) cross-reactivity between *V. ordalii* and *V. anguillarum* serogroup O2 (Chart and Munn, 1984).

Vibrio pelagius

Four isolates were obtained in pure culture, and described as (Angulo *et al.*, 1992):

Vibrio pelagius

The cultures comprise Gram-negative motile fermentative rods, that produce catalase, β-galactosidase (variable), indole (variable) and oxidase, but not arginine dihydrolase, H_2S, or lysine or ornithine decarboxylase, reduce nitrates, degrade alginate (variable), gelatin, starch, Tween 80 and urea (variable), produce acid from mannitol, mannose, trehalose and sucrose (variable) but not L-arabinose, arbutin, inositol, salicin or sorbitol, and are sensitive to the vibriostatic agent O/129, ampicillin and novobiocin. The methyl red test is generally positive. The Voges Proskauer reaction gives a variable response. Growth occurs in 6% (w/v) but not 0% or 10% (w/v) sodium chloride, at 35°C but not 4°C or 42°C. However, the isolates do not agglutinate with O antigens of *V. pelagius* ATCC 25916.

It is noteworthy that the fish isolates differed from the reference culture of *V. pelagius* ATCC 25916 in indole production, urea degradation and whole cell agglutination, which must cast some doubt on the validity of the original identification.

Vibrio salmonicida

An organism, named *V. salmonicida*, has been recovered from diseased salmon (Egidius *et al.*, 1986), and the validity and distinctiveness confirmed (Fig. 4.2; Austin *et al.*, 1997).

Vibrio salmonicida
Cultures contain motile (~9 polar flagella) fermentative Gram-negative curved pleomorphic rods of 2–3 × 0.5 μm in size. Catalase and oxidase are produced, but not arginine dihydrolase, β-galactosidase, H_2S or indole. Nitrates are not reduced, nor is the Voges Proskauer reaction positive. Citrate is not utilized. Neither blood, chitin, gelatin, lipids nor urea are degraded. *N*-acetylglucosamine, glucose, glycerol (slowly), maltose, ribose, sodium gluconate and trehalose are utilized, but not adonitol, amygdalin, D- or L-arabinose, D- or L-arabitol, arbutin, D-cellobiose, dulcitol, erythritol, D- or L-fucose, inositol, β-gentiobiose, lactose, D-lycose, D-mannose, melezitose, melibiose, D-raffinose, rhamnose, salicin, sorbitol, sucrose, L-sorbose, D-tagatose, D-turanose or D- or L-xylose. Growth occurs at 1–22°C, optimally at 15°C but not at 37°C, and in 0–4% but not 7% (w/v) sodium chloride. Sensitivity is displayed to the vibriostatic agent, O/129, but not to novobiocin. The G +C ratio of the DNA is 44 moles % (Holm *et al.*, 1985; Egidius *et al.*, 1986).

Strains have been divided into four different categories on the basis of plasmid profiles (Wiik *et al.*, 1989). The plasmids were 2.6, 3.4 and 24 mDa in size; with the largest plasmid being common to all four groups. There was no apparent difference in biochemical traits among these four plasmid groups. In a separate study, Sørum *et al.* (1990) described plasmids of 2.8, 3.4, 21 and 62 mDa from isolates recovered from Atlantic salmon and cod. These authors reported eleven plasmid profiles for *V. salmonicida*. A similar plasmid composition has been indicated for isolates from the Faroe Islands (Nielsen and Dalsgaard, 1991). Comparing isolates from Canada, Faroe Islands, Norway and Shetland, Sørum *et al.* (1993a) noted a similarity in plasmid profile, with three plasmids of 2.8, 3.4 and 21 mDa revealed. Furthermore, a conclusion has been reached that all strains carry plasmids (Valla *et al.*, 1992). Differences have also been implied by serological studies, which have indicated the presence of two serotypes (Schrøder *et al.*, 1992).

DNA hybridization of four cultures confirmed homogeneity (DNA homology = 82–100%), but low relatedness to *V. anguillarum* (30%), *V. ordalii* (34%) or *V. parahaemolyticus* (40%) (Wiik and Egidius, 1986). Although these data were used to justify the uniqueness of *V. salmonicida*, the relationship to other representatives of the Vibrionaceae was not considered. The validity of the species is not, however, questioned, and in the detailed study of Austin *et al.* (1997), *V. salmonicida* formed a discrete taxon among the other fish-pathogenic vibrios.

Vibrio splendidus

A detailed taxonomic study by Austin *et al.* (1997) of 22 isolates revealed that fish pathogenic isolates (these were non-pathogenic in laboratory-based infectivity experiments with Atlantic salmon and rainbow trout) were markedly heterogeneous, being recovered in seven ribotype clusters, many API 20E profiles and 6 BIOLOG-GN groups (the type strain was recovered as a single member cluster). However, only two serogroups (there were some cross-reactions with *V. pelagius* antiserum) and one LPS profile was recognized. With such heterogeneity, it is difficult to decide whether or not any fish-isolates actually constitute *bona fide V. splendidus*. Nevertheless, an attempt will be made to present a consensus view of the phenotypic traits associated with so-called *V. splendidus*:

> *Vibrio splendidus*
> Cultures contain motile fermentative Gram-negative rods, which produce arginine dihydrolase (some isolates), catalase, β-galactosidase, lysine decarboxylase and oxidase, but not H_2S, indole, ornithine decarboxylase or phenylalanine deaminase, degrade chitin, gelatin, lipids and starch, but not aesculin or urea, reduce nitrates, demonstrate positivity for the methyl red test but not the Voges Proskauer reaction, are sensitive to the vibriostatic agent, O/129, grow at 37°C and in 3% (w/v) but not 0% or 7% (w/v) sodium chloride, utilize citrate but not malonate, and produce acid from glucose (no gas), maltose and mannose but not from arabinose, inositol, mannitol, sorbitol or sucrose. The G +C ratio of the DNA is 45.7–47.8 moles %.

Agglutination was recorded with antisera prepared against whole cells of *V. splendidus* and strongly to O antigen of *V. splendidus* and *V. tubiashi*. Weak agglutination was recorded against O antigen of *V. anguillarum* and *V. ordalii* (Lupiani *et al.*, 1989). It is considered that the pathogen showed similarities to both *V. anguillarum* (including *V. ordalii*) and *V. splendidus*. Differences to *V. splendidus* (biotype II) included arginine dihydrolase, β-galactosidase and indole production, the Voges Proskauer reaction, degradation of gelatin and urea, and acid production from arabinose, mannitol and sorbitol (Baumann *et al.*, 1984). Consequently, an identification as *V. splendidus* can only be regarded as tentative, pending further investigation.

Vibrio trachuri

From an examination of three isolates, homogeneity was apparent (Iwamoto *et al.*, 1995).

> *Vibrio trachuri*
> Cultures contain Gram-negative motile rods, that produce indole, lysine decarboxylase and oxidase, but not arginine dihydrolase, β-galactosidase, H_2S, ornithine decarboxylase or urease. Acid is produced from glucose, lactose, maltose, sorbitol and sucrose, but not from arabinose, inositol, rhamnose. Nitrates are reduced. The Voges Proskauer test is negative. Growth occurs on TCBS, at 30°C, in 3–7% but not 0% or 10% (w/v) sodium chloride. Sensitivity is recorded to the vibriostatic

agent, O/129. Intraspecific DNA homology values are high, i.e. 95.4–100%, whereas interspecific values are low. The G +C ratio of the DNA is 45.4–47.8 moles %. The nearest neighbour is *V. harveyi* (DNA homology of 40.1%), then *V. campbellii* (25.6%) and *V. alginolyticus* (24.4%) (Iwamoto *et al.*, 1995).

Vibrio vulnificus

A group of bacteria was recovered, which appeared to belong to the same taxon. The closest neighbour was thought to be *V. anguillarum* type B (= *V. anguillarum forma anguillicida*), according to the description of Nybelin (1935). However, this group lacks taxonomic meaning, and consequently it was agreed to resurrect the name of *V. anguillicida* (after Brunn and Heiberg, 1935) to accommodate the organisms (Muroga *et al.*, 1976a,b; Nishibuchi *et al.*, 1979). Some excellent detective work by Tison *et al.* (1982) led to the realization that the organisms, in fact, approximated *V. vulnificus* (= lactose-fermenting vibrios). A new biotype, i.e. biogroup 2, was duly established. However, the biotype concept was considered no longer appropriate (Arias *et al.*, 1997a,b), and it was subsequently deemed preferable to refer to the organisms as a serovar rather than biotype (Biosca *et al.*, 1997a).

From an examination of 80 cultures, eel isolates were separated from others by ribotyping (Arias *et al.*, 1997a,b). Moreover, AFLP was very discriminatory (Arias *et al.*, 1997a,b).

Vibrio vulnificus
Isolates comprise short Gram-negative, fermentative rods, motile by means of single polar flagella. Arginine dihydrolase, catalase and oxidase are produced, but not β-galactosidase, indole, or lysine or ornithine decarboxylase. Casein, blood, lecithin and Tween 80 are degraded, but not gelatin or urea. Nitrates are reduced. The Voges Proskauer reaction is positive. Acid is produced from a wide range of compounds including D-amygdalin, cellobiose, D-fructose, D-galactose, D-glucose, glycerol, glycogen, maltose, mannose, melibiose, starch and trehalose, but not from adonitol, arabinose, dulcitol, inositol, inulin, mannitol, raffinose, L-rhamnose, D-sorbitol, sucrose or D-xylose. Tartrate is utilized. Growth occurs at 20°C and 37°C, but not at 5°C or 42°C, and in 0.5–5.0% but not 0% (w/v) sodium chloride. Sensitivity is recorded to the vibriostatic agent, O/129. The G +C ratio of the DNA is 45.7–47.8 moles % (Tison *et al.*, 1982; Amaro *et al.*, 1992).

There are some differences between the species description of *V. vulnificus* and the fish isolates. For example, the latter do not produce indole, ornithine decarboxylase, acid from mannitol or sorbitol, or grow at 42°C. These differences add weight to assigning the fish pathogens to a new biogroup. However, the results of DNA:DNA hybridization, which show 90% homology between *V. vulnificus* and the group of fish pathogens (Tison *et al.*, 1982), confirms taxonomic (species) relatedness.

Results of serology have clearly demonstrated that isolates of *V. vulnificus* contain heterologous surface antigens. It must be emphasized, however, that the fish isolates are distinctly homogeneous (Nishibuchi and Muroga, 1980). There are common soluble intracellular antigens, which are homologous in all isolates. In particular, there is a *V. vulnificus* specific antigen, which may have value in the development of rapid diagnostic tests.

F. MORAXELLACEAE REPRESENTATIVES

Acinetobacter sp.

A bacterium, resembling *Acinetobacter*, was recovered from sexually mature Atlantic salmon, comprising wild stock from the River Surma, Norway (Roald and Hastein, 1980).

Acinetobacter sp.

Cultures comprise round, raised, translucent, mucoid colonies of 1.5 mm in diameter within 48 h incubation at 22°C. Colonies contain fairly unreactive short, plump, facultatively anaerobic, non-motile Gram-negative rods of $1.6-1.8 \times 0.8-1.2$ μm in size. Catalase and oxidase are produced, but not arginine dihydrolase, β-galactosidase, H_2S, indole, lysine or ornithine decarboxylase or tryptophan deaminase. Blood (haemolysis) is degraded, but not gelatin or urea. Nitrates are not reduced. The methyl red test and Voges Proskauer reaction are negative. Sodium citrate is not utilized. Acid is produced from galactose, maltose and mannose, but not from adonitol, amygdalin, arabinose, cellobiose, glucose, inositol, lactose, mannitol, melibiose, raffinose, accharose, salicin, sorbitol or xylose. Unfortunately, the $G + C$ ratio of the DNA has not been determined.

From the available information, Roald and Hastein (1980) considered that the pathogen corresponded to an as-yet-unnamed species of *Acinetobacter*. Although this appears to be a sound decision, *bona fide Acinetobacter* spp. should not produce oxidase (Juni, 1984). In fact, this ability belongs to the morphological similar *Moraxella* (Bøvre, 1984) and *Neisseria* (Vedros, 1984). There is some degree of resemblance between the fish pathogen and *Mor. atlantae* and *Mor. osloensis*, although one of the general traits of moraxellae is an inability to produce acid from carbohydrates. Thus as a compromise solution, it would appear that the fish pathogen should be classified in the area loosely bounded by *Acinetobacter*, *Moraxella* and *Neisseria*.

Moraxella sp.

A bacterium was recovered as pure culture growth from internal organs of juvenile striped bass, and equated with *Moraxella* (Baya *et al.*, 1990b).

Moraxella sp.

Cultures comprise non-motile, short ($0.8-1.0 \times 1.3$ μm in size) non-fermentative paired rods with pronounced bipolar staining. Catalase and oxidase are produced,

> but not so arginine dihydrolase, β-galactosidase, H_2S, indole or lysine or ornithine decarboxylase. Sheep's blood is degraded, but not gelatin or urea. Citrate utilization, nitrate reduction and the Voges Proskauer reaction are negative. Acid is produced weakly from galactose and mannose, but not from amygdalin, arabinose, glucose, inositol, lactose, maltose, mannitol, rhamnose, salicin, sorbitol or sucrose.

Whereas similarities to *Moraxella* were noted (Bøvre, 1984), the organism strongly resembled the *Acinetobacter* described by Roald and Hastein (1980). Indeed, the major differences concerned acid production from maltose. Clearly, the precise taxonomic position of both organisms must await further study.

G. HALOMONADACEAE REPRESENTATIVE

Halomonas (= Deleya) cupida
Homogenates of black sea bream revealed the presence of *Halomonas (= Deleya) cupida (= Alcaligenes cupidus), V. alginolyticus* and *V. nereis* (Kusuda *et al.*, 1986).

> *Halomonas cupida*
> Cultures comprise motile Gram-negative salt-requiring rods, which grow at 10–25°C but not at 4°C and 30°C, and are unreactive in the oxidative-fermentative test. Arginine dihydrolase, catalase and lysine and ornithine decarboxylase are produced, but not H_2S, indole or oxidase. Nitrates are reduced, but the methyl red test and Voges Proskauer reaction are negative. Haemolysis is recorded to eel erythrocytes, but not those of sheep or yellowtail. Sensitivity is recorded to the vibriostatic agent, O/129. Acid is produced from adonitol, galactose, lactose, maltose, mannitol, L-rhamnose, D-sorbitol, salicin, sucrose (weak) and trehalose, but not from fructose or inulin.

From these characteristics, a close relationship to *Alcaligenes cupidus* was noted (see Kersters and De Ley, 1984). The only discrepancy concerned acid production from fructose. However, this taxon was re-classified as *Deleya cupida* (Baumann *et al.*, 1983), and by inference to *Halomonas cupida* on the basis of 16S rRNA sequencing (Dobson and Franzmann, 1996).

H. SMALL, MORPHOLOGICALLY SIMPLE BACTERIA

Mycoplasma mobile
Cell wall defective/deficient bacteria, i.e. L-forms and mycoplasmas, have been recently associated with fish diseases. Since the initial recovery of a motile mycoplasma from fish (Kirchhoff and Rosengarten, 1984), a new species, i.e. *Mycoplasma mobile*, has been described (Kirchhoff *et al.*, 1987). In addition, there is an increasing awareness of L-forms in fish diseases. To date, L-forms have been described for *Aer. salmonicida* and *Y. ruckeri* (McIntosh and Austin, 1990a).

> *Mycoplasma mobile*
> Cultures produce 'fried-egg' colonies of 10–500 μm diameter on Hayflick medium, which contains horse or bovine serum, after incubation at an unstated temperature for 2–6 days. Colonies contain filterable (through the pores of 0.45 μm pore size filters) Gram-negative conical or flask shaped wall-less cells with distinctive terminal structures. The cells demonstrate marked ability to adhere to and glide on glass, plastic, erythrocytes and tissue culture cells. Growth occurs at 4–30°C but not at 37°C. Catalase (weak), oxidase (weak) and phosphatase (weak) are produced. Blood is degraded, but not arginine, casein, gelatin or urea. Acid is produced fermentatively from arabinose, fructose, galactose, glucose, lactose, maltose and mannose. 2,3,5-triphenyltetrazolium chloride and potassium tellurite (weakly) are reduced, but not methylene blue. Gluconate is not oxidized, nor is phenylalanine deaminated. The G +C ratio of the DNA is 22.4–24.6 moles % (Kirchhoff *et al.*, 1987).

Serologically, the tench isolate was distinct from all other validly described species of *Acholeplasma* and *Mycoplasma*. From the traits, listed above, it was deemed that the organism formed a new species, for which the name of *Mycoplasma mobile* was coined (Kirchhoff *et al.*, 1987).

Piscirickettsia salmonis

Degenerate or obligately parasitic bacteria, i.e. chlamydias and rickettsias, have been long established as pathogens of invertebrates, and sporadically mentioned in connection with fish diseases (Wolf, 1981). Yet, firm evidence of their role in fish pathology has not been forthcoming until an upsurge of interest in Chile. Thus since 1989, a disease coined 'coho salmon syndrome', Huito disease (Schäfer *et al.*, 1990) or salmonid rickettsial septicaemia (Cvitanich *et al.*, 1991) has been observed in coho salmon, chinook salmon, Atlantic salmon and rainbow trout, with a spread to Atlantic salmon in Norway (Olsen *et al.*, 1997). Losses fluctuated between 3–7% of stock per week; the cumulative mortalities reaching 90%. The organism was formally recognized as a new taxon, for which the name of *Piscirickettsia salmonis* was proposed (Fryer *et al.*, 1992).

> *Piscirickettsia salmonis*
> The pathogen is a pleomorphic, non-motile Gram-negative predominantly coccoid (and ring forms) organism of variable size (0.5 × 1.5–2.0 μm), occurring intracellularly as individuals, pairs or groups. Electron microscopy reveals that each organism is bound by two membrane layers; a characteristic trait of the Rickettsiales, and possibly the tribe Erlichiae.

A single isolate, designated LF-89, was studied in detail by Fryer *et al.* (1992). The 16S rRNA conformed to the gamma subdivision of the Proteobacteria. Moreover, LF-89 did not show any specific relationship to any of 450 bacterial 16S rRNA sequences held on file. Nevertheless, similarities were apparent with *Wolbachia persica* (similarity = 86.3%)

and *Coxiella burnetii* (similarity = 87.5%) than to representatives of *Ehrlichia*, *Rickettsia* or *Rochalimaea*. In short, it was deemed that the salmonid pathogen was sufficiently novel to warrant description in a new genus of the family Rickettsiaceae.

Rickettsia-like organisms
An increasing number of publications have described rickettsia-like organisms (RLO) as causal agents of disease (e.g. Rodger and Drinan, 1993; Chen *et al.*, 1994; Khoo *et al.*, 1995; Palmer *et al.*, 1997; Jones *et al.*, 1998). Whether or not these organisms correspond with *Piscirickettsia salmonis* has not been always established. For example, a RLO was reported as causing disease in tilapia from Taiwan during October 1992 to February 1993 (Chern and Chao, 1994). The pathogen was described as a Gram-negative rod of $0.86 \pm 0.32 \times 0.63 \pm 0.24$ μm in size, and thought likely to be a representative of the Rickettsiaceae (Chern and Chao, 1994).

I. MISCELLANEOUS PATHOGENS

Janthinobacterium lividum
Pure cultures have been recovered and characterized.

Janthinobacterium lividum
Culture comprise purple-pigmented motile Gram-negative strictly aerobic rods, which produce arginine dihydrolase, catalase and oxidase, but not β-galactosidase, indole or tryptophan deaminase. Nitrates are reduced, and the Voges Proskauer reaction is positive. Gelatin, but not urea, is degraded. Growth occurs at 4–30°C but not 37°C, and in 0–2% but not 3% (w/v) sodium chloride. Caprate, citrate, malate, maltose, mannitol, mannose and phenylacetate are utilized, but not *N*-acetyl glucosamine or adipate.

From these traits, the organisms were identified as typical (Table 4.3; from Northern Ireland) and atypical (Table 4.3; from Scotland) *Janthinobacterium lividum* (Sneath, 1984; Logan, 1989).

Streptobacillus
A possibly unique organism has been recovered from seawater-farmed Atlantic salmon in Ireland. The organism, which occurred intracellularly in tissues, was considered to be related to *Streptobacillus moniliformis* and the fusobacteria on the basis of 16S rRNA analyses (Maher *et al.*, 1995).

Unnamed bacteria
There is an increasing awareness of diseases caused by apparently unique bacteria. For example, Sorimachi *et al.* (1993) and Iida and Sorimachi (1994) described jaundice in yellowtail, attributed to an unknown filamentous bacterium of 4–6 μm in length, which grew only in L15 medium and Eagles MEM medium each supplemented with 10% (v/v) foetal calf serum at 23–26°C.

Table 4.3. Differential characteristics of *J. lividum* recovered from moribund and dead rainbow trout fry. A comparison has been made to taxa which accommodate purple-pigmented bacteria[a]

Character	*Chromobacterium violaceum*[b]	*Iodobacter fluviatile*[b]	*J. lividum*[b]	Atypical *J. lividum*[b]	Isolates from Ireland	Isolates from Scotland
Gelatinous colonies	—	–	v	+	+	+
Oxidative (O)–fermentative (F) metabolism of glucose	F	F	O	O	O	O
Growth at:						
4°C	–	+	+	+	+	+
37°C	+	–	–	–	–	–
Degradation of:						
Aesculin	–	–	+	–	+	–
Arginine	+	–	–	–	–	–
Production of acid from:						
L-arabinose	–	–	+	+	+	+
Gluconate	+	+	–	–	+	+
Glycerol	+	–	+	–	+	–

v = variable result.

[a] From Austin *et al.* (1992b).

[b] From Logan (1989).

The unknown Gram-negative cocco-bacilli of Palmer *et al.* (1994) were characterized, as follows.

Unknown Gram-negative organism of Palmer et al. *(1994)*
Cultures produce small colonies of 0.3 mm diameter after incubation aerobically for 10–14 days. Anaerobic incubation results in larger colonies of 0.6 mm diameter. Improved growth results by the addition of 0.5 mg of L-cysteine hydrochloride per litre. Colonies are off-white, convex and granular, and contain non-motile, Gram-negative β-haemolytic cocco-bacilli. Alkaline phosphatase (weakly positive/variable), arginine dihydrolase, H_2S (weakly positive/variable), indole (weakly positive/variable) and lipases are produced, but not α-glucosidase, *N*-acetyl glucosamine, catalase, β-galactosidase, β-glucuronidase, lysine decarboxylase or oxidase. Nitrates are not reduced. The Voges Proskauer reaction is negative. Neither aesculin, gelatin, Tween 80 nor urea is attacked. Citrate is not utilized. Fructose, glucose and maltose are fermented, but not glycogen, lactose, mannitol, mannose, ribose, sucrose or xylose. Growth occurs at 10°C but not at 37°C, and in 1–4% but not 5% (w/v) sodium chloride.

The organism was linked tenuously to the Neisseriaceae and Pasteurellaceae.

There has been a trend away from the conventional phenotypic approach of characterizing fish pathogens to molecular methods. Whereas the use of new technologies is to be encouraged, an ongoing dilemma remains concerning the authenticity of isolates. Also, too many conclusions seem to result from the examination of too few isolates.

5

Isolation

> There is no single technique suitable for the recovery of all known bacterial fish pathogens. Scientists need to use a combination of methods and incubation conditions to achieve pure cultures.

To an extent, the range of media to be used is governed by personal choice and experience (Table 5.1). The formulae of commonly used media are included in an appendix (Appendix 5.1) at the end of the chapter.

- For marine fish, it is advisable to include media prepared with seawater, e.g. seawater/marine 2216E agar (for example, as supplied by Difco).
- For marine fish with damaged gills, low nutrient agar, such as cytophaga agar (see Appendix 5.1) prepared in seawater, should be used.
- For freshwater fish, the routine use of TSA and BHIA (as supplied by Difco or Oxoid) is recommended.

Although most of the pathogens are aerobic, it is worthwhile remembering that *Cl. botulinum* and *Eu. tarantellae* are anaerobic. For the former, Robertson's meat broth (Appendix 5.1) should be used for isolation, whereas with the latter, BHIA is satisfactory. For some aerobic organisms, i.e. *Acinetobacter*, atypical *Aer. salmonicida* and *Ph. damselae* subsp. *piscicda*, the use of media supplemented with blood aids isolation.

A. ANAEROBES

Clostridium botulinum
Isolation may be achieved by use of straightforward anaerobic techniques. Samples of intestinal contents, which probably support a resident anaerobic microflora, and internal organs should be homogenized in 1% (w/v) peptone phosphate buffer at pH 7.0, and diluted five-fold. These diluted samples should be inoculated into 100 ml or 200 ml aliquots of Robertson's meat broth (see Appendix 5.1), with subsequent anaerobic incubation at 30°C for up to 6 days. Thereupon, the presence of *Cl. botulinum* and its toxicity

Table 5.1. Methods of isolation for bacterial fish pathogens

Medium[a]	Temperature of incubation (°C)	Pathogen
BHIA	20–30°C	*Cit. freundii, Cor. aquaticum, Edw. ictaluri, Edw. tarda, Haf. alvei, Halomonas cupida, Lactococcus garvieae, Lactococcus piscium, Planococcus* sp., *Sal. choloreraesuis* subsp. *arizonae, Ser. liquefaciens, Ser. marcescens, Sta. aureus, Sta. epidermidis, Streptococcus* spp., *Y. ruckeri*
BHIA	22–24°C (anaerobically)	*Eu. tarantellae*
BHIA supplemented with 5–10% (v/v) blood	20–25°C	*Cit. freundii, Haf. alvei, Ph. damselae, Salmonella choleraesuis* subsp. *arizonae*
BHIA supplemented with 3% (w/v) NaCl	15–25°C	*Shewanella putrefaciens*
BHIA supplemented with 10% (v/v) foetal calf serum and 1% (w/v) NaCl	22°C	*Streptobacillus*
Blood agar	15–37°C	streptocococci, *Va. salmoninarum, Y. intermedia*
Blood agar supplemented with 0.5–1.5% (w/v) NaCl	22–25°C	*Acinetobacter* sp., *Ph. damselae* subsp. *piscicida, V. logei, Moritella marina*
CBB	15–25°C	*Aer. salmonicida*
Cytophaga agar	18–20°C	*Fla. branchiophilum, Fl. columnare, Fla. hydatis, Fla. johnsoniae, Fla. ovolyticus, Fla. psychrophilum*
Cytophaga agar prepared in seawater	20°C	*Chrys. scophthalmum, Fle. maritimus, Myxococcus piscicola, Sporocytophaga* sp. (possibly)
Glucose aspargine agar	25–37°C	*Streptomyces salmonis*
Hayflick medium	room temperature	*Mycoplasmaa mobile*
KDM2/SKDM	15°C	*Ren. salmoninarum*
Löwenstein–Jensen medium/Dorset egg medium	15–22°C	*Mycobacterium* spp., *Nocardia* spp. *Haf. alvei, Ps. anguilliseptica, Ps. chloraphis, V. cholerae*

Table 5.1. (*Continued*)

Medium[a]	Temperature of incubation (°C)	Pathogen
Nutrient agar	37°C	*Bacillus* spp.
0.1% (w/v) peptone and agar prepared in seawater	20–25°C	*Fla. piscicida*
Robertson's meat broth	30°C (anaerobically)	*Cl. botulinum*
Seawater agar (marine 2216E agar)	15–25°C	*Ph. damselae* subsp. *piscicida, V. alginolyticus, V. anguillarum, V. cholerae, V. fischeri, V. harveyi, V. ordalii, V. vulnificus*
Skimmed milk agar	15–25°C	*Janthinobacterium* sp., *Micrococcus luteus, Planococcus* sp.
TCBS	15–25°C	*Ph. damselae, V. alginolyticus V. anguillarum, V. fischeri, V. harveyi, V. ordalii, V. pelagius, V. splendidus* and *V. trachuri*
Tissue culture (salmon cell line)	12–21°C	*Piscirickettsia salmonis*
TSA [possibly supplemented with 1–2% (w/v) NaCl]	15–25°C	*Aer. allosaccharophila, Aer. caviae, Aer. hydrophila, Aer. salmonicida, Aer. sobria, Car. piscicola, Cit. freundii, En. faecalis* subsp. *liquefaciens, Ent. agglomerans, Esch. vulneris, Flavobacterium* sp., *J. lividum, Klebsiella pneumoniae, Lactobacillus* spp., *Micrococcus luteus, Moraxella* sp., *Planococcus* sp., *Plesiomonas shigelloides, Pr. rettgeri, Ps. fluorescens, Ps. pseudoalcaligenes, Rhodococcus* sp., *Ser. marcescens, Ser. plymuthica, V. harveyi, V. pelagius, V. salmonicida, V. splendidus, V. vulnificus, Y. ruckeri*
TSA supplemented with 10% (v/v) horse serum	unspecified	*Aquaspirillum*
Yeast extract glucose agar	25°C	*Vag. salmoninarum*

[a] Most of these media may be obtained from Difco and/or Oxoid.

may be assessed (Cann *et al.*, 1965a,b; Cann and Taylor, 1982). In addition, sterile culture filtrates may be injected into mice and/or fish to assess the presence of toxic factors (Cann and Taylor, 1982). However, the recovery of *Cl. botulinum* from intestinal samples should not be used as the sole reason for diagnosing botulism, in view of its possible widespread presence in this habitat (see Trust *et al.*, 1979).

Eubacterium tarantellae
Samples of tissues should be plated on BHIA, whereupon bacterial cultures will develop after 7 days anaerobic incubation at room temperature. Additionally, tissues may be inoculated into Brewer's thioglycollate medium (Appendix 5.1; Udey *et al.*, 1977).

B. GRAM-POSITIVE BACTERIA—THE 'LACTIC ACID' BACTERIA

Carnobacterium piscicola and the lactobacilli
The use of TSA or BHIA with incubation at 22–24°C for 48 h has been advocated (Ross and Toth, 1974; Cone, 1982; Hui *et al.*, 1984).

Gram-positive cocci in chains
Recovery is straightforward, involving use of bovine blood tryptose agar (Naudé, 1975; Roode, 1977; Boomker *et al.*, 1979), Columbia agar (Appendix 5.1), 5% (v/v) defibrinated sheep blood agar (Doménech *et al.*, 1996), Todd–Hewitt broth (Appendix 5.1), nutrient agar supplemented with rabbit blood (Kitao *et al.*, 1981), TSA (Teskeredzic *et al.*, 1993; Michel *et al.*, 1997), 10% (v/v) horse blood in Columbia agar (Oxoid), 10% (v/v) horse serum in Columbia agar (Austin and Robertson, 1993), yeast extract glucose agar (Appendix 5.1; Michel *et al.*, 1997) or BHIA (Minami *et al.*, 1979; Ugajin, 1981; Kusuda *et al.*, 1991; Eldar *et al.*, 1994). Media may be supplemented with 1% (w/v) sodium chloride (Austin and Robertson, 1993). Inoculated media should be inoculated with diseased tissue, notably from the kidney, and incubated at 15–37°C for up to 7 days (usually 48 or 72 h) when 'dull grey' colonies approximately 1–2 mm in diameter develop. These colonies contain cocci in chains. It must be emphasized that an incubation temperature of 37°C is in excess of the normal growth temperature of many fish species, notably salmonids. This indicates that the organisms may well have been derived from warm-blooded animals, and may, therefore, constitute a public health risk.

Vagococcus salmoninarum
Schmidtke and Carson (1994) used Oxoid blood agar base supplemented with 7% (v/v) defibrinated sheep blood with incubation at 25°C for 48 h to recover cultures from brain, kidney, peritoneum, spleen and testes.

C. AEROBIC GRAM-POSITIVE RODS AND COCCI

Bacillus spp.
Oladosu *et al.* (1994) relied on nutrient agar and incubation at the comparatively high temperature of 37°C for an unspecified period to isolate *Bacillus* spp.

Bacillus mycoides

Material from ulcers, brain, kidney, liver and necrotic muscle were inoculated onto a range of media, including 5% (v/v) sheep blood in blood agar base, Mueller Hinton agar (for example, as supplied by Difco or Oxoid) and BHIA with incubation at an unspecified temperature for an unstated duration (Goodwin *et al.*, 1994). Raised, rhizoidal colonies with filamentous swirling patterns developed.

Corynebacterium aquaticum

Brain tissue samples was plated onto BHIA and TSA with incubation at 25°C for 48–72 h (Baya *et al.*, 1992b).

Micrococcus luteus

Cultures were recovered following incubation of swabbed material (kidney, spleen and ascitic fluid) on skimmed milk agar with incubation at 25°C for 48–72 h (Appendix 5.1; Austin and Stobie, 1992a).

Mycobacterium spp.

With many cases of mycobacteriosis, there is no attempt made at isolation of the pathogen. Yet, great scientific conclusions seem to result from the examination of only histological material. Nevertheless, attempts at isolating the aetiological agent often fail, indicating a fastidiousness on the part of the pathogen. Some success occurs by inoculating pieces of infected tissue (especially kidney, liver or spleen) on standard mycobacteria media, including Petragnani, Löwenstein–Jensen, Middlebrook 7H10 and Dorset egg media (see Appendix 5.1), or even blood agar, BHIA or TSA, whereupon growth may occur in 2–28 days at incubation temperatures of 15–30°C in aerobic or microaerophilic, i.e. 3–5% carbon dioxide, conditions (Dulin, 1979; Lansdell *et al.*, 1993). Most difficulty surrounds the recovery of mycobacteria from marine fish species. Clearly, more effort is required to understand the precise nutritional requirements of these organisms. *Myc. abscessus* was not isolated on Löwenstein–Jensen medium. Instead, Middlebrook 7H10 medium was modified by the addition of 10 μm/ml of amphotericin B, 500 μm/ml of chloramphenicol, 5 μm/ml of gentamicin or 30 μm/ml of cephalothin—either singly or in combination. Diseased fish were immersed in the modified Middlebrook 7H10 broth for 1 h at room temperature, before homogenization, and inoculation of modified Middlebrook 7H10 agar with incubation at 25°C for 14–28 days (Teska *et al.*, 1997).

Nocardia spp.

Essentially, the same isolation methods as for mycobacteria, e.g. use of Löwenstein–Jensen medium, have been employed with *Nocardia*. The initial development of colonies usually occurs within 21 days at 18–37°C (Valdez and Conroy, 1963; Conroy, 1964; Snieszko *et al.*, 1964b; Heuschmann-Brunner, 1965; Campbell and MacKelvie, 1968; Ghittino and Penna, 1968).

Planococcus sp.

The pathogen was recovered from kidney swabs following incubation on BHIA, skimmed agar or TSA at 25°C for up to 7 days. Cultures comprised off-white to yellow raised shiny colonies, which were 1–2 mm in diameter after 48 h (Austin *et al.*, 1988).

Renibacterium salmoninarum

Cultivation of the causal agent of BKD *in vitro* from chinook salmon was not achieved until Earp (1950) used a nutrient-rich medium, containing fish extract, glucose, yeast extract and bovine serum/meat infusion, with incubation at 15 or 20°C. An improvement resulted from use of minced chick embryos in 1% (w/v) agar or on Dorset egg medium (Earp *et al.*, 1953). Nevertheless, growth was generally poor, even after prolonged incubation periods of ≥14 days. To prove that the growth was of the pathogen, Earp successfully inoculated the bacterial culture into healthy chinook salmon, and eventually recovered it again from the resultant kidney lesion. Continuing this pioneering work, Ordal and Earp (1956) supplemented Dorset egg medium with 0.05–1.0% (w/v) L-cysteine, tryptone and yeast extract, and succeeded in isolating the pathogen in 3–4 weeks following incubation at 17°C. The outcome of this work was the formulation of cysteine blood agar with which Koch's postulates were fulfilled (Appendix 5.1; Ordal and Earp, 1956). Foetal calf serum was substituted for human blood in a modification proposed by Evelyn *et al.* (1973). This was developed further by removing sodium chloride and substituting peptone for tryptone and beef extract (KDM2; Appendix 5.1; Evelyn, 1977); this medium is now used commonly for growth of the BKD organism. In a parallel development, Wolf and Dunbar (1959) used Mueller–Hinton agar supplemented with 0.1% (w/v) L-cysteine hydrochloride (MHC) to culture the pathogen, although success with this medium did not occur with Smith (1964). However, Bullock *et al.* (1974) confirmed the value of MHC, although this has been subsequently contended by Evelyn (1977). Serum-rich KDM2 was considered to be superior to serum-deficient MHC, indicating the benefit of serum for the cultivation of the BKD organism. This was further supported by Paterson *et al.* (1979), who supplemented MHC with 10% (v/v) foetal calf serum, and successfully used the medium for isolating the pathogen from Atlantic salmon. Daly and Stevenson (1985) proposed replacing serum with charcoal, which serves as a detoxicant (Appendix 5.1, charcoal agar). However, these media, being extremely rich in composition, are generally suitable for the growth of many aerobic, heterotrophic bacteria. Moreover, fast-growing organisms may rapidly outcompete and overgrow the slower-growing BKD organism. A solution was proposed by Evelyn (1977), who advocated the use of a drop-plating technique (essentially, this is analogous to dilution plates, which dilute out potential interference by fast-growing heterotrophs). In a later report, Evelyn (1978) recommended the use of peptone (0.1% w/v)-saline (0.85% w/v) as a diluent to remove any inhibiting factors against the pathogen which may be present in kidney tissue (Evelyn *et al.*, 1981; Austin, 1986). Some inconsistencies in the performance of KDM2 were attributed to variation in the composition of the commercial peptone (Evelyn and Prosperi-Porta, 1989). To overcome this inconsistency when single lots of peptone were not available, two possible modifications were suggested. Firstly, a 'nurse' culture technique was reported. This technique accelerated the growth of the BKD organism, and increased the sensitivity at which the pathogen could be detected. The technique, which was based on satellitism or cross feeding, involved inoculating the nutritionally fastidious pathogen next to a non-fastidious feeder—the nurse organism.

Evelyn *et al.* (1989) placed drops of a dense suspension of a stock culture of the BKD organism (= the nurse organism) onto the centre of KDM2 plates. Samples, suspected of containing the pathogen, were placed as 25 μl drops around the periphery of the nurse culture. With incubation, the nurse culture grew rapidly, presumably modified the conditions in the KDM2, and thus enhanced the growth of the pathogen in the periphery. For example, colonies of the BKD organism were observable after incubation for 19 days, compared to 25 days for the conventional approach. The second modification involved supplemented KDM2 agar with a small amount of spent (KDM2) broth that was previously used for growing the pathogen. In both cases, it seems that an unknown metabolite serves as a growth stimulant (Evelyn *et al.*, 1989, 1990).

Until the advent of selective isolation techniques, initial isolation of the pathogen from fish tissues was an uncertain affair, prone to contamination by fast-growing aerobic heterotrophs. With this in mind, a selective isolation medium, SKDM (Appendix 5.1; Austin *et al.*, 1983a) was devised, which proved to be effective for isolation of the pathogen from dilute samples. SKDM permitted the recovery of the pathogen from seeded river water, and from the kidney and faeces of experimentally infected fish (Embley, 1983; Austin and Rayment, 1985). In contrast, the pathogen was not recovered on corresponding KDM2 plates which were completely overgrown by other bacteria. Clearly, selective media, such as SKDM, should prove useful in further ecological studies on the causal agent of BKD. In a comparison of KDM2, SKDM and the charcoal containing derivative, it was determined that the selective medium (SKDM) was most effective for the primary isolation of the pathogen from Atlantic salmon (Gudmundsdóttir *et al.*, 1991). In this comparison of positive samples, 91%, 60% and 35% were positive on SKDM, the charcoal containing derivative and KDM2, respectively. Clearly, the selective medium enhanced significantly the ability to recover the pathogen. Moreover, serum was more advantageous than charcoal as a medium supplement. However, long incubation periods of 12–19 weeks were necessary to recover colonies on the media from dilute samples (Benediktsdóttir *et al.*, 1991).

The question concerning the necessary growth requirements for the pathogen was addressed in a detailed study by Embley *et al.* (1982). These workers formulated a rich semi-defined medium devoid of serum (Appendix 5.1, semi-defined medium), which was suitable for the cultivation of cells, but not for the initial isolation of cells from infected fish tissues. The semi-defined medium was used, however, to obtain biomass destined for lipid analyses (Embley *et al.*, 1983), and inocula for nutritional and physiological studies. Subsequently, Shieh (1989a) described a complex blood free medium, which permits the growth of the pathogen.

Rhodococcus sp.

Dense growth of two colony types was obtained from diseased tissue following inoculation of blood agar, MacConkey agar (Appendix 5.1) and TSA with incubation at room temperature (19°C) for up to 14 days. Both colony types were also recovered from the kidney and spleen of chinook salmon with ocular lesions (Bachman *et al.*, 1990; Claveau, 1991).

Staphylococcus aureus
Eye and brain tissue revealed the presence of bacteria. However, the precise isolation procedures were not stated (Shah and Tyagi, 1986).

Staphylococcus epidermidis
The pathogen was successfully cultured on BHIA following incubation at 37°C for 24 h (Kusuda and Sugiyama, 1981). However, the use of such a high temperature is puzzling, and it is conceivable that mesophiles, preferring lower growth temperatures, may have been overlooked.

Streptomyces salmonis (= Streptoverticillum salmonis)
The organism may be isolated on glucose asparagine agar, glycerol asparagine agar, yeast extract malt extract agar, Emerson agar, Bennett agar or nutrient agar, following incubation aerobically at 25–37°C for an undisclosed period (Appendix 5.1; Rucker, 1949).

D. AEROMONADACEAE REPRESENTATIVES

Aeromonas allosaccharophila
The precise isolation method was not described. However, it was mentioned that growth occurred after an unstated period on TSA at 4–42°C (Martinez-Murcia *et al.*, 1992).

Aeromonas caviae
Diseased tissues were homogenized in 0.1% (w/v) peptone, before inoculation of TSA and blood agar with incubation at 22°C for 48 h (Candan *et al.*, 1995).

Aeromonas hydrophila
This is quite straightforward, involving use of kidney swabs with non-selective media, such as nutrient agar or TSA, or selective media, namely Rimler–Shotts medium (Appendix 5.1; Shotts and Rimler, 1973) or peptone beef-extract glycogen agar (Appendix 5.1; McCoy and Pilcher, 1974) with incubation at 20–25°C for 24–28 h. Typically, on non-selective media, cream, round, raised, entire colonies of 2–3 mm diameter develop within 48 h at 25°C.

Aeromonas salmonicida
Under ordinary circumstances, i.e. in cases of classical furunculosis caused by 'typical' strains of *Aer. salmonicida*, the pathogen may be readily recovered from diseased fish, especially from surface lesions and the kidney, by use of standard non-selective bacteriological agar media. TSA has been commonly used for this purpose. On TSA, the occurrence of colonies surrounded by a dark-brown water soluble pigment after incubation at 20–25°C for 3–4 days, is considered as indicative of the presence of *Aer. salmonicida*. However, it must be remembered that non-pigmented or slowly pigmented strains of the pathogen occur, and also that some other bacteria produce diffusible brown pigments, e.g. *Aer. hydrophila* and *Aer. media*. In addition, if the fish have succumbed to secondary infection with other micro-organisms, isolation of *Aer. salmonicida* becomes much more

difficult because of overgrowth by other bacteria. Thus, growth of *Aer. salmonicida* may be suppressed or pigment production inhibited. For these reasons, McCarthy and Roberts (1980) recommended that a minimum of six fish should be sampled from any disease outbreak where the presence of *Aer. salmonicida* is suspected. Also, they suggested that samples destined for bacteriological examination should be taken from skin lesions, in all stages of development. In a similar vein, Daly and Stevenson (1985) concluded that it is advisable to sample other organs, i.e. heart, liver and spleen, in addition to the kidney in order to increase the chances of detecting *Aer. salmonicida*. Indeed, these workers found that 45% (14/31) of successful isolations of the pathogen from brown trout, *Salmo trutta*, were from organs other than kidney. Likewise, in a study to detect carrier rates of the pathogen, Rose *et al.* (1989) concluded that sampling of just the kidney might result in an underestimate of the numbers of fish harbouring the pathogen. Thus, they recommended examining the intestine as well as the kidney for the presence of *Aer. salmonicida*. The problem is with the asymptomatic carriers from which recovery of the pathogen is notoriously difficult without stressing the fish (Cipriano *et al.*, 1997).

In an attempt to improve the chances of recovering *Aer. salmonicida* from lake trout in a hatchery experiencing an outbreak of furunculosis, an enrichment procedure using TSB was evaluated by Daly and Stevenson (1985). This entailed placing swabbed material, derived from the organs of diseased animals, into TSB with incubation at 26°C for 48 h. Then, the resulting broth cultures were streaked for single colony isolation onto plates of TSA. The results indicated that the recovery of *Aer. salmonicida* was twice that of using the conventional direct plating of swabbed material onto TSA.

The use of BHIA for isolation and maintenance of *Aer. salmonicida* has been recommended by some groups. In fact, we have observed that a greater proportion of rough-type colonies (this trait is associated with virulence) of *Aer. salmonicida* were recovered on BHIA, than TSA.

In several instances, media supported with blood have been employed for the isolation of the pathogen, especially atypical isolates from cyprinids. McCarthy (1977a), however, stated that unsupplemented media, e.g. TSA, should be used in preference to blood-containing media, but no explanation was given, other than that the observations had resulted from extensive personal experiences.

A more recent addition to the media employed for the isolation of *Aer. salmonicida* is CBB (Appendix 5.1), as developed originally by Udey (1982). CBB was evaluated as a differential and presumptive medium for use in the identification of *Aer. salmonicida* in clinical specimens (Markwardt *et al.*, 1989). The results showed that CBB was effective in differentiating *Aer. salmonicida* among mixed bacterial populations obtained from asymptomatic fish. In laboratory-based experiments, CBB was also successful in differentiating *Aer. salmonicida* colonies from mixed cultures containing *Aer. hydrophila* or *Y. ruckeri*. Here, *Aer. salmonicida* colonies were dark blue. An interesting development concerned the ability to detect *Aer. salmonicida* within 72 h by filtering 100 ml amounts of water through 0.45 μm pore size Millipore cellulose acetate and nitrate filters, and incubating the filters on CBB (Ford, 1994). However, it is apparent that the *Aer. salmonicida* cells must possess the A-layer for the differentiating capacity of CBB to be effective. Also, other aquatic organisms, such as the purple-pigmented *Chromobacterium* and *Janthinobacterium* may produce dark blue colonies on CBB. Nevertheless, CBB is a promising addition to the narrow range of media which may be used for the primary isolation of *Aer. salmonicida.*

The stress-induced furunculosis test to test covertly infected fish has been very successful, and involves intramuscular injection with corticosteroid, namely 20 mg prednisolone acetate/kg of fish followed by increasing the water temperature typically from 12 to 18–20°C. Cultures of *Aer. salmonicida* may be then recovered on TSA or CBB within 3-6 days (Bullock and Stuckey, 1975b; McCarthy, 1977b; Smith, 1991; Bullock *et al.*, 1997). Using such stressed rainbow trout, culturing was the most sensitive method for detecting *Aer. salmonicida* (detected 40 positives out of 80) followed by a direct FAT (detected 6 positives) and then a commercial ELISA system (detected 6 positives) (Bullock *et al.*, 1997). Overall, the culture of gill and mucus was more sensitive (39 positives out of 80 fish examined) than kidney and spleen (18 positives) (Bullock *et al.*, 1997).

In the case of CE and goldfish ulcer disease, the pathogen appears to be located more or less exclusively in the skin lesions. Thus, Bootsma *et al.* (1977) used an inoculating wire, which was plunged below the transparent epidermis at the edge of the ulcer, into the haemorrhagic zone. A loopful of the resulting material was streaked onto agar media. According to Bootsma, satisfactory growth occurred on tryptone-containing media supplemented with serum. Although the enrichment of a culture medium by the addition of serum has been deduced as necessary for the initial recovery of some fastidious strains, notably non-pigmented cultures, Bootsma *et al.* (1977) determined that the fastidiousness of the isolates decreased during maintenance *in vitro*. In our experience, strains associated with non-salmonid fish are extremely difficult to isolate. We have greatest success with blood agar (blood agar base [Oxoid] supplemented with 10% v/v horse, sheep or bovine blood), which is inoculated and incubated at 15–18°C for up to 7 days. Even with this method, *Aer. salmonicida* is recovered from only a small proportion of the clinically diseased fish. This begs the question about the reasons for culturability, when the pathogen is recovered from only a proportion of obviously infected animals. Microscopy will often reveal a greater number of bacterial cells than might be deduced from the results of plating experiments. Perhaps, as has been argued with L-forms, a threshold number of bacterial cells need to be present to enable some to be capable of producing growth in broth or on solid medium. Also, the definition of growth needs to be carefully considered, insofar as the basic criterion reflects observations with the naked eye, i.e. turbidity in broth or clearly visible colonies. The limited growth of micro-colonies may well be missed by classical bacteriological methods.

Aeromonas sobria

Pure culture growth was obtained from kidney, liver and spleen of moribund animals following inoculation of TSA with incubation at 22°C for possibly one or two days (Toranzo *et al.*, 1989).

E. ENTEROBACTERIACEAE REPRESENTATIVES

Citrobacter freundii

Pure culture growth was recovered from kidney on rabbit blood agar (Sato *et al.*, 1982).

Subsequently, similar organisms have been obtained from kidney homogenates spread over the surface of BHIA and TSA with incubation at 25 or 37°C for 24–48 h.

Edwardsiella ictaluri

Isolation has been readily achieved from kidney, liver, spleen, intestine, brain and skin or muscle lesions by inoculation of material into BHIA or blood agar. Following incubation at 26°C for 48 h, smooth circular (2 mm diameter), slightly convex, entire, non-pigmented colonies develop (Hawke, 1979). A selective medium has been described by Shotts and Waltman (1990) (Appendix 5.1).

Edwardsiella tarda

Isolation of *Edw. tarda* from diseased fish is a straightforward procedure involving the use of commonly available media, such as TSA (Meyer and Bullock, 1973) or BHIA (Amandi *et al.*, 1982). On such media, small round (0.5 mm in diameter) raised, transparent colonies develop in 48 h at 24–26°C (Meyer and Bullock, 1973). The use of thioglycollate broth followed by subculturing on BHIA has also been used successfully (Appendix 5.1; Amandi *et al.*, 1982). Indeed, this two-step enrichment procedure has proved to be more sensitive than BHIA used alone. In one experiment, this two-step procedure enabled the recovery of *Edw. tarda* from 19% of a group of chinook salmon, compared to only 2% recovery on BHIA used alone (Amandi *et al.*, 1982).

Enterobacter agglomerans

Pale yellow colonies were recovered from kidney blood on TSA supplemented with 2% (w/v) NaCl following incubation at 16°C for 7 days. Subculturing was possible on marine 2216E agar (Hansen *et al.*, 1990).

Escherichia vulneris

Isolates were recovered on TSA with incubation at 25°C for 48 h (Aydin *et al.*, 1997).

Hafnia alvei

Cultures were obtained from the internal organs and subcutaneous tissues at the base of the pelvic fins on BHIA and nutrient agar supplemented with 10% (v/v) sheep blood following incubation at an unspecified temperature for 48 h, whereupon small, round, smooth colonies developed. Teshima *et al.* (1992) isolated the organism on heart infusion agar following incubation at 30°C for 2 days.

Klebsiella pneumoniae

Pure cultures were obtained after shaking tails and pectoral fins in 0.9% (w/v) saline for 5 min, and thereafter spreading 0.1 ml volumes on TSA with incubation at 15°C for 7 days (Daskalov *et al.*, 1998).

Proteus rettgeri

Pure culture growth was recovered from the heart, kidney and base of the lesions following inoculation of nutrient agar, blood agar and TSA with incubation at 15–35°C for an unspecified period (Bejerano *et al.*, 1979).

Salmonella choleraesuis subsp. *arizonae*

Samples of internal organ were inoculated onto plates of blood agar, BHIA and MacConkey agar with incubation at 25°C for 48 h. Thereupon, an organism was recovered as pure culture growth from heart, kidney, liver and spleen (Kodama *et al.*, 1987).

Serratia liquefaciens

Isolation was readily achieved from kidney by inoculation of swabbed material onto BHIA and TSA with incubation at 25°C for 24–48 h (McIntosh and Austin, 1990).

Serratia marcescens

Pure cultures resulted from inoculation of kidney swabs onto plates of BHIA and TSA with incubation at 22°C for 48–72 h (Baya *et al.*, 1992c).

Serratia plymuthica

Pure culture growth was recovered from the kidney and liver following inoculation of TSA with incubation at 22°C for 7 days (Austin and Stobie, 1992b).

Yersinia intermedia

Blood samples were cultured on Oxoid blood agar base supplemented with 7% (v/v) defibrinated sheep's blood at 18°C for 7 days (Carson and Schmidtke, 1993). After 48 h, the kidneys of two fish revealed the presence of dense growth, equated with *Yersinia intermedia*. In addition, other bacteria were recovered including CLB and *Ps. fluorescens*.

Yersinia ruckeri

The pathogen may be readily recovered from kidney on routine bacteriological media, e.g. BHIA or TSA, following incubation at 20–25°C for 48 h whereupon round, raised, entire, shiny, off-white colonies of 2–3 mm diameter develop (e.g. Ross *et al.*, 1966). Three selective media have been devised (Appendix 5.1; Waltman and Shotts, 1984; Rodgers, 1992; Furones *et al.*, 1993). Tween 80 hydrolysis, which occurred commonly among isolates from the USA, resulted in the precipitation of insoluble calcium salts around colonies on Waltman–Shotts medium. However, according to Hastings and Bruno (1985), there were some limitations insofar as there is some growth variation among *bona fide* strains of *Y. ruckeri*. Moreover, non-motile isolates, which did not degrade Tween 80 have been recovered in Europe, e.g. Germany (Klein *et al.*, 1994). Ribose ornithine deoxycholate medium of Rodgers (1992) appears to have overcome the problems of the earlier formulation, and laboratory use points to its value for the recovery of *Y. ruckeri*. Furone *et al.* (1993) supplemented TSA with 1% (w/v) SDS, 100 μg/ml of coomassie brilliant blue and 100 μg/ml of congo red. On this medium, *Y. ruckeri* produced a creamy deposit around the colonies.

F. *CYTOPHAGA–FLAVOBACTERIUM–FLEXIBACTER* GROUP

Generally, the aetiological agents may be readily recovered from diseased tissues on low-nutrient media, with incubation at 10–25°C for 4–14 days. From 'halibut jelly', which was used to culture *Chrys. balustinum* as yellow-green ('fluorescent') pigmented colonies (Harrison and Sadler, 1929), scientists have devised many suitable media, of which cytophaga agar (Appendix 5.1; after Anacker and Ordal, 1959) has received greatest use. This is suitable for the isolation of *Fla. hydatis*, *Fla. johnsoniae* (specifically with incubation at 27°C for 7 days; Carson *et al.*, 1993), *Fla. psychrophilum*, *Fla. branchiophilum* and *Fla. columnare*. Caution has been suggested over the source of ingredients used to make some media. For example, the importance of the brand of beef extract (the Difco product was reported as superior to Gibco or Oxoid) for the growth of *Fla. psychrophilum* on cytophaga agar was highlighted by Lorenzen (1993). Cytophaga agar, prepared in 70% seawater, is ideal for recovery of marine organism, such as *Fle. maritimus* (Appendix 5.1; Hikida *et al.*, 1979) and *Fle. ovolyticus* (Hansen *et al.*, 1992). Other isolation media have been described, including skimmed milk agar, CP medium (Carlson and Pacha, 1968) and Pacha and Ordal's medium (Pacha and Ordal, 1967) for *Fla. hydatis* (Strohl and Tait, 1978), TSA supplemented with 0.5–3.0% (w/v) sodium chloride for *Flavobacterium* spp. (Acuigrup, 1980a), Bootsma and Clerx's medium for *Fla. columnare* (Appendix 5.1; Bootsma and Clerx, 1976), TCY medium for *Fle. maritimus* (Appendix 5.1; Hikida *et al.*, 1979) and medium K for *Chrys. scophthalmum* (Appendix 5.1; Mudarris and Austin, 1989). Anderson and Conroy (1969) described a seawater-based medium for *Sporocytophaga*-like organisms (Appendix 5.1).

Only two selective media containing antibiotics have been devised (Appendix 5.1). These were formulated for the selective recovery of *Fla. columnare* (Fijn, 1969) and myxobacteria (Hsu *et al.*, 1983), respectively. On these media, the pathogens produce characteristically yellow-orange pigmented colonies.

G. PSEUDOMONADACEAE REPRESENTATIVES

Pseudomonas anguilliseptica

Isolation of *Ps. anguilliseptica* is readily achieved from blood, kidney, liver and spleen samples by use of nutrient agar supplemented with 10% (v/v) horse blood or nutrient agar containing 0.5% (w/v) sodium chloride, and adjusted to a pH of 7.4. Incubation should be at 20–25°C for at least 7 days, when small (≤1 mm in diameter) round, raised, entire, shiny, pale-grey colonies develop (Wakabayashi and Egusa, 1972).

Pseudomonas chlororaphis

This may be accomplished by inoculating homogenates, prepared from the entire fish, onto the surface of nutrient agar plates, with incubation at 25°C for 5 days. It is surprising that this simple method enabled the recovery of pure cultures, because contaminants may be expected from the intestine and body surface (Hatai *et al.*, 1975).

Pseudomonas fluorescens

Ps. fluorescens has been recovered from most organs as pure culture growth on standard bacteriological media, such as Pseudomonas F. agar (Appendix 5.1), blood agar, TSA and nutrient agar, following incubation at 22–28°C for 24–28 h (Csaba *et al.*, 1981; Ahne *et al.*, 1982).

Pseudomonas pseudoalcaligenes
Ps. pseudoalcaligenes was recovered as mixed cultures with *Serratia plymuthica* from the ascitic fluid and surface lesions following incubation on TSA at 22°C for 3–4 days (Austin and Stobie, 1992b).

H. VIBRIONACEAE REPRESENTATIVES

Generally, vibrios may be isolated on marine 2216E agar (supplied by Difco) with incubation at 25°C for 2–7 days (e.g. Ishimaru *et al.*, 1996).

Moritella marina
Benediktsdóttir *et al.* (1998) used 5% (v/v) horse blood agar supplemented with 1.5% (w/v) sodium chloride with incubation at 15°C for 7 days.

Photobacterium damselae
Isolation may be readily achieved by swabbing ulcerative material onto the surface of BHIA supplemented with 5% (v/v) sheep blood, or TCBS (Appendix 5.1) with incubation at 25°C for an unspecified period (probably 2–5 days) (Fujioka *et al.*, 1988; Sakata *et al.*, 1989; Fouz *et al.*, 1991).

Photobacterium damselae subsp. *piscicida*
The organism may be isolated by inoculating swabs of kidney and/or spleen material onto marine 2216E agar (Difco), nutrient agar or blood agar, with incubation at 25°C for 48–72 h. An improved liquid medium has been described, which may be solidified by the addition of 1% (w/v) agar (Appendix 5.1; Hashimoto *et al.*, 1989). On conventional media, shiny, grey-yellow, entire, convex colonies develop, which are approximately 1–2 mm in diameter after 72 h (Kusuda and Yamaoka, 1972). Another approach, which has met with success, has enabled the recovery of *Ph. damselae* subsp. *piscicida* from water in the vicinity of fish. The method involved filtering 250 ml volumes of water through 0.45 μm cellulose nitrate filters, before transfer (of the filters) to 2216E agar supplemented with 1% (w/v) mannitol and 0.5% (w/v) phenol red. The non-fermenting *Ph. damselae* subsp. *piscicida* produced red colonies (Reali *et al.*, 1997). With this method, the pathogen was detected on sea bass 8 days before the outbreak of disease.

Plesiomonas shigelloides
This was accomplished by inoculating samples of kidney and liver onto plates of TSA with incubation at 22–37°C for an unspecified period whereupon round, raised, off-white, circular colonies developed (Cruz *et al.*, 1986).

Shewanella putrefaciens
Bacteria were isolated from the kidney, liver and spleen following inoculation onto BHIA

supplemented with 3% (w/v) sodium chloride, with incubation at an unspecified temperature (presumed to be ≤37°C) for an undetermined period (Saeed *et al.*, 1987).

Vibrio alginolyticus

Isolation may be readily achieved from blood by inoculation onto TSA prepared with sea-water, TCBS, or seawater agar with incubation at 15–25°C for 2–7 days. The precise conditions employed by Colorni *et al.* (1981) were not stated. However, this group succeeded in isolating *V. alginolyticus*, *V. anguillarum* and *V. parahaemolyticus* from blood, and long thin, flexible rods from cases of gill rot. In addition, we have isolated pure culture growth of *V. alginolyticus* from moribund eels.

Vibrio anguillarum

The pathogen may be readily recovered from infected tissue by use of TSA (Traxler and Li, 1972), nutrient agar (Muroga *et al.*, 1976a,b) and BHIA (Tajima *et al.*, 1981) supplemented with sodium chloride at 0.5–3.5% (w/v), seawater agar and TCBS (Bolinches *et al.*, 1988), with incubation at 15–25°C for periods of up to 7 days.

The presumptive identification of *V. anguillarum* has been achieved using a specially designed medium, designated VAM, which combined bile salts with a high sodium chloride concentration, ampicillin, sorbitol and a high pH (Appendix 5.1; Alsina *et al.*, 1994). On VAM, *V. anguillarum* produced bright yellow colonies with yellow haloes. Its usefulness was attested by its ability to recognize the majority, i.e. $197/227 = 87\%$ of *V. anguillarum* isolates. However, some erroneous results occurred insofar as VAM recognized $3/66 = 4\%$ of other vibrios as *V. anguillarum* (Alsina *et al.*, 1994).

Vibrio cholerae (non-O1)

This may be achieved by inoculating kidney material (from swabs) onto the surface of nutrient agar plates, with incubation at 25°C for an undisclosed period (presumably 2–5 days) (Muroga *et al.*, 1979; Kiiyukia *et al.*, 1992).

Vibrio fischeri

By use of TSA supplemented with 2% (w/v) sodium chloride, marine 2216E agar (Difco) and TCBS (Oxoid), virtual pure culture growth of bacteria was recovered from the kidney and liver, following incubation at an unstated temperature for an unspecified duration (Lamas *et al.*, 1990).

Vibrio harveyi

Samples of kidney and liver were inoculated into thioglycollate broth (Appendix 5.1), followed by single colony isolation on plates of TSA supplemented with 1% (w/v) sodium chloride (Grimes *et al.*, 1984a). Yii *et al.* (1997) used TSA supplemented with 2% (w/v) sodium chloride and TCBS with intestinal fluid from diseased grouper. Generally, the use of TSA, MacConkey agar, TCBS and cytophaga agar were advocated, with incubation at 25–35°C for 24–48 h. Thus according to Kraxberger-Beatty *et al.* (1990), the eyes were removed and homogenized in sterile saline; loopfuls of the material being streaked onto the agar-containing media. It should be emphasized that attempts were not made to surface sterilize the eyes. Therefore, the resulting homogenate would have also contained aquatic (saprophytic) bacteria, as well as any potential pathogen.

Vibrio logei

Benediktsdóttir *et al.* (1998) used 5% (v/v) horse blood agar supplemented with 1.5% (w/v) sodium chloride with incubation at 15°C for 7 days.

Vibrio ordalii

As with *V. anguillarum*, isolation involves use of seawater agar and TCBS with incubation at 15–25°C for up to 7 days (Ransom, 1978; Ransom *et al.*, 1984).

Vibrio pelagius

Samples from diseased tissues were inoculated onto TSA supplemented with 2% (w/v) sodium chloride, marine 2216E agar and TCBS with incubation at 25°C for 48 h (Angulo *et al.*, 1992).

Vibrio salmonicida

Microscopy suggests that bacteria are rampant throughout infected fish, and especially in the blood and kidney of moribund and freshly dead specimens. Pure cultures may be readily recovered from blood and kidney samples on TSA supplemented with 1.5% (w/v) sodium chloride following incubation at 15°C for up to 5 days (Holm *et al.*, 1986; Egidius *et al.*, 1986). Colonies are small, i.e. ≤1 mm in diameter, round, raised, entire and translucent. Inexperienced personnel could easily miss the colonies upon cursory glances at inoculated plates. To some extent, the organism is fragile, and will quickly die at supra-optimum temperatures or by failure to carry out regular sub-culturing.

Vibrio splendidus

Kidney (anterior), liver, spleen and fluid from within the peritoneal cavity contained bacterial populations, which grew on TSA supplemented with 2% (w/v) sodium chloride and on TCBS with incubation at 22°C for 48 h (Lupiani *et al.*, 1989).

Vibrio trachuri

It is unclear from the article describing the taxon how the organisms were isolated from fish (Iwamoto *et al.*, 1995). However, it would appear that the growth compares with other vibrios. Therefore, standard techniques, e.g. use of TCBS agar, for marine aerobic heterotrophic bacteria would seem to be appropriate.

Vibrio vulnificus

Use of standard bacteriological media, such as seawater agar and TSA supplemented with sodium chloride, and incubation for up to 7 days at 20–25°C is sufficient to obtain cultures of the pathogen (Muroga *et al.*, 1976a,b; Nishibuchi and Muroga, 1977, 1980; Nishibuchi *et al.*, 1979, 1980).

I. MORAXELLACEAE REPRESENTATIVES

Acinetobacter sp.

Pure cultures were recovered on 5% (w/v) blood agar supplemented with 0.5% (w/v) sodium chloride, following incubation at 22°C for 48 h. Thus, blood, kidney, liver, spleen ulcers in the muscle appeared to contain dense bacterial populations. Pure cultures were obtained, which were capable of reproducing the infection during laboratory-based studies (Roald and Hastein, 1980).

Moraxella sp.

Pure culture growth of round, raised, translucent, mucoid colonies developed in 48 h of incubation at 22°C of kidney, liver and pancreas tissue on TSA (Baya *et al.*, 1990b).

J. HALOMONADACEAE REPRESENTATIVE

Halomonas (= Deleya) cupida

Although the wisdom of using homogenates of whole fish for bacteriological examination may be debated, it is probably an acceptable compromise with fry. Nevertheless, care needs to be taken when interpreting results insofar as bacteria may be derived from the normal surface/intestinal microflora as well as from diseased tissues. In this investigation, cultures were obtained from homogenates following inoculation of BHIA with incubation at 25°C for 24 h (Kusuda *et al.*, 1986).

K. SMALL, MORPHOLOGICALLY SIMPLE BACTERIA

Piscirickettsia salmonis

Isolation of the pathogen from the kidney of infected fish was possible in the cytoplasm of salmon cell lines (including CH5E-214) with incubation at 12–21°C (optimally at 15–18°C), whereupon a cytopathic effect was demonstrated in 5–6 days (Fryer *et al.*, 1990). The cell sheet was completely lysed in 14 days. Growth did not occur on bacteriological media, including BHIA, blood agar, mycoplasma medium, charcoal yeast extract agar, Loeffler medium (Appendix 5.1) or Mueller–Hinton agar.

L. MISCELLANEOUS PATHOGENS

Aquaspirillum sp.

Lio-Po *et al.* (1998) used TSA supplemented with 10% (v/v) horse serum and cytophaga agar at an unspecified temperature and duration to recover *Aquaspirillum*, *Aer. hydrophila* and *Streptococcus* from diseased animals.

Janthinobacterium lividum

Homogenates of whole fish (prepared in quarter-strength Ringer's buffer; Oxoid) and, where possible, loopfuls of kidney, liver, spleen, ascitic fluid and material from surface lesion were spread over the surface of a variety of media, including blood agar (5% v/v bovine blood in Gibco blood agar base), cytophaga agar, KDM2, L-F medium (Appendix

5.1) and TSA, with incubation aerobically at 22°C for up to 14 days. Purple-pigmented colonies were apparent after 3 days.

Streptobacillus
Brain–heart infusion supplemented with 10% (v/v) foetal calf serum and 1% (w/v) sodium chloride with incubation for 10 days at 22°C was used to recover the novel streptobacillus-like organism (Maher *et al.*, 1995).

Unidentified Gram-negative rod
The intracellular Gram-negative cocco-bacilli required serum or blood for growth, which was accomplished on 7% (v/v) horse blood agar and 10% (v/v) foetal calf serum medium after 4–14 days incubation at 15 or 22°C (Palmer *et al.*, 1994).

APPENDIX 5.1 MEDIA USED FOR THE ISOLATION AND GROWTH OF BACTERIAL FISH PATHOGENS

Anderson and Conroy's medium for *Sporocytophaga*-like organisms
5.0% (w/v) enzymic digest of fish muscle
0.1% (w/v) peptone
0.1% (w/v) yeast extract
0.9% (w/v) agar
pH 7.0
This medium is prepared in seawater

Bootsma and Clerx's medium for *Flavobacterium columnare*
0.05% (w/v) casitone
0.05% (w/v) yeast extract
1.0% (w/v) agar
pH 8.0

Brewer's thioglycollate medium
0.1% (w/v) Lab-lemco
0.2% (w/v) yeast extract
0.5% (w/v) peptone
0.5% (w/v) dextrose
0.5% (w/v) sodium chloride
0.11% (w/v) sodium thioglycollate
0.0002% (w/v) methylene blue
0.1% (w/v) agar No. 1
pH 7.2
Sterilize at 121°C/15 min

Charcoal agar; for the growth of *Renibacterium salmoninarum*
1.0% (w/v) peptone
0.05% (w/v) yeast extract

0.1% (w/v) L-cysteine hydrochloride
0.1% (w/v) activated charcoal
1.5% (w/v) agar
pH 6.8
Sterilize at 121°C for 15 min (the charcoal may be placed in dialysis tubing prior to sterilization in order to obtain a clear broth medium)

Columbia agar
2.3% (w/v) special peptone
0.1% (w/v) starch
0.5% (w/v) sodium chloride
1% (w/v) agar No. 1
pH 7.3
Sterilize at 121°C/15 min

Coomassie brilliant blue agar (CBB)
0.01% (w/v) coomassie brilliant blue dye [C.I. 42655]
TSA
Sterilize at 121°C/15 min

Cysteine blood agar
0.3% (w/v) beef extract
0.1% (w/v) L-cysteine hydrochloride
20% (v/v) human blood
0.5% (w/v) sodium chloride
0.05% (w/v) yeast extract
1.5% (w/v) agar

Cytophaga agar (Ancker and Ordal, 1959)
0.05% (w/v) tryptone
0.05% (w/v) yeast extract
0.02% (w/v) sodium acetate
0.02% (w/v) beef extract
0.9% or 1.0% (w/v) agar
pH 7.2–7.4
Sterilize at 121°C for 15 min

Dorset-egg medium
75% (v/v) fresh egg mixture (whites and yolks)
0.25% (w/v) Lab-lemco powder
0.25% (w/v) peptone
0.125% (w/v) sodium chloride
Crystal violet may also be incorporated to suppress contaminants
Inspissated (~80°C for 1 hour)

Emerson agar
0.4% (w/v) beef extract
0.1% (w/v) yeast extract
0.4% (w/v) peptone
1% (w/v) dextrose
0.25% (w/v) sodium chloride
2% (w/v) agar
pH 7.0
Sterilize at 121°C/15 min

***Flavobacterium columnare* selective medium (Fijan, 1969)**
0.02% (w/v) beef extract
0.05% (w/v) tryptone
0.05% (w/v) yeast extract
0.02% (w/v) sodium acetate
0.9% (w/v) agar
pH 7.2–7.4
Sterilize at 121°C/15 min, allow to cool, and add filter sterilized antibiotic solutions:
5 μm/ml of neomycin sulphate and 10 IU/ml of polymyxin B

Kidney disease medium 2 (KDM2)
1.0% (w/v) peptone
0.05% (w/v) yeast extract
0.1% (w/v) L-cysteine hydrochloride
1.5% (w/v) agar
pH 6.5
Sterilize at 121°C for 15 min, cool to 45°C, and add 20% (v/v) sterile foetal calf serum

L-F medium
1.0% (w/v) brain heart infusion
10.0% (w/v) sucrose
0.5% (w/v) yeast extract
1.0% (w/v) agar No. 3
Sterilize at 115°C/20 min
10% (w/v) horse serum, inactivated by heating at 60°C/30 min
Cool medium and warm the horse serum to ~50°C, mix, and pour as plates

Loeffler (serum) medium
75% (v/v) horse serum
0.25% (w/v) Lab-lemco powder
0.25% (w/v) peptone
0.25% (w/v) dextrose
0.125% (w/v) sodium chloride
Distilled water to 1 l

Löwenstein–Jensen medium
0.15% (w/v) potassium hydrogen phosphate
0.15% (w/v) magnesium sulphate hydrated
0.037% (w/v) magnesium citrate
0.22% (w/v) asparagine
0.73% (v/v) glycerol
1.83% (w/v) potato starch
60.97% (v/v) fresh egg mixture (whites and yolks)
0.024% (w/v) malachite green
Inspissated (~80°C for 1 hour)

MacConkey agar
2% (w/v) peptone
1% (w/v) lactose
0.5% (w/v) bile salts
0.5% (w/v) sodium chloride
0.0075% (w/v) neutral red
1.3% (w/v) agar
pH 7.4
Sterilize at 121°C/15 min

Medium K (Mudarris and Austin, 1988)
0.5% (w/v) beef extract
0.6% (w/v) casein
0.2% (w/v) tryptone
0.1% (w/v) yeast extract
0.1% (w/v) calcium chloride
1.5% (w/v) agar
pH 7.0
Sterilize at 121°C/15 min

Middlebrook 7H10 agar
0.05% (w/v) ammonium sulphate
0.15% (w/v) potassium phosphate
0.15% (w/v) sodium phosphate
0.04% (w/v) sodium citrate
0.0025% (w/v) magnesium sulphate
0.00005% (w/v) calcium chloride hydrated
0.0001% (w/v) zinc sulphate
0.0001 (w/v) cupric sulphate
0.05% (w/v) L-glutamic acid
0.5% (v/v) glycerol
0.04% (w/v) ferric ammonium citrate
0.0001% (w/v) pyridozine
0.00005% (w/v) biotin

0.000025% (w/v) malachite green
1.5% (w/v) agar
pH 6.6
Sterilize at 121°C for 15 min, cool to 50–55°C and add supplements of either bovine
albumin–fraction V, glucose and beef catalase (to 0.5% w/v, 0.2% w/v and 0.0003%
w/v, respectively); oleic acid, bovine albumin–fraction V, glucose, beef catalase and
sodium chloride (to 0.005% v/v, 0.05% w/v, 0.2% w/v, 0.0004% w/v and 0.085% w/v,
respectively); or oleic acid, bovine albumin–fraction V, glucose, beef catalase, sodium
chloride and Triton (to 0.005% v/v, 0.05% w/v, 0.02% w/v, 0.0004% w/v, 0.085% w/v
and 0.025% w/v, respectively)

Myxobacterium selective medium (Hsu *et al.*, 1983)
0.3% (w/v) casein
0.2% (w/v) tryptone
0.05% (w/v) yeast extract
0.03% (w/v) calcium chloride
1.0% (w/v) agar
pH 7.0
Sterilize at 121°C/15 min, allow to cool to ~50°C, and add filter sterilized antibiotic
solution: 10 μm/ml of erythromycin, 10 μm/ml of neomycin sulphate or 256 IU/ml of
polymyxin sulphate (this may be replaced by colistin sulphate)

Peptone beef extract glycogen agar (PBG)
1% (w/v) beef extract
0.5% (w/v) glucose
1% (w/v) peptone
0.5% (w/v) sodium chloride
0.004% (w/v) bromothymol blue
1.5% (w/v agar)
2% (w/v agar) for overlay
Sterilize separately at 121°C/15 min. The basal medium is used for pour plates, after
which the water agar is used as an overlay

Petragnani medium
900 ml whole milk
36 g potato flour
500 g potato
1200 ml whole egg (whites and yolks)
70 ml glycerol
1.2 g malachite green
pH 7.2

***Photobcterium damselae* subsp. *piscicida* medium (Hashimoto *et al.*, 1989)**
1% (w/v) casamino acids/polypeptone
0.5% (w/v) yeast extract

0.2% (w/v) galactose
1% (w/v) sodium glutamate
0.5% (w/v) magnesium acetate
Agar may be added to solidify the medium

Pseudomonas F agar
1% (w/v) tryptone
1% (w/v) proteose peptone
0.15% (w/v) dipotassium phosphate
0.15% (w/v) magnesium sulphate
1% (w/v) agar
1% (v/v) glycerol
pH 7.0
Sterilize at 121°C/15 min

Ribose ornithine deoxycholate medium for the isolation of *Yersinia ruckeri* (Rodgers, 1992)
0.3% (w/v) yeast extract
0.1% (w/v) sodium deoxycholate
0.5% (w/v) sodium chloride
0.68% (w/v) sodium thiosulphate
0.08% (w/v) ferric ammonium citrate
0.75% (w/v) maltose
0.375% (w/v) ribose
0.375% (w/v) ornithine hydrochloride
0.1% (w/v) sodium dodecyl sulphate
0.008% (w/v) phenol red
1.25% (w/v) agar
pH 7.4
10 ml of a filtered (0.22 μm) solution containing 0.5 g sucrose/ml should be added after the basal medium has been autoclaved (121°C/15 min) and cooled at 50°C

Rimler-Shotts medium (Shotts and Rimler 1973)
0.05% (w/v) L-lysine hydrochloride
0.65% (w/v) L-ornithine hydrochloride
0.35% (w/v) maltose
0.68% (w/v) sodium thiosulphate
0.03% (w/v) L-cysteine hydrochloride
0.003% (w/v) bromothymol blue
0.08% (w/v) ferric ammonium citrate
0.1% (w/v) sodium deoxycholate
0.0005% (w/v) novobiocin
0.3% (w/v) yeast extract
0.5% (w/v) sodium chloride
1.35% (w/v) agar

pH 7.0
After boiling to dissolve the ingredients, the medium is not sterilized further

Robertson's meat broth (= cooked meat medium)
45.4% (w/v) heart muscle
1% (w/v) peptone
1% (w/v) Lab-lemco powder
0.2% (w/v) dextrose
0.5% (w/v) sodium chloride
pH 7.2
Sterilize at 121°C/15 min

Selective flexibacter medium (Bullock *et al.*, 1986)
0.2% (w/v) tryptone
0.05% (w/v) yeast extract
0.3% (w/v) gelatin
1.5% (w/v) agar
Sterilize at 121°C for 15 min, cool to 45°C and add filter sterilized neomycin sulphate
(0.0004% w/v)

Selective kidney disease medium (SKDM)
1.0% (w/v) tryptone
0.05% (w/v) yeast extract
0.005% (w/v) cycloheximide
1.0% (w/v) agar
pH 6.8
Sterilize at 121°C for 15 min, cool to 50°C, add sterile foetal calf serum to 10% (v/v),
and filter sterilized solutions of L-cysteine hydrochloride (0.1% w/v), D-cycloserine
(0.00125% w/v), polymyxin B sulphate (0.0025% w/v), and oxolinic acid (0.00025%
w/v)

Semi-defined medium for *Renibacterium salmoninarum*

Amount per litre	Ingredient	Preparation
10 g	tryptone	
10 ml	mineral salts solution	contains per 200 ml: EDTA dihydrate (disodium salt), 100 mg; $MgCl_2$, 4 g; $CaCl_2.2H_2O$, 1.4 g; $FeCl_2.6H_2O$, 100 mg; $ZnSO_4.7H_2O$, 100 mg; $MnSO_4.4H_2O$, 100 mg; $CuSo_4.5H_2O$, 50 mg; $CoCl_2.6H_2O$, 50 mg; $Na_2MoO_4.2H_2O$, 50 mg
10 ml	nitrogen compounds	contains per 100 ml: uracil, 50 mg; guanine, 50 mg; adenine, 50 mg; xanthine, 50 mg

Amount per litre	Ingredient	Preparation
1 mg each	vitamins	nicotinic acid, riboflavin, thiamine, calcium pantothenate
2 mg each	vitamins	pyridoxal HCl, pyridoxine HCl (prepared after Rogosa *et al.*, 1961)
20 ml	buffer (pH 6.8)	contains per 100 ml: K_2HPO_4, 15 g; KH_2PO_4, 15 g; sterilized at 121°C/15 min
8 ml	cysteine HCl	prepared immediately prior to addition as a solution (12.5% [w/v]) in N NaOH to give a pH of 6.8; filter sterilized
2 ml	glucose	prepared as a solution (50% [w/v]) in distilled water; sterilized at 115°C/20 min

Note: Tryptone, mineral salts solution and nitrogen compounds are added to distilled water, the pH adjusted to 6.8 with N NaOH, and the medium dispensed in 190 ml amounts in 800 ml capacity Erlenmeyer flasks. Sterilization is at 121°C/15 min. After cooling to ~50°C, the pre-sterilized buffer, glucose, cysteine HCl and vitamin solutions are added to a final volume of 200 ml. Agar to 1% (w/v) may be incorporated, as necessary.

Shotts and Waltman medium for the isolation of *Edwardsiella ictaluri* (Shotts and Waltman, 1990)
1% (w/v) tryptone
1% (w/v) yeast extract
0.125% (w/v) phenylalanine
0.12% (w/v) ferric ammonium citrate
0.0003% (w/v) bromothymol blue
0.1% (w/v) bile salts
1.5% (w/v) agar
980 ml of distilled water
Dissolve by boiling, cool to 50°C and adjust pH to 7.0, sterilise at 121°C for 15 min, cool to 50°C and add mannitol (filter sterilized) to 0.35% (v/v) and colistin sulphate to 10 μg/ml

Skimmed milk agar
0.05% (w/v) yeast extract
1.0% (w/v) skimmed milk powder
1.0% (w/v) agar No. 3
pH 7.2
Sterilize at 115°C/15 min

TCY medium (Hikida *et al.*, 1979)
0.1% (w/v) casamino acids

0.1% (w/v) tryptone
0.02% (w/v) yeast extract
0.1% (w/v) calcium chloride
1.08% (w/v) magnesium chloride
0.07% (w/v) potassium chloride
3.13% (w/v) sodium chloride
pH 7.0–7.2

Thioglycollate broth
0.05% (w/v) L-cystine
0.25% (w/v) sodium chloride
0.5% (w/v) dextrose
0.5% (w/v) yeast extract
1.5% (w/v) pancreatic digest of casein
0.05% (w/v) sodium thioglycollate
pH 7.1
Sterilize at 121°C/15 min

Thiosulphate citrate bile salt sucrose agar (TCBS)
0.5% (w/v) yeast extract
1.0% (w/v) peptone
1.0% (w/v) sodium thiosulphate
1.0% (w/v) sodium citrate
0.8% (w/v) ox bile
2.0% (w/v) sucrose
1.0% (w/v) sodium chloride
0.1% (w/v) ferric citrate
0.004% (w/v) bromothymol blue
0.004% (w/v) thymol blue
1.4% (w/v) agar
pH 8.6
After boiling to dissolve the ingredients the medium will not require further sterilization

Todd–Hewitt broth
1% (w/v) infusion from 450 g of fat-free minced beef
2% (w/v) tryptone
0.2% (w/v) dextrose
0.2% (w/v) sodium bicarbonate
0.2% (w/v) sodium chloride
0.04% (w/v) disodium phosphate, anhydrous
pH 7.8
Sterilize at 115°C/15 min

***Vibrio anguillarum* medium (VAM)**
1.5 (w/v) sorbitol

0.4% (w/v) yeast extract
0.5% (w/v) bile salts
3.5% (w/v) sodium chloride
0.001% (w/v) ampicillin
0.004% (w/v) cresol red
0.004% (w/v) bromothymol blue
1.5% (w/v) agar
distilled water to 1 l
pH 8.6
Heat to dissolve, do not autoclave

Waltman–Shotts medium for the isolation of *Yersinia ruckeri* (Waltman and Shotts, 1984)
0.2% (w/v) tryptone
0.2% (w/v) yeast extract
1.0% (v/v) Tween 80
0.5% (w/v) sodium chloride
0.01% (w/v) calcium chloride hydrated
0.0003% (w/v) bromothymol blue
1.5% (w/v) agar
pH 7.4
Sterilize at 121°C for 15 min, cool to 50°C and add filter sterilize sucrose to 0.5% (w/v)

Yeast extract glucose agar (Michel *et al.*, 1997)
2.5% (w/v) nutrient broth No. 2 [Oxoid]
0.3% (w/v) yeast extract
0.5% (w/v) glucose
1.5% (w/v) agar
pH 6.8

***Yersinia ruckeri* selective medium (Furones *et al.*, 1993)**
TSA
1% (w/v) SDS
100 μm/ml of coomassie brilliant blue
100 μm/ml of congo red

6

Diagnosis

Historically, scientists have seemed loath to make rapid diagnoses, preferring to adopt laborious testing regimes. Why do scientists bother to identify the precise cause of a disease, when the information is often not useful for control purposes? Yet, there have been dramatic improvements in diagnostic improvements, encompassing recent developments in molecular biology and serology.

Diagnosis may often appear to be as much art as science, with a multitude of preferred methods adorning the notebooks of most diagnosticians. Diagnosis is sometimes achieved on purely histological material, with little effort made to isolate the pathogens. This is especially true of some of the acid-fast bacterial fish pathogens. When isolation of the pathogen is attempted, it is a common fault that diagnosis proceeds with emphasis on old-fashioned biochemical tests designed originally for bacteria important to human medicine. Consequently, a superficial glance at many laboratories would suggest that diagnostic techniques for fish pathogens need to be brought up to date. The astute diagnostician synthesizes all available information before reaching a judgement. Useful information includes:

- Gross clinical signs of disease on individual fish.
- Internal abnormalities apparent during post-mortem examination.
- Histopathological examination of diseased tissues.
- Bacteriological examination of tissues (this requires special dexterity to avoid contamination by the normal bacterial microflora present on the surface and in the intestinal tract of fish, and in water; special contamination problems may be encountered with the examination of small fish, such as fry).

A. GROSS CLINICAL SIGNS OF DISEASE

The appearance of every sick fish tells a story, which fits into the proverbial jigsaw pattern of disease diagnosis. Good observation uncovers many useful clues. To an extent, the external appearance of the animal may be ignored by the eager diagnostician in the rush to attack the specimen with scalpel and swabs. Formalin and bacteriological media may be the order of the day. Of course, the same is not true elsewhere in veterinary and human medicine where diagnosis is often achieved by apparently cursory glances at the wretched individual.

Fish may display many behavioural and physical changes, some of which give valuable clues as to the nature of the disease (Table 6.1). It should be emphasized, however, that many symptoms are common to a multitude of bacterial diseases. Consequently, in the following discussion symptoms have been categorized (Fig. 6.1) in clear groups rather than splitting them into a plethora of esoteric detail. Thus, many external signs of disease are recognized which are pertinent to the bacterial fish pathogens (Table 6.1).

- **Sluggish behaviour**
  With some infections, notably those caused by *Bacillus* sp., *Cl. botulinum*, *Cor. aquaticum*, *Fla. johnsoniae*, *Haf. alvei*, *J. lividum*, *Mycobacterium* spp., *Myc. abscessus*, *Nocardia* spp., *Piscirickettsia salmonis*, *Shewanella putrefaciens*, *Str. difficilis*, *Str. iniae*, *V. anguillarum*. *V. ordalii* and *Y. intermedia*, the fish may become very inactive, and often cease feeding. Flatfish, for example turbot, lie listlessly on the bottom of the tank, whereas salmonids float just below the surface of the water. In this state, death may quickly follow.
- **Twirling, spiral or erratic movement**
  This is often indicative of neurological damage. The offending pathogens include *Cit. freundii*, *Cl. botulinum*, *Edw. ictaluri*, *Eu. tarantellae*, *Lactococcus garvieae*, *Str. difficilis*, *Str. iniae* and *V. tracturi*.
- **Faded pigment**
  With this condition, attributed to a reduction in melanin content, the fish become very pale. This may occur with infections caused by *Edw. tarda*, *Mycobacterium,* spp. and *Nocardia* spp.
- **Darkened pigment/melanosis**
  An enhancement of pigment results from infection with *Bac. mycoides*, *Fla. psychrophilum*, *Haf. alvei*, *J. lividum*, *Lactococcus garvieae*, *Lactococcus piscium*, *Micrococcus luteus*, *Piscirickettsia salmonis*, *Rhodococcus* sp., *Streptococcus* spp., *V. alginolyticus*, *V. trachuri*, *Y. intermedia* and *Y. ruckeri*.
- **Eye damage—exophthalmia ('pop-eye')/corneal opacity/rupture**
  The presence of eye damage, i.e. bulging eyes and corneal opacity possibly leading to corneal rupture, occurs with infections by *Aer. caviae*, *Aer. hydrophila*, *Cor. aquaticum*, *Edw. ictaluri*, *Haf. alvei*, *J. lividum*, *Lactococcus garvieae*, *Lactococcus piscium*, *Micrococcus luteus*, *Mycobacterium* spp., *Myc. neoaurum*, *Nocardia* spp., *Ren. salmoninarum*, *Rhodococcus* sp., *Sal. choleraesuis* subsp. *arizonae,* *Shewanella putrefaciens*, *Sta. aureus*, *Sta. epidermidis*, *Streptococcus* spp., *Str. parauberis*, *Vag. salmoninarum*, *V. alginolyticus*, *V. harveyi*, *V. trachuri* and *Y. ruckeri*.
- **Haemorrhaging in the eye**
  This is characterized by the presence of blood spots in the eye. The causal agents include *Ent. agglomerans*, *Haf. alvei*, *Lactococcus garvieae*, *Ps. anguilliseptica*,

Streptococcus spp. *Str. parauberis* (with pus in the eye) and *Y. ruckeri* ('salmonid blood spot')

- **Haemorrhaging in the mouth**
  Essentially blood spots become apparent on the roof of the mouth—a phenomenon attributed to infection by *Ps. anguilliseptica, Shewanella putrefaciens, V. anguillarum, V. ordalii* and *Y. ruckeri.*

- **Erosion of the jaws/mouth**
  This condition occurs principally with infections by *Fla. johnsoniae, Fle. maritimus* and *Y. ruckeri.*

- **Haemorrhaging in the opercula region/gills**
  This occurs with infections of *Chrys. scophthalmum, Lactococcus garvieae, Ps. anguilliseptica, Sta. epidermidis, Streptococcus* spp. and *Vag. salmoninarum.*

- **Gill damage**
  This includes the presence of swollen gill lamellae, anaemia/pale gills, clubbing and gill rot, which are symptoms of gill disease, caused by *Bac. cereus, Bac. mycoides, Chrys. scophthalmum, Esch. vulneris, Fla. branchiophilum, Fla. columnare, Fla. hydatis, J. lividum, Micrococcus luteus, Mycoplasma mobile, Myxococcus piscicola, Piscirickettsia salmonis, Planococcus* sp., *V. alginolyticus, V. anguillarum,* and *V. ordalii.*

- **White nodules on the gills/skin**
  This is one of the characteristics of infections by *Edw. tarda,* and *V. fischeri.*

- **White spots on the head**
  This occurs with infection by *Planococcus* sp.

- **Fin rot/damage**
  The presence of badly damaged (rotted) fins may be a sign of infection by *Aer. hydrophila, Klebsiella pneumoniae, Mycobacterium* spp., *Nocardia* spp., *Ps. fluorescens, Shewanella putrefaciens* and *V. pelagius.*

- **Haemorrhaging at the base of fins**
  The presence of clusters of tiny haemorrhagic lesions at the base of the fins is associated with disease caused by *Aer. hydrophila, Lactococcus garvieae, Ps. fluorescens, Streptococcus* spp., *V. anguillarum, V. ordalii, V. pelagius, V splendidus* and *Y. ruckeri.*

- **Haemorrhaging on the fins**
  This occurs with infections caused by *Sta. epidermidis* and *Str. parauberis.*

- **Tail rot/erosion**
  This occurs with infection by *Fle. maritimus, Klebsiella pneumoniae, Mycobacterium* spp., *Nocardia* spp., *Ps. fluorescens* and *Y. intermedia.*

- **Saddle-like lesions on the dorsal surface (columnaris, saddleback disease)**
  This condition develops with infections by *Fla. columnare, Fla. psychrophilum* and *Sporocytophaga* sp.

- **Distended abdomen**
  This condition, where the abdomen is often filled with ascitic fluid, should not be confused with overfeeding. The presence of a distended abdomen is a sign of many diseases, including those caused by *Aer. hydrophila, Aer. salmonicida, Car. piscicola, Chrys. scophthalmum, Edw. ictaluri, Edw. tarda, J. lividum, Lactobacillus* spp.,

Table 6.1. External signs of disease associated with the bacterial fish pathogens

Pathogen	Disease signs																														
	1	2	3	4	5	6	7	8	9	10	11	12	13	14	15	16	17	18	19	20	21	22	23	24	25	26	27	28	29	30	31
Acinetobacter spp.	–	–	–	–	–	–	–	–	–	–	–	–	–	–	–	–	–	–	–	–	+	–	–	–	–	–	–	–	–	–	–
Aer. caviae	–	–	–	–	+	–	–	–	–	–	–	–	–	–	–	–	–	–	–	–	–	–	–	–	–	–	–	–	–	–	–
Aer. hydrophila	–	–	–	–	+	–	–	–	–	–	–	–	+	+	–	–	–	+	+	–	–	+	–	–	–	–	–	+	–	–	–
Aer. salmonicida	–	–	–	–	–	–	–	–	–	–	–	–	–	–	–	–	–	+	+	–	+	–	+	–	–	–	–	+	–	–	–
Bacillus spp.	+	–	–	–	–	–	–	–	–	–	–	–	–	–	–	–	–	–	–	+	–	–	–	–	–	–	+	–	–	–	–
Bac. cereus	–	–	–	–	–	–	–	–	–	+	–	–	–	–	–	–	–	–	–	–	–	–	–	–	–	–	–	–	–	–	–
Bac. mycoides	–	–	–	+	–	–	–	–	–	+	–	–	–	–	–	–	–	–	–	–	–	+	–	–	–	–	–	+	–	–	–
Car. piscicola	–	–	–	–	–	–	–	–	–	–	–	–	–	–	–	–	–	+	+	–	–	+	–	+	–	–	–	–	–	–	–
Chrys. scophthalmum	–	–	–	–	–	–	–	–	+	+	–	–	–	–	–	–	–	+	+	–	–	–	–	–	–	–	–	–	–	–	–
Cit. freundii	–	+	–	–	–	–	–	–	–	–	–	–	–	–	–	–	–	–	+	–	+	–	–	–	–	–	–	+	–	–	–
Cl. botulinum	+	+	–	–	–	–	–	–	–	–	–	–	–	–	–	–	–	–	–	–	–	–	–	–	–	–	–	–	–	–	–
Cor. aquaticum	+	–	–	–	+	–	–	–	–	–	–	–	–	–	–	–	–	–	–	–	–	–	–	–	–	–	–	+	–	–	–
Edw. ictaluri	–	+	–	–	+	–	–	–	–	–	–	–	–	–	–	–	–	+	+	–	–	–	–	–	–	–	–	–	–	–	–
Edw. tarda	–	–	+	–	–	–	–	–	–	–	+	–	–	–	–	–	–	+	–	–	–	+	–	+	–	–	–	–	–	–	–
En. faecalis subsp. *liquefaciens*	–	–	–	–	–	–	–	–	–	–	–	–	–	–	–	–	–	–	–	–	+	–	–	–	–	+	–	–	–	–	–
Ent. agglomerans	–	–	–	–	–	+	–	–	–	–	–	–	–	–	–	–	–	–	+	–	–	–	–	–	–	–	–	–	–	–	–
Esch. vulneris	–	–	–	–	–	–	–	–	–	+	–	–	–	–	–	–	–	–	+	–	–	–	–	–	–	–	–	–	–	–	–
Eu. tarantellae	–	+	–	–	–	–	–	–	–	–	–	–	–	–	–	–	–	–	–	–	–	–	–	–	–	–	–	–	–	–	–
Fla. branchiophilum	–	–	–	–	–	–	–	–	–	+	–	–	–	–	–	–	–	–	–	–	–	–	–	–	–	–	–	–	–	–	–
Fla. columnare	–	–	–	–	–	–	–	–	–	+	–	–	–	–	–	–	+	–	+	–	–	–	–	–	–	–	–	–	–	–	–
Fla. hydatis	–	–	–	+	–	–	–	–	–	+	–	–	–	–	–	–	–	–	–	–	–	–	–	–	–	–	–	–	–	–	–
Fla. johnsoniae	+	–	–	–	–	–	–	+	–	–	–	–	–	–	–	–	–	–	–	–	–	–	–	–	–	–	+	–	–	+	–
Fla. psychrophilum	–	–	–	+	–	–	–	–	–	–	–	–	–	–	–	–	+	–	–	–	–	–	–	–	–	–	–	–	–	–	–
Fle. maritimus	–	–	–	–	–	–	–	–	+	–	–	–	–	–	–	+	–	–	–	–	–	–	–	–	–	–	–	–	–	–	–
Haf. alvei	+	–	–	+	+	+	–	–	–	–	–	–	–	–	–	–	–	+	–	–	–	–	–	–	–	–	–	–	–	–	–
J. lividum	+	–	–	+	+	–	–	–	–	+	–	–	–	–	–	–	+	–	–	–	–	–	–	–	–	–	–	–	–	+	–
Klebsiella pneumoniae	–	–	–	–	–	–	–	–	–	–	–	–	+	–	–	+	–	–	–	–	–	–	–	–	–	–	–	–	–	–	–

Table 6.1. (*Continued*)

| | Disease signs | | | | | | | | | | | | | | | | | | | | | | | | | | | | | | |
Pathogen	1	2	3	4	5	6	7	8	9	10	11	12	13	14	15	16	17	18	19	20	21	22	23	24	25	26	27	28	29	30	31
Lactococcus garvieae	–	+	–	+	+	+	–	–	+	–	–	–	–	+	–	–	–	+	–	–	–	–	–	–	–	–	–	–	–	–	–
Lactococcus piscium	–	–	–	+	+	–	–	–	–	–	–	–	–	–	–	–	–	+	+	–	–	+	–	+	–	–	–	–	–	–	–
Micrococcus luteus	–	–	–	+	+	–	–	–	–	+	–	–	–	–	–	–	–	+	–	–	–	–	–	–	–	–	–	–	–	–	–
Moraxella spp.	–	–	–	–	–	–	–	–	–	–	–	–	–	–	–	–	–	–	+	–	–	–	–	–	–	–	–	–	–	–	–
Mycobacterium spp.	+	–	+	–	+	–	–	–	–	–	–	–	+	–	–	+	–	+	–	+	–	–	–	–	–	–	+	–	–	–	–
Myc. abscessus	+	–	–	–	–	–	–	–	–	–	–	–	–	–	–	–	–	+	–	–	–	–	–	–	–	–	–	+	–	–	–
Myc. marinum	–	–	–	–	–	–	–	–	–	–	–	–	–	–	–	–	–	–	–	–	–	–	–	–	–	–	–	–	+	–	–
Myc. neoaurum	–	–	–	–	+	–	–	–	–	–	–	–	–	–	–	–	–	–	–	–	–	–	–	–	–	–	–	–	–	–	–
Mycoplasma mobile	–	–	–	–	–	–	–	–	–	+	–	–	–	–	–	–	–	–	–	–	–	–	–	–	–	–	–	–	–	–	–
Myxococcus piscicola	–	–	–	–	–	–	–	–	–	+	–	–	–	–	–	–	–	–	–	–	–	–	–	–	–	–	–	–	–	–	–
Nocardia spp.	+	–	+	–	+	–	–	–	–	–	–	–	+	–	–	+	–	+	–	+	–	–	–	–	–	–	+	–	–	–	–
Ph. damselae	–	–	–	–	–	–	–	–	–	–	–	–	–	–	–	–	–	–	+	–	–	–	–	–	–	–	–	–	–	–	–
Piscirickettsia salmonis	+	–	–	+	–	–	–	–	–	+	–	–	–	–	–	–	–	–	–	–	–	–	–	–	–	–	–	+	–	–	–
Planococcus sp.	–	–	–	–	–	–	–	–	–	+	–	+	–	–	–	–	–	–	–	–	–	–	–	–	–	–	–	–	–	–	–
Plesiomonas shigelloides	–	–	–	–	–	–	–	–	–	–	–	–	–	–	–	–	–	–	–	–	–	–	–	–	+	+	–	+	–	–	–
Pr. rettgeri	–	–	–	–	–	–	–	–	–	–	–	–	–	–	–	–	–	–	+	–	–	–	–	–	–	–	–	–	–	–	–
Ps. anguilliseptica	–	–	–	–	–	+	+	–	+	–	–	–	–	–	–	–	–	–	+	–	–	–	–	–	–	–	–	–	–	–	–
Ps. chlororaphis	–	–	–	–	–	–	–	–	–	–	–	–	–	–	–	–	–	+	+	–	–	–	–	–	–	–	–	–	–	–	–
Ps. fluorescens	–	–	–	–	–	–	–	–	–	–	–	–	+	+	–	+	–	+	+	–	–	–	–	–	–	–	–	–	–	–	–
Ps. pseudoalcaligenes	–	–	–	–	–	–	–	–	–	–	–	–	–	–	–	–	–	–	–	–	–	–	–	–	–	–	–	–	–	+	–
Ren. salmoninarum	–	–	–	–	+	–	–	–	–	–	–	–	–	–	–	–	–	+	–	+	+	–	+	–	–	–	–	–	–	–	–
Rhodococcus sp.	–	–	–	+	+	–	–	–	–	–	–	–	–	–	–	–	–	–	–	–	–	–	–	–	–	–	–	–	–	–	–
Sal. choleraesuis subsp. *arizonae*	–	–	–	–	+	–	–	–	–	–	–	–	–	–	–	–	–	–	–	–	–	–	–	–	–	–	–	–	–	–	–
Ser. plymuthica	–	–	–	–	–	–	–	–	–	–	–	–	–	–	–	–	–	–	–	–	–	–	–	–	–	–	–	–	–	+	–
Shewanella putrefaciens	+	–	–	–	+	–	+	–	–	–	–	–	+	–	–	–	–	–	+	–	–	–	–	–	–	–	–	–	–	–	–
Sporocytophaga sp.	–	–	–	–	–	–	–	–	–	–	–	–	–	–	–	–	+	–	–	–	–	–	–	–	–	–	–	–	–	–	–
Sta. aureus	–	–	–	–	+	–	–	–	–	–	–	–	–	–	–	–	–	–	–	–	–	–	–	–	–	–	–	–	–	–	–

Table 6.1. (*Continued*)

Pathogen	Disease signs																														
	1	2	3	4	5	6	7	8	9	10	11	12	13	14	15	16	17	18	19	20	21	22	23	24	25	26	27	28	29	30	31
Sta. epidermidis	–	–	–	–	+	–	–	–	+	–	–	–	–	–	+	–	–	–	–	–	+	–	–	–	–	–	–	–	–	–	–
Str. difficilis	+	+	–	–	–	–	–	–	–	–	–	–	–	–	–	–	–	–	–	–	–	–	–	–	–	–	–	–	–	–	–
Str. iniae	+	+	–	–	–	–	–	–	–	–	–	–	–	–	–	–	–	–	–	–	–	–	–	–	–	–	–	–	–	–	–
Str. milleri	–	–	–	–	–	–	–	–	–	–	–	–	–	–	–	–	–	–	+	–	–	–	–	–	–	–	–	–	–	–	–
Str. parauberis	–	–	–	–	+	+	–	–	–	–	–	–	–	–	+	–	–	–	+	–	–	–	–	–	–	–	+	–	–	–	–
Vag. salmoninarum	–	–	–	–	+	–	–	–	+	–	–	–	–	–	–	–	–	+	+	–	+	–	+	–	–	–	–	–	–	–	–
Moritella marina	–	–	–	–	–	–	–	–	–	–	–	–	–	–	–	–	–	–	–	+	–	–	–	–	–	–	–	–	–	–	–
V. alginolyticus	–	–	–	+	+	–	–	–	–	+	–	–	–	–	–	–	–	–	–	+	–	–	–	–	–	–	–	–	–	–	–
V. anguillarum	+	–	–	–	–	–	+	–	–	+	–	–	–	+	–	–	–	–	–	–	–	–	–	+	–	–	–	–	–	–	–
V. cholerae	–	–	–	–	–	–	–	–	–	–	–	–	–	–	–	–	–	–	+	–	–	–	–	–	–	–	–	–	–	–	–
V. fischeri	–	–	–	–	–	–	–	–	–	+	–	–	–	–	–	–	–	–	–	+	–	–	–	–	–	–	–	–	–	–	–
V. harveyi	–	–	–	–	+	–	–	–	–	–	–	–	–	–	–	–	–	–	–	+	–	–	–	–	–	–	–	–	–	–	–
V. logei	–	–	–	–	–	–	–	–	–	–	–	–	–	–	–	–	–	–	–	+	–	–	–	–	–	–	–	–	–	–	–
V. ordalii	+	–	–	–	–	–	+	–	–	+	–	–	–	+	–	–	–	–	–	–	–	–	–	+	–	–	–	–	–	–	–
V. pelagius	–	–	–	–	–	–	–	–	–	–	–	+	+	–	–	–	–	–	–	–	–	–	–	–	–	–	–	–	–	–	–
V. salmonicida	–	–	–	–	–	–	–	–	–	–	–	–	–	–	–	–	–	–	+	–	–	–	–	–	–	–	–	–	–	–	–
V. splendidus	–	–	–	–	–	–	–	–	–	–	–	–	–	+	–	–	–	+	–	–	–	–	–	+	–	–	–	–	–	–	–
V. trachuri	–	+	–	+	+	–	–	–	–	–	–	–	–	–	–	–	–	–	–	–	–	–	–	–	–	–	–	–	–	–	–
'V. viscosus'	–	–	–	–	–	–	–	–	–	–	–	–	–	–	–	–	–	–	–	+	–	–	–	–	–	–	–	–	–	–	–
V. vulnificus	–	–	–	–	–	–	–	–	–	–	–	–	–	–	–	–	–	–	+	–	–	–	–	–	–	–	–	–	–	–	–
Y. intermedia	+	–	–	+	–	–	–	–	–	–	–	–	–	–	–	+	–	–	+	–	–	–	–	–	–	–	–	–	–	–	–
Y. ruckeri	–	–	–	+	+	+	+	+	–	–	–	–	–	+	–	–	–	–	–	–	–	–	–	–	–	–	–	–	–	–	–

1 = sluggish behaviour, 2 = twirling, spiral or erratic movement, 3 = faded pigment, 4 = darkened pigment, 5 = eye damage/exophthalmia, 6 = haemorrhaging in the eye, 7 = haemorrhaging in the mouth, 8 = erosion of the jaws/mouth, 9 = haemorrhaging in the opercular area, 10 = gill damage, 11 = white nodules on the gills/skin, 12 = white spots on the head, 13 = fin rot, 14 = haemorrhaging at base of fins, 15 = haemorrhaging on the fins, 16 = tail rot, 17 = saddle like lesions on dorsal surface, 18 = distended abdomen, 19 = haemorrhaging on the surface/muscle, 20 = necrotizing dermatitis, 21 = ulcers, 22 = external abscesses, 23 = furuncles, 24 = blood-filled blisters on flank, 25 = protruded anus, 26 haemorrhaging around the vent, 27 = emaciation, 28 = inappetence, 29 = stunted growth, 30 = sloughing of skin, 31 = dorsal rigidity.

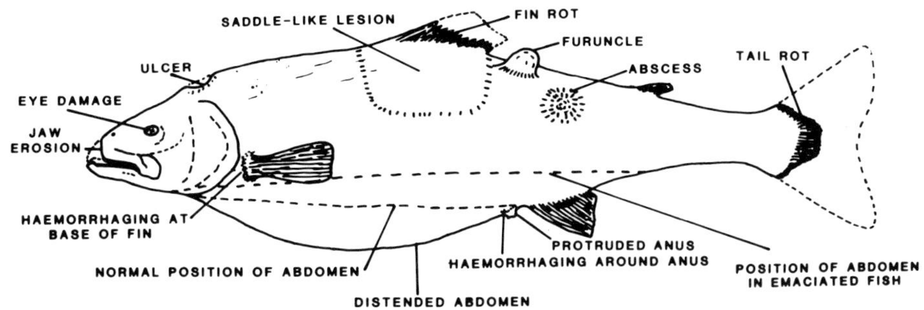

Fig. 6.1. Diagrammatic representation of the possible external disease signs on a fish.

Lactococcus garvieae, *Lactococcus piscium*, *Micrococcus luteus*, *Myc. abscessus*, *Ps. chlororaphis*, *Ps. fluorescens*, *Ren. salmoninarum*, *Streptococcus* spp., *Vag. salmoninarum* and *V. splendidus*.

- **Haemorrhaging on the surface and in the muscle**
  This is a common condition with infections by *Aer. caviae*, *Aer. hydrophila*, *Aer. salmonicida*, *Car. piscicola*, *Chrys. scophthalmum*, *Cit. freundii*, *Esch. vulneris*, *Edw. ictaluri*, *Ent. agglomerans*, *Fla. columnare*, *Haf. alvei*, *Lactobacillus* spp., *Lactococcus piscium*, *Moraxella* sp., *Mycobacterium* spp., *Nocardia* spp., *Ps. anguilliseptica*, *Ps. chlororaphis*, *Ps. fluorescens*, *Shewanella putrefaciens*, *Str. parauberis*, *Vag. salmoninarum*, *V. cholerae*, *V. salmonicida*, *V. vulnificus* and *Y. intermedia*.
- **Necrotizing dermatitis**
  This condition has been associated with infection by *Bacillus* spp.
- **Ulcers**
  The presence of ulcers (or surface erosion) on fish occurs with infections caused by *Acinetobacter* sp., *Aer. salmonicida*, *Bac. mycoides*, *Cit. freundii*, *En. faecalis* subsp. *liquefaciens*, *Moritella marina*, *Mycobacterium* spp., *Nocardia* spp., *Ph. damselae*, *Pr. rettgeri*, *Ren. salmoninarum*, *Sta. epidermidis*, *Str. milleri*, *V. alginolyticus*, *V. fischeri*, *V. harveyi*, *V. logei* and '*V. viscosus*'.
- **External abscesses**
  These may be present with infections by *Aer. hydrophila*, *Car. piscicola*, *Edw. tarda*, *Lactobacillus* spp., *Lactococcus piscium*, *Ren. salmoninarum* and *Vag. salmoninarum*.
- **Furuncles (or boils)**
  These are characteristic of infections by *Aer. salmonicida*.
- **Blood-filled blisters on the flank**
  These have been described in infections by *Car. piscicola*, *Lactobacillus* spp., *Lactococcus piscium*, *Ren. salmoninarum* and *Vag. salmoninarum*.
- **Protruded anus/vent**
  This is symptomatic of infection by *Edw. tarda* and *Plesiomonas shigelloides*.
- **Haemorrhaging around the vent**
  This occurs in infections, caused by *En. faecalis* subsp. *liquefaciens*, *Plesiomonas shigelloides*, *V. anguillarum*, *V. ordalii* and *V. splendidus*.

- **Emaciation** (this should not be confused with starvation)
  This has been described for mycobacteriosis and nocardiosis, and infections by *Bacillus* spp., *Fla. johnsoniae* and *Str. parauberis*.
- **Inappetence**
  This is a common sign of many infections caused by bacterial fish pathogens. Nevertheless, loss of appetite is particularly associated with diseases caused by *Aer. hydrophila*, *Aer. salmonicida*, *Bac. mycoides*, *Cit. freundii*, *Cor. aquaticum*, *Myc. abscessus*, *Piscirickettsia salmonis* and *Plesiomonas shigelloides*.
- **Stunted growth**
  This condition has been reported with *Myc. marinum*.
- **Sloughing off of skin/external surface lesions**
  This condition has occasionally been reported in infections attributed to *Fla. johnsoniae*, *J. lividum*, *Ps. pseudoalcaligenes* and *Ser. plymuthica*.
- **Dorsal rigidity**
  This condition has been reported with infections attributed to *Str. difficilis* and *Str. iniae*.

B. INTERNAL ABNORMALITIES APPARENT DURING POST-MORTEM EXAMINATION

Careful internal observation of diseased fish may reveal the presence of clearly discernible abnormalities (Table 6.2).

- **Skeletal deformities**
  These have been noted with infection by *Myc. neoaurum*.
- **Gas-filled hollows in the muscle**
  These are generally evil-smelling, and characteristic of *Edw. tarda* infections.
- **Opaqueness in the muscle**
  This condition has been described with infections by *Bac. mycoides*.
- **Ascitic fluid in the abdominal cavity**
  This may be responsible for abdominal swelling, a trait common of many bacterial diseases including those caused by *Aer. hydrophila*, *Aer. salmonicida*, *Bacillus* spp., *Car. piscicola*, *Edw. ictaluri*, *Edw. tarda*, *Flavobacterium* sp., *J. lividum*, *Lactobacillus* spp., *Lactococcus garvieae*, *Lactococcus piscium*, *Micrococcus luteus*, *Planococcus* sp., *Ps. chlororaphis*, *Ps. fluorescens*, *Ps. pseudoalcaligenes*, *Ren. salmoninarum*, *Ser. liquefaciens*, *Ser. plymuthica*, *Sta. epidermidis*, *Streptococcus* spp., *Str. parauberis*, *Vag. salmoninarum* and *V. alginolyticus*.
- **Peritonitis**
  Peritonitis is a feature of infections by *Myc. neoarum* and *Vag. salmoninarum*.
- **Petechial (pin-prick) haemorrhages on the muscle wall**
  These are common to many diseases, including those caused by *Acinetobacter* sp., *Aer. hydrophila*, *Aer. salmonicida*, *Edw. ictaluri*, *Lactobacillus* spp., *Lactococcus piscium*, *Piscirickettsia salmonis*, *Ps. fluorescens*, *Ren. salmoninarum*, *Vag. salmoninarum*, *V. anguillarum*, *V. ordalii* and *Y. ruckeri*.

- **Haemorrhaging in the air bladder**
  This has been described for *Acinetobacter*, *Moraxella* and *Piscirickettsia salmonis* infections.
- **Liquid in the air bladder**
  This has been associated with 'vibriosis', caused by *V. alginolyticus*.
- **White nodules (granulomas) on/in the internal organs**
  These have been reported in some diseases, notably those caused by *Cit. freundii*, *Edw. tarda*, *Fla. piscicida*, *Mycobacterium* spp., *Myc. neoaurum*, *Nocardia* spp., *Ph. damselae* subsp. *piscicida*, *Ser. liquefaciens* and the PLO.
- **Yellowish nodules on the internal organs**
  These have been noted with infections by *Ser. liquefaciens*.
- **Nodules in the muscle**
  The presence of nodules has been reported with infections caused by *Myc. neoaurum*.
- **Swollen and/or watery kidney**
  This occurs principally in infections caused by *Aer. caviae*, *Car. piscicola*, *Edw. ictaluri*, *J. lividum*, *Lactococcus piscium*, *Micrococcus luteus*, *Piscirickettsia salmonis*, *Planococcus* sp., *Ren. salmoninarum*, *Ser. liquefaciens* and *Str. parauberis*.
- **False membrane over the heart and/or kidney**
  This has been attributed only to bacterial kidney disease, which is caused by *Ren. salmoninarum*.
- **Haemorrhaging/bloody exudate in the peritoneum**
  This may occur with *Fla. piscicida*, *Lactococcus garvieae*, *Ps. anguilliseptica*, *Ps. fluorescens*, *Sal. choleraesuis* subsp. *arizonae*, *V. alginolyticus*, *V. anguillarum*, *V. ordalii* and *V. splendidus* infections.
- **Swollen intestine, possibly containing yellow or bloody fluid/gastro-enteritis**
  This develops sometimes in infections attributed to *Aer. caviae*, *Cit. freundii*, *Esch. vulneris*, *Lactococcus garvieae*, *Piscirickettsia salmonis*, *Str. parauberis*, *V. alginolyticus*, *V. harveyi*, *V. pelagius*, *V. splendidus* and *Y. ruckeri*.
- **Intestinal necrosis and opaqueness**
  This condition has been documented with *V. ichthyoenteri*.
- **Hyperaemic stomach**
  This develops in infections by *Bacillus* spp.
- **Haemorrhaging in/on the internal organs**
  This is common to most bacterial fish diseases, including those caused by *Acinetobacter* sp., *Aer. hydrophila*, *Aer. salmonicida*, *Bacillus* spp., *Car. piscicola*, *Chrys. scophthalmum*, *Cor. aquaticum*, *Edw. ictaluri*, *Edw. tarda*, *Esch. vulneris*, *Flavobacterium* sp., *Haf. alvei*, *Lactobacillus* spp., *Lactococcus garvieae*, *Lactococcus piscium*, *Ps. anguilliseptica*, *Ps. fluorescens*, *Ps. pseudoalcaligenes*, *Ren. salmoninarum*, *Ser. plymuthica*, *Streptococcus* spp., *V. alginolyticus*, *V. anguillarum*, *V. cholerae*, *V. ordalii*, *V. pelagius*, *V. salmonicida*, *V. trachuri* and *Y. ruckeri*.
- **Brain damage**
  In particular, brain damage has been associated with diseases caused by *Cor. aquaticum*, *Ren. salmoninarum*, *Str. difficilis* and *Str. iniae*.
- **Blood in the cranium**
  This has been reported with infection by *Cor. aquaticum*.

Table 6.2. Internal signs of disease

Taxon	Disease																								
	1	2	3	4	5	6	7	8	9	10	11	12	13	14	15	16	17	18	19	20	21	22	23	24	25
Acinetobacter sp.	–	–	–	–	–	–	+	+	–	–	–	–	–	–	–	–	–	+	–	–	–	–	–	–	–
Aer. caviae	–	–	–	–	–	–	–	–	–	–	–	–	+	–	+	–	–	–	–	–	+	–	+	–	–
Aer. hydrophila	–	–	–	–	+	–	+	–	–	–	–	–	–	–	–	–	–	+	–	–	–	–	–	–	–
Aer. salmonicida	–	–	–	–	+	–	+	–	–	–	–	–	–	–	–	–	–	+	–	–	–	–	–	–	–
Bacillus spp.	–	–	+	–	+	–	–	–	–	–	–	–	–	–	–	–	–	+	–	+	+	–	–	–	–
Car. piscicola	–	–	–	–	+	–	–	–	–	–	–	–	+	–	–	–	–	+	–	–	–	–	–	–	–
Chrys. scophthalmum	–	–	–	–	–	–	–	–	–	–	–	–	–	–	–	–	–	+	–	–	–	–	–	–	–
Cit. freundii	–	–	–	–	–	–	–	–	–	+	–	–	–	–	+	–	–	–	–	–	–	–	+	–	–
Cor. aquaticum	–	–	–	–	–	–	–	–	–	–	–	–	–	–	–	+	+	+	–	–	–	–	–	–	–
Edw. ictaluri	–	–	–	–	+	–	+	–	–	–	–	–	+	–	–	–	–	+	–	–	–	–	–	–	–
Edw. tarda	–	–	–	+	+	–	–	–	–	+	–	–	–	–	–	–	–	+	–	–	–	–	–	–	–
Esch. vulneris	–	–	–	–	–	–	–	–	–	–	–	–	–	–	+	–	–	+	–	–	–	–	–	+	–
Fla. columnare	–	–	–	–	(+)	–	–	–	–	–	–	–	(+)	–	–	–	–	–	–	–	(+)	–	(+)	–	–
Fla. piscicida	–	–	–	–	(+)	–	–	–	–	+	–	–	–	–	–	–	–	+	–	–	–	–	–	–	–
Fla. psychrophilum	–	–	–	–	(+)	–	–	–	–	–	–	–	(+)	–	–	–	–	–	–	–	(+)	–	(+)	–	–
Haf. alvei	–	–	–	–	–	–	–	–	–	–	–	–	–	–	–	–	–	+	–	–	–	–	–	–	–
J. lividum	–	–	–	–	+	–	–	–	–	–	–	–	+	–	–	–	–	–	–	–	+	–	–	–	–
Lactococcus garvieae	–	–	–	–	+	–	–	–	–	–	–	–	–	+	+	–	–	+	–	–	–	–	+	–	–
Lactococcus piscium	–	–	–	–	+	–	+	–	–	–	–	–	+	–	–	–	–	+	–	–	–	–	–	–	–
Micrococcus luteus	–	–	–	–	+	–	–	–	–	–	–	–	+	–	–	–	–	–	–	–	+	–	–	–	–
Moraxella sp.	–	–	–	–	–	–	+	–	–	–	–	–	–	–	–	–	–	–	–	–	–	–	+	–	–
Mycobacterium spp.	–	–	–	–	–	–	–	–	–	–	–	–	+	–	–	–	–	–	–	+	–	–	–	–	–
Myc. neoaurum	–	–	–	–	–	+	–	–	–	+	–	+	–	–	–	–	–	–	–	–	–	–	–	–	–
Nocardia spp.	–	–	–	–	–	–	–	–	–	+	–	–	–	–	–	–	–	–	–	+	–	–	–	–	–
Ph. damselae subsp. *piscicida*	–	–	–	–	–	–	–	–	–	+	–	–	–	–	–	–	–	–	–	–	–	–	–	–	–
Piscirickettsia salmonis	–	–	–	–	–	–	+	+	–	–	–	–	+	–	+	–	–	–	–	–	+	–	–	–	–
Planococcus sp.	–	–	–	–	+	–	–	–	–	–	–	–	+	–	–	–	–	–	–	–	+	–	+	–	–
PLO	–	–	–	–	–	–	–	–	–	+	–	–	–	–	–	–	–	–	–	–	+	–	–	–	–

Table 6.2. (*Continued*)

Taxon	1	2	3	4	5	6	7	8	9	10	11	12	13	14	15	16	17	18	19	20	21	22	23	24	25
Ps. anguilliseptica	−	−	−	−	−	−	−	−	−	−	−	−	−	+	−	−	−	+	−	−	−	−	−	−	−
Ps. chlororaphis	−	−	−	−	+	−	−	−	−	−	−	−	−	−	−	−	−	−	−	−	−	−	−	−	−
Ps. fluorescens	−	−	−	−	+	−	+	−	−	−	−	−	−	+	−	−	−	+	−	−	−	−	−	−	−
Ps. pseudoalcaligenes	−	−	−	−	+	−	−	−	−	−	−	−	−	−	−	−	−	+	−	−	−	−	−	−	−
Ren. salmoninarum	−	−	−	−	+	−	+	−	−	−	−	−	+	−	−	+	−	+	+	−	−	−	−	−	−
Sal. choleraesuis subsp. *arizonae*	−	−	−	−	−	−	−	−	−	−	−	−	−	+	−	−	−	−	−	−	−	−	−	−	−
Ser. liquefaciens	−	−	−	−	+	−	−	−	−	+	+	−	+	−	−	−	−	−	−	−	+	−	−	−	−
Ser. plymuthica	−	−	−	−	+	−	−	−	−	−	−	−	−	−	−	−	−	+	−	−	−	−	−	−	−
Sta. epidermidis	−	−	−	−	+	−	−	−	−	−	−	−	−	−	−	−	−	−	−	−	−	−	−	−	−
Str. difficilis	−	−	−	−	−	−	−	−	−	−	−	−	−	−	−	−	−	−	−	+	−	−	−	−	−
Str. iniae	−	−	−	−	−	−	−	−	−	−	−	−	−	−	−	−	−	−	+	−	−	−	−	−	−
Str. parauberis	−	−	−	−	+	−	−	−	−	−	−	−	+	−	+	−	−	−	−	−	+	−	+	−	−
Vag. salmoninarum	−	−	−	−	+	+	+	−	−	−	−	−	−	−	−	−	−	−	−	−	+	+	−	−	−
V. alginolyticus	−	−	−	−	+	−	−	−	+	−	−	−	−	+	+	−	−	+	−	−	−	−	−	−	−
V. anguillarum	−	−	−	−	−	−	+	−	−	−	−	−	−	+	−	−	−	+	−	−	−	−	−	−	−
V. cholerae	−	−	−	−	−	−	−	−	−	−	−	−	−	−	−	−	−	+	−	−	−	−	−	−	−
V. harveyi	−	−	−	−	−	−	−	−	−	−	−	−	−	−	+	−	−	−	−	−	−	−	−	−	−
V. ichthyoenteri	−	+	−	−	−	−	−	−	−	−	−	−	−	−	−	−	−	−	−	−	−	−	−	−	−
V. ordalii	−	−	−	−	−	−	+	−	−	−	−	−	−	+	−	−	−	+	−	−	−	−	−	−	−
V. pelagius	−	−	−	−	−	−	−	−	−	−	−	−	−	−	+	−	−	+	−	−	−	−	−	−	−
V. salmonicida	−	−	−	−	−	−	−	−	−	−	−	−	−	−	−	−	−	+	−	−	−	−	−	−	−
V. splendidus	−	−	−	−	−	−	−	−	−	−	−	−	−	+	+	−	−	−	−	−	−	−	+	−	+
V. trachuri	−	−	−	−	−	−	−	−	−	−	−	−	−	−	−	−	−	+	−	−	−	−	−	−	−
Y. ruckeri	−	−	−	−	−	−	+	−	−	−	−	−	−	+	−	−	−	+	−	−	−	−	−	−	−

1 = skeletal deformity, 2 = intestinal necrosis, 3 = hyperaemic stomach, 4 = gas-filled hollows in the muscle, 5 = ascitic fluid, 6 = peritonitis, 7 = petechial haemorrhages on muscle wall, 8 = haemorrhaging in the air bladder, 9 = liquid in the air bladder, 10 = white nodules (granulomas) in/on internal organs, 11 = yellowish nodules on the internal organs, 12 = nodules in the muscle, 13 = swollen and/or watery kidney, 14 = haemorrhaging and/or bloody exudate in the peritoneum, 15 = swollen intestine, 16 = false membrane over heart and/or kidney, 17 = blood in the cranium, 18 = haemorrhaging in/on the internal organs, 19 = brain damage, 20 = emaciation, 21 = pale elongated/swollen spleen, 22 = swollen liver, 23 = pale mottled liver, 24 = yellowish liver, 25 = presence of tumours.

(+) = weakly positive.

- **Emaciation**
  This is characteristic of mycobacteriosis and nocardiosis, and with infections by *Bacillus* spp.
- **Pale, elongated/swollen spleen**
  This occurs with infections by *Aer. caviae, Bacillus* spp., *J. lividum, Micrococcus luteus, Planococcus* sp., *Piscirickettsia salmonis*, the PLO, *Ser. liquefaciens, Str. parauberis* and *Vag. salmoninarum*.
- **Pale (possibly mottled) liver**
  This condition occurs with many bacterial fish pathogens, including *Aer. caviae, Cit. freundii, Lactococcus garvieae, Moraxella* sp., *Planococcus* sp., *Str. parauberis* and *V. splendidus*.
- **Yellowish liver (with hyperaemic areas)**
  This occurs with infection by *Esch. vulneris*.
- **Swollen liver**
  This has been noted in infection by *Vag. salmoninarum*.
- **The presence of tumours**
  Although it is unlikely that bacteria are directly involved with tumorous conditions, there is some indication that *V. fischeri* is implicated as a possible secondary invader to viruses.

C. HISTOPATHOLOGICAL EXAMINATION OF DISEASED TISSUES

Although many histological procedures may be routinely used, it is important for the bacteriologist that Gram-stained sections should be prepared. Thus, the presence of any offending bacterial pathogen will quickly be recognized. Microscopic examination of Gram-stained material will enable the determination of the basic staining reaction and micromorphology of the pathogen. Possibilities include the presence of Gram-positive or Gram-negative rods (spore-bearing or asporogenous), cocci and mycelium. For Gram-positive organisms, the acid-fast stain will help in the recognition of *Mycobacterium, Nocardia*, and possibly *Rhodococcus*.

A simplistic approach for the diagnosis of columnaris has involved the observation of wet preparations, prepared from lesion material, which were supposed to demonstrate the presence of 'column'-like masses of bacteria (Snieszko, 1958b). Undoubtedly, use of this technique led to mis-diagnosis with other myxobacteria. However, a recent development, which is meeting with some success, is the microscopic examination of spleen squashes from fish considered to be suffering with RTFS (Sarti *et al.*, 1992). Here, bacteria can be readily seen in the infected animals.

D. BACTERIOLOGICAL EXAMINATION OF TISSUES

Although detailed in Chapter 5, it is worth re-iterating that by use of a comparatively small range of media, a couple of incubators adjusted to 15°C and 20–25°C, an anaerobic jar, and standard aseptic techniques, the majority of the bacterial fish pathogens may be readily recovered as pure culture growth. It should be emphasized, however, that in some asymptomatic 'carrier' states, e.g. BKD, only a scant number of colonies may develop.

Nevertheless, in the majority of clinical cases of disease, it should be possible to obtain dense growth.

Tissues to be sampled
Generally, examination should be made of any damaged tissues. Experience has taught us that it is always advisable to include a kidney sample, which often permits the most satisfactory isolation of the pathogen. Quite simply it is adequate to sample the material by means of swabbing. The swabs should then quickly be used to inoculate the bacteriological media.

Culturing *Aeromonas salmonicida*
Diagnosis of *Aer. salmonicida* is readily achieved by culturing techniques, usually on TSA or BHIA (the preferred medium in the view of many scientists) in which case 'typical' isolates produce a characteristic brown, diffusible pigment. Also, CBB may be employed as a differential medium (Markwardt *et al.*, 1989). However as a primary isolation medium, CBB appears to be less sensitive and gives a poor recovery of the pathogen compared to BHIA (B. Austin, unpublished data).

A special case for diagnosis—BKD
Confusion surrounds which of the available methods is most suitable for detecting and thus diagnosing the presence of BKD. Essentially, the opposing opinions include those who favour serology and those who recommend cultivation. Soon, gene probes may also enter the armoury of the diagnostician (Etchegaray *et al.*, 1991). It is difficult for most microbiologists to determine which of the alternative approaches is correct. Historically, diagnosis of BKD was achieved histologically, by the presence of Gram-positive coccobacilli in kidney tissue. However, the reliability was impaired by the presence of melanin granules and other morphologically similar bacteria (Chen *et al.*, 1974). A subsequent derivative has been a histochemical technique using Lillie's allochrome to stain glycogen in the bacteria (Bruno and Munro, 1982). Again, interference could result from the presence of other morphologically similar glycogen-containing bacteria.

The early spate of interest in culturing techniques improved diagnosis but the effectiveness was marred by the apparently slow growth of the organisms, i.e. up to 6 weeks at 15°C. In fact, recent evidence suggests that 19 weeks may be necessary for the initial incubation period. Therefore, there was widespread attention focused on serological procedures as being the saviour of diagnosticians. This view has been reinforced by the current emphasis on Western blots (e.g. Lovely *et al.*, 1994). But which is more efficient, culturing or serology at detecting renibacterium? Comparing kidney and ovarian fluid from broodstock Atlantic salmon, the selective medium, SKDM (see Austin *et al.*, 1983a) detected a higher number of positive BKD samples than iFAT, which in turn was more sensitive than ELISA or Western blots (Griffiths *et al.*, 1996). Interestingly, renibacterium was found in either kidney or ovarian fluid, but not both. However, the use of culturing in SKDM broth followed by Western blotting increased the sensitivity beyond the maximum level recorded by SKDM alone (Griffiths *et al.*, 1996). This is interesting because the benefit of this broth stage in increasing the detection rate for renibacterium paralleled an observation with peptone water during the 1970s. Here, we found that preincubation of kidney, spleen and more importantly heart tissue in peptone water enhanced the level of resulting colonies, and therefore positivity, on solid medium. Another study concluded that ELISA, using a polyclonal rather than a monoclonal [this was less sensitive] antiserum, was more sensitive than SKDM. Here, SKDM detected *Ren. salmoninarum* in 45% of kidney samples, compared to ELISA which found that 50% of the kidneys were positive (Jansson *et al.*, 1996).

A special case—*Piscirickettsia salmonis*
Isolation in cell culture or detection in acridine orange stained smears or by iFAT has been advocated (Lannan and Fryer, 1991; Lannan *et al.*, 1991).

Identification of bacterial isolates
The most common shortcomings in diagnosis of fish diseases concern the identification of bacterial isolates. There are two schools of thought, namely those that rely on serology *and* those relying on phenotypic tests.

E. SEROLOGY

We will preface further discussion by wholeheartedly endorsing a view that reliable diagnoses occur only with monospecific antisera to assure the homologous reaction between antigen and antibody. The development of monoclonal antibodies has improved diagnoses by standardizing serological tests, i.e. by means of defined reagents (Goerlich *et al.*, 1984), and enhanced the reliability of ELISA, iFAT and immunohistology for the detection of pathogens, such as *Mycobacterium* spp. and *Ren. salmoninarum* (Adams *et al.*, 1995). In contrast, it may be argued that the more conventional polyclonal antibodies have generated contradictory results. Also, the extent of any cross-reactions with polyclonal antibodies has not been adequately determined. Goerlich *et al.* (1984) noted that a monoclonal antibody raised against a typical virulent isolate of *Aer. salmonicida* (i.e. the strain possessed an A-layer) reacted only with virulent cultures, but not with avirulent cells (i.e. those lacking an A-layer).

Nevertheless, tentative diagnoses, especially of asymptomatic infections (Busch and Lingg, 1975; Hansen and Lingg, 1976; Johnson *et al.*, 1974) may result from use of polyclonal antibodies in any of a multiplicity of serological procedures, including FAT, whole-cell (slide) agglutination, precipitin reactions, complement fixation, immuno-diffusion, antibody-coated latex particles (this is akin to human pregnancy testing), co-agglutination using antibody-coated staphylococcal cells, passive haemagglutination, immuno-India ink technique (Geck) or ELISA (e.g. Saeed and Plumb, 1987). The serological techniques are described below.

• Fluorescent antibody technique (FAT)
There are two variations to this test, namely the direct and indirect methods. FAT has found use for the diagnosis of many fish diseases, especially in laboratories (Eurell *et al.*, 1978), for which it is regarded as a highly effective procedure (Kawahara and Kusuda, 1987). For example, α- and β-haemolytic streptococcal isolates may be readily differenti-ated by FAT (Kawahara and Kusuda, 1987). iFAT has a proven track record with the

diagnosis of ERM (Johnson *et al.*, 1974). FAT appears to be useful for detecting *Fle. maritimus* in fish tissues (Baxa *et al.*, 1988a). Moreover, Lorenzen and Karas (1992) detailed an immunofluorescence technique for detecting *Fla. psychrophilum* in the spleen of rainbow trout suffering with RTFS. In addition, iFAT with monoclonal antibodies [and an enzyme immunoassay—also useful for *Edw. tarda* (Rogers, 1981)] have shown promise for the recognition of *Edw. ictaluri* (Rogers, 1981; Ainsworth *et al.*, 1986).

For the direct method (see Bullock *et al.*, 1980 and Smibert and Krieg, 1981 for further details), fluorescein isothiocyanate is conjugated with whole or with the IgG fraction of the antiserum. Two-fold dilutions (1:5 and 1:8) are prepared in PBS (0.1236% (w/v)) Na_2HPO_4, anhydrous; 0.018% (w/v) $NaH_2PO_4.H_2O$; 0.85% (w/v) NaCl; pH 7.6) and used to standardize the 'conjugate'. Then, a bacterial suspension (containing *ca.* 10^8 cells/ml in PBS) is pipetted onto grease-free microscope slides, air-dried and fixed at 60°C for 2 min (or fixed in 95% ethanol for 1 min, and then air-dried). The 'conjugate' is pipetted onto the slide and left in a moist chamber for 5 min to react (room temperature, i.e. 15–20°C, is adequate). Subsequently, the excess antiserum is removed by draining, before the slide is thoroughly rinsed in PBS (for *ca.* 10 min). The slide is air-dried, and the smear covered with a drop of buffered glycerol (Difco; at pH 9.0) before overlayering with a coverslip. This should be quickly examined with a fluorescence microscope. The optimum dilution of conjugate is determined, from the doubling dilutions, by rating the degree of fluorescence from 0 to 4+ (after Jones *et al.*, 1978). The 'use' dilution of the conjugated antiserum is 50% of the highest dilution which gives maximum fluorescence. With this information, the diagnosis may proceed for unknown cultures. However, it is always important to include positive and negative controls.

For iFAT (see Laidler, 1980), bacterial smears are prepared and fixed (as above). Doubling dilutions of rabbit antiserum are prepared, and $20\,\mu l$ aliquots added to the bacterial smears. These are placed in a moist chamber, left for 30 min to react, and washed for 30 min in two changes of PBS. After air-drying, the smears are covered with a suitable dilution of fluorescein-labelled sheep anti-rabbit globulin (Wellcome), incubated for 30 min in the moist chamber, rinsed thoroughly in PBS, air-dried and mounted in buffered glycerol. Examination, with a fluorescence microscope should proceed as quickly as possible.

iFAT (Bullock and Stuckey, 1975b; Mitchum *et al.*, 1979; Paterson *et al.*, 1979; Laidler, 1980) and FAT (Bullock *et al.*, 1980) have been developed for the diagnosis of BKD. Improvements in the FAT included a 60 min staining time with the fluorescent antibody (Cvitanich, 1994). A further refinement involved concentrating samples—in this case coelomic fluid from spawning chinook salmon—on membrane filters, which were used with FAT (Elliott and McKubben, 1997). This modification was regarded as more sensitive than FAT on smears (Elliott and McKubben, 1997). The iFAT has found use for detecting asymptomatic or overt cases of BKD (Bullock and Stuckey, 1975b; Lee and Gordon, 1987), although the technique is not always as sensitive and reliable as culturing (Armstrong *et al.*, 1989). Indeed, Paterson *et al.* (1979) pointed to the enzootic nature of BKD in one Canadian river. These workers reported asymptomatic infections in 33.4% of 456 Atlantic salmon parr and 35.1% of 37 adult salmon in the Margaree River. Seemingly, iFAT was more sensitive than the examination of Gram-stained kidney tissue or cultivation on Mueller–Hinton agar supplemented with 10% foetal calf serum and

0.1% L-cysteine hydrochloride (Paterson *et al.*, 1979). Kawahara and Kusuda (1987) reported that FAT was superior to culturing for the diagnosis of *Aer. hydrophila* (and atypical *Aer. salmonicida*) infections in eels. Moreover, FAT was more successful than culturing for detecting *Aer. hydrophila* in mixed infections. FAT has been found useful for diagnosing *Edw. tarda* and determining the presence of the pathogen in infected fish tissues (Amandi *et al.*, 1982). In contrast to other views, Horiuchi *et al.* (1980) considered FAT as being extremely valuable for field diagnoses, albeit of *Edw. tarda*.

• Whole-cell agglutination

This is a quick and easy technique which provides much useful data. Whole-cell agglutination is used widely as a diagnostic tool, for example with the fish pathogenic streptococci (Kitao *et al.*, 1979; Kitao, 1982a) and *Edw. tarda* (Amandi *et al.*, 1982). Pacha (1968) implied that whole-cell agglutination reactions were effective for differentiating *Fla. psychrophilum* from other (unnamed) myxobacteria. A similar opinion was expressed about *Fla. columnare* (Morrison *et al.*, 1984), and possibly for *Fle. maritimus* (Wakabayashi *et al.*, 1984). Eurell *et al.* (1978) considered the effectiveness of slide agglutination especially for field use in the recognition of *Aer. hydrophila* infections, whereas tube- and macro-agglutination were useful in laboratories. The technique is effective with *Aer. salmonicida*, but only for smooth (non-auto-agglutinating) colonies (Rabb *et al.*, 1964). This is a pity since the majority of isolates recovered from clinical cases of disease are, in fact, rough and auto-agglutinating (McCarthy, 1976). Also, *Ps. anguilliseptica* may be rapidly diagnosed by slide agglutination. However, slight cross-reactions may also occur with *Ps. putida* and *V. anguillarum*. These interfering cross-agglutinations may, however, be effectively eliminated by using diluted antiserum (Nakai *et al.*, 1981). An interesting development concerns the detection of a thermolabile *V. anguillarum* O-antigen, termed the *k*–1 antigen by slide agglutination (Tajima *et al.*, 1987a). So far, the data indicate that this antigen is specific to *V. anguillarum*.

Essentially, a drop of bacterial suspension (*ca.* 10^8 cells/ml, prepared in 0.85% (w/v) saline), is added to a microscope slide. This is followed by adding a drop of antiserum (use a range of dilutions), with gentle mixing for 2 min. A positive response is indicated by clumping of the cells. The reaction may also be carried out in microtitre wells, using serial dilutions of antisera (Toranzo *et al.*, 1987). Of course, it must be emphasized that the reliability reflects the specificity of the antiserum.

• Precipitin reactions and immunodiffusion

The value of these tests for diagnosis has been overshadowed by other techniques, such as FAT and whole-cell agglutination. For detailed discussion, the reader should consult Kimura *et al.* (1978b) and Smibert and Krieg (1981). Nevertheless, an immunodiffusion test for BKD, based on the detection of soluble antigens in infected tissues, was developed by Chen *et al.* (1974) and discussed further by Kimura *et al.* (1978b). Use of this method together with more classical agglutination reactions on ten isolates led to the conclusion by Bullock *et al.* (1974) that the causal agent of BKD was antigenically homogeneous. Following examination of over 50 isolates, we concur with this conclusion. Immunodiffusion was, of course, much quicker than cultivation, with diagnosis, achieved on the basis of specific precipitin lines, taking no more than 24 h.

• Complement fixation
This technique probably has greater value for fish virology than for fish bacteriology; it has been described by Ahne (1981).

• Antibody-coated latex particles
The so-called latex agglutination test has found widespread use for diagnosis of ERM (e.g. Hansen and Lingg, 1976), furunculosis and vibriosis (including Hitra disease). Commercial kits have found success, despite some cross-reactions (Romalde *et al.*, 1995). However, the technique may be readily adapted for most bacterial fish pathogens. As originally described for *Aer. salmonicida* by McCarthy (1975a,b), the test involves the use of globulins from hyper-immune serum (titre = >1:5000) and sensitized latex. The globulins are precipitated by the addition of saturated ammonium sulphate to the antiserum, and the precipitated proteins are sedimented by centrifugation. They are subsequently re-dissolved in 0.9% (w/v) saline, dialysed overnight at 4°C against three changes of saline, and, after centrifugation, the supernatant, which contains the globulins, is stored at –20°C until required. The latex particles (0.81 μm diameter; Difco) are sensitized in globulin solution at 37°C for 2 h. For the test, 200 μl of the antigen (bacterial suspension in glycine-buffered saline, i.e. 0.73% (w/v) glycine and 1% (w/v) NaCl; pH 8.2; supplemented with 1% (w/v) Tween 80) is mixed for 2 min with an equal volume of sensitized latex on a clean glass plate. A positive result is indicated by clumping of the latex. The technique may be used for pure or mixed cultures and tissue. Thus, positive diagnoses may ensue from tissues unsuitable for culturing, e.g. fish stored at –20°C for 14 days, 5°C for 7 days, or from formalin-fixed material (McCarthy, 1975a,b). As before, positive and negative controls are necessary.

• Co-agglutination with antibody-sensitized staphylococci
Reported for *Aer. salmonicida* and *Ren. salmoninarum*, this technique is similar to the latex test (Kimura and Yoshimizu, 1981, 1983, 1984). Essentially, *Sta. aureus* (ATCC 12 598) is suspended in 0.5% formalin-PBS for 3 h at 25°C to inactivate the cells, and washed three times in fresh PBS. The cells are mixed with antiserum in the ratio of 10:1 and incubated at 25°C for 3 h. An equal volume of a boiled bacterial suspension and the sensitized staphylococci are mixed on a glass slide. Following incubation in a moist chamber at room temperature for up to 2 h, a positive response is indicated by clumping of the cells. The advantages of this technique concern its simplicity and reliability. Moreover, it was considered suitable for deployment in field conditions.

The co-agglutination test of Kimura and Yoshimizu (1977) showed considerable promise for rapid detection of BKD, i.e. within 2 hours. The anti-*Renibacterium* antibody coated staphylococcal cells are reacted with the supernatant from heated (i.e. 100°C for 30 min) kidney tissues. Unlike iFAT/FAT, it does not require an expensive fluorescence microscope, and would, therefore, be more suited to field conditions.

• Passive agglutination
For rough colonies of *Aer. salmonicida*, which were unsuitable for use with whole cell agglutination [because of auto-agglutination], McCarthy and Rawle (1975) recommended the mini-passive agglutination test. This technique involves the use of sheep erythrocytes sensitized with *Aer. salmonicida* O-antigen (extracted with hot physiological saline). This

reacts with dilute anti-*Aer. salmonicida* immune serum, assuming that the antigen is present. The obvious advantages of this method concern its application to the detection of both rough and smooth colonies. However, McCarthy and Rawle (1975) cautioned that false negative results may sometimes be obtained with cultures that have been maintained in laboratory conditions for prolonged periods. Hence, old cultures may not be suitable for use in serological studies (or, for that matter, vaccine production!).

• **Immuno-India ink technique (Geck)**

Another rapid technique, which allows diagnosis within 15 min, is the India ink immunostaining reaction as developed initially by Geck (1971). This is a microscopic technique, in which the precise mode of action is unknown, although Geck suggested that it could be regarded as an immuno-adsorption method. The technique has been described only for use with *Aer. salmonicida* (McCarthy and Whitehead, 1977). A drop of bacterial suspension is smeared onto a clean (de-fatted) microscope slide, air-dried, and heat-fixed. The smear is covered with a 1:1 mixture of India ink and antiserum, before incubation in a moist chamber for 10 min at room temperature. Subsequently, the mixture is removed by washing with ferric chloride (0.00001% w/v), and the slide air-dried prior to microscopic examination. A positive result is indicated by the presence of cells, clearly outlined with India ink.

• **Enzyme-linked immunosorbent assay (ELISA)**

This is a technique which is becoming widely adopted for the detection and diagnosis of bacterial fish pathogens, some commercial kits having been developed. This is a useful technique which has already gained widespread use in human and veterinary medicine. Essentially, there is an requirement for a specific antiserum, an enzyme, e.g. alkaline phosphatase or horseradish peroxidase, and a substrate, e.g. *o*-phenylenediamine (for use with alkaline phosphatase) (see Austin *et al.*, 1986). A positive result is indicated by a colour change, which may be recorded quantitatively with a specially designed reader.

Hsu *et al.* (1991) described a monoclonal antibody based ELISA which appears to be effective for the diagnosis of BKD. This system detected 0.05–0.1 μg of antigen/ml within a few hours. An ELISA was developed, which successfully identified *Ph. damselae* subsp. *piscicida* albeit in artificially infected fish tissue within 4 h. By visually recording the ELISA the threshold for positive was 10^5 cells/ml. However, use of a reader cut this level to only 10^3 cells/ml (Bakopoulos *et l.*, 1997a). A sensitivity limit of 10^4–10^5 cells/well was detailed for the *Aer. hydrophila* system devised by Sendra *et al.* (1997). Rapid identification of *Fla. psychrophilum* and *Fla. branchiophilum* was achieved by ELISA, which detected $\geq 1 \times 10^4$ cells/ml from infected spleen (Rangdale and Way, 1995) and 1×10^3 cells/ml from gills [= the threshold value] (MacPhee *et al.*, 1995), respectively. Similarly, an ELISA was developed for *V. vulnificus* and field tested, the results of which indicated a sensitivity of 10^4–10^5 cells/well, and an ability to detect non-culturable cells (Biosca *et l.*, 1997b). We have successfully married monoclonal antibodies to *Aer. salmonicida* with ELISA for a test which has proven suitable for use on fish farms. Indeed, experiments demonstrated that reliable diagnoses were achieved within 30 min (Austin *et al.*, 1986). It is noteworthy that ELISA systems appear to be more sensitive than culturing for the detection of *Aer. salmonicida* (Hiney *et al.*, 1994).

An indirect ELISA has been effective at detecting *Y. ruckeri* (Cossarini-Dunier, 1985) and for determining the presence of antibodies to *Edw. ictaluri* in fish serum (Waterstrat *et al.*, 1989). A development of this approach involved the use of tissue homogenization (using 0.5% v/v Triton X-100 in 0.05 M PBS [pH 7.2]), filtration and then the ELISA (Earlix *et al.*, 1996). This approach was used successfully to detect asymptomatic carriers, and permitted live bacteria to be filtered from 1 g quantities of tissue slurries, with a sensitivity of <10 colony forming units/g of tissue. The filter-ELISA system detected *Edw. ictaluri* in 80% of 98 channel catfish compared to a detection of 24% by culturing (Earlix *et al.*, 1996).

• **Immunohistochemistry**
Immunohistochemistry, based on avidin–biotin complexes, have identified *V. salmonicida* in fixed tissues (Evensen *et al.*, 1991). Also, a peroxidase–antiperoxidase immunohistochemical technique appears to be useful for differentiating *Y. ruckeri* (Jansson *et al.*, 1991).

• **Immunomagnetic separation of antigens**
A comparatively novel approach concerns the recovery of *Aer. salmonicida* cells with immunomagnetic beads coated with monoclonal antibodies to LPS coupled with culturing (Nese and Enger, 1993). Thus, the cells are recovered by serological procedures for culturing on routine bacteriological media.

Which method is best?—The saga of BKD
Pascho *et al.* (1987) compared five techniques for the detection of *Ren. salmoninarum* in coho salmon. The conclusion was that the ELISA (Dixon, 1987) was most sensitive, followed by FAT, filtration-FAT, culturing, counterimmuno-electrophoresis and immunodiffusion. This view was echoed by results from Meyers *et al.* (1993), who regarded ELISA as more sensitive than FAT. Specifically, FAT did not detect *Ren. salmoninarum* in 80% of the samples positive by ELISA. However, a complication was that in the same study FAT detected *Ren. salmoninarum* in 28% of the samples negative by ELISA (Meyers *et al.*, 1993). The comparative benefit of a membrane-filtration FAT over ELISA was illustrated when the former detected *Ren. salmoninarum* in 66/103 (= 64%) ovarian samples compared to 40/103 (= 39%) of positives with the latter (Pascho *et al.*, 1998). The membrane-filtration FAT was capable of detecting ≥ 25 renibacterial cells/ml of ovarian fluid, whereas the ELISA was not consistent in detection at levels of $\leq 1.3 \times 10^4$ cells/ml (Pascho *et al.*, 1998). Yet both methods were inferior to a nested PCR amplifying a 320 bp fragment of the p57 antigen. This PCR detected *Ren. salmoninarum* in all of the ovarian samples (Pascho *et al.*, 1998).

Following an examination of 1239 kidney samples, Gudmundsdóttir *et al.* (1993) considered that a double sandwich ELISA was more sensitive than culturing on SKDM. Yet, Bandin *et al.* (1996) reported the comparatively high cutoff for ELISA of $\sim 10^6$ bacteria/g of tissue. An amount of 0.3 μg of antigen/ml was noted by Olea *et al.* (1993). Sakai *et al.* (1989b,c) favoured the indirect dot blot assay (Sakai *et al.*, 1989b,c; Sakai and Kobayashi, 1992) [involving peroxidase and diaminobenzidine tetrahydrochloride as enzyme and substrate, respectively], which detected 10^2 cells/g of kidney tissue, over iFAT, co-agglutination, microscopy by Gram-stain, immunodiffusion and latex agglutination. Confirmatory diagnoses were made using dot blot and Western blot assays (Sakai *et al.*, 1990; see also Olivier *et al.*, 1992). Certainly, Griffiths *et al.* (1991) highlighted the value of Western blots over FAT and culturing for the detection of renibacteria. Immunohistochemistry is another approach which is gaining popularity for the efficient detection of BKD. In particular, the indirect peroxidase technique, as applied to tissue sections, was deemed to be more sensitive than iFAT or Gram-staining (Hoffman *et al.*, 1989). Further work has also highlighted the value of the peroxidase–antiperoxidase immunohistochemical technique for the detection of *Renibacterium* (Jansson *et al.*, 1991).

The developments of serological methods for the detection and/or diagnosis of BKD must be examined sceptically because the reports preceded detailed taxonomic study of the organisms. It is unclear how workers knew that the aetiological agent possessed a unique antigenic profile, distinguishing it from other Gram-positive organisms. Unfortunately, the reliability of serological methods may now be questioned insofar as cross-reactions with apparently unrelated organisms have been recognized. Bullock *et al.* (1980) observed large bacteria, in faecal samples of brook trout, which fluoresced with antiserum to *Renibacterium*. After studying authentic representatives of 44 Gram-positive bacterial taxa and 101 cultures from fish and water, Austin and Rayment (1985) reported false positive reactions with coryneform bacteria obtained from fish, a fish pathogenic *Mycobacterium* spp., and *Rothia dentocariosa*. Yoshimizu *et al.* (1987) noted a cross-reaction between a *Pseudomonas* and *Ren. salmoninarum* in iFAT. A 60 kDa heat-shock protein (hsp60) of *Chlamydia psittaci* migrated with the 57 kDa protein of *Ren. salmoninarum*, and may explain the cross-reactivity of polyclonal renibacterium antiserum (Wood *et al.*, 1995). Of course, antisera can be made more specific by cross-absorbing with these organisms. Reinforcing a separate article (Toranzo *et al.*, 1993) by using Western blots, Bandín *et al.* (1993) reported a common antigen, i.e. the 57 kDa protein, between *Cor. aquaticum*, *Car. piscicola* and *Ren. salmoninarum*. Also, it was noted that some isolates of *Ren. salmoninarum* did not produce the 57 kDa protein (Bandín *et al.*, 1993). This is interesting because using the same strains, McIntosh *et al.* 1996) could not find the 57 kDa protein in *Cor. aquaticum* or *Car. piscicola*. Also in contrast to Bandín and co-workers, *Ren. salmoninarum* strain K57 was found to produce the 57 kDa protein. Brown *et al.* (1995) produced evidence that bacteria other than renibacterium could cross-react with antiserum to *Ren. salmoninarum*. Moreover, these workers used PCR and confirmed the conclusion of McIntosh *et al.* (1996) that *Cor. aquaticum* and *Car. piscicola* lacked the p57 antigen.

Investment may be placed in the development of monoclonal antibodies, which should be totally specific for *Renibacterium* (Arakawa *et al.*, 1987; Wiens and Kaattari, 1991). Evelyn (1978) contradicted the Utopian opinion if serology, by reporting that culturing was more sensitive than fluorescent antibody techniques for the detection of renibacteria in kidney tissue by a factor of 10:1. This theme was continued in a later study (Evelyn *et al.*, 1981) when experiments were undertaken to determine whether or not there was correlation between culturing and fluorescent antibody based diagnoses of the BKD carrier state. Again, culturing was reported as more sensitive than fluorescent antibody methods (Evelyn *et al.*, 1981). Nevertheless, from the work of Paterson and Colleagues, it could not be explained what was present in the fish which gave a positive fluorescence test but which could not be cultured. Explanations include the presence of dead cells which retain the ability to fluoresce, anaerobes which would require specialized isolation procedures, fastidious aerobes, damaged, dormant or inhibited cells of renibacteria, or even inanimate particles which microscopically could be mistaken for bacteria. Obviously, caution is needed in interpreting serological diagnoses. Whenever possible, culturing should be used for confirmation. Certainly, the selective medium (SKDM; Austin *et al.*, 1983a) has proved useful in isolating *Renibacterium* from mixed cultures. In addition, during a comparative exercise with KDM2, SKDM consistently enabled a greater recovery of cells from infected fish. In some case, scant growth of only one or two colonies were recovered from kidney tissues on SKDM, although the parallel KDM2 plates were devoid of any growth (Austin *et al.*, 1983a). Suspect cultures of renibacteria were subsequently confirmed by the characteristic profile on API-zym and other phenotypic traits, namely catalase production and inability to produce oxidase.

Whereas debate has centred over the most effective means of detecting BKD, it may be concluded that effective diagnosis should encompass a multiplicity of methods. These include isolation and characterization, and serology on infected tissue. It is proposed that clinical cases of disease should be examined by FAT and culturing methods. Asymptomatic cases should be the subject of full bacteriological examination.

Which method is best?—Furunculosis

In a comparative study of serodiagnostic techniques, Sakai *et al.* (1986) reported that iFAT and the peroxidase–antiperoxidase enzyme immunoassay (PAP) were more sensitive (capable of detecting 10^3 colony forming units [cfu]/ml) than the latex agglutination and co-agglutination techniques. These required 10^7 cfu/ml for positive results to be recorded. Nevertheless with latex agglutination and co-agglutination techniques, more (15/15 = 100%) positive samples were detected than by iFAT (10/15 = 67%) or PAP (1/15 = 73%).

F. MOLECULAR TECHNIQUES

There is evidence that molecular techniques have been used for bacterial pathogens. A timely overview of PCR with emphasis on validation of the techniques and problems relating to diagnosis has been published (Hiney and Smith, 1998). A noteworthy advance resulted from the use of PCR technology to identify *Mycobacterium* spp. in sea bass (Colorni *et al.*, 1993; Knibb *et al.*, 1993) and *Myc. chelonei* in a chichlid oscar (*Astronotus ocellatus*) (McCormick *et al.*, 1995).

A PCR using a 1100 bp fragment has distinguished *Lactococcus garvieae* from *Lactococcus lactis* (Zlotlin *et al.*, 1998). In terms of sensitivity, the PCR detected *Lactococcus garvieae* in 1 μl of fish plasma.

León *et al.* (1994a,b) published details of a PCR assay using a 149 base pair (bp) DNA sequence, which was sensitive enough to detect 22 renibacterial cells even in tissue, and of sufficient specificity to recognize *Ren. salmoninarum* but not *Aer. hydrophila*, *Aer. salmonicida*, *Car. piscicola*, *Fla. columnare*, *V. anguillarum*, *V. ordalii* or *Y. ruckeri*. Then, this group detailed a 2282 bp DNA fragment that appeared to be responsible for internalization of renibacteria, at least into CHSE-tissue culture cells (Maulén *et al.*,

1996). A nested reverse transcription PCR of 16S rRNA sequences successfully detected 1–10 renibacterial cells in ovarian fluid, but was unreliable with kidney (Magnússen *et al.*, 1994). Although sensitive, this system took 1–2 days to carry out. In a further development, McIntosh *et al.* (1996) devised a simplified PCR invoking the 376-bp region of the gene encoding the 57 kDa surface antigen. This system had a minimum detection limit of 5×10^3 renibacterial cells/ml in rainbow trout lymphocytes. Two 24-base oligonucleotide primers used to amplify a 501-bp region of the gene encoding the 57 kDa soluble protein (p57) formed the basis of a PCR, which was capable of detecting two renibacterial cells within individual salmonid eggs (Brown *et al.*, 1994). This PCR was considered to have value for the screening of broodstock for the presence of BKD (Brown *et al.*, 1994). A nested PCR amplifying a 320 bp fragment of the p57 antigen was suitable for detecting *Ren. salmoninarum* in ovarian samples (Pascho *et al.*, 1998).

A clone, pRS47, of 5.1 kb, was used to develop a specific DNA probe (Hariharan *et al.*, 1995). In a dot blot assay, the biotinylated pRS47/*Bam*HI insert probe hybridized with only DNA from three strains of *Ren. salmoninarum* but not with *Arthrobacter protophormiae, Aer. salmonicida, Cor. aquaticum, Car. piscicola, Micrococcus luteus, Ps. fluorescens, V. anguillarum, V. ordalii* or *Y. ruckeri*. With kidney tissue from fish challenged with *Ren. salmoninarum*, the dot blot assay was regarded as sensitive as culture and FAT. With the latter technique, samples negative by dot blot assay and culturing revealed the presence of $\leqslant 1$ fluorescing object, presumed to be a bacterium)/50 microscope fields. Consequently, this probe was regarded as having potential for the diagnosis of BKD (Hariharan *et al.*, 1995).

DNA probe technology has extended to *Aer. salmonicida* (e.g. Mooney *et al.*, 1995; Miyata *et al.*, 1996; Oakey *et al.*, 1998). Barry *et al.* (1990) suggested that such probes have the potential to detect the pathogen in environmental and clinical samples. These workers found that specific probes for micro-organisms could be developed, even if only two base pair differences existed in the target sequence. Hiney *et al.* (1992) continued with developmental work leading to the isolation of a DNA fragment specific to *Aer. salmonicida*, which when incorporated into a polymerase chain reaction technique enabled a sensitivity of detection of approximately two cells of *Aer. salmonicida*. Mooney *et al.* (1995) examined the blood from 61 wild Atlantic salmon from three rivers in Ireland, and recorded 87% positives, with 100 genome equivalents/fish, using a specific DNA probe for *Aer. salmonicida*. Høie *et al.* (1997) designed primers and probes from 16S rRNA and plasmid DNA; the former of which amplified *Aer. hydrophila, Aer. salmonicida* subsp. *achromogenes, Aer. salmonicida* subsp. *masoucida* and atypical isolates, whereas the latter detected only *Aer. salmonicida* subsp. *achromogenes* and *Aer. salmonicida* subsp. *salmonicida*. Based on an examination of 100 ml volumes of kidney suspension and gill swabs, the PCR detected 20 and 200 colony forming units in 10 μl of PCR template by 16S rRNA and plasmid primers, respectively. The numbers corresponded to 10^3 and 10^4 colony forming units in 100 ml of kidney suspension, respectively (Høie *et al.*, 1997). A conclusion was reached that the PCR detected *Aer. salmonicida* more often than culturing. A possible drawback for use of DNA probes, however, has been proposed by Hennigan *et al.* (1989). Using four DNA probes in combination with seven restriction enzymes and seven strains of *Aer. salmonicida*, their data suggested that the DNA sequences of the species is very strongly conserved. They emphasized that the use of DNA probe technology to identify different strains of *Aer. salmonicida* may be limited.

In a comparison of the sensitivity of culturing with DNA probes, the former detected *Aer. salmonicida* from the kidney of only dead or moribund farmed Atlantic salmon smolts in Ireland whereas probe technology allied to a PCR assay was capable of recognizing the pathogen in water, faeces and effluent (O'Brien *et al.*, 1994).

Specific DNA hybridization probes for *Ph. damselae* subsp. *piscicida* offer promise for the future. Already, initial studies with a ^{32}P-labelled DNA probe indicated a minimum detection limit of 3.9 ng of DNA or 10^5 bacterial cells (Zhao and Aoki, 1989). Aoki *et al.* 1997) emphasized the value of a 629 bp DNA fragment from the subspecies-specific plasmid pZP1 for PCR.

Molecular methods have been invoked to improve the identification of *V. anguillarum*. Using partial 16S rRNA sequences, a specific 16S rRNA oligonucleotide probe detected a minimum of 5×10^3 cells/ml in culture or tissue extracts (Rehnstam *et al.*, 1989). Powell and Loutit (1994a) used a 310 bp DNA fragment as a probe for *V. anguillarum*. This system detected 100 (but not 10) ng of purified genomic GNA of most serogroups (not serogroup O7) of *V. anguillarum* but did not react with other vibrios. Using the species-specific probe in combination with membrane filtration, *V. anguillarum* could be detected in water (Powell and Loutit, 1994b). In a parallel development, an oligonucleotide (VaV3) detected 150 ng by DNA:DNA slot blot hybridization. The system did not cross-react with other species, and was capable of detecting eight out of 10 serogroups of *V. anguillarum* (Martínez-Picardo *et al.*, 1994).

PCR technology offers promise for the detection of *Piscirickettsia salmonis*, insofar as a system has already been designed which is capable of detecting one tissue culture infectious dose (Mauel *et al.*, 1996). Subsequently, a PCR was described that was effective with a few millilitres of serum (Marshall *et al.*, 1998). The benefit of this system is that the test could be carried on live fish, such as valuable broodstock.

PCR targeted 16S rRNA has been successful with *Fla. branchiophilum*, *Fla. columnare* and *Fle. maritimus* (Toyama *et al.*, 1994, 1996). In particular, a PCR was sensitive enough to detect 1.5 colony forming units (per PCR reaction tube) of *Fla. psychrophilum* from apparently healthy coho salmon eggs and juvenile ayu (Izumi and Wakabayashgi, 1997).

Also, there is evidence that molecular techniques are finding use for the identification of infections caused by *Y. ruckeri* (e.g. Argenton *et al.*, 1996).

G. PHENOTYPIC TESTS

For many pathogens emphasis has been placed on conventional phenotypic tests for diagnosis. For example, Boulanger *et al.* (1977) highlighted the value of confirming the Gram-staining reaction, fermentative metabolism of glucose, and production of arginine dihydrolase, catalase and oxidase but not of lysine or ornithine decarboxylase. Caution is advocated should consideration be given to using commercially available diagnostic kits. The API 20E and API-zym systems, and more recently the API 20NE API 50CH, API 50L, Biolog-GN, Enterotubes and RapidID 32 systems (Meyer and Bullock, 1973;

Amandi *et al.*, 1982), have made an inroad into routine diagnostic laboratories. Whereas some systems, e.g. Biolog-GN, are clearly adaptable to and useful for environmental isolates, others have been developed specifically for a given use, usually medically important bacteria. The consequence is that the supporting identification schemes may be inappropriate. Many workers have used the API 20E system for the identification of fish pathogenic bacteria, and the technique has been considered useful by some (e.g. Santos *et al.*, 1993). For example, by inference the API 20E rapid identification system may be useful for diagnosing *Flavobacterium* sp. (Acuigrup, 1980a). Kent (1982) reported that *V. ordalii* produces a characteristic profile in API 20E, i.e.

$$- - - - - - - - - - + + + - - - + - - - - +$$

Kent (1982) reported the API 20E profile for *V. anguillarum* as:

$$+ v - - + - - - - v + + + + - + - + - + v +$$

whereas Maugeri *et al.* (1983) published a slightly different pattern, i.e.

$$v + - - v - - - - v v v + + - - - - v - + v +$$

where 'v' indicated a variable result. Kent (1982) reported that *Ph. damselae* subsp. *piscicida* gave positive responses in the API 20E rapid identification system for arginine dihydrolase and weak acid production from glucose; all other tests being negative. Of course, it is necessary to modify the protocol for use with marine bacteria. Thus, it was essential to suspend cultures in 2–3% (w/v) saline rather than distilled water, and the inoculated test strips were incubated at 25°C (not 37°C) for up to 48 h. Maugeri *et al.* (1983) considered that it was essential to carry out some additional tests with putative *V. anguillarum*, namely motility and sensitivity to 0/129, in order to confirm that the isolates, were indeed, motile and were inhibited by the vibriostatic agent. However, some cultures may be resistant to the action of 0/129 (Muroga *et al.*, 1979) and appear to be non-motile. The latter phenomenon may result from exposure to the partial inhibitory activity of some antimicrobial compounds.

The API 20E system weighs heavily on the use of sugar fermentation reactions, which may be influenced by the presence of plasmids. The profile(s) for *Aer. hydrophila* are similar to those of *Aer. allosaccharophila* and *Aer. sobria*, whereas *Fle. maritimus* and *Ps. anguilliseptica* may be indistinguishable by the API 20E rapid identification system. *Y. ruckeri* may be confused with *Haf. alvei*. Toranzo *et al.* (1986) compared the API 20E rapid identification system for *Aer. hydrophila* with Kaper's medium (Kaper *et al.*, 1979) and conventional biochemical tests. Kaper and co-workers formulated a single tube medium, which was suitable for determining motility, inositol and mannitol fermentation, ornithine decarboxylase and deamination, and the production of H_2S and indole. Thus, *bona fide* isolates of *Aer. hydrophila* gave an alkaline reaction on the top of the medium, acid production in the butt, motility, and indole but not H_2S production (H_2S production may occur on the top). Toranzo and colleagues pointed to shortcomings of the API 20E system, insofar as many environmental isolates were mis-identified or not listed by the published profile index. In contrast, Kaper's medium was effective for fast, presumptive identification. Problems were encountered with the reliability of some conventional biochemical tests, notably the Voges Proskauer reaction, fermentation and gas production from arabinose, gelatinase production, and the lysine decarboxylase test. Ironically, these tests have also been considered to be correlated with virulence in motile aeromonads. Nevertheless, it is accepted that the API 20E and API 20NE systems have a role in the diagnosis of bacterial fish pathogens. Consequently, diagnostic schemes based on these systems are included in Tables 6.3 and 6.4. In addition, API-zym is very useful for diagnosing *Ren. salmoninarum* (see Table 6.5), which gives a characteristic profile:

$$- + - + - + - - + - + + - - - + - - + / - -$$

A diagnostic scheme for use with Biolog-GN has been included as Table 6.6.

Diagnoses may also be achieved by means of diagnostic schemes based on reactions in conventional phenotypic tests. The procedure may be automated, as with the Abbot Quantum II system (Teska *et al.*, 1989), and involves spectrophotometric readings at 492.6 nm, with a sample cartridge containing 20 inoculated biochemical chambers. Alternatively, specially constructed diagnostic tables may be used, such as presented in Tables 6.7 and 6.8.

Diagnosis of botulism has been accomplished by isolation of *Cl. botulinum* from diseased tissues, and more importantly, by demonstrating the presence of circulating toxin (particularly to *Cl. botulinum* type E) in the blood of moribund fish (Cann and Taylor, 1982).

Presumptive identification of *Lactococcus garvieae*-like organisms has been made following growth on bile (40%)-aesculin agar (see Facklam and Moody, 1979), with hydrolysis of aesculin and by the characteristic growth on eosin-methylene blue agar (lactose is not fermented).

Bacillus mycoides could be distinguished from other bacilli, as follows (Goodwin *et al.*, 1994):

	Presence of parasporal crystals	Rhizoidal growth	Motility	Growth at 45°C
Bacillus mycoides	−	+	−	−
Bacillus anthracis	−	−	−	+
Bacillus cereus	−	−	+	+
Bacillus thuringiensis	+	−	+	+

A simple chemical technique has been described, which may readily delineate *Nocardia* from *Mycobacterium* (Kanetsuna and Bartoni, 1972). Assuming that pure cultures are available, the bacteria are saponified in 2.5% (w/v) potassium hydroxide in a 1:1 (v/v) mixture of methanol and benzene at 37°C for 24 h. Crude mycolic acids from *bona fide* mycobacteria may be subsequently precipitated by addition of an equal volume of ethanol to an ethereal solution of the extracted lipids. Mycobacteria give rise to copious quantities of white precipitate of melting point between 45 and 70°C, whereas nocardias produce negligible amounts, which do not melt below 150°C (Kanetsuna and Bartoli, 1972).

Aer. allosaccharophila isolates may be identified by the examination of key phenotypic characters. In particular, the utilization of L-arabinose and L-histidine as sole carbon

Table 6.3. Profiles of fish pathogens obtained with the API 20E rapid identification system. Results were recorded after incubation at 15–37°C for 24–48 h

Taxon	API 20E test no																				
	1	2	3	4	5	6	7	8	9	10	11	12	13	14	15	16	17	18	19	20	21
Acinetobacter–Moraxella spp.	–	–	–	–	–	–	–	–	–	–	–	–	–	–	–	–	–	–	–	–	+
Aer. allosaccharophila	+	v	+	v	v	–	–	–	+	–	+	+	+	–	–	v	+	v	–	v	+
Aer. hydrophila	+	+	v	–	v	–	–	–	+	v	+	+	+	–	–	–	+	–	v	v	+
Aer. salmonicida	–	+	v	–	–	–	–	–	–	–	+	+	+	–	–	–	–	–	–	–	+
Aer sobria	+	+	+	–	–	–	–	–	+	+	+	+	+	–	–	–	+	–	–	–	+
Cit. freundii	+	–	–	–	+	–	–	–	–	–	–	–	–	–	+	+	+	+	–	+	–
Edw. ictaluri	+	–	+	+	–	–	–	–	–	–	–	+	–	–	–	–	–	–	–	–	–
Edw. tarda	–	–	+	+	–	+	–	–	+	–	–	+	–	–	–	–	–	–	–	–	–
Ent. agglomerans	+	–	–	–	+	–	–	–	–	+	+	+	+	–	–	–	+	–	–	–	–
Fla. branchiophilum	–	–	–	–	–	–	–	–	–	–	+	+	–	–	–	–	+	+	–	–	–
Fla. columnare	–	–	–	–	–	+	–	–	–	–	+	–	–	–	–	–	–	–	–	–	+
Fla. hydatis	–	–	–	–	–	–	–	–	–	–	+	+	+	–	–	–	+	–	–	+	–
Fla. psychrophilum	–	–	–	–	–	–	–	–	–	–	+	–	–	–	–	–	–	–	–	–	(+)
Fle. maritimus	–	–	–	–	–	–	–	–	–	–	+	–	–	–	–	–	–	–	–	–	+
Haf. alvei	–	–	+	+	+	–	–	–	–	–	–	+	+	–	–	–	–	–	–	+	–
Halomonas cupida	–	+	+	+	–	–	–	–	–	–	–	–	+	–	+	+	(+)	–	–	–	–
J. lividum	–	+	–	–	+	–	–	–	–	+	+	(+)	+	–	–	–	–	–	–	–	+
Klebsiella pneumoniae	+	–	v	–	+	–	v	–	–	+	–	+	+	+	+	+	+	+	+	+	–
Photobacterium damselae	–	+	–	–	–	–	+	–	–	+	–	+	–	–	–	–	–	–	–	–	+
Ph. damselae subsp. *piscicida*	–	–	–	–	–	–	–	–	–	(+)	–	(+)	–	–	–	–	–	–	–	–	+
Plesiomonas shigelloides	+	+	+	+	–	–	–	–	+	–	–	+	–	+	–	–	–	–	–	–	+
Pr. rettgeri	–	–	–	–	–	–	+	+	+	–	–	+	+	+	–	+	–	–	–	–	–
Ps. anguilliseptica	–	–	–	–	–	–	–	–	–	–	+	–	–	–	–	–	–	–	–	–	+
Ps. fluorescens	–	+	–	–	+	–	–	–	–	–	+	(+)	+	+	+	–	+	–	–	+	+

Table 6.3. (*Continued*)

Taxon	API 20E test no																				
	1	2	3	4	5	6	7	8	9	10	11	12	13	14	15	16	17	18	19	20	21
Sal. choleraesuis subsp. *arizonae*	+	−	+	−	+	+	−	−	−	−	+	+	−	+	−	+	−	−	−	−	−
Ser. liquefaciens	+	−	+	+	+	−	v	−	−	v	+	+	+	v	+	−	+	v	+	−	−
Ser. plymuthica	+	−	−	−	+	−	−	−	−	+	+	+	+	+	−	−	+	−	−	+	−
Shewanella putrefaciens	−	−	−	+	+	+	−	−	−	−	+	−	−	−	−	−	−	−	−	−	+
V. alginolyticus	−	−	+	−	v	−	−	−	+	−	v	+	+	−	−	−	+	−	−	−	+
V. anguillarum	+	+	−	−	+	−	−	−	+	+	+	+	+	−	+	−	+	−	+	+	+
V. cholerae	+	−	+	+	v	−	−	−	+	v	+	+	+	−	−	−	+	−	−	−	+
V. harveyi	−	−	+	+	−	−	−	−	+	−	+	+	−	−	−	−	+	−	−	−	+
V. ordalii	−	−	−	−	−	−	−	−	−	−	+	+	+	−	−	−	+	−	−	−	+
V. salmonicida	−	−	−	−	−	−	−	−	−	−	−	+	−	−	−	−	−	−	−	−	+
V. splendidus	−	+	−	−	−	−	+	−	−	+	−	+	−	−	−	−	−	−	−	−	+
V. vulnificus	+	−	v	−	+	−	−	−	−	−	+	+	+	−	−	−	−	−	+	−	+
Y. intermedia	+	−	−	+	−	−	+	−	+	−	−	+	+	v	+	+	+	+	+	v	−
Y. ruckeri	+	−	+	+	+	−	−	−	−	−	+	+	+	−	−	−	−	−	−	−	−

1 = β-galactosidase, 2 = arginine dihydrolase, 3 = lysine decarboxylase, 4 = ornithine decarboxylase, 5 = citrate utilization, 6 = H_2S production, 7 = urease production, 8 = tryptophan deaminase, 9 = indole production, 10 = Voges Proskauer reaction, 11 = gelatin hydrolysis, 12 = acid from glucose, 13 = acid from mannitol, 14 = acid from inositol, 15 = acid from sorbitol, 16 = acid from rhamnose, 17 = acid from sucrose, 18 = acid from melibiose, 19 = acid from amygdalin 20 = acid from arabinose, 21 = oxidase production.

+, − and v correspond to >80, <20 and 21–79% of positive results, respectively.

( ) indicates weakly positive results.

Table 6.4. Differential characteristics of some fish pathogens obtained with the API 20NE rapid identifica
tion system. Results were recorded after incubation at 30°C for 24–48 h

Taxon[a]	API 20NE test no																				
	1	2	3	4	5	6	7	8	9	10	11	12	13	14	15	16	17	18	19	20	21
Acinetobcter calcoaceticus	–	–	–	–	–	–	–	–	–	–	–	–	–	–	–	+	–	–	+	+	–
Aer. hydrophila	+	v	+	+	–	+	+	+	+	v	+	+	+	+	+	+	–	+	v	–	+
Aer. salmonicida subsp. achromogenes/masoucida	+	v	–	–	–	v	v	–	v	–	v	v	–	–	–	–	–	–	–	–	+
Aer. salmonicida subsp. *salmonicida*	+	v	v	v	–	+	+	–	+	–	v	+	+	+	+	–	–	+	–	–	+
Aer. sobria	+	v	+	+	–	–	+	+	+	–	+	+	+	+	+	+	–	+	+	–	+
Ph. damselae	+	–	+	+	+	–	–	–	–	–	–	–	–	–	–	–	–	v	–	–	+
Plesiomonas shigelloides	+	+	+	+	–	–	–	+	v	–	–	–	v	v	v	v	–	v	–	–	+
Ps. chlororaphis	–	–	–	–	–	v	–	–	+	+	+	+	–	–	+	v	–	+	+	–	v
Ps. fluorescens	v	–	–	v	–	–	v	–	+	v	+	+	v	–	+	+	–	+	+	–	+
Shewanella putrefaciens	+	–	–	–	–	v	+	–	–	–	–	–	+	–	–	v	–	+	–	–	+
V. alginolyticus	+	+	+	–	–	v	+	–	v	–	v	v	v	v	v	–	–	+	–	–	+
V. cholerae	+	+	+	–	–	–	+	+	+	–	v	v	v	+	+	–	–	+	+	–	+

[a] +, – and v correspond to ≥80, ≤20 and 21–79% of positive results, respectively.

Table 6.5. Distinguishing profiles of Gram-positive bacteria as obtained with API-zym[a]

Taxon (and source of strains)	API-zym test no.																			
	1	2	3	4	5	6	7	8	9	10	11	12	13	14	15	16	17	18	19	20
Actinomyces viscosus (ATCC 15987)	–	–	–	+	–	+	+	–	–	–	+	+	+	+	–	–	+	–	–	+
Aerococcus viridans subsp. *homari* (NCIMB 1119)[b]	–	+	+	+	–	+	–	–	–	–	–	+	–	–	–	–	–	–	–	–
Arthrobacter aurescens (NCIMB 8912)	–	–	–	+	–	+	–	+	+	–	+	+	–	+	+	+	+	–	+	–
Arthrobacter crystallopoietes (ATCC 15481)	–	–	–	+	–	+	+	–	+	–	–	+	–	–	–	–	–	–	–	–
Arthrobacter globiformis (NCIMB 8907)	–	–	–	+	–	+	+	+	+	–	+	+	–	+	–	+	+	–	+	–
Arthrobacter nicotianae (NCIMB 9458)	–	–	–	+	–	+	+	–	+	–	–	+	–	–	–	+	–	+	+	–
Bacillus cereus (CCM 2010)	–	+	+	+	–	+	+	–	+	+	+	+	–	–	–	+	–	–	–	–
Bacillus licheniformis (ATCC 9945)	–	+	+	+	–	–	–	–	–	–	+	+	–	–	–	+	–	–	–	–
Bacillus megaterium (CCM 2007)	–	+	+	+	–	+	–	–	–	+	+	+	+	+	–	–	–	–	–	–
Bacillus polymyxa (ATCC 12321)	–	+	+	+	–	+	–	–	–	+	–	+	+	+	–	–	–	–	–	–
Bacillus sphaericus (ATCC 10208)	–	+	+	+	–	+	–	–	–	+	+	+	–	–	–	–	–	–	–	–
Brevibacterium flavum (ATCC 13826)	–	+	–	+	–	+	+	–	+	–	+	+	–	–	+	–	+	–	–	–
Cor. acnes (NCTC 737)	–	–	–	+	–	–	–	–	(+)	–	–	+	–	–	–	–	–	–	–	–
Cor. pyogenes (NCTC 5224)	–	+	–	+	–	–	–	–	+	–	–	+	–	–	–	–	–	–	–	–
Cor. xerosis (NCTC 7929)[b]	–	+	+	+	–	+	–	–	–	–	–	+	–	–	–	–	–	–	–	–
En. faecalis (CCM 1875)	–	–	–	+	–	+	–	–	–	+	+	+	–	+	–	+	–	–	–	–
En. faecium (CCM 2801)[b]	–	+	+	+	–	+	–	–	(+)	–	–	+	–	–	–	–	–	–	–	–
Kurthia zopfii (CCM 3478)	–	–	–	(+)	–	+	–	–	–	+	+	+	–	–	–	–	–	–	–	–
Lactobacillus casei (CCM 1753)	–	–	+	+	–	+	+	–	–	–	+	+	–	+	–	–	+	–	–	–
Lactobacillus curvatus (NCIMB 9710)	–	–	–	–	–	+	+	–	–	–	–	+	–	–	–	–	–	–	–	–
Lactobacillus brevis (NCDO 1749)	–	–	(+)	+	–	+	+	–	–	–	+	+	+	+	–	+	+	–	–	–
Lactobacillus sp. (pseudokidney disease, 3 isolates)	–	–	(+)	+	–	+	(+)	–	–	(+)	+	–	(+)	–	–	+	+	–	–	–
Listeria denitrificans (ATCC 14870)	–	–	–	+	–	+	–	–	–	–	–	+	+	+	+	+	+	–	–	–
Listeria grayi (CCM 5887)	–	–	(+)	+	–	(+)	(+)	–	–	–	–	+	–	–	–	–	–	–	–	–
Listeria murrayi (CCM 5990)	–	–	+	+	–	+	–	–	–	(+)	–	+	–	–	–	–	–	–	–	–

Table 6.5. (*Continued*)

Taxon (and source of strains)	API-zym test no.																			
	1	2	3	4	5	6	7	8	9	10	11	12	13	14	15	16	17	18	19	20
Microbacterium lacticum (NCIMB 8450)	−	−	−	+	−	+	+	−	−	−	+	+	−	+	−	−	−	−	−	−
Micrococcus luteus (NCIMB 9278)	−	+	−	+	−	+	−	−	−	+	+	+	−	−	−	+	−	−	−	−
Myc. aquae (Körmendy)	−	+	+	+	+	+	−	−	−	−	+	+	−	−	−	−	−	−	−	−
Myc. fortuitum (Körmendy)	−	+	+	+	−	+	+	−	+	−	+	+	−	−	−	−	−	−	−	−
Myc. marinum (Körmendy)	−	+	+	+	−	+	−	−	−	−	+	+	−	−	−	−	−	−	−	−
Myc. smegmatis (Körmendy	−	+	+	+	−	+	−	−	−	−	+	+	−	−	−	−	−	−	−	−
Mycobacterium sp. (Ashburner, SC 744)	−	+	+	+	−	+	−	−	−	−	+	+	−	−	−	+	−	−	−	−
Noc. asteroides (ATCC 14759)	−	+	+	+	−	−	+	+	−	−	+	+	−	−	−	−	−	−	−	−
Noc. corallina (ATCC 4273)	−	+	+	+	−	+	+	+	−	−	−	+	−	−	−	−	−	−	−	−
Planococcus citreus (NCIMB 1493)	−	−	−	+	−	+	+	+	(+)	(+)	−	+	−	+	−	+	−	−	−	−
Propionibacterium acnes (CCM 3343)	−	−	−	(+)	−	+	−	−	−	−	−	+	−	+	−	−	−	−	+	−
Ren. salmoninarum (48 isolates)	−	+	−	+	−	+	−	−	+	−	+	+	−	−	−	+	−	−	(+)	−
Rothia dentocariosa (ATCC 17931)	−	−	−	+	−	+	+	−	+	−	−	+	−	−	−	+	−	−	−	−
Sta. epidermidis (NCIMB 2699)	−	+	(+)	+	−	−	−	−	−	−	+	+	−	−	−	+	−	−	−	−
Str. agalactiae (CCM 6187)	−	+	−	+	−	−	−	−	−	−	+	+	−	−	−	(+)	−	−	−	−
Streptomyces griseus (ATCC 23345)	−	+	−	+	−	+	−	−	−	+	−	+	−	−	−	−	−	−	−	−
Presumptive coryneform (laboratory isolate 198)	−	−	+	+	−	+	−	−	−	−	+	+	−	−	−	+	−	−	−	−

1 = control, 2 = alkaline phosphatase, 3 = esterase (butyrate), 4 = esterase (caprylate), 5 = lipase (myristate), 6 = leucine arylamidase, 7 = valine arylamidase, 8 = cystine arylamidase, 9 = trypsin, 10 = chemotrypsin, 11 = acid phosphatase, 12 = phosphoamidase, 13 = α-galactosidase, 14 = β-galactosidase, 15 = β-glucuronidase, 16 = α-glucosidase, 17 = β-glucosidase, 18 = N-acetyl-β-glucosaminidase, 19 = α-mannosidse, 20 = α-fucosidase.

[a] 15°C/18 h.

[b] Distinguish by results of the Gram-staining reaction.

Sources: ATCC, American Type Culture Collection; CCM, Czechoslovak Collection of Micro-organisms; NCDO, National Collection of Dairy Organisms; NCIMB, National Collection of Industrial and Marine Bacteria; NCTC, National Collection of Type Cultures; Dr L. D. Ashburner, Freshwater Fisheries Research Station, Victoria, Australia; Dr B. Körmendy, Central Veterinary Institute, Hungary.

Table 6.6. Characteristics of selected taxa by Biolog-GN

Taxon	1	2	3	4	5	6	7	8	9	10	11	12	13	14	15	16	17	18	19	20	21	22	23	24	25	26	27	28	29	30	31	32
Aer. caviae	−	v	+	+	+	v	−	+	−	+	v	+	−	+	−	+	+	+	−	v	v	+	+	+	−	+	+	−	−	+	+	+
Aer. hydrophila	−	v	+	+	+	+	+	+	−	v	−	−	−	+	−	+	v	+	−	v	v	+	+	+	−	+	+	−	−	+	+	+
Aer. salmonicida	−	−	+	+	+	+	−	+	−	−	−	−	−	−	+	−	+	−	+	−	−	−	+	v	+	−	−	+	−	−	v	−
Aer. sobria	−	−	+	+	+	+	−	+	−	−	−	−	−	−	+	−	+	v	+	−	v	v	+	+	+	−	+	+	−	−	+	+
Cit. freundii	−	−	+	+	+	+	+	+	−	+	v	v	−	+	+	+	v	+	v	+	v	+	+	+	v	+	+	v	+	+	v	+
Edw. tarda	−	−	+	+	−	−	+	+	−	−	−	−	−	+	+	+	+	−	−	−	+	−	+	−	v	+	−	−	−	−	−	−
Ent. agglomerans	−	−	+	+	v	v	v	+	v	+	+	v	−	+	v	+	+	+	+	v	v	+	+	+	v	+	+	v	+	+	+	+
Esch. vulneris	−	−	+	+	v	v	−	+	−	+	v	+	−	+	−	+	+	+	v	+	v	+	+	+	+	+	+	+	+	+	v	+
Haf. alvei	−	v	+	+	+	+	+	+	v	+	v	v	−	+	+	+	+	+	v	v	v	+	+	+	−	v	+	v	v	+	v	+
Klebsiella pneumoniae	−	−	+	+	v	v	−	+	+	v	+	v	−	+	+	+	+	+	+	v	v	+	+	v	+	+	+	+	+	+	+	+
Ph. damselae	−	−	+	+	−	−	v	+	−	−	−	v	−	+	−	+	v	+	−	−	−	+	−	+	−	−	+	−	−	−	−	v
Plesiomonas shigelloides	−	−	−	v	−	−	v	+	−	−	−	−	−	−	−	v	−	+	+	−	−	+	−	v	v	+	−	−	−	−	−	+
Pr. rettgeri	−	−	v	+	v	v	v	+	v	−	v	−	v	+	−	+	v	+	+	−	−	+	v	+	−	v	v	−	−	v	v	v
Ps. fluorescens	−	−	v	v	v	v	−	v	v	v	v	−	−	+	−	v	v	+	+	−	−	v	+	+	−	−	v	−	−	v	v	v
Ps. pseudoalcaligenes	−	−	−	+	+	−	−	−	−	−	−	−	+	−	−	−	−	+	−	−	−	−	−	−	−	+	−	−	−	−	−	−
Ser. liquefaciens	−	−	+	+	v	v	+	+	v	+	v	v	−	+	+	+	+	+	+	+	+	+	+	+	+	+	+	v	+	+	+	+
Ser. marcescens	−	v	+	+	+	+	+	+	+	+	+	+	v	+	+	+	+	+	+	+	+	+	+	v	+	+	v	v	+	+	+	+
Ser. plymuthica	−	−	+	+	+	−	+	−	+	v	+	−	+	+	+	+	+	+	+	+	+	+	+	+	+	+	+	v	+	+	+	+
V. alginolyticus	−	+	+	+	+	v	−	+	−	−	v	−	−	+	−	−	+	+	−	−	−	+	+	+	−	v	+	−	−	+	+	+
V. harveyi	−	+	v	+	+	−	+	+	−	−	−	+	−	+	−	+	−	+	−	−	−	+	+	+	−	−	+	−	−	v	+	+
V. vulnificus	−	−	v	+	−	v	v	+	−	−	−	+	−	+	−	v	+	+	−	−	−	+	v	+	−	v	+	−	−	v	v	+
Y. intermedia	−	−	+	+	v	v	+	+	−	+	+	+	−	+	+	+	+	+	+	−	−	+	+	+	−	+	+	−	−	+	+	+
Y. ruckeri	−	−	+	+	−	−	v	+	−	−	v	−	−	+	−	+	v	+	−	−	−	+	+	+	−	v	+	−	−	+	−	+

+, − and v correspond to ≥80%, ≤20% and 21–79% of positive responses, respectively.

1 = water, 2 = α-cyclodextrin, 3 = dextrin, 4 = glycogen, 5 = Tween 40, 6 = Tween 80, 7 = N-acetyl-D-galactosamine, 8 = N-acetyl-D-glucosamine, 9 = adonitol, 10 = L-arabinose, 11 = D-arabitol, 12 = cellobiose, 13 = i-erythritol, 14 = D-fructose, 15 = L-fucose, 16 = D-galactose, 17 = gentiobiose, 18 = α-D-glucose, 19 = m-inositol, 20 = α-lactose, 21 = lactulose, 22 = maltose, 23 = D-mannitol, 24 = D-mannose, 25 = D-melibiose, 26 = β-methyl glucoside, 27 = psicose, 28 = D-raffinose, 29 = L-rhamnose, 30 = D-sorbitol, 31 = sucrose, 32 = D-trehalose,

Table 6.6. (*Continued*)

Taxon	33	34	35	36	37	38	39	40	41	42	43	44	45	46	47	48	49	50	51	52	53	54	55	56	57	58	59	60	61	62	63	64
Aer. caviae	+	−	+	+	+	−	−	+	−	v	+	−	−	+	v	−	−	−	−	+	−	+	−	v	−	−	−	+	+	+	−	+
Aer. hydrophila	+	−	+	+	v	v	−	v	−	−	+	−	−	v	−	−	−	−	−	v	−	v	−	−	−	−	−	+	+	+	−	v
Aer. salmonicida	−	−	+	v	−	−	−	−	−	−	−	−	−	−	−	−	−	−	−	−	−	−	−	−	−	−	−	v	−	v	−	v
Aer. sobria	−	−	+	v	v	v	v	−	−	−	+	−	−	−	−	−	−	−	−	v	−	v	−	−	−	−	−	+	+	+	−	v
Cit. freundii	−	−	+	+	+	v	+	+	−	+	+	v	+	+	+	−	−	−	+	v	−	+	−	+	−	+	−	+	+	+	+	+
Edw. tarda	−	−	+	+	+	v	v	+	−	−	+	−	+	+	−	−	−	−	v	+	−	+	−	−	−	−	−	+	+	+	+	−
Ent. agglomerans	v	v	+	+	v	+	v	v	v	+	+	v	+	v	v	−	−	−	−	v	−	+	v	v	v	+	−	+	+	+	+	−
Esch. vulneris	v	−	+	+	+	−	−	+	+	+	+	v	+	v	−	−	−	−	−	−	−	+	v	−	v	+	−	+	+	+	+	−
Haf. alvei	v	v	+	+	+	−	−	+	+	+	+	+	+	+	−	−	v	−	v	v	−	+	v	−	−	−	−	+	+	+	+	v
Klebsiella pneumoniae	−	−	v	+	+	v	+	+	v	+	+	−	+	v	−	−	v	−	−	v	−	+	v	−	v	+	−	+	+	+	+	v
Ph. damselae	−	−	+	+	v	−	−	−	−	−	−	−	−	−	−	−	−	−	−	−	−	+	−	−	−	−	−	+	−	v	−	−
Plesiomonas shigelloides	−	−	+	+	v	−	−	−	−	−	+	−	−	−	−	−	−	−	−	v	−	−	−	−	−	−	−	+	+	v	−	−
Pr. rettgeri	−	v	+	+	+	+	+	+	v	−	+	v	−	v	−	−	+	−	v	+	−	v	−	v	−	−	−	+	+	+	−	v
Ps. fluorescens	−	v	+	v	v	+	+	+	v	v	+	v	v	v	+	v	v	+	v	+	v	+	+	+	+	v	−	+	v	+	v	v
Ps. pseudoalcaligenes	−	−	+	+	+	v	v	+	−	−	−	−	−	+	+	+	−	+	+	+	v	+	−	+	−	−	v	+	+	+	−	+
Ser. liquefaciens	+	v	+	+	+	+	+	+	+	+	+	+	+	v	v	−	v	−	−	+	−	+	v	−	v	−	−	+	+	+	v	+
Ser. marcescens	+	+	+	+	+	+	+	+	v	+	+	+	+	+	+	v	+	−	+	+	−	+	v	+	v	v	v	+	+	+	+	+
Ser. plymuthica	+	−	+	+	+	+	+	+	v	+	+	+	+	−	−	−	+	−	−	+	−	+	−	−	+	−	−	+	+	−	v	−
V. alginolyticus	v	−	+	+	+	−	−	v	−	−	+	−	−	v	−	−	−	−	v	v	−	+	−	+	−	−	−	+	v	+	−	v
V. harveyi	−	−	+	−	−	−	−	−	−	−	+	−	+	−	−	−	−	−	−	−	−	+	−	−	−	−	−	−	−	+	v	−
V. vulnificus	−	−	+	v	v	−	−	−	−	−	v	−	−	−	−	−	−	−	−	−	−	v	−	−	−	−	−	+	−	v	−	−
Y. intermedia	−	−	+	v	+	−	v	v	+	+	+	−	+	−	−	+	−	−	−	v	−	v	−	−	−	−	−	+	v	v	+	+
Y. ruckeri	−	−	+	v	+	−	−	−	−	+	+	−	v	v	−	−	−	−	−	−	−	+	−	−	−	−	−	+	v	v	v	−

+, − and v correspond to ≥80%, ≤20% and 21–79% of positive responses, respectively.

33 = turanose, 34 = xylitol, 35 = methyl pyruvate, 36 = mono-methyl-succinate, 37 = acetic acid, 38 = cis-aconitic acid, 39 = citric acid, 40 = formic acid, 41 = D-galactonic acid lactone, 42 = D-galacturonic acid, 43 = D-gluconic acid, 44 = D-glucosaminic acid, 45 = D-glucuronic acid, 46 = α-hydroxy butyric acid, 47 = β-hydroxy butryric acid, 48 = χ-hydroxy butyric acid, 49 = p-hydroxy phenylacetic acid, 50 = itaconic acid, 51 = α-keto-butyric acid, 52 = α-keto-glutaric acid, 53 = α-keto-valeric acid, 54 = D,L-lactic acid, 55 = malonic acid, 56 = propionic acid, 57 = quinic acid, 58 = D-saccharic acid, 59 = sebacic acid, 60 = succinic acid, 61 = bromosuccinic acid, 62 = succinamic acid, 63 = glucuronamide, 64 = alaninamide,

Table 6.6. (*Continued*)

Taxon	65	66	67	68	69	70	71	72	73	74	75	76	77	78	79	80	81	82	83	84	85	86	87	88	89	90	91	92	93	94	95	96
Aer. caviae	+	+	+	+	+	+	+	+	+	−	v	+	−	+	−	+	+	+	−	v	+	+	+	+	−	−	−	−	+	+	−	−
Aer. hydrophila	+	+	+	+	+	+	+	+	+	v	−	−	v	v	−	+	+	v	−	−	+	+	+	+	−	−	−	−	+	+	+	+
Aer. salmonicida	v	+	+	+	v	+	v	+	−	−	v	−	−	−	−	v	+	−	−	−	+	+	+	+	−	−	−	−	+	−	−	−
Aer. sobria	v	+	+	+	+	+	+	+	v	−	v	−	−	v	−	+	+	v	−	−	+	+	+	+	−	−	−	−	+	+	+	+
Cit. freundii	+	+	+	+	+	+	+	+	+	−	−	v	−	+	−	+	+	v	−	−	−	+	+	+	−	v	+	−	+	+	+	+
Edw. tarda	−	−	v	+	+	+	+	+	−	−	−	−	−	+	−	+	+	+	−	−	−	+	+	+	−	−	−	−	+	+	+	+
Ent. agglomerans	+	+	+	+	+	+	+	+	+	−	−	v	−	+	−	v	+	v	−	v	+	+	+	+	−	−	−	−	+	+	+	+
Esch. vulneris	+	+	+	+	+	+	+	+	−	−	−	−	−	v	v	v	+	+	−	−	v	+	+	+	−	−	−	−	+	+	+	+
Haf. alvei	v	+	+	+	+	+	+	+	+	−	v	+	+	+	−	+	+	+	−	v	v	+	+	+	v	+	−	−	+	+	+	+
Klebsiella pneumoniae	+	+	+	+	+	+	+	+	+	−	−	v	v	+	−	+	+	+	−	v	v	+	+	+	−	v	−	−	+	+	+	+
Ph. damselae	−	−	−	+	+	+	+	v	−	−	−	−	−	−	−	v	−	−	−	−	−	+	+	+	v	−	−	−	+	+	+	+
Plesiomonas shigelloides	−	−	−	+	+	+	v	v	−	−	−	−	−	v	−	−	v	−	−	−	−	+	+	+	v	−	−	−	+	+	−	−
Pr. rettgeri	v	v	v	+	+	+	+	+	+	+	+	v	+	+	+	+	+	v	v	−	+	+	+	+	−	−	−	−	+	+	+	+
Ps. fluorescens	+	+	+	+	+	+	−	v	+	+	v	+	−	+	+	−	+	+	−	v	v	v	−	−	−	+	+	−	v	−	−	−
Ps. pseudoalcaligenes	+	+	v	+	v	+	−	−	v	v	v	v	v	+	+	−	+	+	−	v	v	v	−	−	−	+	+	−	v	−	−	−
Ser. liquefaciens	+	+	+	+	+	+	+	+	+	+	−	+	v	+	v	+	+	+	−	v	+	+	+	+	−	+	−	−	+	+	+	+
Ser. marcescens	+	+	+	+	+	+	+	+	+	+	−	+	v	+	v	+	+	+	−	+	+	+	+	+	−	+	−	−	+	+	+	+
Ser. plymuthica	+	+	+	+	+	+	+	+	+	v	−	−	−	+	−	+	+	v	−	−	+	+	+	+	−	v	−	−	+	+	+	+
V. alginolyticus	+	+	+	+	+	+	+	+	+	v	v	v	−	+	−	v	+	+	−	−	+	+	+	+	v	−	v	−	+	v	+	+
V. harveyi	−	+	+	−	+	+	+	v	−	−	−	−	−	−	−	+	+	+	+	+	−	+	+	+	−	−	−	−	+	+	+	+
V. vulnificus	−	v	v	v	v	+	v	v	−	−	−	−	−	−	−	−	v	−	−	−	v	+	+	v	−	−	−	−	v	−	v	v
Y. intermedia	v	+	+	+	v	+	+	+	v	−	−	−	−	v	−	+	+	v	−	−	v	+	+	+	v	−	−	−	+	+	+	+
Y. ruckeri	−	+	+	+	v	+	v	+	−	−	−	−	−	+	−	−	+	+	−	−	−	+	+	+	−	−	−	−	+	v	−	−

+, − and v correspond to ≥80%, ≤20% and 21–79% of positive responses, respectively.

65 = D-alanine, 66 = L-alanine, 67 = L-alanyl-glycine, 68 = L-asparagine, 69 = L-aspartic acid, 70 = L-glutamic acid, 71 = glycyl-L-aspartic acid, 72 = glycyl-L-glutamic acid, 73 = L-histidine, 74 = hydroxy-L-proline, 75 = L-leucine, 76 = L-ornithine, 77 = L-phenylalanine, 78 = L-proline, 79 = L-pyroglutamic acid, 80 = D-serine, 81 = L-serine, 82 = L-threonine, 83 = D,L-carnitine, 84 = χ-aminobutyric acid, 85 = urocanic acid, 86 = inosine, 87 = uridine, 88 = thymidine, 89 = phenylethylamine, 90 = putrescine, 91 = 2-aminoethanol 92 = 2,3-butanediol, 93 = glycerol, 94 = D,L-α-glycerol phosphate, 95 = glucose-1-phosphate, 96 = glucose-6-phosphate.

Table 6.7. Diagnostic traits of the Gram-positive bacterial fish pathogens

Taxon	1	2	3	4	5	6	7	8	9	10	11	12	13	14	15	16	17	18	19	20	21	22	23	24
Bacillus spp.	c	r	−	+	−	−	−	+	F	?	?	?	?	−	?	−	?	?	?	+	?	?	?	?
Bacillus mycoides	w	r	−	+	−	−	−	+	?	?	?	?	?	−	?	+	?	?	?	+	?	?	?	?
Car. piscicola	w	r	−	−	−	−	−	+	F	−	?	?	−	−	−	−	?	−	+	+	+	+	?	−
Cl. botulinum[b]	w[a]	r	−	+	−	+	−	−	−	−	?	?	−	−	−	+	?	−	?	+	?	?	?	?
Cor. aquaticum	y	r	−	−	−	−	−	+	−	+	?	+	−	−	−	+(β)	?	−	+	+	+	?	?	−
Eu. tarantellae	w	r	−	−	−	−	−	−	−	?	?	?	−	−	+	?	?	?	+	?	?	?	?	?
Lactococcus garvieae	w	c	−	−	−	−	−	+	F	−	?	?	−	−	−	+	?	?	+	+	+	+	?	+
Lactococcus piscium	w	c	−	−	−	−	−	+	F	−	?	?	?	−	−	?	+	?	+	−	?	?	?	−
Micrococcus luteus	y	c	−	−	−	−	−	+	O	+	−	−	+	−	−	?	?	?	?	?	+	?	?	?
Myc. abscessus	−	r	−	−	+	−	−	+	O	+	?	?	?	−	?	?	?	+	?	?	?	+	?	?
Myc. chelonei subsp. *piscarium*	w	r	−	−	+	−	−	+	O	+	?	?	−	−	?	?	?	+	+	−	?	−	+	−
Myc. fortuitum	w	r	−	−	+	−	−	+	O	+	?	?	−	−	?	?	?	+	?	+	?	?	+	?
Myc. marinum	w	r	−	−	+	−	−	+	O	−	?	?	−	−	?	?	?	+	?	v	−	−	−	?
Myc. neoaurum	y	r	−	−	+	−	−	+	O	?	?	?	?	−	?	?	?	?	?	−	+	−	?	?
Noc. asteroides	w	c/r	(+)	−	+	−	−	+	O	+	?	?	−	−	−	?	+	−	?	+	?	?	−	−
Noc. seriolae	w	c/r	−	−	+	−	−	+	O	+	?	?	−	−	+	?	−	−	?	?	?	?	?	?
Planococcus sp.	y	c	−	−	−	+	−	+	O	+	?	+	+	−	−	?	?	?	?	+	+	+	?	?
Ren. salmoninarum[c]	w	r	−	−	−	−	+	+	−	+	−	−	−	−	−	?	−	?	+	−	+	−	?	−
Rhodococcus sp.	re	r	−	−	+	−	−	+	F	−	?	?	−	−	?	?	?	+	?	−	+	?	?	+
Sta. epidermidis	w	c	−	−	−	−	−	+	F	+	?	+	−	−	−	+(β)	−	+	?	+	+	+	?	?
Str. difficilis	w	c	−	−	−	−	−	+	F	−	−	−	?	−	?	−	−	?	−	−	?	−	?	−
Str. iniae	w	c	−	−	−	−	−	+	F	−	−	−	?	−	?	+(α)	+	?	−	+	?	−	?	−

Table 6.7. (*Continued*)

Taxon	1	2	3	4	5	6	7	8	9	10	11	12	13	14	15	16	17	18	19	20	21	22	23	24
Str. milleri	w	c	−	−	−	−	−	+	F	−	−	+	−	−	−	+(β)	−	−	?	?	+	?	?	−
Str. parauberis	w	c	−	−	−	−	−	+	?	−	+	?	?	−	?	+(α)	?	?	+	+	?	−	?	?
Streptomyces salmonis	o/r	m	+	−	−	+	−	+	?	?	?	?	?	−	−	?	+	?	?	+	−	?	−	?
Vag. salmoninarum	w	c/r	−	−	−	−	−	+	F	−	?	?	?	−	+	+(α)	?	−	+	−	?	?	?	−

1 = colony pigmentation, 2 = rods-cocci, 3 = presence of aerial hyphae, 4 = presence of endospores, 5 = acid-fast staining reaction, 6 = motility, 7 = requirement for L-cysteine hydrochloride, 8 = growth in air, 9 = oxidative-fermentative metabolism of glucose, 10 = catalase production, 11 = α-galactosidase production, 12 = β-galactosidase production, 13 = oxidase production, 14 = coagulase production, 15 = H_2S production, 16 = blood degradation (haemolytic activity), 17 = starch degradation, 18 = urea degradation, 19 = growth at 10°C, 20 = growth at 37°C, 21 = growth in 0% (w/v) sodium chloride, 22 = growth in 6.5% (w/v) sodium chloride, 23 = growth in 0.001% (w/v) crystal violet, 24 = acid production from sorbitol.

(+) = weakly positive result; v = variable results; ? = not stated; c = cocci; r = rods; m = mycelium; O = oxidative metabolism; F = fermentative metabolism.
[a] Colony pigmentation: w, y, re and o correspond to off-white/white, yellow, red and orange, respectively.
[b] Confirm by presence of toxin.
[c] Confirmatory profile with API-zym.

Table 6.8. Diagnostic traits of the Gram-negative bacterial fish pathogens

Taxon	1	2	3	4	5	6	7	8	9	10	11	12	13	14	15	16	17	18	19	20	21
Acinetobacter sp.	w	–	–	–	–	–	–	–	–	–	+	–	–	–	+	–	?	–	?	?	?
Aer. allosaccharophila	w	–	–	–	–	+	F	?	+	+	+	+	?	–	?	+	+	+	+	+	+
Aer. caviae	w	–	–	–	–	+	F	+	+	+	+	–	+	–	+	+	?	+	+	+	+
Aer. hydrophila	w	–	–	–	–	+	F	v	+	+	+	–	–	v	+	+	+	+	+	+	+
Aer. jandaei	w	–	–	–	–	+	F	+	?	+	+	+	?	+	?	+	+	?	+	+	+
Aer. salmonicida subsp. *achromogenes*	w	–	–	–	–	–	F	+	+	+	+	–	–	–	–	+	+	+	–	–	+
Aer. salmonicida subsp. *masoucida*	w	–	–	–	–	–	F	+	+	+	+	+	–	+	+	+	+	+	+	–	+
Aer. salmonicida subsp. *salmonicida*	b	–	–	–	–	–	F	+	+	–	+	–	–	–	+	+	+	+	+	–	+
Aer. salmonicida subsp. *smithia*	b	–	–	–	–	–	F	–	+	–	+	–	–	–	–	+	+	–	–	–	+
Aer. sobria	w	–	–	–	–	+	F	+	+	+	+	+	?	+	+	+	+	?	+	+	+
Chrys. balustinum	y	–	–	–	–	–	O	–	–	+	+	+	?	?	?	+	–	?	+	+	?
Chrys. scophthalmum	o	–	–	–	+	–	F	?	?	–	+	–	?	–	?	+	–	?	?	?	?
Cit. freundii	w	–	–	–	–	+	F	–	+	–	–	–	?	–	?	–	?	?	+	+	+
Edw. ictaluri	w	–	–	–	–	+	F	–	–	–	–	+	+	–	+	–	?	+	+	+	+
Edw. tarda	w	–	–	–	–	+	F	–	–	+	–	+	+	–	(+)	–	?	+	+	+	+
Ent. agglomerans	y	–	–	–	–	+	F	–	+	–	–	–	+	+	–	+	–	?	+	+	?
Esch. vulneris	w	–	–	–	–	+	F	+	?	–	–	+	+	–	+	–	–	+	+	+	?
Fla. branchiophilum	y	–	–	–	–	–	O	?	?	–	+	?	?	?	?	+	+	?	+	–	+
Fla. columnare	y	–	–	–	+	–	O	?	?	–	+	–	–	–	?	+	–	?	+	(+)	+
Fla. hydatis	y	–	–	–	+	–	O	?	?	–	–	+	–	–	+	+	+	?	+	+	+
Fla. johnsoniae	y	–	–	–	+	–	?	?	+	–	+	?	?	?	?	+	+	?	+	–	–
Fla. psychrophilum	y	–		–	+	–	–	?	?	–	–	–	?	–	?	+	–	?	–	–	+
Fle. maritimus	y	–	–	–	+	–	O	?	?	–	+	?	?	–	?	+	–	?	+	–	–
Fle. ovolyticus	y	–	–	–	+	–	O	–	–	–	+	–	?	?	?	+	–	?	–	–	–
Haf. alvei	w	–	–	–	–	+	F	–	+	–	–	+	+	+	?	–	?	?	+	+	+

Table 6.8. (*Continued*)

Taxon	1	2	3	4	5	6	7	8	9	10	11	12	13	14	15	16	17	18	19	20	21
Halomonas cupida	w	–	–	–	–	+	–	+	?	–	–	+		–	+	?	?	?	–	–	–
J. lividum	p	–	–	–	–	+	O	+	–	–	+	?	?	+	?	+	?	+	+	–	+
Klebsiella pneumoniae	w	–	–	–	–	+	F	–	?	–	–	–	?	?	?	–	?	+	+	+	+
Moraxella sp.	w	–	–	–	–	–	O/–	–	–	–	+	–	?	–	+	–	?	+	?	?	?
Moritella marina	w	–	–	–	–	+	F	–	?	?	+	?	+	–	+	+	+	?	?	–	–
Mycoplasma mobile	?	–	+	–	+	–	?	?	?	?	+	?	?	?	+	–	?	+	+	–	?
Myxococcus piscicola	y	+	–	–	+	?	?	?	?	?	?	?	?	?	?	?	?	?	?	?	+
Ph. damselae	w	–	–	–	–	+	F	+	–	+	+	–	+	+	?	–	+	+	+	+	–
Ph. damselae subsp. *piscicida*	w	–	–	–	–	–	F	–	–	–	+	+	+	+	–	–	–	–	+	–	–
Piscirickettsia salmonis	?	?	–	+	–	–	–	?	?	?	?	?	?	?	?	?	?	?	?	?	?
Plesiomonas shigelloides	w	–	–	–	–	+	F	+	+	+	+	+	?	–	?	–	?	?	+	+	+
Pr. rettgeri	w	–	–	–	–	+	F	–	–	+	–	–	+	–	?	–	–	–	+	+	+
Ps. anguilliseptica	w	–	–	–	–	+	O	–	–	–	–	–	?	?	–	+	–	?	+	–	+
Ps. chlororaphis	g/fl	–	–	–	–	+	O	+	?	?	+	?	?	?	?	+	?	?	+	?	+
Ps. fluorescens	w/fl	–	–	–	–	+	O	v	–	–	+	–	?	–	?	+	–	?	+	+	+
Ps. pseudoalcaligenes	w	–	–	–	–	+	O(alk)	+	–	–	+	–	?	–	?	+	–	–	+	–	+
Sal. choleraesuis subsp. *arizonae*	w	–	–	–	–	+	F	?	+	–	–	+	+	–	–	–	–	+	+	+	+
Ser. liquefaciens	w	–	–	–	–	+	F	–	+	–	(–)	+	–	–	+	+	–	+	+	+	+
Ser. marcescens	r	–	–	–	–	+	F	–	+	–	–	+	–	+	+	+	+	+	+	+	?
Ser. plymuthica	r	–	–	–	–	+	F	–	+	–	–	–	–	+	–	+	?	+	+	+	+
Shewanella putrefaciens	w	–	–	–	–	+	–	–	–	–	+	–	?	–	?	–	?	–	+	+	–
Sporocytophaga sp.	y	+	–	–	+	–	O	?	?	?	?	?	?	?	?	?	?	?	?	?	–
V. alginolyticus	w	–	–	–	–	+	F	–	+	+	+	+	+	+	+	+	+	+	+	+	–
V. anguillarum	w	–	–	–	–	+	F	–	+	v	+	–	–	+	?	+	+	+	+	+	–
V. cholerae	w	–	–	–	–	+	F	–	+	+	+	+	+	+	+	+	+	?	+	+	+

Table 6.8. (*Continued*)

Taxon	1	2	3	4	5	6	7	8	9	10	11	12	13	14	15	16	17	18	19	20	21
V. harveyi	w	–	–	–	–	+	F	–	–	+	+	+	?	–	+	+	–	–	+	+	–
V. ichthyoenteri	w	–	–	–	–	+	F	–	–	–	+	?	?	?	?	–	–	+	+	–	+
V. logei	w	–	–	–	–	+	F	–	?	–	+	–	?	–	+	–	–	+	–	–	–
V. ordalii	w	–	–	–	–	+	F	–	–	–	+	–	–	–	?	+	–	+	(+)	–	–
V. pelagius	w	–	–	–	–	+	F	–	v	v	+	–	+	v	?	+	+	?	+	+	?
V. salmonicida	w	–	–	–	–	+	F	–	–	–	+	?	?	–	–	–	?	?	–	–	+
V. splendidus	w	–	–	–	–	+	F	+	–	–	+	–	+	+	?	–	+	+	+	+	–
V. trachuri	w	–	–	–	–	+	F	–	–	+	+	+	?	–	?	?	?	?	+	?	–
V. vulnificus	w	–	–	–	–	+	F	–	–	–	+	+	+	–	+	+	+	+	+	+	–
Y. intermedia	w	–	–	–	–	+	F	–	+	+	–	–	+	+	?	?	?	?	+	+	?
Y. ruckeri	w	–	–	–	–	+	F	–	+	–	–	+	+	–	?	+	?	+	+	+	+

1 = colony pigmentation, 2 = presence of microcysts, 3 = filterability through 0.45 μm pore filters, 4 = growth only occurs in tissue culture, 5 = gliding, 6 = motility by flagella, 7 = oxidative–fermentative metabolism of glucose, 8 = arginine dihydrolase production, 9 = β-galactosidase production, 10 = indole production, 11 = oxidase production, 12 = lysine decarboxylase production, 13 = methyl red test, 14 = Voges Proskauer reaction, 15 = degradation of blood, 16 = gelatin degradation, 17 = starch degradation, 18 = acid production from maltose, 19 = growth at 30°C, 20 = growth at 37°C, 21 = growth in 0% (w/v) sodium chloride.

b, fl, g, p, r, w and y correspond to brown, fluorescent, green, purple, red, cream/white and yellow pigment, respectively.

+, –, v, ?, (+) and (–) correspond to positive ($\geq$80% of positive responses), negative ($\leq$20% of positive responses), variable (21–79% of positive responses), unstated, weakly positive, and weakly negative result, respectively.

O, F, Alc = oxidative or fermentative metabolism, or production of alkali, respectively.

sources, acid production from D-mannitol, D-melibiose, D-raffinose, L-rhamnose, salicin and sucrose, and the Voges Proskauer reaction were considered differential (Martinez-Murcia *et al.*, 1992). However, a word of caution is necessary, insofar as the organisms which clearly demonstrated genetic homogeneity were markedly heterogeneous phenotypically. This would complicate diagnoses.

Aer. salmonicida may be distinguished from other fish pathogens on the basis of a small number of phenotypic tests, notably the Gram-staining reaction (small Gram-negative rods), motility (usually appears to be non-motile), growth at 37°C (usually a negative response), fermentative metabolism, catalase and oxidase production (both positive) and acid production from sucrose and xylose (both negative; recently, acid production from sucrose has been attributed to some isolates (Wiklund *et al.*, 1992)). These tests will result in a provisional identification of *Aer. salmonicida* (McCarthy, 1976). In addition, it is recommended that pathogenic isolates should be examined for degradation of gelatin (positive), starch (positive) and urea (negative), arginine dihydrolase (positive), gluconate oxidation (negative) and ornithine decarboxylase production (negative). Unfortunately, this apparently simple state of affairs may be complicated by the increasing presence of 'atypical' isolates, particularly in non-salmonid fish. In particular, these may be non- or slowly-pigmenting.

Diagnosis of *Y. ruckeri* may be achieved by isolation of the pathogen, such as on the selective media of Waltman and Shotts (1984) or Rodgers (1992), and thence identification. According to Waltman and Shotts (1984), 53/60 isolates hydrolysed Tween 80 but none fermented sucrose. Therefore, typically on the selective medium, *Y. ruckeri* colonies were green with a zone of hydrolysis (indicated by the presence of insoluble calcium salts) around them. Unfortunately in our experience with this medium, UK isolates rarely hydrolysed Tween 80. Therefore, interpretations should be made carefully.

Wakabayashi and Egusa (1972) proposed an identification scheme for *Ps. anguilliseptica* based on a small number of phenotypic traits, principally motility, growth at 37°C, presence of soluble pigment, production of H_2S, indole and oxidase, nitrate reduction, gelatin degradation, susceptibility to the vibriostatic agent (0/129), and the ability to attack glucose. According to these workers, the tests were sufficient to differentiate *Ps. anguilliseptica* from *Ps. fluorescens*, *Ps. alcaligenes*, *V. anguillarum*, *Aer. liquefaciens* (= *Aer. hydrophila*), *Ph. damselae* subsp. *piscicida* and *H. piscium*.

A simplified diagnostic test for *V. anguillarum*, involving 'glucose motility deeps' (GMD) has been reported (Walters and Plumb, 1978). Essentially, GMD is a much modified version of the oxidation-fermentation test medium, comprising:

phenol red broth base (Difco)	1.6% (w/v)
glucose	1.0% (w/v)
yeast extract	0.3% (w/v)
agar	0.3% (w/v)

Stab-inoculated media are incubated at 25°C for 24–48 h, when acid production and motility (indicated as a carrot-like diffuse growth around the stab mark) are recorded. It remains for further work to confirm the specificity of the reaction for *V. anguillarum*.

It was proposed to separated *V. trachuri* from other vibrios by means of a small number of phenetic tests (Iwamoto *et al.*, 1995):

Test	V. trachuri	V. anguillarum	V. parahaemolyticus
Arginine dihydrolase	–	+	–
Lysine decarboxylase	+	–	+
Voges Proskauer reaction	–	+	–

- **Colony morphology and pigmentation**
  This should be recorded from 'young' colonies, i.e. shortly after growth is initially detected. The presence of aerial hyphae may be assessed with a stereo-microscope. The presence of pigment should be assessed from basal medium supplemented with 5–10% (w/v) skimmed milked powder (Oxoid).
- **The Gram-staining reaction**
  With smears from young cultures, this reaction serves also to determine the presence of rods, cocci, mycelia, microcysts and endospores. For convenience, we recommend the use of commercially available staining and decolorizing solutions, such as those marketed by Difco. Heat-fixed smears should be stained for 1 min with crystal violet, washed in tap water, covered with Grams-iodine for 1 min, re-washed, decolorized by a few seconds in acetone-alcohol, and counterstained for 30 sec in safranin. The smears are washed thoroughly, and gently blotted dry, prior to microscopic examination preferably at a magnification of × 1000.
- **The acid-fast staining reaction**
  This reaction highlights the presence of *Mycobacterium*, *Nocardia* and possibly *Rhodococcus*. Heat-fixed smears may be flooded with carbol fuchsin, and heated until the steam rises by means of wafting a source of heat (from a Bunsen burner or cotton wool plug soaked with alcohol) underneath the slide. After 5 min, the stain is washed away with tap water, and the smear decolorized with acid-alcohol until only a faint pinkish tinge remains. The slide is re-washed, before applying a methylene blue counterstain for 30 s. Following re-washing with tapwater, the slide is gently blotted dry, and examined by oil-immersion (Doetsch, 1981).
- **Motility**
  In our experience, wet-preparations prepared from barely turbid suspensions are most satisfactory when viewed by phase contrast microscopy at × 400 magnification.
- **Gliding motility**
  This may be assessed from the development of spreading growth on low-nutrient (cytophaga) agar. It should be differentiated from locomotion by means of flagella.
- **Filterability through the pores of 0.45 μm pore size porosity filters**
  The ability of cells to pass through the pores of 0.45 μm pore size porosity filters is indicative of the presence of L-forms and mycoplasmas. Thus, the bacterial suspension is filtered, and the filtrate applied to a suitable growth medium. Growth within 7 days is indicative of filterability.
- **The ability to grow only in fish cell cultures**
  Viruses and rickettsias are only capable of growth in suitable cell cultures.
- **Aerobic or anaerobic requirements for growth**
  These are apparent after incubating inoculated media aerobically and anaerobically.

- **Catalase production**
  This is recorded by effervescence within 1 min from 3% (v/v) hydrogen peroxide following application of a bacterial colony. Quite simply, the 'young' colony may be scraped with a thin glass rod and transferred to a drop of hydrogen peroxide on a glass slide.
- **Fluorescent (fluorescein) pigment production**
  This is assessed by the presence of a fluorescent green pigment seen under ultraviolet light, after 7 days' incubation on the medium of King *et al.* (1954).
- **Growth at 10, 30 and 37°C**
  Growth at 30 and 37°C should be recorded within 72 h incubation on basal medium. At 10°C, the media should be retained for up to 14 days.
- **Growth on 0% and 6.5% (w/v) sodium chloride and on 0.001% (w/v) crystal violet**
  This is reported after 7 and 14 days's incubation on suitably modified basal medium.
- **Requirement for 0.1% (w/v) L-cysteine hydrochloride**
  This is essentially a requirement for the growth of *Ren. salmoninarum*. Inoculated media should be incubated at 15°C, and examined at weekly intervals for up to 16 weeks.
- **Oxidation-fermentation test**
  This involves the measurement of acid production from glucose metabolism under aerobic and/or anaerobic conditions in the basal medium of Hugh and Leifson (1953). The production of an alkaline reaction is indicated by a deep blue colour which develops, usually in the open tube. For marine organisms, it is necessary to use the modified medium of Leifson (1963). The presence of acid, indicated by a colour change to yellow, should be recorded after incubation for 1, 2 and 7 days.
- **Indole production**
  This is recorded after 7 days incubation in 1% (w/v) peptone water. For marine organisms, this should be prepared MSS (2.4% (w/v) NCl; 0.7% (w/v) $MgSO_4.7H_2O$; 0.075% (w/v) KCl; after Austin *et al.*, 1979). A positive response is indicated by a red coloration following the addition of a few drops of Kovacs reagent.
- **α-galactosidase production**
  One of the most reproducible methods is to record α-galactosidase production from the API-zym system after incubation for 48 h at 15 or 25°C.
- **β-galactosidase production**
  This involves use of the medium of Lowe (1962). Inoculated medium is incubated for 7 days, whereupon a positive response is indicated by a yellow coloration. For marine organisms, the medium should be prepared in MSS.
- **Production of arginine dihydrolase and lysine decarboxylase**
  We recommend use of the medium described by Møller (1955). Essentially inoculated medium is incubated for 7 days, when a positive reaction is indicated by a purple coloration. With marine organisms, the medium should be prepared in MSS.
- **Urease production**
  Using the medium of Stuart *et al.* (1945), a positive response develops as a reddish coloration, within 28 days. For marine organisms, it is suggested that the medium is supplemented with 2.4% (w/v) sodium chloride.

- **Methyl red test and Voges Proskauer reaction**
  These may be recorded after 7 days incubation in MRVP broth (Difco). Following the addition of a few drops of methyl red, a bright red coloration indicates a positive methyl red test. The Voges Proskauer reaction is recorded after use of commercially available reagents. A positive reaction is indicated by a red coloration which develops within 18 h (usually within 1 h) after the addition of the reagents. As before with marine organisms, the medium may be prepared in MSS.

- **Degradation of blood**
  This should be recorded within 7 days as zones of clearing around colonies on basal medium supplemented with 5% (v/v) defibrinated sheep's blood.

- **Degradation of gelatin**
  This is detected after 7 days incubation by the addition of saturated ammonium sulphate solution to the medium of Smith and Goodner (1958). A positive result is indicated by zones of clearing around the bacterial growth. For marine organisms, the medium should be supplemented with MSS.

- **Degradation of starch**
  Basal medium supplemented with 1% (w/v) soluble starch is streaked, and incubated at 15–25°C. After 7 days, the starch plates are flooded with an iodine solution (e.g. Difco Gram's iodine). The degradation of starch is indicated by a clear area surrounded by a blue/black background.

- **Acid production from maltose and sorbitol**
  The use of Andrade or phenyl red-peptone water supplemented with maltose or sorbitol is advocated (see Cowan, 1974). This medium contains 1% (w/v) bacteriological peptone, 0.5% (w/v) sodium chloride (for marine organisms this amount should be increased to 2%), 1% (w/v) maltose and Andrade or phenyl red indicator. The filter sterilized (0.22 μm pore size porosity filter) maltose solution should be added to the basal medium after autoclaving, and the completed medium dispensed into test tubes. The production of acid is indicated by the development of a pink colour within 48 h at 25–37°C.

- **Production of hydrogen sulphide**
  Many methods have been developed to detect the production of hydrogen sulphide. We have found success with triple sugar iron agar (Oxoid), which should be prepared as slopes in test tubes. Following incubation of the inoculated media at 15–25°C for up to 7 days, the production of hydrogen sulphide is indicated by blackening of the agar.

- **Coagulase test**
  We recommend a simple test using citrated plasma (or rabbit, sheep, donkey or ox). The bacterial culture should be emulsified [to form a dense suspension of ~5×10^8 cells/ml] in a drop of 0.9% (w/v) saline on a clean grease-free microscope slide. This suspension is then carefully mixed with one drop of citrated plasma. A positive result, which is indicated by clumping of the bacterial cells, is apparent within 2–3 min.

Most of the above-mentioned phenotypic tests have been derived from medical microbiology. Nevertheless, careful attention to detail will generate useful data about bacterial fish pathogens. Undoubtedly, more modern methods will eventually enter the realms of fish microbiology. These methods may include the development of highly reliable rapid techniques, such as offered by high-pressure liquid chromatography and mass-spectrometry. Moreover, lipid analyses could be adapted further for fisheries work. Serological techniques, such as those involving ELISA and monoclonal antibodies, are steadily entering the domain of the fish disease diagnosticians. In addition, molecular genetic techniques, notably gene probe technology, are under evaluation in several laboratories.

H. OTHER TECHNIQUES

A novel diagnostic approach concerns determination of plasmid profiles for *Edw. ictaluri* (Lobb and Rhodes, 1987; Speyerer and Boyle, 1987).

Development of a bacteriophage typing scheme may be of considerable value for diagnosis in the future. Reference is made here to a collection of tailed icosahedral bacteriophages which are specific to *Y. ruckeri* (Stevenson and Airdrie, 1984b).

The use of microwave radiation [700 w energy from a domestic microwave] has been suggested for *Pis. salmonis* (Larenas *et al.*, 1996).

7

Epizootiology: Gram-positive bacteria

The reservoir of many Gram-positive bacterial fish pathogens is unknown. Whereas some groups, e.g. streptococci, occur in polluted waters, other organisms, e.g. *Ren. salmoninarum*, seem to be restricted to fish. How do such organisms spread between separate fish populations?

A. ANAEROBES

Clostridium botulinum

Cl. botulinum is widespread in soil, marine and freshwater sediments and in the gastro-intestinal tract of man and other animals, including fish (Bott *et al.*, 1968; Cato *et al.*, 1986). In one study of 530 trout in Danish earth ponds, *Cl. botulinum* type E was discovered to occur in 5–100% of the fish in water, and in 85–100% of the population in late summer (Huss *et al.*, 1974a). It was supposed that the principal source of contamination with this organism was from minced trash fish used as feed, although soil and water could also be involved (Huss *et al.*, 1974a). Moreover, it was considered likely that clostridia became established in the mud and bottom-living invertebrates in trout ponds (Huss *et al.*, 1974b). In Britain, it has been determined from an examination of 1400 trout collected from 17 fish farms, that the incidence of *Cl. botulinum* in whole fish and viscera was 9.4% and 11.0% respectively. Nevertheless, *Cl. botulinum* lingers in the fish farm environment for considerable periods following outbreaks of disease. Thus, at the English trout farm which experienced botulism, the organism (possibly as endospores) was recovered for a year after the outbreak of disease. The numbers ranged from 1 to 800 organisms/g of sediment, compared to <1/g at an unaffected control site (Cann and Taylor, 1982). There are no data available to assess the level of contamination in wild fish stocks (Cann *et al.*, 1975). Similarly, there is no evidence to suggest that trout contaminated with *Cl. botulinum* could comprise a human health hazard (Bach *et al.*, 1971).

Eubacterium tarantellae

So far, the organism has only been recovered from the brain of mullet and ten other unnamed species of estuarine fish caught in Biscayne and Florida Bays. It has not been found outside this area. Moreover, the inability to grow in 2% (w/v) sodium chloride implies that the organism is likely to be restricted to estuarine environments (Udey *et al.*, 1977). These authors consider that isolates recovered from moribund fish, caught off the Texas coast and tentatively identified as *Catenabacterium* (Henley and Lewis, 1976), also belong in *Eubacterium*, as *Eu. tarantellae*. Therefore, the range would appear to be restricted to the warmer waters of the southern part of the USA. It is uncertain whether or not the organism occurs in water, or indeed as part of the resident microflora of fish, although Trust *et al.* (1979) isolated eubacteria from the intestinal tract of three fish species. Therefore, it is conceivable that *Eu. tarantellae* could comprise part of the anaerobic microflora of the digestive tract, although it will be necessary for further study to clarify this point.

B. GRAM-POSITIVE BACTERIA—THE 'LACTIC ACID' BACTERIA

Carnobacterium piscicola and the lactobacilli

So far, it would appear that the disease is confined to Europe and North America. However, it is unclear whether fish are the natural hosts for *Car. piscicola* and other lactobacilli, or if they comprise part of the natural aquatic microflora.

Gram-positive cocci in chains

Streptococcicosis (= streptococcosis) was initially described among populations of rainbow trout farmed in Japan (Hoshina *et al.*, 1958). Since then, the disease has increased in importance, with outbreaks occurring in yellowtails (Kusuda *et al.*, 1976a; Kitao *et al.*, 1979), coho salmon (Atsuta *et al.*, 1990), Jacopever (*Sebastes schlegeli*) (Sakai *et al.*, 1986), Japanese eels (Kusuda *et al.*, 1978b), ayu and tilapia (Kitao *et al.*, 1981). The disease, also known as 'pop-eye', has assumed importance in rainbow trout farms in Australia, Israel, Italy and South Africa (Barham *et al.*, 1979; Boomker *et al.*, 1979; Carson and Munday, 1990; Ceschia *et al.*, 1992; B. Austin, unpublished data) and in Atlantic croaker (*Micropogon undulatus*), blue fish (*Pomatomus saltatrix*), channel catfish, golden shiner (*Notemigonous chrysoleuca*), hardhead (sea) catfish (*Arius felis*), menhaden (*Brevoortia patronus*), pinfish (*Lagodon rhomboides*), sea trout (*Cynoscion regalis*), silver trout (*Cynoscion nothus*), spot (*Leiostomus xanthurus*), stingray (*Dasyatis* sp.), striped bass (*Morone saxatilis*) and striped mullet (*Mugil cephalus*) in the USA (Robinson and Meyer, 1966; Plumb *et al.*, 1974; Cook and Lofton, 1975; Baya *et al.*, 1990c). The disease may also have occurred sporadically in Great Britain and Norway. There is good evidence that streptococcicosis is problematical in both farmed and wild fish stocks.

There is evidence that the pathogens abound throughout the year in the aquatic environment, occurring in water, mud and in the vicinity of fish pens (Kitao *et al.*, 1979). Some seasonality has been recorded, with higher numbers present in seawater during the summer. In contrast, greatest numbers were isolated in mud during autumn and winter (Kitao *et al.*, 1979). This is interesting information, but unfortunately the authors do not comment further about the reasons for the presence of streptococci in the aquatic environment. Conceivably, the organisms may have been released from infected fish and were being merely retained in the water and underlying sediment. Alternatively with the inconclusive taxonomic status of the fish pathogenic streptococci, it is difficult to conclude that any environmental isolates correspond precisely to the description of the pathogens. Therefore, any environmental isolates could be merely indicators of an unsanitary condition and not necessarily imply the presence of fish pathogenic strains. However, this evades the question about the precise source of infection. Minami (1979) determined that streptococci, with similarities to the fish pathogens, were present in fresh and frozen fish used for yellowtail diets. This worker reported that the isolates were pathogenic, and could survive for over 6 months in the frozen state. The suggestion was made, therefore, that the contaminated diets served as an important source of infection. The importance of food-borne infection was further highlighted by Taniguchi (1982a,b; 1983).

It is recognized that streptococcicosis may be transmitted by contact with infected fish. In this context, Robinson and Meyer (1966) transmitted the disease by cohabiting an infected golden shiner with healthy specimens of the same species. The healthy fish succumbed to streptococcicosis, and died within 5 days.

Bacteriophages of *Lactococcus garvieae* have been found in seawater and sediment, but after defining 14 phage types (among 111 isolates), it was concluded that there was not any correlation between phage type and geographical source of the isolates (Park *et al.*, 1998).

Although the source of *Str. parauberis* was not determined, it may be relevant that previously the organism has been associated with raw milk and bovine mastitis (Doménech *et al.*, 1996).

C. AEROBIC GRAM-POSITIVE RODS AND COCCI

Corynebacterium aquaticum
Apart from fish, the organism was recovered from water and the 'scum' forming at the air–water interface on the tank walls (Baya *et al.*, 1992b).

Mycobacterium spp.
Very little is known about the epizootiology of fish pathogenic mycobacteria. Undoubtedly, the reservoir for the organism is the aquatic environment, although the factors that lead to the development and spread of the disease condition are unknown. To illustrate the potential spread of the disease, in Oregon as many as 26% of hatchery fish may be infected (Arakawa and Fryer, 1984). Possibly, transmission may be by ingesting contaminated food or debris (Dulin, 1979). Infection, via the intra-ovarian route, has been demonstrated for the Mexican platyfish (*Xiphophorus maculatus*) (Conroy, 1966). However, other investigators have ruled out vertical (i.e. egg) transmission as a means of spreading the disease (Ross and Johnson, 1962; Wood, 1974).

Nocardia spp.
Nocardia spp. (including *Noc. asteroides*) occur in freshwater and soil. With the observation that *Noc. kampachi* grows in up to 4.5% (w/v) sodium chloride, it was suggested

that the normal habitat is terrestrial or limnetic. Moreover, in survival experiments, the organism remained viable in clean seawater for only a few days, although in polluted conditions viability was considerably extended (Kariya *et al.*, 1968). The inference, therefore, is that fish become infected from the natural environment. Presumably these diseased animals serve as a reservoir for further infection.

Renibacterium salmoninarum

To date, there has been no evidence to suggest that *Renibacterium* is a component of the normal aquatic microflora. Indeed in one study, water and sediment from 56 fish farms were examined for the presence of renibacteria, but to no avail (Austin and Rayment, 1985). Survival experiments confirmed that *Renibacterium* could survive in fish tank sediment/faecal material for up to 21 days in the absence of any fish. However, the organism was not at any time recovered from the overlying water, suggesting that renibacteria have an affinity with organic matter. The question regarding survival of the pathogen in water was the topic of detailed experimentation. This confirmed that laboratory-grown cultures were short-lived in river water. In the absence of indigenous water-borne organisms, i.e. using filter-sterilized river water, renibacterial cells survived for 28 days, after which there was a rapid decline in numbers. In the absence of indigenous water-borne organisms, i.e. using filter-sterilized river water, renibacteria survived for 28 days, after which there was a rapid decline in numbers (Fig. 7.1). Essentially, these data show that renibacteria have the potential to survive outside of fish for limited periods, although in water it is probably unable to compete with members of the normal aquatic microflora (Austin and Rayment, 1985). In one study, the pathogen was found only in association with asymptomatic and clinically diseased fish (Austin and Rayment, 1985). In another investigation, it was determined that the blue mussel (*Mytilus edulis*) cleared and killed most *Ren. salmoninarum* cells from seawater (Paclibare *et al.*, 1994). However, some renibacterial cells could be found in mussel faeces during settling. Yet, it was conceded that mussels were unlikely to pose a realistic threat to fish farms regarding the survival and spread of renibacterium. But what if infected mussels are transferred to clean seawater? The answer according to Paclibare *et al.* (1994) was that the mussel cleared *Ren. salmoninarum* from within them upon transfer to clean sites. Clearly, early studies may have been hampered by lack of a suitable selective medium. Nevertheless, the use of SKDM has not, as yet, produced any definite evidence to suggest a non-fish reservoir for the organism (Austin *et al.*, 1983a; Embley, 1983; Austin and Rayment, 1985). The precise source of infection is unclear, but may include clinically or asymptomatically diseased fish (Wood and Wallis, 1955; Wolf, 1966; Bucke, 1978; Mitchum *et al.*, 1979; Paterson *et al.*, 1979; Fryer and Sanders, 1981). The organism has been recovered from faeces of both cultured and wild salmonid stocks. According to Balfry *et al.* (1996), renibacterium is shed from faeces, and may survive in seawater for a week. Attention has also been focused on the role of eggs in the transmission of BKD ('vertical' transmission) (Allison, 1958; Wolf, 1966; MacLean and Yoder, 1970; Mitchum *et al.*, 1979; Lee and Evelyn, 1989). Allison (1958) indicated the involvement of eggs when BKD occurred following transfer of ova from an infected site. Similarly, Bullock *et al.* (1978b) implicated disinfected eggs of chinook salmon in the spread of the disease. Moreover, the preliminary data of Paterson *et al.* (1981) pointed to the presence of

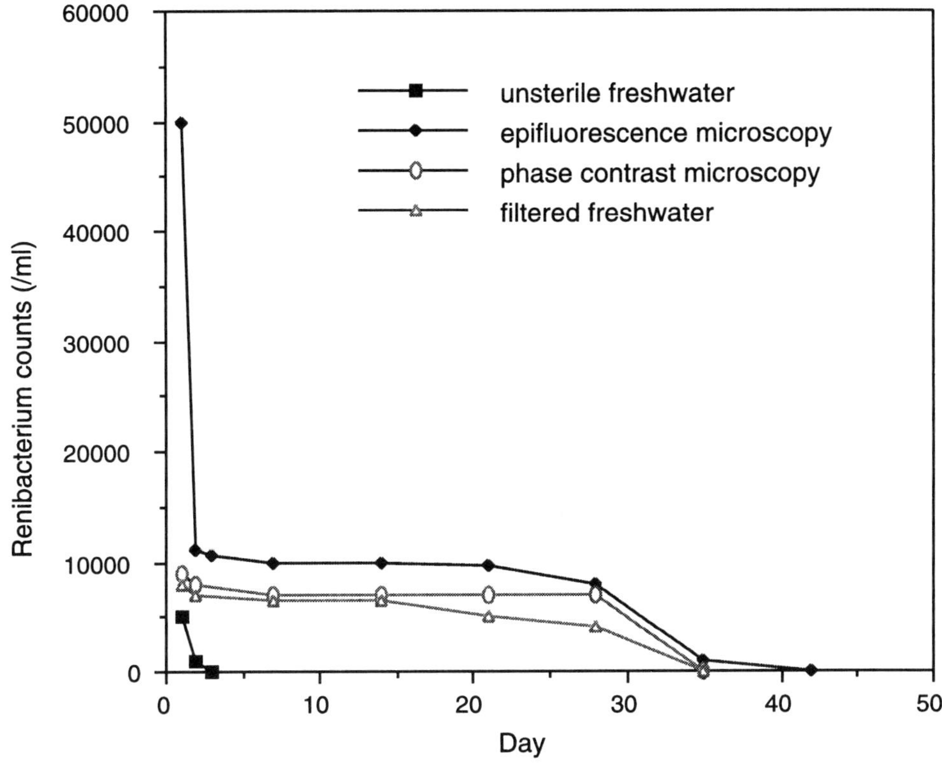

Fig. 7.1. Survival of *Ren. salmoninarum* in river water at 15°C (based on the publication of Austin and Rayment, 1985). Essentially, there was a dramatic decline in numbers of bacterial cells in unsterile and filtered freshwater within 48 h, as assessed by epifluorescence and phase contrast microscopy. In unsterile freshwater, the populations declined to negligible numbers shortly afterwards. However, in filtered freshwater, a residual population of *Ren. salmoninarum* remained for 40+ days.

Renibacterium within fertilized eggs. Evelyn *et al.* (1984) demonstrated the presence of renibacteria in 11.6–15.1% of eggs from a coho salmon which was infected with BKD, such that the coelomic fluid was cloudy because of high numbers of the organism. These authors suggested that *Ren. salmoninarum* was present in the yolk of the eggs, even after treatment with erythromycin (Evelyn *et al.*, 1986a). Of greater significance was the finding that iodophors were ineffective at preventing intra-ovum infections. Clearly, this has profound implications for control of the disease.

The manifestation of the disease is complicated by certain environmental factors, including water hardness (Warren, 1963), temperature, salinity and diet. Belding and Merrill (1935) were the first workers to describe the seasonal nature of BKD, with a correlation between water temperature and level of mortality. Earp *et al.* (1953) found that BKD occurred over a wide range of water temperatures from 8–18°C. Most epizootics occurred in the autumn and winter, i.e. during periods of declining water temperatures. However, most mortalities occurred at higher temperatures, a conclusion which has been echoed by Austin (1985), although the reverse has also been reported (Sanders *et al.*, 1978). At low temperatures, the effect was the continual loss of small numbers of fish (Snieszko and Griffin, 1955). This is quite a feat for a supposedly unreactive organism!

BKD has been diagnosed in fish following movement from freshwater to seawater (Earp *et al.*, 1953; Bell, 1961). Indeed, the disease may be of paramount importance in the ability to acclimatize to seawater (Frantsi *et al.*, 1975) and the survival of salmonids in the oceanic environment (Fryer and Sanders, 1981).

Data obtained with ELISA have shown that *Ren. salmoninarum* occurs commonly, in the absence of pathological signs of BKD, in wild fish, i.e. Arctic charr and brown trout in Iceland (Jónsdóttir *et al.*, 1998). Thus, the route of transmission to aquaculture may be from wild fish.

The effect of nutritional (dietary) status on the development of BKD is only partially understood. Some diets, notably those containing corn gluten (Wedemeyer and Ross, 1973) or lipid (Austin, 1985), enhanced the disease. Nutritional studies with Atlantic salmon have shown that levels of vitamin A, iron and zinc are lowered in BKD infected fish (Paterson *et al.*, 1981). Subsequent experiments in which fish were administered diets rich in trace elements resulted in reduced incidences of BKD. This theme should be exploited further for control purposes.

In fish culture, *Ren. salmoninarum* appears to be a most unaggresssive organism, generally devoid of much production of exoenzymes (exotoxins). Yet, it causes such a severe problem in salmonids. With lack of evidence to the contrary, it is our hypothesis that the organism is a normal resident of some fish, in (or on) which it exists, probably in fairly low numbers. Conceivably, it may be a normal resident of kidney tissue, forming a synergistic or controlled parasitic relationship with the host, possibly in the macrophages. Alternatively, it may be a normal resident of the digestive tract (Austin, 1986). To continue the scenario, we postulate that at times of stress to the host, such as sub-clinical infections, damage to the digestive tract, starvation, kidney damage or temperature shock, the organism is able to migrate to the kidney (if not already there) and multiply. This would lead ultimately to the condition known as BKD. The problems with recovery of the organism, particularly from asymptomatic fish, may be explained within the realms of this concept. Evelyn and co-workers (Evelyn, 1978; Evelyn *et al.*, 1981) have considered the presence of inhibitors in the kidney which suppress the development of *Renibacterium* on solid medium. Could these unnamed compounds control the growth and development of the organism in healthy fish? This remains a possibility. However, there are other equally plausible explanations, namely dormancy, damage, or the presence of altered— osmotically fragile—cells. The renibacteria may normally be in a dormant or altered phase within the fish and, thus, would require to be triggered back into activity in order to produce colonies. Alternatively, renibacteria may be in some way damaged in the fish, and require repair before being able to produce colonies. This parallels the problem of damaged coliform bacteria in the aquatic environment (Olson, 1978). Of course, it is also possible that the media are deficient in certain essential nutrients, necessary for the replication of *Renibacterium*. A long lag phase, which has been suggested by Embley (1983), would be necessary for the organism to adjust to the new environment of the laboratory medium, prior to replication. Any or all these possibilities could apply to

Renibacterium. Careful thought is necessary to unravel many of the mysteries still surrounding the biology of this pathogen.

Rhodococcus sp.
It was considered that feeding with crude fish offal may have been the cause of infection (Claveau, 1991).

Staphylococcus epidermidis
Sugiyama and Kusuda considered that the bacteria originated from water (or fish) rather than from human beings, because of the pronounced antigenic differences to human—Tachikawa—strains (Sugiyama and Kusuda, 1981b). This seems to be a likely possibility in view of other ecological studies which have clearly demonstrated the presence of *Sta. epidermidis* in the aquatic environment (e.g. Gunn *et al.*, 1982).

8

Epizootiology: Gram-negative bacteria

> The value of survival experiments based on laboratory cultures is questionable. Cultures, grown on laboratory media, tend to be larger and less aggressive, than their 'natural' counterparts. So when laboratory cultures are added to experimental microcosms, the outcome need not represent the true fate of the organism in the aquatic environment.

The source of many bacterial fish pathogens has not been considered. Details concerning the remainder are given below:

A. AEROMONADACEAE REPRESENTATIVES

Aeromonas hydrophila

The epizootiology of *Aer. hydrophila* has not been considered in any great detail, although it has been concluded that the organism is rife in freshwater (Heuschmann-Brunner, 1978; Allen *et al.*, 1983), aquatic plants and fish (Trust and Sparrow, 1974; Ugajin, 1979) and fish eggs (Hansen and Olafsen, 1989), and may be associated with invertebrates, such as the ciliated protozoan *Tetrahymena pyriformis* (King and Shotts, 1988), from where it will be readily available for infection of fish. Some isolates have been demonstrated to exhibit chemotactic responses to the mucus of freshwater fish (Hazen *et al.*, 1982). The chemotaxic substance, which is heat stable at 56°C, has been reported to have a molecular weight of 100 kDa.

The evidence points to a stress-mediated disease condition (Bullock *et al.*, 1971), in which mortalities, if indeed they occur at all (Huizinga *et al.*, 1979), are influenced by elevated water temperatures (Groberg *et al.*, 1978; Nieto *et al.*, 1985). Thus, Groberg and co-workers determined that deaths among fish, which were challenged via i.p. injection, occurred only at water temperatures in excess of 9.4°C. This implies that the disease is not associated with cold water. In addition to water temperatures, the presence of pollutants, notably nitrate at 6 mg/l, increased the susceptibility of channel catfish to infection (Hanson and Grizzle, 1985). It is interesting to note that survivors possess high serum titres of IgM-like antibody (Hazen *et al.*, 1981).

During an examination of likely portals of entry of the pathogen into walking catfish, Lio-Po *et al.* (1996) found evidence of localization in the muscle.

Aeromonas salmonicida

Historical consideration

The origin of the pathogen will probably never be definitively resolved, and it remains uncertain whether the disease was spread by importation of fish stocks from Europe to North America or vice versa. Circumstantial evidence exists to support either interpretation. For instance, the observation that furunculosis was not confirmed in North America until 1902 (Marsh, 1902) compared to 1894 in Europe could indicate that the disease originated in Europe and was introduced into North America with imports of brown trout (*Salmo trutta*) (McCarthy, 1975a). An alternative suggestion has been that exports of sharta rainbow trout from North America to Europe were responsible for spreading the disease. In support of this contention is the fact that rainbow trout are relatively resistant to furunculosis, which is taken to imply that they were the original host, but had developed resistance to the disease (Fish, 1937). It appears that awareness of the pathogen was brought about through the initial reports, so that by the early 1900s, furunculosis was found to occur in a variety of European countries, notably Austria, Belgium, France, Great Britain, Ireland and Switzerland. It took until 1964 for the disease to be recognized in Norway; the spread being due to the importation of infected rainbow trout from Denmark. In 1985, furunculosis was recognized in Norwegian marine Atlantic salmon farms; the cause being infected smolts from Scotland (Johnsen and Jensen, 1994). Although at first it was believed that furunculosis was confined to hatcheries, this concept was dispelled when extensive studies by Plehn (1909; 1911) revealed the presence of the disease in wild trout taken from 25 rivers and streams in Bavaria. In Great Britain, the occurrence of furunculosis was first reported in the literature during 1909. The following discussion on the history of the disease in Great Britain is taken from Lund (1967), who noted that the first report of furunculosis did not attract significant attention, other than to cause an awareness of the disease among a few trout farm owners. However, in 1911, a conspicuously high mortality among salmonids in the rivers Wye, Exe, Teign and Dart occurred. This disease outbreak was investigated by Masterman and Arkwright (1911), who proved it to be furunculosis. By 1912, the disease had appeared in the rivers Test and Itchen, and eventually spread to other trout streams in the south of England. A host of reports followed, with the occurrence of furunculosis in the river Conway in 1917 or 1918, the river Kennett in 1924 and 1925; the Welsh rivers Teify, Towy, Usk and Elway in 1925, the river Coquet in Northumberland during 1926, and the Welsh river Dee in 1925. The disease also reached Scotland in 1926 where high mortalities were reported in salmon in the river Doon in southwest Scotland. By 1929, furunculosis had encroached upon the rivers of northern Scotland, and was confirmed in the Aberdeen Dee. As the seriousness of the situation became increasingly apparent, the first organized attempt to tackle the problem was made by the Kennett Valley Fisheries Association in 1925, who raised a fund for studying the disease. The Fishery Board of Scotland followed suit in

1927, and in 1928 the efforts of the two groups were united. Thus in 1929, the Furunculosis Committee was formed. This committee reviewed information about the spread of the disease, and compiled all the facts on the occurrence of furunculosis in other countries. In Great Britain by the end of 1926, it was observed that furunculosis was proving to be less troublesome in some areas where the infection had initially been reported, but fresh outbreaks had been documented in other, additional salmon rivers. These new outbreaks also caused great consternation, due not only to a decline in fish numbers, especially valuable breeding stock, but also to loss of revenue from fishing rights (Lund, 1967). There seem to be no doubt that the disease had become enzootic in Great Britain.

In North America, the first report of furunculosis was made by Marsh (1902) who described an organism, which was named as *Bacillus truttae*. This caused an epizootic in hatchery fish in Michigan, USA. Later, Belding (1927) reported the loss of an entire stock of brook trout due to furunculosis at the Massachusetts State Hatchery. Davis (1929) contended that serious mortality had been restricted to brown and brook trout in hatcheries and rearing ponds. However, the extent of the problem was reflected in the calculations of Smith (1942), who reckoned that the disease had occurred in more than 25% of the hatcheries in the USA. However, as in Great Britain, furunculosis in North America was eventually shown to be not solely a hatchery disease; a study by Fish (1937) revealed the presence of *Aer. salmonicida* in wild stock of adult Loch Leven trout (*Salmo trutta levensis*) in Wyoming, USA. In Canada, Duff and Stewart (1933) investigated disease outbreaks among salmonids in British Columbia. Several fish species, such as wild Rocky Mountain whitefish (*Prosopium williamsoni*), Dolly Varden (*Salvelinus marme* Walbaum) and cutthroat trout (*Salmo clarki*), were reported in 1929 and 1930 to be dying from disease, a major symptom of which was the appearance of red 'blisters' on the body. In 1931, *Bacillus salmonicida* was isolated from diseased specimens. Nevertheless, despite the occurrence of furunculosis in wild stock in North America, it does not appear to have assumed epizootic status in rivers.

At present, the geographical distribution of the pathogen is worldwide, including Australia (Trust *et al.*, 1980b) and the mainland of Asia, i.e. Korea (Fryer *et al.*, 1988) where it was formerly believed to be absent. The pattern formed by the locations from where the disease has occurred since the initial report of Emmerich and Weibel (1894), suggests that it spread from a focal point in Europe. However, according to McCarthy and Roberts (1980), this may reflect progress in bacteriological diagnostic facilities as well as an actual dissemination of the pathogen.

Investigations into the epizootiology of *Aer. salmonicida* have not provided unequivocal answers for several crucial questions concerning the factors which control or determine the dissemination of the pathogen. Although *Aer. salmonicida* has been recognized for approximately 100 years, the precise route of transmission has not been conclusively resolved. Some studies indicate that the pathogen is widespread among wild salmonids. For example, in a study using DNA probe technology of the blood from 61 fish caught from three rivers in Ireland, Mooney *et al.* (1995) reported a widespread low-level infection. Controversy persists as to whether or not the organism is capable of a free-living existence in the natural environment, away from the fish host. The site(s) of entry into the fish also remain uncertain. All of these questions must be answered before the gaps in the understanding of the epizootiology of diseases of *Aer. salmonicida* aetiology can be filled. Unfortunately, epizootiological studies have not been aided by modern scientific approaches, such as plasmid profiling (Austin *et al.*, 1998) or ribotyping (Nielsen *et al.*, 1994).

The ecology of Aeromonas salmonicida
McCarthy (1980) performed detailed experiments concerned with the ecology of *Aer. salmonicida* and also reviewed the work carried out by others. According to his report, contact with infected fish or contaminated water and fish farm materials, and transovarian transmission have all been cited as probable routes of infection. Also, carrier fish, which show no overt signs of disease but harbour the pathogen in their tissues, appear to be implicated in horizontal or vertical transmission. Such carrier fish are presumed to provide a reservoir which retains the pathogen in fish populations.

To understand how *Aer. salmonicida* is transmitted both among and within fish populations it is necessary to know the source of the pathogen and its capacity to survive in the environment. In fact, most of the work done on epizootiological aspects of fish diseases caused by *Aer. salmonicida* has focused on investigations of potential sources of infection. The role of water, mud and detritus, carrier fish, (i.e. salmonids as well as non-salmonids) as potential sources of infection with *Aer. salmonicida* have been examined. The popular approach to the study of this subject has been to determine the presence of and survival capabilities of *Aer. salmonicida* in the variety of habitats listed above.

Aeromonas salmonicida—survival studies
The survival of *Aer. salmonicida* in water has been thoroughly examined by numerous investigators (Williamson, 1929; Smith, 1962; Lund, 1967; McCarthy, 1980; Sakai, 1986a,b; Rose *et al.*, 1990a,b; Morgan *et al.*, 1991; Effendi and Austin, 1991; 1994; Table 8.1). Unfortunately, caution must be used in the interpretation of some of the data as many of the studies employed pre-sterilized water or types of water in which *Aer. salmonicida* would not normally be present, e.g. distilled or tap water. Thus, the information gleaned from such studies does not necessarily reflect the behaviour of the pathogen in the natural aquatic environment. However, enough work has been done to allow tentative conclusions to be drawn. Based on the survival data accumulated, it appears that *Aer. salmonicida* is capable of surviving for a prolonged period in fresh, brackish and seawater, although contradictory results as to the exact time interval involved abound. Thus for unsterilized freshwater, including river water, recovery of the pathogen from as little as 24 h to as long as 19 days had been reported by different studies. Survival in unsterilized brackish water was between 16 and 25 days (Smith, 1962; McCarthy, 1980). In unsterilized seawater, the organism could be recovered from between 24 h and 8 days (McCarthy, 1980). If sterilized water samples were used, survival time in the absence of competing—antagonistic—organisms was invariably greatly increased. For example, survival times in freshwater of up to 63 days were reported (Cornick *et al.*, 1969), and in seawater up to 24 days (Lund, 1967). Using whole cells and DNA released into lake water microcosms, with media and PCR for detection, Deere *et al.* (1996a) cultured *Aer. salmonicida* for <4 weeks, but found the DNA remained intact for >13 weeks. This

discrepancy between the results of culturing and other techniques opens a veritable Pandora's box. Why should intact DNA be found 2+ months after culturing techniques indicated that the population of *Aer. salmonicida* had disappeared?

Using laboratory-based microcosms and culturing, direct counts, respiratory activity (the reduction of tetrazoliums to coloured formazans; after Effendi and Austin, 1993), iFAT, epifluorescence microscopy and the direct viable count techniques (by incorporating yeast extract and nalidixic acid; after Kogure *et al.*, 1979), Effendi and Austin (1994) confirmed the emerging view that cells of *Aer. salmonicida* remained after plates counts reached zero (Fig. 8.1). These workers determined that survival was maximal in brackish conditions, i.e. salinity = 25% (Fig. 8.2), notably on substrates—especially on wood (Fig. 8.3) but also in sediment—rather than in the water column (Fig. 8.4). Similarly, using an oxidase-negative atypical isolate, sterilized microcosms and culturing techniques, Wiklund (1995a) deduced that survival was better at 4°C than 15°C in brackish rather than seawater or freshwater, and in the presence of particulates, i.e. sand. The addition of nutrients did not resuscitate cells after colony counts declined to zero. This is important, insofar as workers have been generally unsuccessful at retrieving culturable cells after plate counts declined to zero, regardless of the method that indicated cells or cellular components remained.

McCarthy attributed the discrepancies among the various investigations to the technical difficulty of isolating *Aer. salmonicida* from mixed cultures. The temperature at which the experiments were run may well have also influenced the results. For instance, McCarthy (1980) conducted his experiments between 11 and 13°C, and reported longer survival times for the pathogen in fresh, brackish and seawater (17, 24 and 8 days, respectively), than had been recorded by most other investigators (Table 8.1). Also, the differences may reflect inherent variations between cultures.

In the majority of reports on survival of *Aer. salmonicida*, a large initial inoculum of the bacterium, usually 10^6 to 10^7 cells/ml of sample, was used. It is unlikely, however, that the pathogen would occur in these numbers even in the event of a free-living existence in the natural environment, except perhaps during epizootics where moribund and dead fish were releasing large numbers of *Aer. salmonicida* into the immediate vicinity. Therefore, studies were undertaken using a low number of cells, *ca.* 10^1 to 10^2/ml, as an inoculum (Allen, 1982). When the pathogen was placed in sterilized reservoir water in such minimal numbers, and incubated at 15°C, the organism underwent a severe reduction in numbers within 72 h, such as to be virtually unrecoverable by plating methods on solid non-selective medium. From these results, which are in contrast to other studies, it becomes apparent that several factors, including the size of the inoculum and the temperature at which the experiments are conducted, are crucial in determining the outcome of survival studies. In contrast, when an inoculum of approximately 10^5 to 10^6 cells/ml was placed into sterilized reservoir water, *Aer. salmonicida* survived and multiplied with up to 10^8 cells/ml in the system in 72 h. It could still be recovered in substantial numbers (10^7 cells/ml) at 55 days, when the experiment was concluded. In addition, nutrient conditions appeared to have an effect as it was observed that supplementation with low concentrations of nutrient, e.g. 0.005% (w/v) brain heart infusion broth, caused an increase in the number of *Aer. salmonicida* cells within 24 h. This increase was maintained until the end of the sampling period. The addition of nutrient,

Table 8.1. Experimental data concerning the survival of *A. salmonicida* in water

Type of experimental system	Temperature (°C)	Survival time	Reference
Sterilized water:			
Distilled water	20	35 days	Lund (1967)
Distilled water	20	<7 days	Sakai (1986b)
Lake water	10	8 days	Morgan *et al.* (1991)
	10	21 days	Morgan *et al.* (1991)
	20	60 days	Deere *et al.* (1996a)
Physiological saline	20	<7 days	Sakai (1986b)
Tap water	—	5 days	Arkwright (1912)
	—	3 days	Williamson (1929)
	20	12 h	Lund (1967)
Reservoir water, low inoculum of *A. salmonicida* $(10^1\text{–}10^2$ cells/ml), in mixed culture with other aquatic bacteria	15	*ca.* 3 days	Allen (1982)
Reservoir of water, as above, supplemented with 0.005% (w/v) brain heart infusion	15	>3 days	Allen (1982)
River water	—	5 day	Williamson (1929)
	20	8 day	Lund (1967)
	10	63 days	Cornick *et al.* (1969)
	20–25	28 days	Cornick *et al.* (1969)
Seawater	—	19 h	Arkwright (1912)
	—	3 days	Williamson (1929)
	20	24 days	Lund (1967)
	—	<10 days	Rose *et al.* (1990b)
	20	>24 days	Effendi and Austin (1991)
	5–25	≤28 days	Effendi and Austin (1994)

Table 8.1. (*Continued*)

Type of experimental system	Temperature (°C)	Survival time	Reference
Unsterilized water			
Brackish water	—	16–25 days	Smith (1962)
	11–13	24 days	McCarthy (1980)
Distilled water	—	14 days	Horne (1928)
	—	4 days	Williamson (1929)
	—	7 days	Duncan (1932)
	20	9 days	Lund (1967)
Freshwater	—	24–30 h	Duncan (1932)
	11–13	17 days	McCarthy (1980)
River water	—	2 days	Williamson (1929)
Sterilized water	—	7–19 days	Smith (1962)
	20	2 days	Lund (1967)
10^1–10^2 A. *salmonicida* cells/ml	15	*ca.* 3 days	Allen (1982)
10^5–10^6 A. *salmonicida* cells/ml	15	>3 days	Allen (1982)
Sea water	—	2 days	Williamson (1929)
	—	24–30 h	Duncan (1932)
	20	5–6 days	Lund (1967)
	11–13	8 days	McCarthy (1980)
	—	<10 days	Rose *et al.* (1990b)
	20	6 days	Effendi and Austin (1991)
Tap water	—	3–4 days	Horne (1928)
	—	3 days	Williamson (1929)
	—	4 days	Duncan (1932)
	20	3 days	Lund (1967)
Mixtures of water type:			
50% seawater + 50% tap water	—	19 h	Arkwright (1912)
25% seawater + 75% tap water	—	45–67 h	Arkwright (1912)

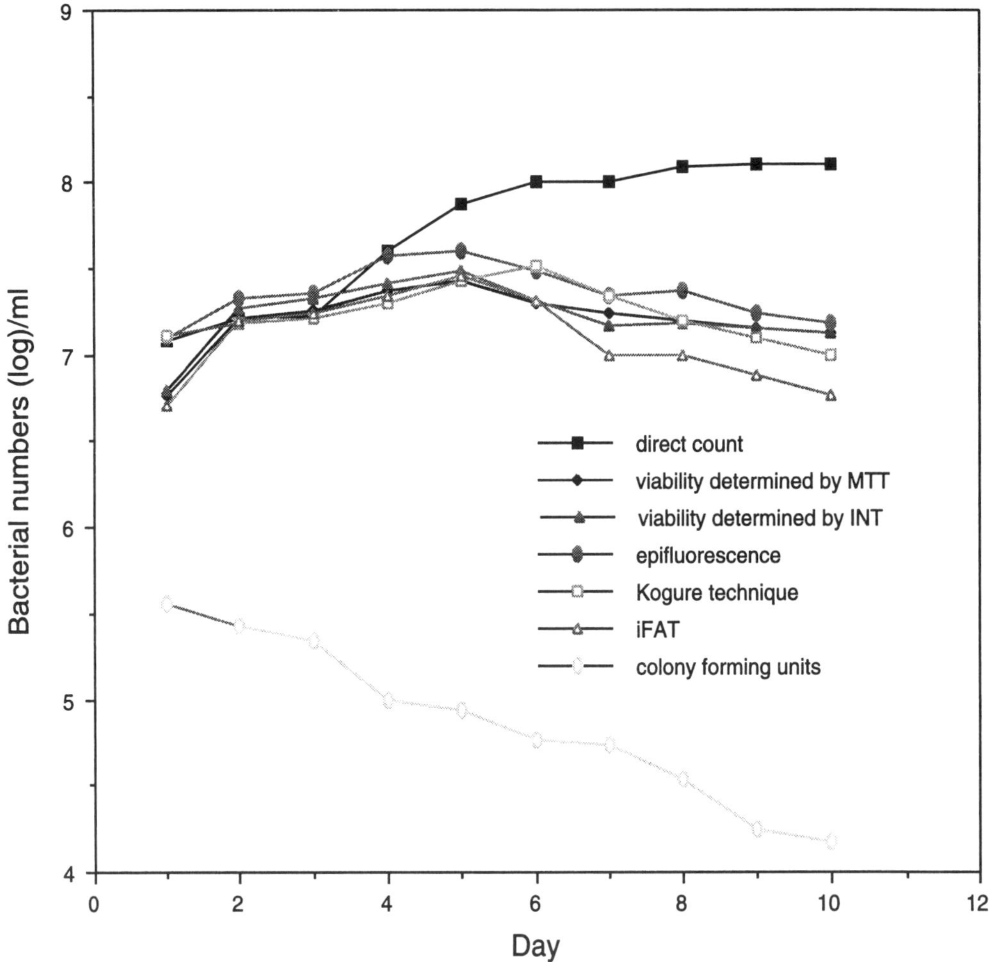

Fig. 8.1. Longevity of cells of *Aer. salmonicida* in a sterile marine microcosm, as determined by culturing, phase-contrast microscopy, respiratory activity, epifluorescence microscopy, iFAT and the direct viable count (= Kogure) technique (after Effendi and Austin, 1994). Whereas culturing revealed a progressive decline in the population of *Aer. salmonicida*, most of the other techniques pointed to an initial increase in cell numbers followed by a slow decrease over the 10-day sampling period.

moreover, caused the increase in numbers of the pathogen regardless of whether the initial inoculum of cells was large or small. Supplementation with nutrient was also reported to increase survival time by McCraw (1952), who observed that the addition of 0.1% (w/v) peptone to seawater enabled *Aer. salmonicida* to survive up to 80 days, a much longer time than recorded for unsupplemented seawater.

The ability of *Aer. salmonicida* to persist in mud (sediment) or detritus in the fish farm environment has also been examined. McCarthy (1980) demonstrated that the pathogen

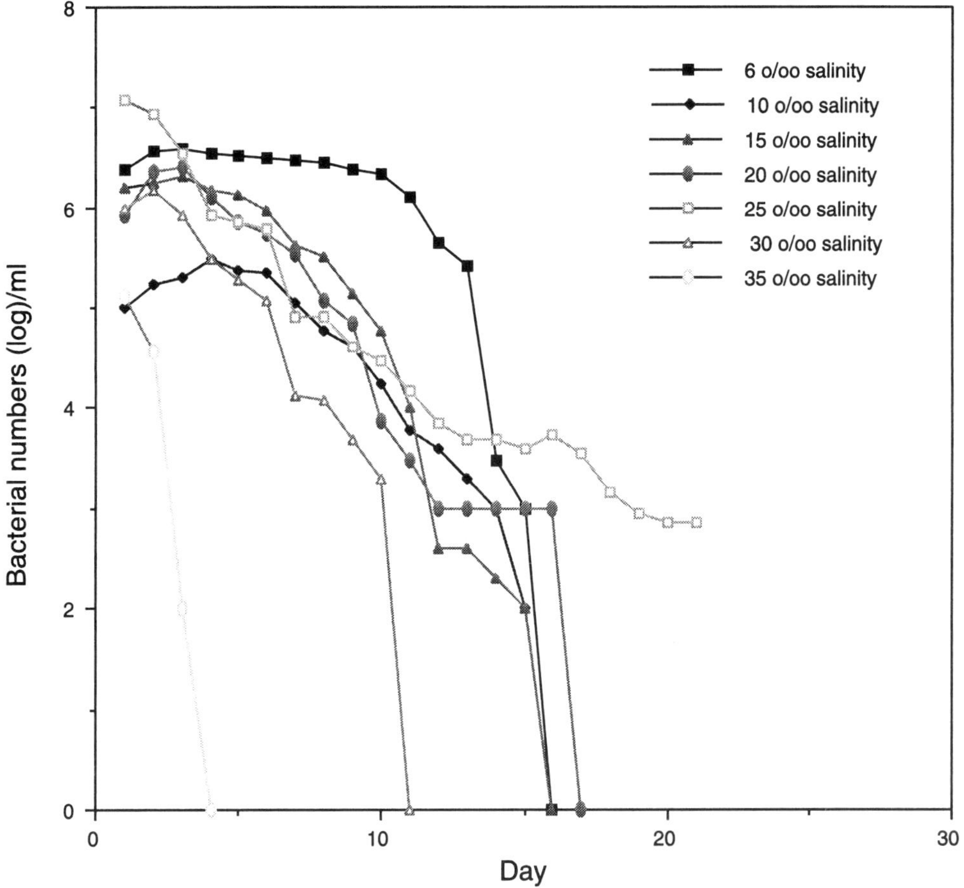

Fig. 8.2. Survival of *Aer. salmonicida* at 15°C in salinities of 6–35‰ (after Effendi and Austin, 1994). The data indicated maximal survival at 25‰. In contrast, the cells rapidly declined in full seawater, i.e. 35‰. The lowest salinity, i.e. 6‰, revealed the maximal number of cells for two weeks, thereafter there was a rapid decline in the population.

was able to survive in numbers of *ca.* 10^5 viable cells in fish pond mud and detritus at least up to 29 days, and probably longer as the experiments had to be terminated prematurely due to decomposition of the dialysis bags containing the bacteria. He further pointed out that the 10^5 viable cells remaining after 29 days are significant, since from his studies it was shown that if this number of cells was released into fresh water, their survival time would be 14 days. Michel and Dubois-Darnaudpeys (1980) investigated the persistence of *Aer. salmonicida* in sediments and reported that the pathogen survived and grew in sterilized river sediments for over 10 months. However, pathogenicity of the two isolates tested was lost after 8 or 9 months. They concluded that in natural conditions such a length of time would enable the pathogen to be released from sediment into the water, and that the behaviour of bottom-feeding fishes would allow direct contamination

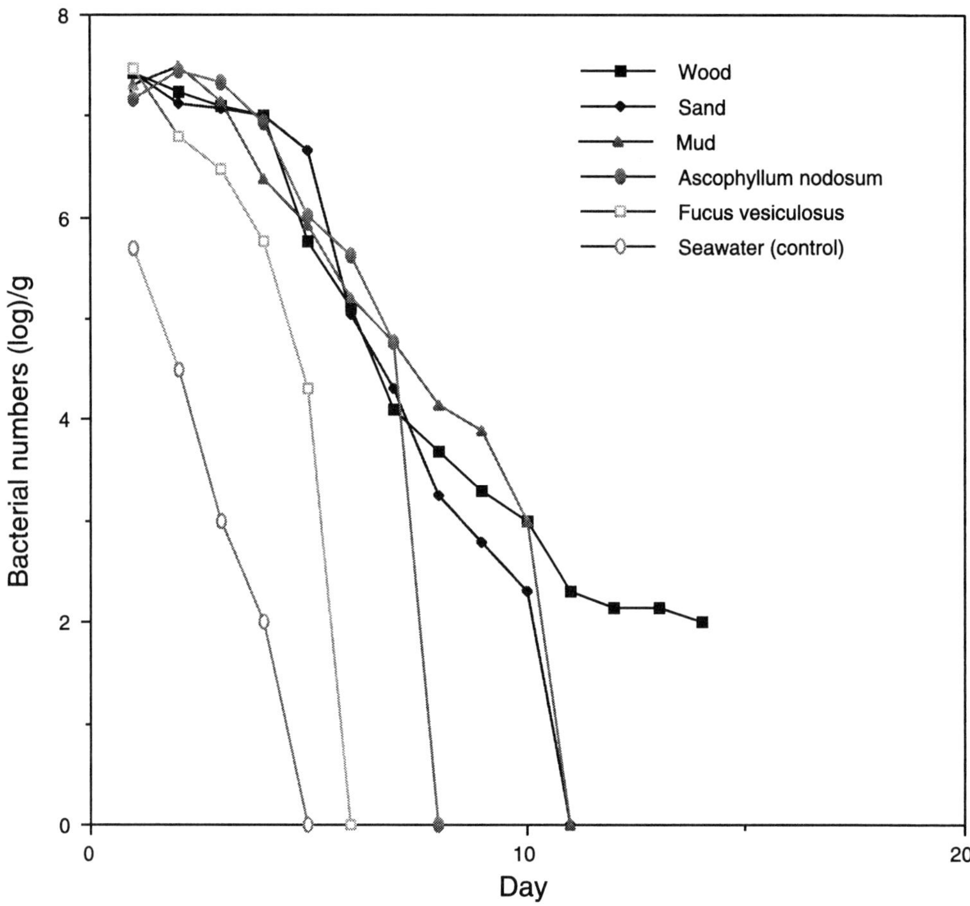

Fig. 8.3. Survival of *Aer. salmonicida* in a non-sterile marine microcosm (after Effendi and Austin, 1994), indicating the benefit of substrates compared to seawater. Although there was a steady decline in bacterial numbers on all substrates, *Aer. salmonicida* survived for longest on wood fragments, which were present in the sediment.

of fish, possibly becoming carriers. A reduction in pathogenicity, subsequent to prolonged incubation in river sediments, was also noted by Sakai (1986a,b). He offered an explanation whereby avirulent cells (with a positive electrical charge), which originate from virulent cells (negatively charged) attached to sediment, spontaneously detach from the sediment particles (river sand in survival experiments) thus decreasing the number of virulent cells recovered. Michel and Dubois-Darnaudpeys (1980) conceded that competition of *Aer. salmonicida* with large numbers of other bacteria in streams, some with an ability to synthesize bacteriocins, may act as a regulatory mechanism and limit the proliferation of the pathogen. However, previous work by Dubois-Darnaudpeys (1977b) supports the concept that *Aer. salmonicida* is genuinely capable of survival and multiplication in natural sediments, hence providing a reservoir of infection, even though direct

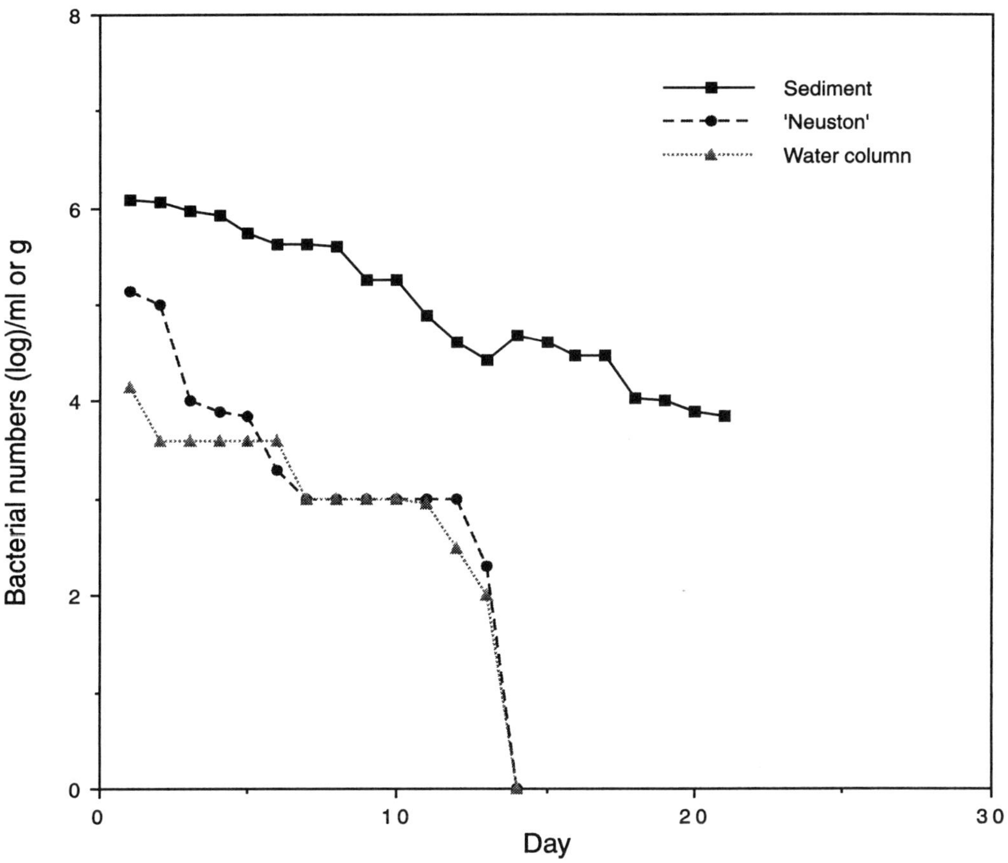

Fig. 8.4. Survival of *Aer. salmonicida* in sterile (121°C/15 min) seawater and sediment (after Effendi and Austin, 1994). The data revealed that the pathogen survived longest in sediment rather than the surface layer of seawater in the water column.

contact with diseased fish is likely to remain the primary route of transmission. In addition, the regular detection of bacteriophages specific for *Aer. salmonicida* in samples of river sediments was taken as an indirect demonstration that the pathogen was present throughout the year (Dubois-Darnaudpeys, 1977b). Sakai (1986b) reported extended survival times (>15 weeks) for virulent cultures of *Aer. salmonicida*; if placed in the presence of dilute humic acid (10 μg/ml), tryptone (10 μg/ml), and cleaned river sand (100 g/100 ml of medium). Without the addition of the sand, detection of viable cells ceased within five weeks. However, avirulent strains of the pathogen did not survive more than two weeks regardless of whether or not sand was included in the experimental system. Sakai (1986b) determined that humic acid and amino acid–humic acid complexes were absorbed onto the sand, which led to a build-up of 30–50 times the environmental concentration of amino acids on the surface of the sand particles. This, in turn, allowed only colonization of/attachment by bacterial cells with net negative electrical charges (virulent cells of *Aer. salmonicida* in this instance), which resulted in their enhanced survival in the presence of the sand. Thus, Sakai (1986b) concluded that the electrostatic interrelationship occurring among humic acid, river sand and the bacteria explains the ability of virulent *Aer. salmonicida* strains to survive for extended periods in river sediments.

McCarthy (1980) contended that during epizootics of furunculosis there existed a strong possibility that fish farm implements could become contaminated with *Aer. salmonicida*. In a study emphasizing survival of *Aer. salmonicida* on fish nets which would be used both to remove dead infected fish and to move healthy fish, it was established that the pathogen survived up to 6 days on both dry and wet contaminated nets. In addition, wet and dry contaminated netting was disinfected using three compounds, i.e. acriflavine, Teepol-sodium hydroxide and hypochlorite solutions. *Aer. salmonicida* was not recovered from either wet or dry netting disinfected with the acriflavine or Teepol-sodium hydroxide solutions, but the hypochlorite solution failed to disinfect dry nets. McCarthy (1980) concluded from these results that the use of contaminated and improperly disinfected nets is potentially dangerous to healthy stock as it is known that netting abrades fish to some extent, and such abrasions can facilitate bacterial invasion.

More recently, it has been found that wrasse (these are small inshore benthic fish which have gained popularity as a means of controlling sea lice populations among infested Atlantic salmon) are also susceptible to furunculosis (Treasurer and Cox, 1991). These investigators reported the recovery of typical *Aer. salmonicida* from the liver and kidney, and the presence of skin lesions reminiscent of chronic furunculosis, in golsinny, rock cook and cuckoo wrasse of fish farm origin. But could salmon lice harbour and transmit *Aer. salmonicida*? By means of recovery techniques with immunomagnetic beads coated with monoclonal antibodies to LPS and culturing techniques, it was determined that *Aer. salmonicida* was recoverable from lice (~10^4 *Aer. salmonicida* cells/louse) and also marine marine plankton 600 *Aer. salmonicida* cells/g of homogenized plankton) (Nese and Enger, 1993). Perhaps more worrisome is the report by Frerichs *et al.* (1992) of the recovery of atypical *Aer. salmonicida* from apparently healthy wild wrasse captured in the open sea. Fortunately in this instance, the isolates were proven to be non-pathogenic to Atlantic salmon smolts. Consequently, both groups have cautioned against the arbitrary stocking of wrasse in fish farms without first checking for the possible presence of *Aer. salmonicida* in the fish. Furthermore, Treasurer and Cox (1991) recommended that wrasse should not be released back into the wild, or transferred between fish farms at the end of the production cycle.

Only a few investigators have examined animals, other than fish, as a potential source of infection. The extensive study carried out on this topic was that of Cornick *et al.* (1969); a total of 2954 vertebrate and invertebrate specimens, collected from fish ponds during an epizootic of furunculosis, were examined for the presence of *Aer. salmonicida*. No isolates of the pathogen were recovered despite this heroic attempt. This study is cited time and time again as evidence against the likelihood of animals, other than fish, acting as reservoirs of infection. Williamson (1928) was also unsuccessful in isolating the pathogen from water snails under similar conditions. Allen (1982) examined macroscopic algae and zooplankton taken from fish tanks prior to, during, and after a furunculosis epizootic at a fish-rearing unit in Essex, UK, in an unsuccessful attempt to recover *Aer. salmonicida* from these organisms. In contrast, King and Shotts (1988) determined that viable cells of *Aer. salmonicida* survived and, indeed, multiplied (twofold) within the digestive tract of the ciliated protozoan, *Tetrahymena pyriformis*. It may be concluded from some of these studies that the pathogen may be found in association with other aquatic animals, but it is apparent that many of the existing methods available for detection of the pathogen are inadequate.

Data have pointed to the possibility that *Aer. salmonicida* may be disseminated in aerosols (Wooster and Bowser, 1996). In particular, experiments demonstrated that the pathogen travelled 104.1 cm (the limit of the test chamber), via the airborne route (Wooster and Bowser, 1996). Thus, another possible means of spreading the pathogen needs to be considered.

Difficulties in recovering Aeromonas salmonicida from the aquatic environment
It is relevant to digress, at this point, from the discussion of sources of infection, in order to comment upon the difficulties besetting the isolation of *Aer. salmonicida* from environmental samples, other than fish. A dependable isolation procedure for the pathogen is of critical importance to an understanding of the epizootiology of diseases caused by *Aer. salmonicida*. For example, if the pathogen is capable of a free-living existence outside a fish host, the prevention and control of diseases of *Aer. salmonicida* aetiology would be rendered much more difficult if not impossible. However, currently *Aer. salmonicida* is defined as an obligate fish pathogen not found in surface waters (Popoff, 1984). This definition has no doubt been formulated due to the paucity of conclusive evidence for a free-living existence of the pathogen. The organism, for instance, often cannot be isolated from water on fish farms even during an epizootic of the disease (Cornick *et al.*, 1969; Kimura, 1970; Allen, 1982). Several reasons have been put forward to explain this disconcerting phenomenon. One is that *Aer. salmonicida* is notoriously difficult to isolate from mixed microbial populations as it is quickly outcompeted in growth by most other commonly occurring aquatic bacteria. In addition, pigment production on agar plates, heavily relied upon as a first indication that *Aer. salmonicida* is present, is inhibited by the close proximity of colonies of other bacterial types. Therefore, there is the perceived problem of recognizing *Aer. salmonicida* in large mixed microbial communities. Many of the problems with habitat and survival studies on *Aer. salmonicida* are blamed on lack of adequate methodology. Both McCarthy (1980) and Michel and Dubois-Darnaudpeys (1980) stress contamination difficulties when employing non-selective media, e.g. TSA, for isolation of the pathogen. Cornick *et al.*(1969) reported that *Aer. salmonicida* was isolated most frequently from environments containing few, if any, other bacterial species, particularly representatives of the genus *Pseudomonas*. Their preliminary experiments suggested that some pseudomonad taxa obtained from water and fish inhibited the growth of *Aer. salmonicida* in liquid and on solid culture media. Both cell-free filtrates and extracts of disrupted cells of the pseudomonads caused the inhibition, believed to be due to antimicrobial activity. Dubois-Darnaudpeys (1977a) also examined the effects of the bacterial flora commonly

occurring in surface water, such as the *Pseudomonas–Achromobacter* and *Flavobacterium* groups, on the survival and growth of *Aer. salmonicida*. She found that the pathogen was inhibited at all temperatures if the experiments were run in the presence of the other bacteria. The ability of other micro-organisms, e.g. *Acinetobacter*, *Aer. hydrophila*, *Chromobacterium*, *Esch. coli*, *Flavobacterium* and *Pseudomonas*, and their metabolites, to inhibit the survival of *Aer. salmonicida* in non-sterile seawater was also reported by Effendi and Austin (1991). Thus, because the isolation and study of the viability of the pathogen is complicated by competition and inhibition by other organisms, it is not surprising that investigations into survival of the pathogen, which could help to establish whether or not it is capable of a free-living existence outside of fish, have invariably retreated to laboratory-based experiments using sterilized water. A filtration method tested by Maheshkumar *et al.* (1990) attempted, with some degree of success, to overcome the difficulties of isolation of *Aer. salmonicida* posed by the presence of other bacteria, i.e. the possible overgrowth of small numbers of cells of the pathogen. In their studies, up to 5 l of hatchery water was seeded with *Aer. salmonicida* and passed through 1^{-MDS} electropositive filters. This technique was used in combination with removal of the filters after backwashing, soaking them in a small volume of 3% beef extract solution, and followed by the scraping of the filters to remove trapped bacteria. When all the eluates from the filters were combined, recovery of *Aer. salmonicida* was determined to be 35%. Thus, Maheshkumar *et al.* (1990) concluded that the filtration technique demonstrated greater sensitivity than the direct examination of water. Also, the enumeration of *Aer. salmonicida* was not overly effected by the presence of other bacteria because in the water samples the pathogen retained the ability to produce a brown pigment. However, these authors noted that biochemical/serological tests would be necessary for the detection of atypical non-pigmenting isolates. Nonetheless the classic dilemma of many ecological studies persists, i.e. how does the response of an organism in laboratory-based experiments relate to its performance in the natural environment where it needs to interact in a diverse and heterogeneous community? It is a problem still in search of a reliable solution.

It has often been stated that ecological investigations of *Aer. salmonicida* are hampered by the lack of an effective selective isolation medium specifically formulated for the pathogen. It was, for example, the opinion of Cornick *et al.* (1969) that the development of a selective medium for *Aer. salmonicida* could quite possibly change the present views on the habitat and viability of the pathogen. McCarthy (1980) also believed a selective medium would greatly assist ecological work. No doubt such a medium would be extremely useful; however, it is unlikely that its existence alone would cause all the remaining difficulties concerning ecological work on *Aer. salmonicida* to evaporate. It should be noted, however, that although a selective medium for *Aer. salmonicida* has not yet been formulated, CBB serves well as a differential growth medium, and is increasingly employed for this purpose. This medium is especially useful in the detection of *Aer. salmonicida* in fish tissues. For instance using CBB, Cipriano *et al.* (1992) recovered the pathogen from 56% of mucus samples, but interestingly from only 6% of kidney material, taken from salmonids. This was reinforced by a later study which pointed to the presence of the organism in gills and as well mucus, i.e. external carriage (Cipriano *et al.*, 1996a,b). The data has been summarized, as follows:

Number of samples revealing *Aer. salmonicida*

15/100 gill samples revealed the pathogen at $6.3 \times 10^2 - 1 \times 10^4/g$
19/100 mucus samples revealed the pathogen at $9.1 \times 10^2 - 1.7 \times 10^4/g$

Using CBB, Cipriano *et al.* (1996b) reported *Aer. salmonicida* in higher numbers from the mucus than kidney of six salmon. Thus, the populations of *Aer. salmonicida* were in the range of 1.1×10^3 to $1.8 \times 10^7/g$ and 1.0×10^3 to $1.4 \times 10^7/g$ for mucus and kidney, respectively. Hiney *et al.* (1994) reiterated the view that *Aer. salmonicida* may colonize mucus, gills and also fins. Also, these workers considered that the intestine may well be the primary location of *Aer. salmonicida* in Atlantic salmon with asymptomatic infections. In addition, there are other considerations that may contribute to the problems experienced. For instance, parallels might be drawn from studies on the isolation methods used for other micro-organisms, i.e. notably coliforms and enteric pathogens such as *Salmonella* spp., where it has been observed that a pre-enrichment step must be employed prior to use of selective media that impose too stringent conditions on these organisms which have been stressed, injured or are too sensitive to selective agents, and, thus, are rendered unrecoverable by selective methods alone (Geldreich, 1977; Kaper *et al.*, 1977; Olson, 1978). Alternatively, it can be postulated that *Aer. salmonicida* follows the pattern of certain other types of micro-organisms which are extremely difficult to detect in the natural environment by means of routine bacteriological procedures (i.e. plate counts) and, hence, have been assumed to be absent from these environments. Stevenson (1978) commented on adjustments made by bacteria, which enable the organisms to survive in the variable and often stressful conditions imposed upon them by existence in natural aquatic systems. He suggested that bacteria surmount changes in their environment, including varying degrees of solar input, temperature, availability of nitrogen and dissolved oxygen, by entering a state of dormancy defined as 'any rest period or reversible interruption of the phenotypic development of an organism' (Sussman and Halvorson, 1966). Thus, the possibility that *Aer. salmonicida*, in the natural environment outside a fish host, assumes a physiological state such that it cannot be recovered on agar plates used for primary isolation should not be discounted. There is tentative evidence that the situation for *Aer. salmonicida* is similar to that of a related pathogen, *V. cholerae*. In the survival of *V. cholerae* in aquatic microcosms, Singleton *et al.* (1982a,b) reported that certain combinations of environmental parameters, i.e. sub-optimal salinities and low nutrient concentrations, not only affected multiplication of *V. cholerae* populations but also recoverability of the cells. Thus, these investigators found that *V. cholerae* cells were observed using acridine orange staining in conjunction with epifluorescence microscropy when culturable cells were not detected. Further work on the theme of a 'non-recover-able' stage of existence for bacterial populations which, however, remain viable was done by Xu *et al.* (1982), also for *V. cholerae*. They used direct viable counting, a procedure allowing estimation of substrate-responsiveness, i.e. viable cells, using microscopy. This method revealed that a significant proportion of the non-culturable cells were, in fact, viable.

Survival of Aeromonas salmonicida in the absence of culturing
The observations for *V. cholerae* were of especial interest because we found that cells of *Aer. salmonicida* could be enumerated by microscopic direct counts when no colony-forming units were recovered on agar plates, raising the possibility that problems concerning the recovery of *Aer. salmonicida* may bear similarities to *V. cholerae*. Our experiments had shown that if low numbers of *Aer. salmonicida* were inoculated into sterilized reservoir water the organism could not be recovered using plating methods, after 72 h. However, when microscopic direct counts were done the number of bacteria underwent an initial increase 6 h post-inoculation, and maintained these numbers until the conclusion of the experiment. Unfortunately, it was not possible to know from these results whether the cells observed by microscopy were, indeed, viable. Thus, in a follow up study, the hypothesis that *Aer. salmonicida* may enter a non-recoverable but viable state was tested. However, these data bore some similarity to survival studies with other pathogens, namely that there is often an initial increase in the number of cells as visualized by microscopic methods despite culturing indicating a progressive decline in numbers.

The non-culturable but viable (NCBV) state
It was confirmed by Allen-Austin *et al.* (1984) that small inocula of cells rapidly declined in filter-sterilized river water, when enumerated using total viable count procedures on agar plates, such that the system appeared to be totally devoid of viable cells by day 17. However, the microscopic procedures showed that, after an initial decrease, the number of *Aer. salmonicida* cells remained constant at approximately 8.0×10^2/ml. TSB to 0.01% (v/v) was added to the experimental system 7 days after the plate counts reached and remained at zero, and the sample was split into three equal volumes, incubated at 22 and 18°C in addition to 15°C. At 22°C, 150 colony-forming units of *Aer. salmonicida*/ml of sample were recovered on TSA 6 h after supplementation with the nutrient. There was no apparent increase in the direct microscopic count at 6 h, but by 24 h the microscopic fields contained too many cells to count. This result demonstrated that 6 h after nutrient addition, a proportion of the cells had regained the ability to produce colonies on TSA, despite the fact that for the previous 7 days none had been capable of colony formation. At 18°C, the response to the added nutrient was much slower, insofar as colonies were not detected until after 4 days had elapsed. There was, however, a pronounced increase in the direct counts at 24 h after nutrient was added. This continued up to 4 days, when the first colonies were cultured on solid medium. Similarly, there was a lag of 5 days before colonies were recoverable at 15°C. This coincided with an increase in the direct count. In comparison, it is emphasized that plate counts remained at zero in unsupplemented river water at 15, 18 and 22°C. The factors involved in triggering the return of *Aer. salmonicida* to a culturable state need careful evaluation. As demonstrated in the experiment reported here, temperature and nutritional changes appear to be responsible for reactivating cells of *Aer. salmonicida* (Allen-Austin *et al.*, 1984).

Sakai (1986b) also proposed a mechanism for the long-term survival of *Aer. salmonicida* in the aquatic environment based on electrostatic charge differences on individual cells, with net negative charges reported on virulent, agglutinating cells, and net positive charges on avirulent non-agglutinating strains. He suggested that the negatively charged virulent form of *Aer. salmonicida* is able to persist, albeit under starvation conditions, retaining viability in river sediments. It was also proposed that the decline in negatively charged virulent cells in sediments over prolonged periods, also noted by other investigators (e.g. Michel and Dubois-Darnaudpeys, 1980), may be caused by the spontaneous occurrence of positively charged avirulent free-living cells of *Aer. salmonicida*. These cells originate from the virulent ones, attached to sediment particle surfaces, and subsequently detach from the sediment/sand particles. This free-living form could be considered to enter a dormant phase, according to Sakai (1986b), because the viability of these bacteria declines due to a lack of nutrients. It was further proposed that the free-living cells represent a transitional life stage of the pathogen which would ultimately lose viability (Sakai, 1986b).

Subsequently, Rose *et al.* (1990a) re-examined the possibility that *Aer. salmonicida* may enter a dormant state in water, using methods modified from the work of Allen-Austin *et al.* (1984), as described above. However, in their experiments, the addition of 0.1% (w/v) TSB to aliquots withdrawn from microcosms after viable counts of *Aer. salmonicida* had reached zero, did not result in renewed growth of the organism. Thus, Rose *et al.* (1990a) concluded that the most probable explanation for the results obtained in the previous study (when there appeared to be resuscitation of dormant cells by added nutrients) was the presence of small numbers of viable culturable cells, which were too few in quantity to be detected by the sampling protocol employed. This conclusion was based on the observation that the addition of 0.1% (w/v) TSB to microcosms after the viable count had reached zero resulted in the re-appearance of viable culturable cells within 48 h of incubation at 22°C. However in both studies, bacteria enumerated by microscopic techniques remained at levels of approximately 10^4/ml in water samples retrieved from the experimental microcosms containing *Aer. salmonicida*, even after viable counts had apparently reached zero. It is curious that Rose *et al.* (1990a) proffered no explanation which account for the level of bacteria that were observed microscopically (were the cells alive or could they have been dead?). In a later study, which again addressed the issue of dormancy/NCBV for *Aer. salmonicida*, Morgan *et al.* (1991) assessed the survival of the pathogen in lake water, employing an extensive range of techniques, including epifluorescence microscopy, respiration, cell culture, cell revival, flow cytometry, plasmid maintenance and membrane fatty acid analysis. These workers found that *Aer. salmonicida* became unculturable in sterile lake water, but microscopic and flow cytometric methods revealed the continued presence of cells. However, attempts to revive these cells by the addition of TBS were unsuccessful. Despite this, it was found that both genomic and plasmic DNA, and also RNA, were maintained in the cells, even though they could not be cultured on conventional media. Morgan *et al.* (1991) concluded that morphologically the cells remained intact, although their viability could not unfortunately be definitely demonstrated. In addition, they commented (and we strongly agree) that non-culturability of some bacteria from environmental samples maybe as much a function of the ignorance of the parameters necessary for their recovery, as the occurrence of a truly non-culturable, specialized survival state. Subsequently by means of flow cytometry with rhodamine 123, they established a NCBV state in sterile lake water (Morgan *et al.*, 1993). However, flow cytometry indicated that cellular properties related with viability was lost shortly after culturability disappeared in distilled water, but not so in lake water (Deere *et al.*, 1996b). Additionally, these workers marked an isolate of *Aer. salmonicida* with the *xylE* gene, using the plasmid pLV1013. This isolate was culturable on TSA from normal (non-sterile) lake water for three weeks, after which non-culturability developed, with the NCBV cells retaining chromosomal and plasmid DNA. The NCBV state can be postponed (60 days was mentioned by Pickup *et al.*, 1996) by the addition of high levels of nutrient, especially 125 μM quantities of the amino acids arginine and methionine, to experimental microcosms. *Aer. salmonicida* decreased in size and became rounded, but were still culturable (Pickup *et al.*, 1996).

The development of a dormant, nonculturable state of *Aer. salmonicida* in seawater at 4°C was apparent from the work of Ferguson *et al.* (1995). These workers incorporated a luciferase gene, *luxAB*, from *V. fischeri* into *Aer. salmonicida* and followed the fate of the cells. As before, intact non-culturable cells could not be resurrected.

Ecology of Aeromonas salmonicida—an explanation
To develop some previous points, there is tentative evidence to support the possibility that *Aer. salmonicida* undergoes sufficient modifications to its morphology in seawater so as to be only recoverable on specialized media. Thus while conducting experiments on the survival of *Aer. salmonicida* in seawater, Effendi and Austin (1991) found that samples where the pathogen was believed to be absent (or unculturable) actually contained cells which passed through 0.22 and 0.45 μm pore size porosity filters. These isolates grew on specialized media designed for the recovery of L-forms, and showed agreement with the characteristics of *Aer. salmonicida* L-forms as reported by McIntosh and Austin (1988; 1990; 1991b). Subsequently, *Aer. salmonicida* colonies developed on basal marine agar (BMA) plates inoculated with material from turbid L-form broth medium. On this basis, Effendi and Austin (1991) recorded populations of *ca.* 10^3 *Aer. salmonicida* cells/ml in the microcosms after corresponding enumeration of colonies on BMA had reached zero. Thus, they suggested that the existence of specialized forms, e.g. L-forms, of *Aer. salmonicida*, may be a factor in the difficulties previous researchers have experienced in attempts to recover the pathogen from environmental samples. Continuing this theme, Effendi and Austin (1995a) examined the characteristics of the so-called NCBV cells. Using a marine microcosm, it was observed that these NCBV cells became much smaller and coccoid while retaining respiratory activity as measured by the reduction of tetrazoliums to insoluble formazans. There was not any alteration in the LPS composition of the cells, but an alteration in the protein composition was recorded, with a reduction in some (15, 17, 22, 30 and 70 kDa proteins) and an increase in a 49 kDa protein. This was accompanied by a loss in DNA. That these cells were still alive was indicated by the development of large bizarre shapes following the addition of yeast extract and nalidixic acid (after Kogure *et al.*, 1979).

To summarize, the question as to whether *Aer. salmonicida* is able to persist in a free-living form outside a fish host is still outstanding. An equally important corollary concerns the pathogenicity of such forms, assuming that they exist, i.e. can they retain the ability to infect fish? Traditionally, *Aer. salmonicida* has been defined as an obligate fish pathogen. However, there appears to be an increasing trend in ecological studies to at least consider the possibility that this definition may no longer holder true at all times. On the basis of the available data obtained from numerous survival studies, it may be stated that *Aer. salmonicida* has the ability to persist in the aquatic environment for protracted periods. It is the mechanism of this survival and its effects on the organism in the natural environment around which the debate now centres. It is possible that *Aer. salmonicida* could exist in a so-called non-culturable stages (perhaps due to the presence of an altered morphological state, such as L-forms). After all, the reason for the occurrence of explosive outbreaks of furunculosis among fish populations, particularly salmonids, which have not been previously exposed to the disease, has yet to be explained. The answer to this important question of facultative versus obligate pathogen still awaits the development of methods for the more efficient and refined detection of the organism in the natural environment. As one astute scientist has said, 'absence of evidence is not evidence of absence'; hence further efforts to breach this gap in our understanding of *Aer. salmonicida* epizootiology are essential.

Aeromonas salmonicida—transmission by fish
Fish undoubtedly play a major role in the transmission of disease among themselves. One early study by Blake and Clark (1931) stated that furunculosis was only spread by infected fish or by material which has come into contact with them. Fortunately, the recovery of *Aer. salmonicida* from fish tissues is generally less troublesome than seeking it in other environmental sources. Fish may act as a source of infection in two ways: those which have died or are ill with furunculosis are heavily contaminated with the pathogen, alternatively they may be carriers which, although appearing healthy, harbour the pathogen in their organs where it can be released if the fish eventually succumb to the disease. Both these aspects have received attention. McCarthy (1980) established that material from a furuncle could contain up to 10^8 viable cells/ml of necrotic tissue, and was interested to assess the viability of *Aer. salmonicida* in dead fish and the degree to which they could contaminate pond water. He found that *Aer. salmonicida* remained viable in fish (muscle) tissue for 32 days, and for 40 days in the tank water where the dead fish had been kept, thus providing a possible source of infection for healthy fish. Another study showed that *Aer. salmonicida* remained viable in infected trout tissues and internal organs (heart, liver, spleen and kidney) for 49 days when the fish were stored at −10°C (Cornick *et al.*, 1969). However, the pathogen was isolated only from the kidney of infected trout held at 4°C for 28 days. McCarthy (1980) pointed out that the prolonged survival of *Aer. salmonicida* in dead diseased fish shows the risk of using scrap fish, which might be infected with a chronic form of furunculosis, as food, particularly in view of the finding of Cornick *et al.* (1969) on the survival of the pathogen at low temperatures. Some investigations, which have examined the survival of *Aer. salmonicida* in water, have also commented on the potential of the non-culturable cells, detected microscopically in the experimental systems, to infect fish. However, these studies have generally concluded that *Aer. salmonicida* in this form does not appear to be pathogenic, as re-infection of fish does not occur (Rose *et al.*, 1990a; Morgan *et al.*, 1991).

Aeromonas salmonicida—L-forms
Another line of investigation examined L-forms (spherical, filterable cells) of *Aer. salmonicida*. The L-forms were induced experimentally, and were found to be unculturable by conventional plating methods. Subsequently, a stable L-form was

induced with benzylpenicillin, and determined to contain more OMP of ~40 kDa molecular weight than its parental cell (Gibb *et al.*, 1996). This stable L-form did not require specialized media, and could grow on BHIA at 0–5°C (Gibb and Austin, 1994). It could be argued that the development of an L-form state could enable it to persist in tissues of infected fish, although not causing clinical disease (McIntosh and Austin, 1991b). However, attempts to infect fish using L-forms did not result in recovery of the pathogen from fish, even after the administration of immunosuppressants. In addition, it was reported that small numbers of 'natural' L-form colonies were observed, albeit infrequently, on specialized medium (containing horse serum and high quantities of sucrose), which had been inoculated with kidney and spleen samples taken from farmed Atlantic salmon suffering with furunculosis. This suggests that L-forms may have implications in the disease process. Such findings certainly merit further investigation to more conclusively establish the role of this form of *Aer. salmonicida* in disease processes. Obviously, if *Aer. salmonicida*, in a dormant or NCBV or indeed in any other altered state it may undergo outside a fish host, is genuinely unable to transmit infection, then control of the diseases is vastly simplified and rendered less difficult. It is not possible, on the basis of the limited data available, to draw grand, definitive conclusions about this crucially important aspect of the pathogen's behaviour. Further work will, hopefully, clarify the situation.

Aeromonas salmonicida—the role of carriers
The second mechanism by which fish may provide a source of infection is by becoming carriers of *Aer. salmonicida*. The existence of such fish and their role in transmission of the disease was recognized early in the history of furunculosis. In a study of furunculosis in trout in the River Kennett, Horne (1928) found that *Bacterium salmonicida* was recovered in the blood of 17% (3 out of 18) of trout examined, and it was concluded that carriers provided a source of infection in the river. He also sampled fish from fish farms and found them to be generally healthy and thus contended that these results indicated that trout farms were not harbouring the disease. Horne (1928) additionally commented that knowledge of carrier rates in fish populations before, during and after the furunculosis season would be of great epizootiological value.

Because of the obvious importance carrier fish have in the epizootiology of furunculosis, it is essential that the methods used to detect their presence are effective. However, research into the carrier state has been hampered by technical difficulties concerned with detecting such fish with certainty, as present culture isolation methods appear to be too insensitive (McCarthy, 1980). Blake and Clark (1931) reported that raising the temperature of the water in which suspected carriers (usually survivors which had been previously exposed to infection) were maintained from 5 to 18°C induced the disease. At present, a combination of increasing the water temperature to 18°C and the injection of corticosteroids is employed to activate the carrier state. This method is based on work of Bullock and Stuckey (1975b), who tested corticosteroid injection and heat stress as means of producing overt furunculosis in carrier trout. They reported that although heat stress alone produced mortality, the pathogen could not be recovered from the majority of test fish which died. Direct kidney cultures to isolate *Aer. salmonicida* were found to be ineffective for carrier detection. McCarthy (1980) reported success with the method of

Bullock and Stuckey (1975b), although he stated that prednisolone acetate was the most effective corticosteroid of the several he tested, for inducing furunculosis in fish. In further experiments, he assessed the prevalence of carriers in fish populations, examining 1-year-old brown trout from four different commercial fish farms. The fish had a high carrier rate of 40–80%; however, similar populations of rainbow trout tested had a very low incidence of <5% or were possibly free of carriers. The technique had also been applied by McCarthy (1975a) to non-salmonids, e.g. silver bream (*Blicca bjoerkna*), infected with a non-pigmented aberrant *Aer. salmonicida* strain.

The ELISA technique has also been assessed for its effectiveness in the detection of carrier fish. Rose *et al.* (1989) compared a commercial ELISA kit (obtained from Stirling Diagnostics Ltd.), the commonly used corticosteroid/heat protocol of McCarthy (1980) and plating of rectum and kidney samples from Atlantic salmon onto BHIA and TSA. The Atlantic salmon were obtained from a site where outbreaks of furunculosis had previously occurred. From the results, it was apparent that the ELISA was the most successful technique, with 56.17% (14/26) of the fish shown to be carriers. This compared with a carrier rate of 26.4% as determined by means of the corticosteroid/heat test. Yet, plating techniques failed to reveal *Aer. salmonicida* in any of the fish. However, some disadvantages of ELISA are apparent, namely the inability to distinguish living from dead cells, and, for that matter, to differentiate viable, pathogenic cells from those resulting from use of living, attenuated or dead, virulent cells from vaccines. Moreover, ELISA does not enable the provision of cultures, which could be used for additional investigation, such as the determination of antibiograms. Notwithstanding such drawbacks, ELISA is a valuable tool for the rapid detection of *Aer. salmonicida* in clinically diseased and asymptomatic carrier fish.

It has not been conclusively established which organs of carrier fish serve as the site of carriage. From experiments using immunofluorescence techniques, Klontz (1968) concluded that the intestine is a primary site of infection, leading to the establishment of asymptomatic carriers. There is also evidence to suggest that the kidney is involved, but the intestinal tract has again been mentioned (McCarthy, 1980). This is a subject which requires further work to clarify it.

Another problematic aspect of the carrier state is that antibiotic therapy to control furunculosis outbreaks does not necessarily completely remove the bacterium from the tissues of fish, at least when sulphamerazine, furazolidone and potentiated sulphonamide are used. McCarthy (1980), however, reported that fish treated with tetracycline hydrochloride (i.m. injection of 50 mg/kg) survived attempts to induce the disease. On the basis of such findings, McCarthy (1980) warned that it must be anticipated that fish populations treated for furunculosis will remain carriers.

However, more recently, other antibacterial compounds have been investigated for their ability to reduce or eliminate the carrier state of *Aer. salmonicida* in fish. Examples include erythromycin phosphate (Roberts, 1980), flumequine, and an aryl-fluoroquinolone (Scallen and Smith, 1985) in combination with 0.01% Tween 80 to enhance the assimilation of the antimicrobial compound into fish (Markwardt and Klontz, 1989). The last mentioned group of workers found that the aryl-fluoroquinolone/surfactant combination was effective in eliminating the asymptomatic carrier state of *Aer. salmonicida* within 48 h of treatment. Markwardt and Klontz (1989) also pointed out the advantages of the use of the aryl-fluoroquinolone, such as its broad spectrum activity, low *in vitro* minimal inhibitory concentration for many bacterial pathogens (Stamm *et al.*, 1986), and lack of occurrence of resistant strains (Fernandes *et al.*, 1987).

Aeromonas salmonicida—transmission; what does it all mean?
Several possible routes for the transmission of furunculosis have been propounded and investigated. It is commonly accepted that the disease is disseminated by lateral transmission of *Aer. salmonicida* which includes contact with contaminated water and infected fish in addition to possible infection via the gastro-intestinal tract. Also, vertical transmission has been considered in several investigations. A water-borne route where water contamination with *Aer. salmonicida* has occurred initially from moribund infected fish or from overtly healthy carriers shedding the pathogen is favoured as the common means of transmission. Once released into the aquatic environment, the organism is then able to persist for a prolonged period of time and the disease spread in this way. Early studies demonstrated that trout placed into water which had contained diseased fish contracted the infection (Emmerich and Weibel, 1894; Horne, 1928; Blake and Clark, 1931). The Furunculosis Committee of the UK concluded, on the basis of available data, that both food and diseased fish could constitute sources of infection. McCarthy (1980) has examined in detail the likely transmission mechanisms, specifically with regard to the ability of *Aer. salmonicida* to penetrate fish tissues (invasiveness), a prerequisite for the occurrence of infection. Transmission by contaminated water was tested by McCarthy (1980) in laboratory-based experiments by seeding water in a tank containing six brown trout with a suspension of *Aer. salmonicida* to a final concentration of 10^6 cells/ml. Five of the six fish died of furunculosis, and the sixth succumbed when given an injection of prednisolone acetate. In a subsequent large-scale experiment using 50 brown trout placed in a pond on a fish farm experiencing a summer epizootic of furunculosis, 41 fish had died within 28 days, and the remaining 9 succumbed after injection with corticosteroid. McCarthy (1980) concluded from these experiments that the disease is readily disseminated through water and also that brown trout surviving infection probably become carriers. However, both Blake and Clark (1931) and McCarthy (1980) reported failure in attempts to infect rainbow trout by co-habitation with infected brown trout or the addition of *Aer. salmonicida*, respectively. It is known that rainbow trout are more resistant to the disease than are brown trout. In experiments, which examined different routes of exposure to *Aer. salmonicida* subsp. *salmonicida* in Atlantic salmon in seawater, Rose *et al.* (1989) noted that a minimum dose of 10^4 colony forming units/ml by bath was required to initiate infection.

Aeromonas salmonicida—uptake into fish
Another unresolved aspect of the transmission of furunculosis is the uptake of *Aer. salmonicida* into a fish host. It is possible that the pathogen may gain entry to a new host through the gills, lateral line, mouth, anus or surface injury (e.g. Effendi and Austin, 1995b). McCarthy (1980) demonstrated that rainbow trout that resisted the disease subsequently died from furunculosis after their flanks had been abraded with sandpaper. Also, Lund (1967) found that infection was acquired by fish which had been scarified and experimentally challenged with the pathogen. However, these injuries were artificially

induced. Effendi and Austin (1995b) evaluated many different routes for the possible uptake of *Aer. salmonicida* into fish. The data may be summarized, as follows:

Route	Recovery of *Aer. salmonicida* from
Gill	blood and kidney
Oral	blood
Lateral line	blood and spleen (but not kidney)
Ventral surface	blood and spleen (but not kidney)
Flank	blood and spleen (but not kidney)
Anus	blood (but not kidney or spleen)

Generally, *Aer. salmonicida* remained at the site of administration for 24 h. The most effective route of uptake leading to mortalities was the gill and anus. In contrast, fewer deaths resulted from challenge via the lateral line, flank or ventral surface (Effendi and Austin, 1995b). Yet, despite all this work, the natural mode of uptake remains unresolved.

In a study of uptake of *Aer. salmonicida* by rainbow trout, it was observed that the pathogen could be detected in the blood and kidney within 5 min of immersion in a suspension containing 10^5 cells/ml (Hodgkinson *et al.*, 1987). Interestingly, it was also found that uptake of the pathogen was enhanced by the addition of particulates, e.g. latex, to the bacterial suspension. If latex was indeed added, *Aer. salmonicida* was isolated from blood at 12 min, and from kidney, spleen and faeces at 4 h post-challenge. The organism was also cultured from the skin, gills, blood and faeces for up to 48 h. In the absence of latex, the pathogen could be again recovered at 12 min, but from a wider range of sites including kidney, spleen and the lower intestine. However by 24 h, the pathogen was no longer recovered from the fish. From culturing methods alone, it may not, however, be assumed that *Aer. salmonicida* had been totally removed from the animals. In fact, cultures of the pathogen were isolated from kidney, spleen and faeces within 1 to 4 h of immunosuppression of the fish at 7 days post-challenge. In addition, the method of challenge yielded different results. Thus, when the entire fish was immersed in the bacterial suspension, superior uptake occurred compared to exposing only the head or tail regions. The explanation of this phenomenon is unclear, but such results suggest that uptake may occur through several locations rather than a single site, e.g. mouth, nares, gill or anus. It is possible that the pathogen gains entry via all these sites or additionally through the lateral line and/or skin (Hodgkinson *et al.*, 1987). Perhaps, the most significant observation resulting from these experiments was the rapidity by which *Aer. salmonicida* entered the rainbow trout. Other investigators have not sampled so close to the initial time of challenge. McCarthy (1980) reported uptake to occur from the oral route within 5 h, with the organism found in the kidney. Tatner *et al.* (1984) first sampled the fish at 24 h post-challenge. However, all these studies indicate the localization of *Aer. salmonicida* principally within the reticulo-endothelial systems of fish (Tatner *et al.*, 1984).

Another route of infection which has been proposed is via the gastro-intestinal tract, due to intake of contaminated food. However, there is disagreement as to whether or not this in fact occurs. Plehn (1911) and Blake and Clark (1931) reported success in experimentally infecting fish by feeding contaminated food. However Krantz *et al.* (1964a,b) and McCarthy (1980) failed to infect brown trout by feeding with food containing the pathogen. Despite the contradictory information available on the subject, infection via the gastro-intestinal tract cannot be dismissed as a possibility, particularly in view of a report by Klontz and Wood (1972) concerning clinical furunculosis in the sable fish, apparently caused by ingestion of carrier coho salmon. In addition, in laboratory-based experiments that compared various methods designed to induce the carrier state of *Aer. salmonicida* in juvenile spring chinook salmon, Markwardt and Klontz (1989) observed that gastric intubation (of *ca.* 1×10^8 bacteria) resulted in a 65% carrier state. This result was a significantly higher percentage than those recorded for exposure to a broth culture as a bath, ingestion of broth culture coated food, and exposure to intraperitoneally injected fish (40%, 20% and 10% carrier rates, respectively). Markwardt and Klontz (1989) also commented that the exposure to clinically diseased fish and bathing in broth cultures most probably simulated the natural routes of infection (Paterson, 1982).

Aeromonas salmonicida—transmission; the role of eggs
Vertical or transovarian transmission of *Aer. salmonicida* as a possible route of infection has been widely studied with inconclusive results. Smith (1939) claimed the pathogen could be carried on the egg surface, in contrast to Plehn (1911) and Mackie *et al.* (1930), who were of the opinion that the organism was unable to infect fish eggs. The possibility that the pathogen could be transmitted at fertilization was examined by the Furunculosis Committee in the UK. The experimental evidence gathered for their report indicated that furunculosis was not transmitted in such a fashion. Lund (1967), however, contended that since the conclusions were based on the results of a sole set of experiments, further work was necessary to confirm this point. In a detailed series of experiments aimed at clarifying the situation regarding transmission of *Aer. salmonicida* at fertilization, Lund (1967) examined ovaries, testes and ova of experimentally infected fish for the presence of the pathogen. *Aer. salmonicida* was recovered in pure culture from the ovaries and testes of infected fish. Results of isolation of the pathogen from ova were decisive as, of 500 ova sampled, only three obtained from the same fish yielded the bacterium in pure culture from the interior of the ovum. In further experiments, using wild spawners, ova were contaminated with *Aer. salmonicida* on the external surface at the time of fertilization was effected, and the eggs then planted out in a river bed. It was observed that these ova died quickly and were subjected to *Saprolegnia ferax* infection. *Aer. salmonicida* was not isolated from dead or living ova, and Lund (1967) concluded that the experiment had been unsuccessful. Ova taken from parents experimentally infected with *Aer. salmonicida* also failed to yield the bacterium. Continuing the study of vertical transmission, McCarthy (1980) examined the fertilized ova of mature brood stock brown trout taken from a known carrier population (5/8 proved to be carriers when challenged with prednisolone acetate). However, *Aer. salmonicida* was not recovered from the fertilized eggs sampled. When artificially infected brood stock were stripped as soon as signs of clinical furunculosis had developed, both fish organs and fertilized eggs were positive for *Aer. salmonicida*. However, the high numbers of viable cells initially present, i.e. 10^6 cells/ml of egg macerate, rapidly declined and could not be detected 5 days after incubation

began. Based upon these experiments, McCarthy (1980) concluded that vertical transmission of *Aer. salmonicida* was not a significant means of disseminating the disease, and, moreover, in the improbable event that overtly infected fish were used for stripping, the pathogen was unlikely to survive to the eyed-egg stage at which the eggs are marketed. Neither Lund (1967) nor McCarthy (1980) recovered *Aer. salmonicida* from fry derived from experimentally infected parents or of known carrier brood stock, respectively. However, both these authors pointed out the possibility that the negative results obtained may have been due to the inadequacy of techniques used for detection and isolation of the pathogen in the face of inhibition or overgrowth by commensal bacteria, or the presence of only small numbers of *Aer. salmonicida*.

Aeromonas salmonicida—transmission in seawater
A remaining aspect of the epizootiology of *Aer. salmonicida* diseases which requires consideration is the transmission of the infection in seawater. This is an important topic for the aquaculture industry, as salmonids are not infrequently placed in seawater for on-growing. In addition, early in the study of the pathogen, the possibility that migratory Salmonidae could spread the infection was considered. Lund (1967) investigated the possibility that the disease could be carried by salmon or sea trout smolts (previously infected in freshwater) when they migrated to the sea. The Furunculosis Committee had not agreed with this theory because examination of large numbers of smolts taken from the River Coquet in 1928 and 1929 had given no evidence for the presence of *Aer. salmonicida*, although the reverse process, i.e. salmon or sea trout contracting the infection upon migration into rivers containing infected trout, had become generally accepted. In a examination of 234 smolts from the River Coquet, Lund (1967) isolated and confirmed *Aer. salmonicida* from four smolts (two salmon and two sea trout), and believed the findings to be significant as such fish would possibly develop the disease on exposure to suitable conditions or remain resistant, possibly transmitting the infection upon contact with healthy fish in sea or brackish waters. Lund (1967) could not offer a definitive reason for the results differing from those of Williamson and Anderson (see Mackie *et al.*, 1930), who examined 1339 smolts taken from the Coquet without recovering any isolates of *Aer. salmonicida*. Certainly, mortalities attributed to *Aer. salmonicida* in anadromous fish in seawater and in trout grown-on in seawater have been reported (Evelyn, 1971a; Håstein and Bullock, 1976; Novotny, 1978). However, it has not been determined whether the disease outbreaks resulted from stress experienced by fish carrying a latent infection initially contracted in freshwater, or whether they represented a case of lateral transmission of the pathogen via seawater. Smith (1962), for example, had established that *Aer. salmonicida* survived in seawater for a prolonged period of time. It had also been demonstrated that *Aer. salmonicida* is capable of infecting sea and brown trout by contact with infected fish in sea and brackish waters (Scott, 1968). She found that the infection was transmitted between salinities of 2.54 and 3.31% (w/v) at water temperatures ranging from 5.6 to 14.5°C. Smith *et al.* (1982) reported on mortalities of Atlantic salmon from two marine fish farms in Ireland, presenting evidence for the lateral transmission of *Aer. salmonicida* in seawater to a group of fish not known to be carriers. They also provided data suggesting that subsequent to the stocking in spring 1978, and removal of carrier fish in summer 1979, at a marine fish farm, the pathogen became

established and persisted in the fish farm environment for at least 6 months after the removal of the carrier fish. Thus, a carrier-free population placed on the site in the spring of 1980 was infected. Unfortunately, it was not determined whether the pathogen persisted in feral fish outside the cages or in the sediments under the cages. To lend support to a seawater transmission of furunculosis, Evelyn (1971a) has documented isolation of *Aer. salmonicida* from a strictly marine host, the sable fish, although probably the route of infection was by ingestion of moribund or dead salmonid carrier fish (Klontz and Wood, 1972). Obviously, *Aer. salmonicida* has wider potential for causing disease problems than has been hitherto suspected. The study of the epizootiology of *Aer. salmonicida* diseases in aquatic environments other than freshwater, again, as in so many other aspects of the pathogen, demands additional attention to unravel its complexities.

Aeromonas salmonicida—the summary
In summary, although substantial efforts have been made to reach an understanding of the natural disease cycle in the environment, there still remain many issues to be resolved. Some of the data provide partial answers, some are contradictory. Lack of appropriate methodology appears to constitute a major hindrance to progress. In the face of the several unresolved questions concerning *Aer. salmonicida* epizootiology, McCarthy (1980) has valiantly put forth several hypotheses aimed at deriving an overall view of *Aer. salmonicida* epizootiology. The first of these is that furunculosis is introduced to a fish farm by importation of healthy carrier fish or that non-carrier resident fish may be infected from a water supply contaminated by wild or imported carrier fish. Accordingly, if contact with carriers of *Aer. salmonicida* can be prevented, furunculosis should not arise in the population. The second hypothesis McCarthy (1980) advanced is that the mechanisms by which carriers disseminate the disease to non-carriers is dependent upon environmental parameters at a given fish farm site. He postulated that if stressful conditions are absent, resident fish may become carriers without developing clinical signs of disease, and that under such conditions a high carrier rate could be maintained without clinical evidence of furunculosis. This, in the opinion of other workers, would be an arguable premise, as some believe all carriers will eventually succumb to the disease. It is reasonable to suppose that once carrier fish are present on a fish farm they are responsible for initiation of epizootics, although infection also occurs by the water-borne route. The third hypothesis is that the bacterium in the carrier state provides a measure of protection to the fish in return for shelter from the ravages of the aquatic environment. Were this the case, the observation by McCarthy (1980) that non-carrier and carrier fish exhibit marked differences in their susceptibility to furunculosis would be explained.

Finally, it is our opinion that while a firm understanding of the interactions of the pathogen in its milieu and in its fish host is lacking, fisheries scientists will remain at a serious disadvantage when tackling the problems of infectious diseases of *Aer. salmonicida* aetiology that currently beset fish cultivation.

Aeromonas sobria
It may be assumed that the natural reservoir is eutrophic freshwater.

B. ENTEROBACTERIACEAE REPRESENTATIVES

Citrobacter freundii
The organism is common in eutrophic freshwater, from which spread to fish is possible (Allen *et al.*, 1983). In addition, the organism was widespread in the seawater in the Japanese aquarium (Sato *et al.*, 1982).

Edwardsiella ictaluri
The natural reservoir for infection has not yet been firmly established. It is known that the organism survives for limited periods, i.e. up to 8 days, in sterile pond water (Hawke, 1979). This suggests that *Edw. ictaluri* has only limited ability to survive in the aquatic environment. A carrier state in channel catfish has been documented (Klesius, 1992). Also, data have shown that cells can pass from dead fish to non-infected channel catfish (Klesius, 1994). Of course, the results of such laboratory-based survival experiments need to be treated cautiously. It is conceivable that the organism comprises part of the normal microflora of fish, lurking perhaps in the digestive tract.

Edwardsiella tarda
The precise source of infection is often unknown. Nevertheless, it has been speculated that snakes or faecal contamination from humans or other animals may have been involved in the initial outbreak (Meyer and Bullock, 1973). Environmental parameters, namely water temperature and the quantity of organic matter in the water, undoubtedly influence the severity of outbreaks. In particular, it is noteworthy that most disease outbreaks occur at high water temperature, i.e. 30°C, when high levels of organic matter are present. However, data showed that the incidence of disease in catfish ponds rarely exceeded 5%, except when infected fish were moved to holding tanks. During these periods, edwardsiellosis became rampant with total mortalities approaching 50% of the population (Meyer and Bullock, 1973).

There is some evidence that *Edw. tarda* occurs in sediment (Rashid *et al.*, 1994a,b) and water within the vicinity of fish farms (Minagawa *et al.*, 1983; Park *et al.*, 1983; Rashid *et al.*, 1994a,b). It has been reported to inhabit aquatic invertebrates, such as the freshwater aquarium snail (*Ampullaria* sp.; Bartlett and Trust, 1976) and sea-urchin (Sasaki and Aita, 1975), and has been associated with vertebrates, namely snakes (Iverson, 1971), frogs (Sharma *et al.*, 1974), turtles (Otis and Behler, 1973), gulls (Berg and Anderson, 1972) and human beings (Koshi and Lalitha, 1976). Any one of these hosts may serve as a reservoir for infections of fish. However, it is unclear whether *Edw. tarda* should be regarded as a primary or opportunistic pathogen of fish. Indeed, it may comprise part of the normal microflora of fish surfaces, insofar as Wyatt *et al.* (1979) have indicated the widespread occurrence of the micro-organisms in and around channel catfish. A suggestion has been made that the organism becomes non-culturable in the aquatic environment (Sakai *et al.*, 1994).

The presence of sub-lethal concentrations of copper (100–250 μm/l) in water leads to increased susceptibility of Japanese eels to infection (Mushiake *et al.*, 1984). This increased susceptibility seems likely to be attributed to a reduction in lymphocytes and granulocytes in the blood, leading to lowered phagocytosis (Mushiake *et al.*, 1985).

Enterobacter agglomerans
The source of the pathogen was unknown (Hansen *et al.*, 1990).

Proteus rettgeri
Inference was made that the source of infection was from poultry faeces, which were used extensively to fertilize the carp ponds (Bejerano *et al.*, 1979). In this connection, it is relevant to note that *Pr. rettgeri* has been associated with the digestive tract of poultry.

Salmonella choleraesuis subsp. *arizonae*
Presumably, the pathogen had been derived from the aquarium water or from fish (carriers) (Kodama *et al.*, 1987).

Serratia liquefaciens
The natural reservoir of the organism was probably polluted waters (McIntosh and Austin, 1990b).

Serratia marcescens
Seemingly, the organism may be associated with fish, and possibly comprise part of the 'normal' microflora of eutrophic waters (Baya *et al.*, 1992c).

Serratia plymuthica
Possibly, the pathogen comprises a component of the freshwater bacterial community, particularly where pollution with organic material is rife (Austin and Stobie, 1992b).

Yersinia intermedia
The organism has been associated with water and the intestines of apparently healthy fish (e.g. Kapperud, 1981; Shayegani *et al.*, 1986; Zamora and Enriquez, 1987), which are likely to be the source.

Yersinia ruckeri
In rainbow trout, ERM most commonly affects fish of approximately 7.5 cm in length. The disease is less severe but more chronic in larger fish, i.e. of 12.5 cm in length. Severity peaks with a water temperature of 15–18°C, and decreases when it drops to 10°C or below. Overly fat or debilitated (stressed) fish are thought to be more susceptible to severe epizootics (Rucker, 1966). Moreover, occurrence of ERM may be cyclic, suggesting the presence of asypmtomatic carrier fish in the population (Busch and Lingg, 1975). Such fish would be capable of shedding the organism in the faeces. Alternatively, it may be argued that the pathogen is present normally in the water and/or sediment (Rucker, 1966). From the results of some fascinating experiments, Hunter *et al.* (1980) demonstrated that unstressed carrier fish did not transmit *Y. ruckeri* to recipient fish. In fact, it was necessary to stress these carriers, such as by heat (i.e. a water temperature of 25°C), in order to enable release and thus transmission, of the pathogen to other fish. Even under these conditions, colonization of the lower intestine took place in the recipient fish without any mortalities. However, the handling of fish, which may appear to be extremely healthy, and crowded conditions causing excess ammonia and metabolic

waste products in the water and, thus, a decrease in oxygen levels, may precipitate outbreaks of clinical disease (Bullock and Snieszko, 1975). Nevertheless, outbreaks generally occur only after the fish have been exposed to large numbers of the pathogen (Ross *et al.*, 1966). Mortalities tend to start 5–19 days after exposure, depending on the size of the inoculum, and last for 30–60 days (Rucker, 1966; Busch, 1973). Cumulative losses due to ERM may account for as much as 30–35% of the rainbow trout population (Klontz and Huddleston, 1976).

There is controversy over some aspects of the ecology of *Y. ruckeri*, insofar as the organism has been alternatively considered to be well adapted as a normal aquatic saprophyte (McDaniel, 1972) and, secondly, as not capable of free-living for any extended period in water (Klontz and Huddleston, 1976). Nevertheless, these authors conceded that survival for up to 2 months was possible in mud. Further experiments indicated that the pathogen could survive for 4 months in unsupplemented water at a salinity of 0–20‰; survival being reduced in seawater (salinity = 35‰) (Thorsen *et al.*, 1992). Romalde *et al.* (1994) reported survival for 3 months. Therefore, the implication is that *Y. ruckeri* is capable of surviving in the (fresh) water column for a long time after an outbreak of disease has occurred. Perhaps, the organism is a normal inhabitant of the digestive tract of fish or in contaminated waters. In this respect, *Y. ruckeri* has been isolated from sewage sludge (Dudley *et al.*, 1980) and a bacteriophage specific for *Y. ruckeri* found in sewage (Stevenson and Airdrie, 1984b). Moreover, aquatic invertebrates, notably crayfish, and even terrestrial mammals, namely muskrats (Stevenson and Daly, 1982) may harbour large numbers of the pathogen, thereby serving as a reservoir of infection. It is noteworthy that McArdle and Dooley-Martin (1985) recovered *Y. ruckeri* from the digestive tract of an apparently healthy goldfish following inspection of a consignment upon importation into Ireland. The isolate was subsequently demonstrated to be pathogenic for rainbow trout. Thus it would appear that ornamental fish could pose a reservoir of infection into salmonids. If Dulin *et al.* (1976) are correct in asserting that *Y. ruckeri* is not ubiquitous in nature, it is puzzling why ERM should be found in remote fish stocks, without prior history or association with the disease (Janeke, personal communication). Under these conditions, it is doubtful if movement of infected fish could account for the spread of ERM. Yet, isolations have been made from wild salmonids. In one example, *Y. ruckeri* type I was recovered from a mature wild Atlantic salmon in freshwater [in Scotland] (Petrie *et al.*, 1996). Whether or not this suggests a natural reservoir of the pathogen in wild fish or reflect a transfer from aquaculture remains to be established.

There is some evidence that *Y. ruckeri* may well be capable of surviving in the environment in a form (the so-called dormant or non-culturable state) that is not readily culturable on conventional media (Romalde *et al.*, 1994). These workers found that the number of culturable cells increased for the first 15 days after *Y. ruckeri* was seeded into an experimental system. Thereafter, a decline in numbers was recorded over a 100-day period. Culturable cells persisted in sediment more so than water at 6 and 18°C. Yet, by means of acridine orange staining and fluorescence microscopy, a so-called dormant state was considered to develop. There were slight changes in the LPS of the dormant cells, but not in membrane proteins or plasmid composition. Interestingly, virulence was maintained during the period of non-culturability. Unlike the situation with *Aer. salmonicida*, cells were resuscitated to a culturable state following the addition of nutrients (Romalde *et al.*, 1994).

A serotyping scheme, based on O-antigens, has been proposed (Davies, 1990), and may have value in epizootiological investigations.

C. *CYTOPHAGA–FLAVOBACTERIUM–FLEXIBACTER* GROUP

The precise reservoir of most of these fish pathogens remains unclear. Yellow and orange pigmented bacteria occur in large numbers in freshwater (Allen *et al.*, 1983) and seawater (Austin, 1982a), and comprise part of the normal microflora of gills of healthy salmonids (Trust, 1975) and possibly even eggs (see Hansen *et al.*, 1992). Therefore, it seems likely that the taxa comprising the fish pathogens occur naturally in the aquatic environment. Rucker *et al.* (1953) isolated highly virulent strains of *Fla. columnare*, which were capable of killing fish within 24 h, from the water in the upper Columbia River basin. Furthermore, it has been demonstrated that large numbers of *Fla. columnare* cells are present in water during epizootics (McCarthy, 1975c), with good survival occurring over a wide range of pH and hardness values (Fijan, 1968). Moreover, in one study, Collins (1970) reported a relationship between eutrophication and the numbers of *Fla. columnare* cells in lake water. Although the precise reservoir of *Fla. psychrophilum* has not been established, we have recovered some organisms, with the key characteristics of *Fla. psychrophilum*, from freshwater in England. However, it is uncertain whether or not the organisms were native, or merely transient, in the aquatic environment. Some evidence has been presented that *Fla. psychrophilum* may be transmitted within salmonid eggs (Brown *et al.*, 1997).

Colonization of fish may be a forerunner for the development of disease. For example, colonization of eyed rainbow trout ova with *Fla. psychrophilum* has been considered to lead to the development of RTFS (Rangdale *et al.*, 1997b). Using a modified Anacker and Ordal's medium, the pathogen was recovered from ovarian fluid of 2/15 hen fish, egg surfaces 14 days after fertilization, but not from milt (Rangdale *et al.*, 1996).

Ribotyping and plasmid profiling appear to be useful for epizootiology. In particular, several ribotypes have been recognized among 85 isolates, and determined to be associated with the fish species from which the cultures were obtained (Chakroun *et al.*, 1998).

There is no dispute that these organisms are capable of causing severe losses to fish populations. Coldwater disease may cause losses of up to 50% of the fish population, as determined for coho salmon fry (Rucker *et al.*, 1953). This disease is especially troublesome at low water temperatures (Holt *et al.*, 1989), i.e. ≤15°C. *Fle. maritimus* may also be responsible for heavy losses, i.e. 20–30% of the population, among juvenile red sea bream and black sea bream (Hikida *et al.*, 1979). Similarly, *Fla. branchiophilum* may be fatal to small fish following infection via water, i.e. adding a bacterial suspension to tank water (Wakabayashi *et al.*, 1980). Certainly, the pathogen is present in water at times of disease outbreaks (Heo *et al.*, 1990).

Concerning columnaris, the consensus of opinion is that the disease is problematic only in the warmer periods of the year. Generally, epizootics occur when the water temperature is in the region of 18–22°C, and the disease is rarely troublesome at <15°C

(Amend, 1970). Thus, most outbreaks of columnaris occur between May and October (Bowser, 1973; Kuo *et al.*, 1981). To illustrate the dramatic effects of water temperature on the level of mortalities, the investigation of Holt *et al.* (1975) is especially relevant. This team challenged steelhead trout, chinook salmon and coho salmon with *Fla. columnare*, via the water-borne route. At a water temperature of 9.4°C there were not mortalities attributable to columnaris. By increasing the temperature to 12.2°C, 4–20% mortalities ensued; whereas at 20.5°C all the steelhead trout and coho salmon died, together with 70% of the chinook salmon.

The level of mortality may be extremely high, and figures of 60–90% are not uncommon. For example, columnaris was considered to be the most important contributing factor for 72.3–97.6% and 75.4–95.4% mortality among populations of adult sockeye salmon and adult chinook salmon, respectively (Fish and Hanavan, 1948). Similarly, Chen *et al.* (1982) reported 77% and 88.3% losses among groups of carp and goldfish, respectively. Certainly, there are good data demonstrating the seasonal effects of mortalities due to columnaris, insofar as Bowser (1973) reported 60% infection of bullheads in mid May whereas only a few months later the incidence had dropped to only 10%.

In addition to water temperature, the severity of columnaris is influenced by a multiplicity of environmental (stress) and host-related factors. Chen *et al.* (1982) described the highest eel mortality levels to be associated with stagnant running water, whereas the lowest losses occurred in running water. Interestingly with aeration, the total losses fell between these two extremes. In this respect, the mortality rate has been inversely correlated with the level of dissolved oxygen in the water. Moreover, with adequate dissolved oxygen, deaths increased with a concomitant rise in the level of ammonia.

The susceptibility of juvenile chinook salmon and rainbow trout has been correlated with the age of the fish and the level of crowding, as well as water temperature (Fujihara *et al.*, 1971). These workers concluded that rainbow trout (1 g average weight) and chinook salmon (average weight = 3 g) were less susceptible to the rigours of columnaris than smaller fish. Therefore, it was concluded that age is more important than weight in determining susceptibility to infection.

Chrys. scophthalmum was found in the water, from which it is surmised that spread to turbot occurred (Mudarris and Austin, 1989).

D. PSEUDOMONADACEAE REPRESENTATIVES

Pseudomonas anguilliseptica
Red spot disease, also known as 'Sekiten-byo', was first recognized in pond-cultured eels (*Anguilla japonica*) in Japan (Wakabayashi and Egusa, 1972). Since then, it has developed into one of the most destructive diseases of eels in Japan (Nakai and Muroga, 1979). During 1981, the disease was recognized in European eels (*A. anguilla*) within Scotland (Nakai and Muroga, 1982; Ellis *et al.*, 1983; Stewart *et al.*, 1983). Conceivably, the disease is spreading, and this may reflect the natural migratory patterns of wild eel populations, or the rapid increase in intensive eel cultivation during the mid-1970s. A spread to other fish groups has now occurred. The pathogen was recovered from striped jack at water temperatures of 14–16°C during February to April, 1993 (Kusuda *et al.*,

1995), diseased rainbow trout, sea trout and whitefish (*Coregonus* sp.) in Finland during 1986–1991 (Wiklund and Lönnström, 1994), wild Baltic herring (*Clupea harengus membras*) (Lönnström *et al.*, 1994), sea bass, sea bream and turbot in France (Berthe *et al.*, 1995), and as a winter disease outbreak in sea bream in Spain (Doménech *et al.*, 1997). However, in Finland, there was some evidence also for the presence of *V. anguillarum* and *Aer. salmonicida.*

Very little has been published about the epizootiology of Sekiten-byo. The disease is recognized, however, to be prevalent in brackish water, when the water temperature is between 20 and 27°C (Muroga *et al.*, 1977b). Indeed, temperature is considered to be the major factor in influencing epizootics. All the available information points to the seriousness of outbreaks in terms of high mortalities. Thus in one outbreak of the condition in Scotland, 67000 elvers (96% of the total) plus 154 adult eels (3.9% of the total) died (Stewart *et al.*, 1983). These authors observed that generally the large adult eels fared worse than smaller adults. Nevertheless, the greatest losses occurred with elvers. It is interesting to note that the disease eliminated 14% of the total weight of the farm stock. This represents a significant investment; therefore, Sekiten-byo has the potential to be a severe economic problem.

Ps. chlororaphis
Ps. chlororaphis occurs in water (Palleroni, 1984), which is assumed to be the reservoir of infection. No other information is available.

Ps. fluorescens
In view of its widespread occurrence in the aquatic environment, *Ps. fluorescens* is likely to be spread through water, which will serve as the primary reservoir of infection. The disease is especially troublesome at low water temperatures, i.e. at approximately 10°C. In one experiment, Ahne *et al.* (1982) achieved 100% mortality in tench fry within 10 days at a water temperature of 10°C. Challenge was by i.p. injection of a bacterial suspension.

Ahne and co-workers noted that initial occurrence of disease, with 30% of the population developing skin haemorrhages, was 7 days after transfer of the tench to laboratory aquaria. Mortalities began on day 14, and by 2 weeks later, 90% of the fish had died. It is noteworthy that the problem in silver carp and bighead developed after a stressful period during winter, when the water temperature fluctuated around freezing. In this outbreak, the mortality rate was at 5% of the population per day (Csaba *et al.*, 1981).

Pseudomonas pseudoalcaligenes
Pseudomonads occur in polluted/eutrophic freshwater, which is considered to be the source of *Ps. pseudoalcaligenes* (Austin and Stobie, 1992b). Moreover, it was apparent that the fish farm waters received sewage from a neighbouring septic tank (B. Austin, unpublished data).

E. VIBRIONACEAE REPRESENTATIVES

Photobacterium damselae

The organism was associated initially with ulcerative lesions along the flank of blacksmith (*Chromis punctipinnis*), one of the damselfish. These ulcers were noted in summer and autumn among fish populations in the coastal waters of southern California. Surveys of wild fish populations led to a conclusion that the ulcers were restricted to species of damselfish. Additional information pointed to a role in human pathogenicity, insofar as the organism has been isolated from human wounds (Love *et al.*, 1981). Subsequent work demonstrated this organism in sharks (Grimes *et al.*, 1984; Fujioka *et al.*, 1988), turbot (Fouz *et al.*, 1991; 1992), yellowtail (Sakata *et al.*, 1989) and nurse sharks and rainbow trout [in Denmark] (Pedersen *et al.*, 1997a).

Ulcerated fish, presumably the result of *Ph. damselae*, accounted for 10–70% of the population in King Harbor, Redondo Beach, California, during August to October, and at a second site (Ship Rock, Catalina Island) during June to October. This suggests a seasonal distribution in the incidence of disease, and possibly coincides with warmer water temperatures and lower resistance caused by physiological changes in the host during sexual maturity. Conceivably, *Ph. damselae* occurs normally in the marine environment, where it could pose a constant threat to susceptible fish species.

Photobacterium damselae subsp. *piscicida*

There is no doubt that pasteurellosis is a serious condition of both farmed and wild fish populations (Snieszko *et al.*, 1964a; Kusuda and Yamaoka, 1972; Ohnishi *et al.*, 1982; Yasunaga *et al.*, 1983). Heavy mortalities, in the range of 40–50% of the stock, have occurred during summer months. Unfortunately, the reasons for these outbreaks are largely unknown. Likewise, the precise nature of the pathogenicity mechanism remains to be elucidated.

It is thought that infection takes place in seawater at temperatures of approximately 25°C (Yasunaga *et al.*, 1983). Toranzo *et al.* (1982) devised a series of survival experiments, and concluded that *Ph. damselae* subsp. *piscicida* was short-lived in freshwater and estuarine conditions. Thus in freshwater, the organism could not be cultured after 48 h at 20°C. Survival in estuarine water (salinity = 12‰) was slightly longer, i.e. 4–5 days. These results support the earlier findings of Janssen and Surgalla (1968) that the organism does not appear to survive well away from fish. However, there is some evidence that has demonstrated the discharge of viable cells from experimentally infected yellowtail for 1–2 days before death (Matsuoka and Kamada, 1995). The survival of starved cells in seawater needs to be clarified (Magariños *et al.*, 1997a). Furthermore, it was speculated that transmission of the disease is likely to be fish to fish. Although this reasoning is reminiscent of the debate about the spread of furunculosis, it should not be overlooked that *Ph. damselae* subsp. *piscicida* may survive in water, albeit in a non-culturable, dormant or altered form. Indeed, a view was expressed that the pathogen may well exhibit a dormant phase (Magariños *et al.*, 1994a). Plate count data revealed that *Ph. damselae* subsp. *piscicida* could survive in seawater and sediment for 6–12 days, with metabolism being reduced by 80%. Indeed, in terms of numbers, culturable cells persisted in sediment better than in seawater. However, when culture techniques inferred a reduction in bacterial numbers, microscopy using acridine orange suggested that the populations remained at 10^5 (Magariños *et al.*, 1994a).

A possibly useful tool for epizootiology is ribotyping, which has already successfully discriminated between European and Japanese isolates ((Magariños *et al.*, 1997b).

Plesiomonas shigelloides

The outbreak in 1984 affected mostly 1–2-year-old fish, in which the total mortality approximated 40% of the population. There was a correlation with water temperature, which increased in June from 10 to 17°C. Moreover, there was a pronounced increase in organic matter from food within the water. These factors would undoubtedly contribute to stress in the fish. It has been noted that *P. shigelloides* may be normally resident in the gastro-intestinal tract of warm-water fish (Vandepitte *et al.*, 1980), from where it could serve as a reservoir of infection.

Shewanella putrefaciens

It is assumed that the organism was derived from the coastal marine environment. Possibly, the organism comprises part of the normal microflora of fish (see Lee *et al.*, 1977; Gillespie, 1981).

Vibrio alginolyticus

Vibrios abound in the marine and estuarine environments (see Kaneko and Colwell, 1974), and therefore present a constant threat for any susceptible host. In particular, *V. alginolyticus* has been recovered from the water in marine fish tanks (Gilmour, 1977).

Vibrio anguillarum

Vibriosis had gained considerable notoriety in mariculture, where it has become a major limiting factor in the successful rearing of salmonids (Mahnken, 1975). To cite one example, in Denmark, the disease has resulted in cumulative losses of 30% among eel populations (Bruun and Heiburg, 1935). This represents a significant economic loss. An authoritative publication reported that vibriosis occurs in more than 14 countries, where it has ravaged approximately 48 species of marine fish. Vibriosis appears to have been confined initially in European waters. North America escaped the ravages of the disease until 1953 (Crosa *et al.*, 1977). Its arrival in Japan in 1975 may have resulted from the importation from France of contaminated eels (Muroga *et al.*, 1976a,b). Evidence is also accumulating that the disease may occur in freshwater conditions (Muroga, 1975; Ghittino and Andruetto, 1977). This suggests that vibriosis is an extremely widespread problem. Consequently, the literature abounds with reports of 'new' isolations and titbits of gossip, which slowly contribute to an overall saga. However, it would appear that vibriosis is, in fact, a syndrome caused by a multiplicity of vibrios (see Schiewe, 1981). Here, emphasis will be placed on *V. anguillarum*.

The causal agent of 'red-pest' in eels was first isolated by Canestrini (1893), who designated the organism as *Bacterium anguillarum*. A further case among eels in Sweden during 1907 was investigated by Bergman (1909), and it was directly attributable to this scientist that the name of *V. anguillarum* was coined.

V. anguillarum constitutes part of the normal microflora of the aquatic environment (e.g. West and Lee, 1981; Muroga *et al.*, 1986), particularly associated with rotifers (Tatani *et al.*, 1985; Muroga and Yasunobu, 1987), with maximal and minimal numbers in summer and winter, respectively (Larsen, 1982). Experiments have suggested that the pathogen survives for prolonged periods in seawater. Thus, Hoff (1989) reported survival for >50 months in a seawater microcosm. The organism may also constitute part of the normal microflora of marine fish (Oppenheimer, 1962; Mattheis, 1964). Some elegant work, albeit with only one isolate, has addressed the precise changes to the organism, i.e. starvation-stress responses, in the marine environment (Nelson *et al.*, 1997). When starved of carbon, nitrogen and phosphorus, the number of colony-forming units (note: this is a dubious measure of viability) dropped rapidly over an initial 5–7 day period, and then gradually declined over 3–4 weeks. Some cells became small and spherical, corresponding to the notion of ultramicrobacteria (see Austin, 1988), whereas others elongated to short spirals. Protein synthesis, as measured by incorporation of [^{35}S]-methionine declined during the first 6 h of starvation, and increased to >70% of the rate in exponentially growing cells by 5 days into the starvation regime (Nelson *et al.*, 1997).

The precise origin of an isolate has importance for epizootiology. In this respect, Olsen and Larsen (1990) detailed a seemingly useful method, namely restriction fragment length polymorphism of the 65–70 kb plasmid. This method should have value for epizootiological investigations.

The exact mode of infection is unclear, but undoubtedly involves colonization of (attachment to) the host, and thence penetration of the tissues. Ransom (1978) postulated that infection probably begins with colonization of the posterior gastro-intestinal tract and rectum. This conclusion resulted from the observation that *V. anguillarum* was seen initially in these sites. Horne and Baxendale (1983) reported adhesion of *V. anguillarum* to intestinal sections derived from rainbow trout. All regions of the intestine were colonized (approximately 10^3 cells/cm^2), with maximum attachment occurring within 100 min. The skin appears to become colonized within 12 h of immersion in a virulent culture (Kanno *et al.*, 1990). Then, invasion of the liver, spleen, muscle, gills and intestine follows (Muroga and De La Cruz, 1987).

It has been well documented that epizootics occur in the warm summer months when water temperatures exceed 10°C, the water is depleted of dissolved oxygen, and the fish stressed by overcrowding and poor hygiene (Anderson and Conroy, 1970). There are exceptions to the norm insofar as outbreaks have been documented in fresh water (e.g. Rucker, 1959) and at low temperature, i.e. 1–4°C (Olafsen *et al.*, 1981). It is perhaps ironic that isolates recovered from rainbow trout in freshwater have an obvious salt requirement for growth (Rucker, 1959). Perhaps, the organisms were contained in a protected ecological niche such as within the fish body, prior to the manifestation of the disease. However, it should be remembered that the pathogen has been recovered sporadically from freshwater (West and Lee, 1982). The determination of plasmid profiles may have value for epizootiological investigations (Wiik *et al.*, 1989).

The presence of heavy metals, notably copper and iron, contributes to an exacerbation of vibriosis. Yet sub-lethal concentrations of chlorine do not appear to promote the development of infections (Hetrick *et al.*, 1984). Levels of only 30–60 μg copper/ml and

10 μg of iron/ml have caused severe problems (Rødsaether *et al.*, 1977; Nakai *et al.*, 1987). Further investigation demonstrated the susceptibility to vibriosis was dependent upon concentration and time of exposure to copper (Baker *et al.*, 1983). The debilitating effect has been attributed to coagulation in the mucus layer of the gills, and thus the inhibition of oxygen transport leading to respiratory stress (Westfall, 1945). The practical outcome from this information is that fish holding facilities should not be coated with copper-containing anti-fouling compounds, which could trigger vibriosis.

V. cholerae (non-O1)

V. cholerae survives in the aquatic environment. For example at 25°C, strain PS-7701 survived for 32 days in fresh water, saline, Ringers buffer, and normal strength and diluted seawater. Survival was considerably reduced at 2°C (Yamanoi *et al.*, 1980); an observation which coincides with the findings of Singleton *et al.* (1982a,b). It is likely that infection occurs via the water-borne route, insofar as *V. cholerae* appears to inhabit the aquatic environment (Lee *et al.*, 1982; West and Lee, 1982).

Moritella marina, V. fischeri, V. furnissii, V. harveyi, V. ichthyoenteri and *V. logei*

It may be assumed that the source of the bacteria was seawater.

V. ordalii

V. ordalii appears to have a more restricted niche than *V. anguillarum*, insofar as isolates are rarely recovered from outside of fish. It has been postulated that infection (colonization) begins in the rectum and posterior gastro-intestinal tract. Alternatively, its presence on skin suggests that entry may proceed by direct invasion of the integument (Ransom, 1978).

V. pelagius

This was not considered by Angulo *et al.* (1992).

V. salmonicida

It has been demonstrated that *V. salmonicida* survives for >14 months in laboratory-based experiments with seawater, when seeded at ~10^6 cells/ml (Hoff, 1989). Thus, there is the potential for long-term survival in the vicinity of fish farms, as confirmed by Husevåg *et al.* (1991). Moreover, the pathogen has been detected in the sediment (12–43 cells/ml) below fish farms, several months after an outbreak of Hitra disease. In addition, *V. salmonicida* has been detected in the sediments from fish farms which were not experiencing clinical disease (Enger *et al.*, 1989; 1991). Clearly, there will be a reservoir of the pathogen around farmed fish, from which further infections may occur.

V. splendidus

It seems likely that the organism is a component of the normal aquatic bacterial micro-flora, with survival of >114 days recorded (Lopez and Angulo, 1995).

V. vulnificus

V. vulnificus is ubiquitous in the coastal marine and estuarine environment, where it occurs routinely in low numbers (Oliver *et al.*, 1983). So, the reservoir is almost certainly the aquatic, especially seawater, environment (Høi *et al.*, 1998). It has been documented to survive in brackish water and on the surfaces of eels for 14 days (Amaro *et al.*, 1995). It is feasible that fish are constantly exposed to the potential vagaries of this organism. Moreover, it is capable of entering eels through the skin (Amaro *et al.*, 1995).

F. MORAXELLACEAE REPRESENTATIVES

Acinetobacter sp.

Unreactive Gram-negative rods that are difficult to identify and which resemble *Acinetobacter* are common inhabitants of freshwater (Allen *et al.*, 1983) and marine ecosystems (Austin, 1982a). In particular, the organisms populate the skin and gills (Horsley, 1973) and digestive tract (Shewan, 1961; Trust and Sparrow, 1974; Roald, 1977) of salmonids. Therefore, a ready inoculum of cells is likely to be in continual contact with fish. Conceivably any break in the integument of the host may lead to colonization of the nutrient-rich tissues by components of the water-borne or, indeed, fish microflora. This may lead to the start of a disease cycle.

Moraxella sp.

It may be assumed that the organism constitutes part of the aquatic microflora.

G. HALOMONADACEAE REPRESENTATIVE

Halomonas (= Deleya) cupida

H. cupida is a water-borne organism (Baumann *et al.*, 1972). Therefore, the source of infection was undoubtedly the marine environment.

H. SMALL, MORPHOLOGICALLY SIMPLE BACTERIA

Mycoplasma mobile

Mycoplasma-like bacteria occur on fish (Kirchhoff and Rosengarten, 1984), from which infection probably occurs.

Piscirickettsia salmonis

It is presumed that the pathogen is passed directly between fish, or via invertebrate vectors. Data have shown that physical contact may be necessary for horizontal transmission between salmonids (Almendras *et al.*, 1997). Certainly, the pathogen appears to survive extracellularly (Lanan and Fryer, 1994).

I. MISCELLANEOUS PATHOGENS

Janthinobacterium lividum

Typical and atypical forms of *J. lividum* are regarded as part of the normal microflora of fresh water (Sneath, 1984) and soil (Moss and Ryall, 1980). Therefore, there would be a ready inoculum of the pathogen in the environment around fish.

9

Pathogenicity

Many publications about pathogenicity mechanisms have resulted from the examination of single isolates, often of questionable authenticity. The usefulness of such approaches to the understanding of pathogenicity of bacterial species is doubtful. Also, the value of studies involving bacterial subcellular components produced on agar plates or in broth cultures at explaining disease mechanisms *in situ* is unclear.

The pathogenicity of some supposed bacterial fish pathogens has not been considered. Such organisms have not been included in this chapter.

A. ANAEROBES

Eubacterium tarantellae

Invasion of the body may occur through wounds or as a result of damage inflicted through parasites, weak pathogens or stress. Once inside the body tissues, further damage may be inflicted as a result of exo- or endotoxins. The organism produces haemolysins and lecithinase, which may harm the fish. Nevertheless, it should be emphasized that the precise pathogenicity mechanisms have yet to be elucidated (Udey *et al.*, 1976).

B. GRAM-POSITIVE BACTERIA—THE 'LACTIC ACID' BACTERIA

Carnobacterium piscicola and the lactobacilli

Small-scale experiments with rainbow trout maintained in fresh water at 18°C have shown that death may result within 14 days of i.p. injection of 10^5 cells/fish. Dead and moribund fish had swollen kidneys, and ascitic fluid accumulated in the abdominal cavity. However, adverse effects were not recorded following injection of cell-free extracts. This suggests that exotoxins did not exert a significant role in pathogenicity. It remains for further work to elucidate the effect, if any, of endotoxins (Ross and Toch, 1974; Cone, 1982; Hui *et al.*, 1984).

Gram-positive cocci in chains

Experimental infections with organisms likely to correspond with *Lactococcus garvieae* have been achieved by injection of 10^4 to 10^5 cells (Cook and Lofton, 1975), and by exposure of fish for 10 min to 10^6 bacteria (Robinson and Meyer, 1966). Thereafter, disease becomes established, and death ensues. Some host specificity to Gram-positive cocci in chains exists, insofar as trout suffer heavy mortalities whereas Mozambique bream (*Sarotherodon mossambicus*), banded bream (*Tilapia sparramanii*), carp (*Cyprinus carpio*) and largemouth bass (*Micropterus salmoides*) do not (Boomker *et al.*, 1979). It has been established that challenge with low-virulence isolates or low doses of high-virulence isolates together with cell-free culture supernatants are sufficient to establish infection (Kimura and Kusuda, 1979). The toxic activity of supernatants was further researched, and two fractions were demonstrated to have a significant effect on pathogenicity (Kimura and Kusuda, 1982). These were recovered in Todd-Hewitt broth after incubation at 30°C for 48 h. The fraction, although not toxic by oral administration (presumably the compounds were digested), produced damage, i.e. exophthalmia and petechial haemorrhages, following percutaneous injection of yellowtails.

Evidence has been presented that a cell capsule may be involved with the resistance of Gram-positive cocci in chains to opsonophagocytosis in yellowtail (Yoshida *et al.*, 1997).

An isolate of *Str. milleri* (G3K) injected at 5×10^6 cells/fish caused 20% mortalities in Atlantic salmon. Interestingly, all the fish darkened, albeit with negligible signs of internal or external abnormalities. With rainbow trout, there was evidence of kidney liquefaction (Austin and Robertson, 1993).

Laboratory infections with *Vag. salmoninarum* were achieved using a comparatively high dose of 1.8×10^6 cells/rainbow trout (Michel *et al.*, 1997).

C. AEROBIC GRAM-POSITIVE RODS AND COCCI

Bacillus sp.

Oladosu *et al.* (1994) infected *Clarias gariepinus* via the oral and subcutaneous routes with a comparatively low dose of 0.5 ml, which contained 1.8×10^3 cells/ml. Thus, 60% and 30% mortalities were achieved over a 3-week period by oral and subcutaneous challenge, respectively.

Bacillus mycoides

Injection of 1.6×10^4 cells intramuscularly led to lesions in channel catfish, as described in the original outbreak (Goodwin *et al.*, 1994). Intraperitoneal and subcutaneous injections did not lead to the development of any lesions in the infected fish.

Corynebacterium aquaticum

The fish isolate, RB 968 BA, killed rainbow trout and striped bass, with LD_{50} doses calculated as 5.8×10^4 and 1.0×10^5, respectively (Baya *et al.*, 1992a). Experimentally infected fish developed haemorrhaging in the cranial cavity, but did not develop any external signs of disease. ECP, which contained caseinase and gelatinase activity, was harmful to fish, with an LD_{50} dose equivalent to 1.2 μg of protein/g of fish.

Coryneforms

As a result of pathogenicity experiments with rainbow trout (average weight = 8 g) maintained in freshwater at 18°C, it was established that 1.25×10^6 cells, administered by i.p. injection, were capable of killing fish within a few days (Austin *et al.*, 1985).

Micrococcus luteus

Injection of 10^5 cells, via the i.m. and i.p. routes, led to 54% mortalities in rainbow trout fry within 14 days (Austin and Stobie, 1992a).

Mycobacterium spp.

Only *Myc. chelonei* subsp. *piscarium* has been studied in any detail. At a water temperature of 12°C, experimental infections developed in rainbow trout which were injected, via the i.p. route, with approximately 10^7 cells. Accumulative mortalities ranged from 20% to 52%. With juvenile chinook salmon, 98% mortalities were recorded within 10 days at a water temperature of 18°C (Arakawa and Fryer, 1984).

Evidence has indicated that ECP may well be involved with the pathogenic process. For example, Chen *et al.* (1997) determined the LD_{50} of ECP from *Mycobacterium* spp. as >400 μg of protein/fish to rainbow trout and nile tilapia.

Nocardia spp.

Experimental infections have been established in Formosa snakehead (*Chanos maculata*) and largemouth bass (*Micropterus salmoides*) (Chen, 1992). Thus, typical granulomatous lesions and mortality followed in 14 days of i.p. or i.m. injection of 8 mg of suspensions of *Noc. asteroides* (Chen, 1992).

Planococcus sp.

Fish, injected intraperitoneally with 10^5 cells displayed erratic swimming within 48 h. At this time, the gills were pale, the anus was protruded and abdomen was swollen. The intestine became swollen and haemorrhagic. Slight kidney liquefaction was noted. Approximately 30–40% of the infected fish died (Austin *et al.*, 1988; Austin and Stobie, 1992a).

Renibacterium salmoninarum

Pathogenicity experiments have met with varying degrees of success. Mackie *et al.* (1933; 1935) succeeded in transmitting 'Dee disease' to brown trout by subcutaneous and i.m. injections of emulsified spleen from Atlantic salmon. In these experiments, death followed in 5 weeks, although typical lesions, as found in field situations, did not occur. A similar observation was made by Belding and Merrill (1935) who injected, intramuscularly, brook trout with purulent material collected from kidney abscesses in the same species. Death followed in 18 to 25 days, but characteristic BKD lesions did not occur. This was, however, achieved by Earp (1950) following the injection of chinook salmon with a pure culture of the BKD organism. Koch's postulates were finally satisfied by Ordal and Earp (1956) following the establishment of BKD in chinook salmon after i.p. injection of an organism obtained from sockeye salmon. Mortalities started after 12 days, and continued until day 23, when all the fish were dead. At this point, the organism was re-isolated. Sakai *et al.* (1989c) found mortalities began 17 days after rainbow trout were injected with 4×10^8 cells. In comparison, carp (*Cyprinus carpio*) were markedly resistant. Failure greeted the attempt by Snieszko and Griffin (1955) to transmit BKD to brook trout by co-habiting with diseased fish for 21 days, followed by feeding with infected viscera. However using feeding, success was achieved by Wood and Wallis (1955) with 100% infection of 993 chinook salmon fingerlings. Later, Wolf and Dunbar (1959) achieved success by immersing experimentally wounded brook trout into a suspension of the pathogen. Murray *et al.* (1992) succeeded in inducing BKD in chinook salmon by immersion (10^4–10^6 cells/ml for 15–30 min) and co-habitation with other experimentally infected fish. However, the time to death was much longer than in most experimental models. By co-habitation and immersion, the average periods leading to mortalities were 145 and 203 days, respectively. Transmission from wild to cultured fish has been reported (Mitchum and Sherman, 1981) and vice versa (Frantsi *et al.*, 1975). Prior infection with *Ren. salmoninarum* may well contribute to the poor survival of coho salmon upon transfer from freshwater to seawater (Moles, 1997).

Evidence has pointed to the ability of *Ren. salmoninarum* becoming internalized in macrophages in which putative virulence factors are produced (McIntosh *et al.*, 1997). Within the phagocytic cells, renibacterium exhibits a slow rate of division, and survives certainly for 10 or more days (Gutenberger *et al.*, 1997). Conversely, the macrophages may well inhibit the growth of and kill renibacterium by the live bacterial cells generating respiratory burst products (Hardie *et al.*, 1996; Campos-Pérez *et al.*, 1997). With this scenario, exposure to *Ren. salmoninarum* would enhance the killing activity of the macrophages (Hardie *et al.*, 1996).

There is a divergent opinion as to the presence of biological activity in ECP of *Ren. salmoninarum*. One view is that the ECP is generally devoid of extracellular enzymes; haemolytic and cytolytic activity being absent (Bandin *et al.*, 1991). Yet in other investigations, proteases (Sakai *et al.*, 1989c) and haemolysins (Grayson *et al.*, 1995a) have been detected. ECP at 0.1 mg/ml and 1.0 ml/ml inhibited respiratory burst but not phagocytic activity in brook trout splenic phagocytes (Densmore *et al.*, 1998). Hydrophobicity, haemagglutination and haemolysin activity to rabbit and trout erythrocytes have been recorded from water soluble extracts (proteins) (Bandin *et al.*, 1989; Daly and Stevenson, 1987; 1990; Evenden *et al.*, 1990). In particular, hydrophobicity and auto-aggregation have been linked with virulence (Bruno, 1988). *Ren. salmoninarum* has agglutinated spermatozoa from salmonids and goldfish (Daly and Stevenson, 1989). Shieh (1989b) reported an unidentified toxin from *Renibacterium*, which was lethal to fingerling Atlantic salmon. Also, an iron acquisition mechanism has been found (Grayson *et al.*, 1995b).

There is some evidence that fish respond to infection with renibacterium by the production of stress proteins. Thus, a 70 kDa stress protein (HSP70) was recognized in coho salmon with BKD (Forsyth *et al.*, 1997).

Rhodococcus sp.

Intraperitoneal injection of Atlantic salmon smolts with a very high dose of 5×10^8 cells resulted in severe peritoneal granulomatous reactions, with a low accompanying mortality rate, within 21 days (Speare *et al.*, 1995). Unlike the natural disease where the most severe pathological changes occurred in the renal interstitium, experimental challenge resulted in damage in the direct vicinity of the injection site. Yet, the development of large bacterial colonies was common to both natural and artificial infections.

Negligible information is available about the pathogenicity of other Gram-positive aerobic rods and cocci, as included in Table 1.1.

D. AEROMONADACEAE REPRESENTATIVES

Aeromonas allosaccharophila
It was not concluded that the organisms were indeed pathogenic to fish. Yet, the recovery from diseased elvers suggests a pathogenic role for the organisms (Martinez-Murcia *et al.*, 1992).

Aeromonas hydrophila
Most of the information concerning pathogenicity mechanisms of *Aer. hydrophila* appertains to isolates of medical importance and will not be considered further here. The value of using cultures, grown on nutrient-rich media, has been cast into doubt following the observation that starved cells (N.B. this is akin to the natural state of bacteria in the aquatic environment) are more virulent that their counterparts from nutrient rich cultures (Rahman *et al.*, 1997). Nevertheless as a general rule, it is apparent that the pathogen has considerable exo-enzyme potential, some of which has relevance in fish pathology. The precise function of these 'toxins', which number at least six (Bernheimer and Avidad, 1974; Donta and Haddow, 1978; Cumberbatch *et al.*, 1979) in fish pathology, has yet to be fully elucidated.

Surface structures
Recent studies have emphasized the surface structures of *Aer. hydrophila*, which appear to be involved in autoaggregation/hydrophobicity and haemagglutination (e.g. Paula *et al.*, 1988). There is some evidence that a capsule may be produced *in vivo* (Mateos and Paniagua, 1995). The presence or absence of lateral flagella (as opposed to the more typical polar pattern) was demonstrated by electron microscopy on three isolates from catfish in Nigeria (Nzeako, 1991). Del Corral *et al.* (1990) demonstrated the presence of pili/fimbriae, regardless of virulence. These workers considered that there was not a direct correlation between virulence and haemagglutination.

The surface array matrix, i.e. the S-layer, has been considered to influence the interaction between the bacterial cell and its environment. A major function is believed to be the provision of physical protection from lytic components, including serum proteins and bacteriophages (Dooley *et al.*, 1988). Work also links the presence of an S-layer with invasive disease in humans and mice (but not fish!) (Murray *et al.*, 1988). As a result of studying one isolate, i.e. TF7 (isolated from a lesion on trout in Quebec), it was determined that the S-layer did not confer any increase in surface hydrophobicity or any enhanced association with macrophages, and did not specifically bind porphyrin or immunoglobulin (Murray *et al.*, 1988). Nevertheless in *Aer. salmonicida*, the S-layer has indeed been shown to be a prerequisite for virulence, by increasing hydrophobicity and enhancing macrophage association (Murray *et al.*, 1988).

The detailed structure of the S-layer has been revealed in an excellent series of publications (Dooley *et al.*, 1986; 1988; Dooley and Trust, 1988). After studying eight isolates of a serogroup with a high virulence to fish, Dooley and Trust (1988) concluded that the S-layer was tetragonally arrayed. SDS-PAGE revealed a protein of 52 kDa molecular weight, which was the major surface (protein) antigen. This protein effectively masked the underlying OMP.

Ascencio *et al.* (1991) investigated extracellular matrix protein binding to *Aer. hydrophila*. In particular, binding of ^{125}I-labelled collagen, fibronectin and laminin is common to isolates from diseased fish. Moreover, the binding property was specific, with cultural conditions influencing expression of the bacterial cell surface binding structures. Experiments showed that calcium (in the growth medium) enhanced expression of the bacterial extracellular matrix protein surface receptors. The conclusion was reached that success in infecting/colonising a host depended on the ability of the pathogen to bind to specific cell surface receptors of the mucus layer, epithelial cells and subepithelial basement membranes.

'Adhesins'

So, it appears that the pathogen has the ability to attach to selected host cells, e.g. erythrocytes, and tissue proteins, i.e. collagen, fibronectin, serum proteins and glycoproteins, via the action of 'adhesins' (Trust *et al.*, 1980a,b,c; Toranzo *et al.*, 1989; Ascencio *et al.*, 1991; Lee *et al.*, 1997) and become internalized (Tan *et al.*, 1988). The adhesins appear to be extremely selective, recognizing D-mannose and L-fucose side chains on polymers located on the surface of the eukaryotic cells. The specificity was further highlighted by the observation that a human isolate of *Aer. hydrophila* failed to bind (or bound poorly) to fish tissue culture cells (Krovacek *et al.*, 1987). Indeed using tissue culture cells from rainbow trout liver and chinook salmon embryos, Krovacek *et al.* (1987) demonstrated that some (~33%) isolates of *Aer. hydrophila* from fish adhered to the tissue culture cells and glass surfaces coated with rainbow trout mucus. Adhesion and adsorption were time dependent; and the activities were lost after treatment of the bacteria with heat, proteolytic enzymes or ultrasound.

Invasion of fish cells

A 43 kDa protein, possibly an outer membrane porin, has been regarded as important for the invasion of epithelial cells, *in vitro* (Lee *et al.*, 1997). Other workers have pointed to the relevance of capsular polysaccharides, which appear to enhance slightly adherence to fish cells but contribute more significantly to cell invasion (Merino *et al.*, 1997). With attachment, the host cell will be at the mercy of the pathogen. Although the precise mechanism of cell damage and tissue damage remains unproven, the available evidence points to the involvement of both endo- and exo-toxins. Experiments with fish epidermal cells revealed that *Aer. hydrophila* could survive internally (Tan *et al.*, 1998). Here, a role for tyrosine phosphorylation in the internalization process was suggested (Tan *et al.*, 1998).

Extracellular products
In comparison with *Aer. salmonicida*, fish pathogenic strains of *Aer. hydrophila* produce
ECP, which contains considerable enzymatic activity (Shotts *et al.*, 1984; Santos *et al.*,
1987), including haemolysins and proteases (Angka *et al.*, 1995; Khalil and Mansour,
1997), with optimum production [of protease] at 27.6 ± 4.9°C (Uddin *et al.*, 1997). The
relevance of the ECP was highlighted by Allan and Stevenson (1981) and Stevenson and
Allan (1981), who succeeded in causing a pathology in fish as a result of injection of the
material. Yet, the role of ECP is debatable with contrasting views of the importance of
'haemolysins' in virulence (Thune *et al.*, 1986; Toranzo *et al.*, 1989; Karunasagar *et al.*,
1990; Paniagua *et al.*, 1990). Stevenson and colleagues reported haemolytic (heat-labile)
and proteolytic activity, the former of which was concluded to be of greater importance
in pathogenesis. Kanai and Takagi (1986) recovered an α-type haemolysin which was
deemed to be heat-stable and stable at pH 4–1.2, but inactivated by EDTA, trypsin and
papain. The crude preparation caused swelling and reddening of the body surface
following injection into carp. Previously, Boulanger *et al.* (1977) isolated two types of
haemolysins. The reasons then for the conclusion about the importance of haemolysins
were based upon work with protease-deficient mutants, the ECP from which was more
toxic to recipient fish than from wild-type cultures. Conversely, Thune *et al.* (1982a,b)
obtained a fish-toxic fraction, which possessed proteolytic but not haemolytic activity.
Moreover, in a comparison of ECP from virulent and weakly virulent isolates, Lallier *et
al.* (1984) noted that both were haemolytic, enterotoxigenic and dermonecrotic, but the
weakly virulent isolate produced twenty-fold more haemolysin than the virulent
organism. Yet, only cell-free supernatants from virulent isolates produced toxic
(oedematous) effects in fish. Following detailed chemical analyses, this heat-labile toxic
factor was separated on Sephacryl S-200 from the haemolysin. These data suggest that
factors other than haemolysins and proteases may be relevant in fish pathology. Indeed,
after studying numerous isolates, Hsu *et al.* (1981; 1983), Shotts *et al.* (1985) and
Paniagua *et al.* (1990) correlated virulence with extracellular proteolytic enzymes,
notably caseinase and elastase. Santos *et al.* (1987) reported a relationship between
virulence in fish and elastase and haemolysin (of human erythrocytes) production and
fermentation of arabinose and sucrose. On this theme, Hsu *et al.* (1983) associated
virulence with gas production from fructose, glucose, mannitol, mannose, salicin and
trehalose, and the possession of resistance to colistin.

Extracellular metallo- and serine-proteases of *Aer. hydrophila* (strain B_5) have been
characterized, and deemed to be heat (to 56°C) (Leung and Stevenson, 1988) and cold-
stable (to –20°C) (Nieto and Ellis, 1986). Most activity was inhibited by EDTA. Overall,
there were many differences in the proteases (four or five were present) described by
Nieto and Ellis (1986) to reports from other workers. This may be explained by the work
of Leung and Stevenson (1988), who examined the proteases from 47 *Aer. hydrophila*
isolates. Of these isolates, 27 produced both metallo- and serine-proteases, 19 produced
only metallo-proteases, and ATCC 7966 produced only a serine-protease. The differences
in these 47 isolates may well explain the apparent conflicting reports which result from
the examination of only single isolates. Certainly, it seems that there are pronounced
differences in the characteristics of the ECP and thus protease composition between
strains.

It has been suggested that the proteases may be involved in protecting the pathogen against serum bacteriocidal effects, by providing nutrients for growth following the destruction of host tissues, and by enhancing invasiveness (Leung and Stevenson, 1988). Also, proteases may be involved with the activation of haemolysin (Howard and Buckley, 1985).

A further study identified acetylcholinesterase (a 15.5 kDa polypeptide) in the ECP, and regarded the enzyme as a major lethal factor, possibly with neurotoxic activity (Nieto *et al.*, 1991; Rodriguez *et al.*, 1993a,b; Pérez *et al.*, 1998). The minimal lethal dose of the compound was given as 0.05 μg/g of fish.

Precipitation of Aeromonas hydrophila
The importance of precipitation after boiling is a debatable issue in screening of *Aer. hydrophila* isolates for virulence. Santos *et al.* (1988) considered that precipitation was not an important indicator, whereas Mittal *et al.* (1980) and Karunasagar *et al.* (1990) reported that settling after boiling was indeed an important measure of virulence.

Scavenging for iron
The ability of a potential pathogen to scavenge successfully for iron (in iron-limited conditions) will influence the outcome of the infection process. The haemolysins of *Aer. hydrophila* are iron-regulated, and access to iron in the haemolytic destruction of the host cells may be necessary (Massad *et al.*, 1991). The acquisition of iron from iron-transferrin in serum is dependent on the siderophore amonabactin. Many aeromonads use haem as a sole source of iron for growth. Some have evolved both siderophore-dependent (iron-transferrin) and independent mechanisms (haem compounds) for the acquisition of iron from host tissues (Massad *et al.*, 1991).

Enterotoxigenicity
Strains have been attributed with enterotoxigenicity, as assessed by the rabbit ileal-loop technique, and cytotoxicity (Boulanger *et al.*, 1977; Jiwa, 1983; Paniagua *et al.*, 1990), and correlated with lysine decarboxylase production (Santos *et al.*, 1987). Enterotoxigenic strains have been shown to produce two types of enterotoxins, which appear to be antigenically related, although the mode of action differs (Boulanger *et al.*, 1977). This was an interesting observation because de Meuron and Peduzzi (1979) isolated two types of antigen, of which the K-antigen (this was thermolabile at 100°C) was considered to represent a pathogenicity factor. Possibly, this corresponded to the enterotoxin or cytotoxin as described by Boulanger *et al.* (1977). However, the O (somatic)-antigen, which was heat stable, may have greater relevance, insofar as most virulent isolates share a common O-antigen (Mittal *et al.*, 1980). In an excellent study, Dooley *et al.* (1985) used SDS-PAGE to analyses LPS (considered to constitute an O-antigen) from virulent strains, which auto-agglutinated in static broth culture. The LPS contained O-polysaccharide chains of homogeneous chain length. Two strains produced a surface protein array, which was traversed by O-polysaccharide chains and thus exposed to the cell surface. Antigenic analysis revealed that the polysaccharide of the LPS carried three antigenic determinants.

Clearly, the evidence indicates the involvement of both endo- and exo-toxins in the pathogenesis of *Aer. hydrophila* infections. It still remains for further work to elucidate the precise mechanism of action.

Aeromonas jandaei

Esteve (1995) and Esteve *et al.* (1995b) reported a high LD_{50} dose of ~10^6 cells for eel. Possibly, the ECP activity, which was equated with production of caseinase, collagenase, elastase, protease, lipase and haemolysin, caused pathogenicity (Esteve *et al.*, 1995b).

Aeromonas salmonicida

The spread of the pathogen

Historically, *Aer. salmonicida* was regarded as a risk primarily to salmonids (e.g. Mackie *et al.*, 1930; 1933; 1935). Then, cyprinids followed by other freshwater and marine fish became recognized to be vulnerable to infection (e.g. Herman, 1968; Austin *et al.*, 1998).

- Could farmed salmonids pose a realistic risk to native marine fish species?

The data on this topic are confusing. Certainly, marine fish larvae have been infected with *Aer. salmonicida* subsp. *salmonicida*, with turbot regarded as being more susceptible than halibut (Bergh *et al.*, 1997). Using co-habitation and injection challenges, experiments suggested that *Aer. salmonicida* subsp. *salmonicida* could be transmitted rarely from Atlantic salmon to Atlantic cod, halibut and wrasse (Hjeltnes *et al.*, 1995).

- Could atypical isolates, which are appearing with increasing frequency in wild fish, pose a threat to cultured salmonids?

Wiklund (1995a,b) using an atypical isolate from ulcerated flounder concluded that there was not any risk to rainbow trout.

- What about the risk of transferring *Aer. salmonicida* from the freshwater to seawater stage of salmonids?

Eggset *et al.* (1997) concluded that the susceptibility of Atlantic salmon to furunculosis in seawater possibly reflected the overall quality of the smolts.

Pathogenicity—historical aspects

Although the factors conferring pathogenicity on *Aer. salmonicida* strains have been the subject of speculation since early in the study of the pathogen, it is only relatively recently that the details concerning pathogenesis and virulence have begun to be elucidated. The initial investigations, carried out in the 1930s, resulted in several key observations, notably that prolonged laboratory maintenance of *Aer. salmonicida* isolates was frequently responsible for a loss of virulence, and that histopathological examinations of infected fish suggested the occurrence of leucopenia and proteolysis in certain tissues. Among the first studies concerned with virulence mechanisms of the organism was the extensive work of the Furunculosis Committee in the UK (Mackie *et al.*, 1930; 1933, 1935). This group did not detect any toxin production by *Aer. salmonicida* when either ultra-filtrates of broth cultures or diseased fish tissue was injected into healthy fish. Based on their failure to demonstrate toxin production, they hypothesized as a result of detailed clinical observations, that the pathogenic processes caused by *Aer. salmonicida* could be explained by the prolific growth in the blood and tissues of its host which, in turn, interfered with blood supply resulting in anoxic cell necrosis and ultimately death. Additional evidence for a possible contribution to virulence, in the form of a leucocytolytic component, was provided by Blake (1935), who described the presence of 'free' bacteria and little phagocytosis in the blood of diseased fish, with no definite leucocytic infiltration at the foci of infection. Mackie and Menzies (1938) confirmed the production of a leucocytolytic substance, as did Field *et al.* (1944), who determined the absence of leucocytosis by performing repeated blood counts on experimentally infected carp. Perhaps, a more significant finding of their study, however, was the rapid decline in blood sugar levels resulting in hypoglycaemic shock, which was sufficient in some instances to cause acute mortalities. They suggested that the hypoglycaemic shock was the outcome of rapid utilization of blood glucose by the multiplying pathogen. Regarding virulence mechanisms of *Aer. salmonicida*, Griffin (1953) theorized that leucocidin production *in vivo* by *Aer. salmonicida* would account for the observations by previous workers that marked cytolytic tissue necrosis did not seem to be accompanied by leucocytic infiltration. Another aspect of *Aer. salmonicida* pathogenicity, which eventually proved to be extremely important, was discussed by Duff (1937). He reported a loss in pathogenicity among strains after 6 or more months of maintenance on artificial culture in the laboratory. The loss was accompanied by a change in the appearance of colonies on nutrient agar from glistening, convex and translucent to strongly convex, distinctly opaque and cream-coloured. Because of such observations, Duff further investigated this phenomenon of dissociation into different colony types. Subsequently, he discovered that dissociation could be induced by culturing the pathogen in nutrient broth with the addition of either 0.25% lithium chloride or 0.1% phenol. Use of this procedure gave rise to several distinct colony forms. One of these resembled the original stock culture, a second corresponded to the 'new' type, and a third was intermediate between the other two forms. The colonies resembling those of the original stock culture were described as opaque, strongly convex, cream-coloured and friable, whereas the new colony form appeared translucent, slightly convex, and a bluish-green in colour with a butyrous consistency. When the two different colony types were inoculated intraperitoneally into goldfish, the blue-green translucent dissociant caused the deaths of the fish and was accompanied by lesions typical of the disease. In contrast, the original type of colony did not adversely affect fish, which survived for the 30-day duration of the experiment without any signs of illness. Thus, Duff concluded that the cream-opaque form which produced friable colonies was non-pathogenic, and more stable on prolonged storage. Duff designated this colony type as 'rough'. The 'smooth' form (i.e. the blue-green-translucent dissociant, which produced butyrous colonies on agar media) was pathogenic, but less stable in prolonged storage. In the subsequent study, Duff (1939) also reported the presence of an extra antigen in the rough strains. Although Duff (1937) was the first worker to report the ability of *Aer. salmonicida* to dissociate into several distinct colony types with differences in pathogenicity, a phenomenon which is now widely accepted, it is curious that he ascribed pathogenicity to the smooth colony type. This is in contrary to the view currently held that the rough colony type is, in fact, virulent. Interestingly, the Furunculosis Committee had also reported a variation in colony morphology among isolates (it may be assumed that these corresponded to the rough and smooth variants) but contended that this phenomenon was not accompanied by a difference in virulence. It is regrettable that this initial confusion over dissociation occurred, preventing an earlier realization of its significance. In fact, the relevance of dissociation of *Aer. salmonicida* colonies and the relationship to virulence was not made apparent until the work of Udey (1977), almost 40 years later. Early studies provided tentative evidence for a variety of possible pathogenic mechanisms, but there is no doubt that progress in the understanding of *Aer. salmonicida* pathogenesis and virulence has been accelerated by rapid advances in the knowledge of cell biology and the development of sophisticated biochemical techniques. It is the application of such techniques which continues to yield considerable new information about the manner in which *Aer. salmonicida* may effect its disease processes in fish.

Pathogenicity—modern developments (intraperitoneal chambers)
An intriguing and significant development concerned the description of intraperitoneal chambers, which could be implanted into fish (Garduño *et al.*, 1993a). These chambers could be filled with pathogens—or for that matter a range of other objects—implanted into fish, and measurements made with time. Garduño and colleagues placed *Aer. salmonicida* into a chamber, and studied its fate in the peritoneal cavity of rainbow trout. In one set of investigations, these workers observed that when the pathogen was contained in the chamber killing occurred rapidly as a result of host-derived lytic activity (in the peritoneal fluid). In contrast, free cells had a better chance of survival (Garduño *et al.*, 1993a). Moreover, within the peritoneal chamber, *Aer. salmonicida* produced novel antigens, as determined by Western blots (Thornton *et al.*, 1993). In another publication using the peritoneal chamber, evidence was presented that the capsular layer around *Aer. salmonicida* permitted the pathogen to resist host-mediated bacteriolysis, phagocytosis and oxidative killing (Garduño *et al.*, 1993b).

Pathogenicity—cell-associated versus extracellular components
A variety of pathogenicity mechanisms and virulence factors have been proposed for diseases caused by *Aer. salmonicida*, namely possession of an extracellular (A) layer (= the surface or 'S' layer) and the production of ECP, although there is confusion and even contradiction about the relative merits of the various components in pathogenicity (see Ellis *et al.*, 1988b). Yet, ironically fish may mount an antibody response during infection (Hamilton *et al.*, 1986). Indeed, complement and non-α_2 m-antiprotease activity have been considered important host defence mechanisms against *Aer. salmonicida* (Marsden *et al.*, 1996c).

Munro (1984) has grouped the virulence/pathogenicity factors into cell-associated and extracellular components, a division which is convenient for the purpose of this narrative. The best-studied cell-associated factor is the additional layer, external to the cell wall, termed the A-layer.

The A-layer
The A-layer is now thought to be the product of a single chromosomal gene (Belland and Trust, 1985), is produced *in vivo* (Ellis *et al.*, 1997) and contributes to survival in

macrophages (Daly *et al.*, 1996). First reported by Udey and Fryer (1978), and resulting from detailed electron microscopic studies, the A-layer was determined to be correlated with virulence. Thus, it was observed that virulent strains possessed the A-layer, whereas avirulent isolates did not. In addition, the presence of the A-layer was found to correspond with strong auto-agglutinating properties of the organism, and to the adhesion to fish tissue culture cells. The auto-agglutination trait has been found to be influenced by temperature, with weak and strong auto-agglutination at 25 and 15–20°C, respectively (Moki *et al.*, 1995). The presence of the A-layer may confer protection against phagocytosis and thus destruction by macrophages (Olivier *et al.*, 1986; Graham *et al.*, 1988). Essentially, these workers noted that avirulent cells, i.e. those without an A-layer, were phagocytosed and destroyed when virulent cells with A-layer were more resistant. Moreover, the bacteriocidal activity of macrophages was stimulated by prior exposure to low doses of *Ren. salmoninarum*, but inhibited by high amounts of living or dead renibacterial cells or the p57 antigen (Siegel and Congleton, 1997). Interestingly, it was deduced that living and formalized virulent cells, in the absence of serum, attracted macrophages more readily than avirulent cells after a period of 90 min (Weeks-Perkins and Ellis, 1995). The surface layer may inhibit growth at 30°C, enhance cell filamentation at 37°C, and enhance uptake of the hydrophobic antibiotics streptonigrin and chloramphenicol (Garduño *et al.*, 1994).

For its formation, Belland and Trust (1985) reasoned that the A-layer subunits pass through the periplasm and across the outer membrane for assembly on the cell surface. A requirement for the presence of O-polysaccharide chains, for which the AbcA protein is involved in biosynthesis (Noonan and Trust, 1995) of the LPS, was reported as necessary for the assembly of A-layer (Dooley *et al.*, 1989). These virulent, auto-agglutinating forms produce characteristic deep blue colonies on CBB agar (Wilson and Horne, 1986; Bernoth, 1990). Sakai (1986a,b) postulated that a possible mechanism for auto-agglutination and adhesion could be attributed to the presence of net negative electrical charge in the interiors or on the surfaces of cells. In particular, pathogenic cultures were highly adhesive (Sakai, 1987). It should be emphasized that Udey and Fryer (1978) determined that strains maintained for long periods in laboratory conditions were not auto-agglutinating, and demonstrated reduced virulence. Conversely, it was observed that fresh isolates, obtained from epizootics, were of the aggregating type. From the results of experiments, Udey and Fryer (1978) concluded that the presence of the A-layer was necessary for virulence. However, they contended that more work was needed to establish whether or not the A-layer alone could confer virulence. The discovery of the A-layer generated much interest, resulting in further study of its chemical composition and its specific role in fish pathology. Kay *et al.* (1981) succeeded in purifying the A-layer from virulent isolates, and concluded that it was composed of a surface-localized protein with a molecular weight of 49 kDa. Phipps *et al.* (1983) continued with work on purification and characterization of the substance, determining that it was hydrophobic in nature, present on the entire cell surface, did not possess any enzymic activity, but instead constituted a macromolecular refractive protein barrier which was essential for virulence. Meanwhile, an independent parallel investigation of Evenberg *et al.* (1982) highlighted the relationship between auto-agglutination and the presence of the A-layer. This group examined the cell envelope protein patterns of a variety of isolates obtained from a wide range of geographical locations and different fish species (i.e. carp, minnow, goldfish and salmonids). These fish were suffering from either furunculosis, CE or ulcer disease. A major protein (molecular weight = 54 kDa) was found in all auto-aggregating strains, but little or no trace occurred in isolates which were not auto-agglutinating. When examination for the presence of the protein were carried out after a change of growth medium, i.e. replacement of horse serum by synthetic sea salt, it was observed that an almost complete loss of the additional cell envelope and the auto-agglutinating ability of the isolate had occurred. Using gel immunoradio assays, it was also determined that the extra cell envelope proteins of all the isolates, irrespective of fish host, type of infection or geographical source, were immunologically related. Evenberg and Lugtenberg (1982) pursued this topic, and described the protein as water insoluble with an amino acid composition similar to those of the additional surface layers of other bacteria, e.g. the adhesive K88 fimbriae of enteropathogenic strains of *Esch. coli.* It is particularly relevant that the findings of Evenberg *et al.* (1982), concerning the auto-agglutinating ability of 'atypical' strains from cases of CE and ulcer disease, and the presence of the A-layer, were in excellent agreement with the work of Trust *et al.* (1980c) and Hamilton *et al.* (1981). These earlier studies deduced the presence of an outer layer protein, which was estimated to have a molecular weight of 50 kDa. Evidence was provided by Ishiguro *et al.* (1981) that loss of the A-layer and loss of auto-agglutinating properties resulted in decreased virulence. After examining the effects of temperatures on the growth of *Aer. salmonicida*, it was shown that in cells, cultured at 30°C (the generally accepted upper limit for the organism), virulence was restricted to <10% of the population. The avirulent attenuated cells which resulted from use of the higher growth temperature, did not auto-agglutinate and, for that matter, did not possess the A-layer. It is interesting to note that higher maximum growth temperatures were recorded for the attenuated strains, in comparison to their virulent counterparts. Perhaps, this is explained by their selection at high temperatures. Because of this observation, Ishiguro *et al.* (1981) hypothesized that the A-layer is important in determining physical properties of the cell envelope, and that these properties undergo a change when the A-layer is lost, permitting growth at higher than normal temperatures. If the A-layer is a prerequisite for virulence, it may be assumed that its presence confers advantages on the bacterial cell in its role as a pathogen. Indeed, several prime functions for the A-layer have been proposed. Thus, evidence exists that the extracellular layer protects *Aer. salmonicida* cells from the action of protease (Kay and Trust, 1991) and bacteriophage, by shielding its phage receptors (Ishiguro *et al.*, 1981). In addition, the layer may protect the cell from serum complement, insofar as Munn and Trust (1984) demonstrated that virulent strains (with A-layer) were resistant to complement bacteriocidal activity in the presence (and indeed absence) of specific antibody in rainbow trout serum. Other investigations have revealed that hydrophobicity is conferred upon the bacterial surface by the A-layer (Trust *et al.*, 1983; Van Alstine *et al.*, 1986). These workers reported that the hydrophobic A-layer provided *Aer. salmonicida* cells with an affinity for fatty acid esters of polyethylene glycol and an enhanced ability to associate with rainbow trout and mouse phagocytic monocytes (macrophages), in the absence of opsonizing antibody. Although Trust *et al.* (1983) conceded that the advantages to the pathogen of the increased association with macrophages remained to be determined, they suggested as a tentative explanation, the possibility that *Aer.*

salmonicida is a facultative intracellular pathogen able to survive within phagocytes. Indeed, Munn and Trust (1984) demonstrated that A-layer[+] bacteria (i.e. bacteria with A-layer) were able to multiply within the principal phagocytic organs, i.e. the spleen, following experimental infection. Subsequently, it has become established that *Aer. salmonicida* is capable of replication in macrophages, where the pathogen is presumed to be able to resist reactive radicals (Garduño *et al.*, 1997). It has been argued that the surface layer constitutes the first line of defence for *Aer. salmonicida*, with an inducible catalase and manganese superoxide dismutase as second defensive systems against macrophage-mediated killing via reactive oxygen species.

The A-layer has also been implicated in a role concerning adhesion to fish tissues. By means of *in vitro* experiments, Parker and Munn (1985) examined the ability of avirulent (A-layer[-]) cells to adhere to cells of baby hamster kidney and rainbow trout gonad in tissue culture. Attachment of A-layer[+] *Aer. salmonicida* to both types of cells was greater than for the A-layer[-] derivative. As a result, Parker and Munn (1985) proposed that since attachment to epithelial cells may be the primary step in the pathogenic process, their observations could account for the association of virulence with the presence of an extra outer membrane layer.

To summarize, the accumulating body of evidence indicts the A-layer as a principal virulence determinant, even though its precise functions and the mechanism of action obviously require further clarification. However, blithe acceptance of an absolute relationship between virulence and possession of an A-layer must unfortunately be cautioned against. This is in view of reports by Johnson *et al.* (1985) and Ward *et al.* (1985) on the occurrence of virulent, auto-agglutinating strains that have no detectable A-layer. Conversely, Olivier (1990) recovered non-virulent A-layer[+] isolates. Thus, the association between presence of the extracellular layer and virulence, but not between auto-agglutination and virulence, appears to be open to question. It is important that the extent of this problem should be determined, particularly because the use of A-protein as an antigenic component of a potential vaccine for control of diseases caused by *Aer. salmonicida* has been advocated. This is due to the apparent immunological relatedness of the A-protein among isolates from different locations and a variety of fish hosts (Evenberg *et al.*, 1982). In view of the existence of virulent, auto-agglutinating *Aer. salmonicida* strains apparently lacking the A-layer, the effectiveness of such a vaccine would possibly be subject to severe limitations.

Capsules

Capsular polysaccharides have been found to develop around cells of *Aer. salmonicida* in the presence of glucose, phosphate, magnesium chloride and/or trace elements. Production of this material was improved in the presence of yeast extract (Bonet *et al.*, 1993). Interestingly, a striking difference in cells grown *in vitro* or *in vivo* reflected the presence or absence of capsules. Using intraperitoneal chambers, it was observed that *Aer. salmonicida* produced capsules with virulence functions (Garduño and Kay, 1995). Adherence to fish cell cultures was slightly higher in cultures of *Aer. salmonicida* grown in conditions to promote capsule formation. Also, invasion of fish cells was more pronounced in the capsulated cells (Merino *et al.*, 1996). Another role for the capsule concerns resistance to complement. Thus, it was recorded that when grown under conditions promoting capsule development, *Aer. salmonicida* were partially resistant to complement (Merino *et al.*, 1997).

Agglutination of fish cells

Another cell-associated factor, possibly relevant to virulence of *Aer. salmonicida*, is the ability of the pathogen to agglutinate trout and mammalian erythrocytes (Mellergaard and Larsen, 1981). The haemagglutination capability is proported to be related to the presence of adhesins, which are structures on the bacterial surface that mediate the attachment of the pathogen to the host's cell surface. Thus, the interest in haemagglutination is due primarily to its use to provide semi-quantitative information on the adhesive potential of a bacterial strain, while the sugar inhibition of haemagglutination has allowed adhesive specificity to be demonstrated (Duguid and Old, 1980). There is good agreement in the literature on the haemagglutinating ability of *Aer. salmonicida*. For instance, Jiwa (1983) reported mannose-resistant agglutination of bovine, chicken, human group A and guinea-pig erythrocytes by two *Aer. salmonicida* isolates which were recovered from diseased brown trout. It has also been demonstrated that smooth strains were unable to agglutinate erthrocytes, whereas rough strains showed a broad spectrum of haemagglutinating activity. Similarly, Parker and Munn (1985) observed that virulent, auto-agglutinating A-layer[+] cells agglutinated trout erythrocytes as well as a range of mammalian erythrocytes. They also noted that the process was not inhibited by specific sugars, and thus concluded that adhesion was a relatively non-specific process, attributable to the hydrophobic properties of the A-layer.

Extracellular products

Researchers interested in the biology of *Aer. salmonicida* have been aware for some time that extracellular substances (produced by the organism) presumably exerted a role in virulence. These ECP have been the focus of numerous investigations. Unfortunately, the work has been rendered more difficult by the complexity of the substance(s), which, at present, is known to include acetylcholinesterase (an ichthyotoxin with neurotoxic activity; Pérez *et al.*, 1998), several proteases, phospholipase, haemolysins and a leucocidin (Munro *et al.*, 1980; Sheeran and Smith, 1981; Shieh and MacLean, 1975; Titball and Munn, 1981; Cipriano *et al.*, 1981; Fuller *et al.*, 1977; Rockey *et al.*, 1988; Huntly *et al.*, 1992; Lygren *et al.*, 1998), as well as LPS. Work with monoclonal antibodies has shown heterogeneity in the LPS (Rockey *et al.*, 1991). Ellis *et al.* (1981) reported that ECP of the pathogen, prepared by a cellophane overlay method, reproduced the lesions normally associated with the chronic form of furunculosis, e.g. muscle necrosis and oedematous swelling at the site of injection. This suggests that the toxins and aggressins released by the bacteria *in vivo* are responsible for much of the pathology of the disease. In addition, when injected intraperitoneally into rainbow trout, the ECP proved to be fatal for the fish (Munro *et al.*, 1980). In preparations of fish cells, the ECP exhibited cytotoxic effects, and at higher concentrations was leucocytolytic and haemolytic. These investigators concluded that most of the virulence factors were produced extracellularly, with most strains of *Aer. salmonicida* producing similar compounds, although the quantities varied. However, a detailed chemical analysis was not carried out.

Other studies have also indicated that injection of ECP closely reproduce the pathological condition attributed to furunculosis (Sakai, 1977; Cipriano *et al.*, 1981). Cipriano *et al.* (1981) attempted to determine the role between ECP and virulence by extracting the compounds from culture supernatants. The ECP was resolved into four fractions by ion-exchange chromatography. It was deduced that fraction II possessed leucocytolytic activity, although this fraction was not associated with virulence. Rather, a link between virulence and the toxicity of crude material, and fractions II and III, to cultured rainbow trout was observed. In this experiment, the extracted material from virulent isolates was more toxic to tissue culture cells than preparations derived from the avirulent strains. Fraction II also demonstrated proteolytic activity. Furthermore, results of *in vivo* toxicity studies revealed that three of the fractions were toxic to fish, although their activities varied according to the nature of the fish species used. Thus, mortalities, accompanied by haemorrhaging at the vent and fins, and inflammation at the site of injection, occurred in Atlantic salmon and brook trout which received fractions I and II. Fraction III also caused haemorrhaging at the base of the fins, injection site, and in the mouth; however, the majority of fish administered with this fraction survived. In contrast, rainbow trout were relatively resistant to the effects of all four fractions, insofar as no mortalities resulted. However, administration of fraction II resulted in the development of characteristic furuncle-like lesions at the inoculation site. Fractions I, III and IV did not cause any obvious pathology. The results of Cipriano *et al.* (1981) supported the previous findings of Sakai (1977) who, on the basis of work with crude ECP preparations, considered that a protease was the most pathogenic substance produced by *Aer. salmonicida*. The effects ascribed to proteolytic activity by Sakai (1977) were analogous to those noted by Cipriano *et al.* (1981) for fraction II. In fact, the muscle necrosis and degeneration of connective tissue associated with furunculosis indicates proteolytic enzyme activity. Yet, Fyfe *et al.* (1986) recorded that protease preparations were less effective than equivalent amounts of ECP (with similar amount of proteolytic activity) at causing lesions, i.e. furuncles, following i.m. injection of juvenile Atlantic salmon. This team identified three major components with molecular weights of 70 kDa (a serine protease; Ellis *et al.*, 1997), 56 kDa (a haemolysin) and 100 kDa (unidentified protein) in the ECP (Fyfe *et al.*, 1987a); the first-mentioned of which was produced in greater quantities after incubation for 18 h at 25°C compared to 125 h at 10°C (Fyfe *et al.*, 1987b). Haemolysin production was similar at both temperatures, but tenfold more of the 100 kDa protein was produced at the lower temperature.

Proteases, as prime candidates for exerting a significant role in disease pathogenesis, have aroused substantial interest as a research topic. Indeed, a variety of investigators have performed detailed analyses, and suggested heterogeneity among isolates. In particular, Gudmundsdóttir (1996) described six protease groups, but this information might have greater value for taxonomy than an understanding of pathogenicity. Thus by examining five typical and 25 atypical isolates, it was determined that the proteases produced by the type strains of *Aer. salmonicida* subsp. *achromogenes* (this produced a metallocaseinase = AsaP1) and *Aer. salmonicida* subsp. *salmonicida* were different to those of the fresh isolates. Moreover, all the typical isolates belonged to one protease group with proteolytic activities comparable to P1 and P2 isolates, whereas the atypical cultures were different. With the exception of three atypical oxidase-negative isolates, which secreted a protease reminiscent of P1, the others produced metallo-gelatinase. Ten of the atypical isolates produced AsaP1 (Gudmundsdóttir, 1996).

Shieh and MacLean (1975) purified a proteolytic enzyme, which was determined to have a molecular weight of 11 kDa, and an optimum pH range of 8–11. Because the enzyme was found to be inhibited by PMSF, these workers concluded that it was a serine protease. Mellergaard (1983) also isolated and purified a proteolytic enzyme (molecular weight = 87.5 kDa; optimum pH of 9.0), as did Tajima *et al.* (1984), who reported the presence of an extracellular protease with a molecular weight of 71 kDa, and a pH range of 5–10. This enzyme was deduced to be an alkaline serine protease. Sheeran *et al.* (1984) described two extracellular proteolytic activities that differed in their susceptibility to inhibitors and substrate specificity. One of the enzymes, designated P1, hydrolysed casein, elastin and gelatin, and showed a low non-specific activity against collagen. The second enzyme (P2) hydrolysed collagen and gelatin, but not casein or elastin; a pattern which suggested that it is a specific collagenase. Also, Rockey *et al.* (1988) described two proteases, coined P1 and P2, and a haemolysin (T-lysin) in the ECP. P1 and T-lysin were shown to work separately in the complete lysis of (rainbow trout) erythrocytes. T-lysin interacted with the outer membrane of the erythrocytes, whereas P1 destroyed the nuclear membrane. A role for P2 was not described. Hastings and Ellis (1985) recorded differences in the pattern of extracellular protein production, according to the origin of the bacterial isolates. For example, isolates from Iceland (achromogenic) and the USA lacked caseinase and gelatinase activity in the ECP. Indeed, an isolate of *Aer. salmonicida* subsp. *achromogenes* from Iceland has been credited with the production of a novel metalloprotease (Gudmundsdóttir *et al.*, 1990). For other strains, caseinase and gelatinase activities were inhibited by PMSF and EDTA. These data suggested that both enzyme activities could be attributed to a single serine protease, which depends upon divalent cations for activity. An extracellular metallocaseinase, AsaP1, has been linked with lethal toxicity of atypical *Aer. salmonicida* in Atlantic salmon, with furuncles being produced by ECP with AsaP1 (Gunnlaugsdóttir and Gudmundsdóttir, 1997).

Using the 'P1' and 'P2' terminology, Lygren *et al.* (1998) discussed differences in protease secretion according to the age of the culture of *Aer. salmonicida* subsp. *salmonicida*. Essentially, two different proteolytic activities were found in early and late exponential phase of growth. The P2 activity, found in culture supernatants in the early exponential phase was described as a metalloprotease (molecular weight = 30–40 kDa) with activity against casein and gelatin. This caseinase activity was regarded as novel for metalloproteases of *Aer. salmonicida* subsp. *salmonicida*. The second and major protease, P1, appeared in the culture supernatant during late exponential phase of growth. This protease was regarded as identical to the 70 kDa serine protease (Lygren *et al.*, 1998).

Intramuscular injection of the proteases into brown trout resulted in the development of gross symptoms similar to those occurring in natural outbreaks of furunculosis, i.e. muscle liquefaction along the flanks adjacent to the lesion and swelling at the site of injection (Sheeran *et al.*, 1984). From these data, Sheeran and co-workers concluded that their observations differed from those of other workers. They stated that the absence of significant mortalities, or haemorrhaging of the fins and anus in the experimentally infected fish contrasted with the reports of others, suggesting that, in previous work, enzyme preparations may have contained toxic material other than proteases. Alternatively, it seems possible that Sheeran used insufficient quantities to achieve the pathological changes in question. However, the latter probability seems to be unlikely insofar as the levels of P1 enzyme injected by Sheeran *et al.* (1984) were equivalent to those of Sakai (1978), who reported furuncle formation, haemorrhaging and mortalities in kokanee salmon (*Oncorhynchus nerka*) following administration via i.m. injection. Further evidence for the role of protease production in the pathogenesis of furunculosis was provided by Sakai (1977), who reported a reduction in virulence for a proteolytically deficient mutant strain of *Aer. salmonicida* compared with its isogenic wild-type. In his later article, Sakai (1985) considered a role for proteases in reproduction of the pathogen by making available small peptides and amino acids from proteolysis. Of course, this would considerably benefit the nutrition of the pathogen. It should be emphasized that serum from salmonids is capable of neutralizing lethal doses of proteases (Ellis *et al.*, 1981), possibly through the action of an α-migrating antiprotease (Grisley *et al.*, 1984). Nevertheless, it is apparent that proteases play an important role in the pathogenicity process. A complication concerns the recovery of a pathogenic non-protease secreting strain (Tajima *et al.*, 1987b). Therefore, much work is still required to clarify the precise mode of action of proteases in the pathogenic process.

The greater susceptibility of brown trout (compared to rainbow trout) to furunculosis has long been recognized (e.g. McCarthy, 1975a,b). An explanation for this difference has been provided by Ellis and Stapleton (1988), who found that at low ratios of exotoxin to serum, brown trout serum considerably enhanced *Aer. salmonicida* proteolytic activity. Yet at similar ratios, rainbow trout serum demonstrated some inhibition of the bacterial protease activity. The interpretation of these data is that during the initial stages of infection, *Aer. salmonicida* would have greater potential to multiply in brown trout rather than rainbow trout. Furthermore, Rockey *et al.* (1988) determined that serum from rainbow trout protected the erythrocytes from the haemolysins of *Aer. salmonicida*.

A LPS-free phospholiphase has also been recovered from the ECP, and demonstrated to cause disease signs upon injection into Atlantic salmon (Wong *et al.*, 1989; Huntly *et al.*, 1992). Death resulted overnight following injection of 10 μg/g body weight of fish. Disease signs included lethargy, melanosis, and other defects characteristic of furunculosis. In addition, erythema was noted on the undersurfaces, particularly around the vent, at the bases of the pectoral and pelvic fins, and head (Huntly *et al.*, 1992). Erythrocyte membranes were degraded (= haemolytic activity). It was concluded that this phospholipase exhibited glycerophospholipid:cholesterol acyltransferase activity.

Production of haemolysins by *Aer. salmonicida* may also contribute to the pathogenesis of furunculosis, insofar as it has been established that ECP contain components with pronounced haemolytic activity for trout erythrocytes (Munro *et al.*, 1980). Titball and Munn (1981) carried out the first extensive study of haemolysin production by *Aer. salmonicida*. These authors reported the existence of two distinct haemolytic activities. Essentially, they determined that the supernatant from unshaken broth cultures contained haemolytic activity against erythrocytes from a diverse range of vertebrate species, with maximal activity against horse red blood cells. Titball and Munn termed this 'H' activity. If cultures were shaken, however, the resulting supernatant yielded an activity against trout erythrocytes only (this was designated the 'T' activity).

Furthermore, the H lysin was reported as unstable in culture supernatants, sensitive to heat after exposure to 56°C for 5 min, and became membrane-bound when solutions were filtered. In contrast, the T lysin was stable in supernatants, and was inactivated by normal rainbow trout serum. Nomura and Saito (1982) also studied the extracellular haemolytic toxin, which was recorded as cytotoxic for sheep and salmonid erythrocytes. These investigators observed that the production of haemolysin was stimulated by the addition of enzymic hydrolysates of protein, but suppressed by carbohydrates, such as glucose or sucrose. Moreover, bivalent metal ions, e.g. Ca^{2+}, Co^{2+} and Mn^{2+}, and phosphate ion ($(HPO_4)^{2-}$) were necessary for production of the haemolysin. The optimum pH range and optimum temperature for toxin production was 7.5–8.0 and 20°C, respectively. Nomura and Saito (1982) concluded that the haemolysin was produced during the stationary phase of the growth cycle, and was relatively heat labile, being inactivated at 60°C. These observations coincided with those of Titball and Munn (1981).

In continued studies of the T and H lysins, Titball and Munn (1983; 1985a) purified the components, and examined properties of the haemolytic activity. Thus the T lysin activity was separated into two factors, namely a caseinase and another, apparently membrane-associated (T_1) activity, which by itself caused only incomplete lysis. In fact, complete lysis of trout erythrocytes occurred only in the presence of both T_1 activity and the caseinase (also see Rockey *et al.*, 1988). Titball and Munn (1983) believed that this phenomenon was due to the co-operative effect of both activities on the red blood cell membrane, rather than the conversion of T_1 to T lysin by caseinase. This opinion was reached because the inhibition of caseinase resulted in the loss of complete lytic potential from supernatant fluids containing T lysin. Titball and Munn (1985a) regarded the H lysin to be a proteinaceous substance, on the basis of results of the ultraviolet absorption spectrum. Additionally, they observed that the partially purified H lysin contained detectable levels of GCAT. This enzyme possesses some similarities to the H lysin, e.g. molecular weight of 23.2 and 25.9 kDa, respectively (Buckley *et al.*, 1982). Yet, the molecular weights were much smaller than the 200 kDa size of 'salmolysin', the haemolytic toxin described by Nomura *et al.* (1988). The ionic strengths needed for the elution of GCAT and H lysin from ion-exchange gels were similar. These factors may complicate the isolation of pure H lysin, assuming that GCAT and the haemolysins are separate entities, which appears to be the case. Thus, GCAT has not been reported to possess haemolytic activity, and is stable at room temperature (Buckley *et al.*, 1982). Membrane filtration of the preparation failed to remove GCAT, whereas H lysin activity was lost after the procedure (Titball and Munn, 1981). Other observations of H lysin activity have indicated that haemolysis of horse erythrocytes occurs in two steps, namely a first stage in which there is no detectable cell lysis (this was coined the pre-lytic stage), followed by a second phase involving haemoglobin release and disruption of the cell membrane. Binding of the H lysin to the erythrocytes during the pre-lytic stage does not occur. Together with the observation of an optimum temperature of 25–33°C for lysis, this suggests that the H lysin has enzymic action on the erythrocyte membrane. Nevertheless, fish injected with H lysin appeared to be unaffected, despite an apparent toxicity to rainbow trout gonad tissue cell lines. Titball and Munn (1981) concluded that the failure of H lysin to elicit a response in the fish experiments was explained by the use of an unsuitable route of administration or the injection of too low a quantity of the material. In addition, these authors argued that possibly H lysin is non-toxic to fish, with no important role in the pathological process.

Several investigators have explained the relationship of haemolysins and proteases to virulence by using different strains of *Aer. salmonicida*. For example, Hackett *et al.* (1984) studied the possibility of a plasmid-encoded origin for these extracellular enzymes. Significantly, the team concluded that the loss of proteolytic and haemolytic activity in variants of wild-type *Aer. salmonicida*, obtained by treatment with ethidium bromide, did not correlate with loss of plasmid DNA. Moreover, there was no apparent change in the LD_{100} between the virulent wild-type strain and its protease-haemolysin deficient variant. This implied that the extracellular activities were not essential for virulence or, indeed, pathogenicity, at least with regard to the acute form of furunculosis in rainbow trout. Two clones derived from another virulent strain, one of which was negative for protease and haemolysin production whereas the second derivative was positive for these attributes, were avirulent (LD_{50} increased by greater than four orders of magnitude). This was an important observation, denoting that attenuation of a virulent strain occurred without loss of the A-layer, plasmids or extracellular proteolytic and haemolytic activities. For this reason, Hackett *et al.* (1984) concluded that virulence was attributable to other, as yet unknown, factors. Titball and Munn (1984b) also studied the effects of quantitative differences in virulence on the production of potential toxins by *Aer. salmonicida*. The release of ECP, i.e. proteases and haemolysins, by virulent strains and their avirulent attenuated derivatives (differing only in the presence or absence of the A-layer) could not be linked directly to virulence. Insofar as no appreciable differences were recorded between the levels of ECP from virulent (possessing an A-layer) and avirulent (no A-layer) cells, it appears that these compounds are not virulence determinants.

Hastings and Ellis (1985) reported that there was a marked variation in the production of haemolysins and proteases among different strains of *Aer. salmonicida*. In their study, four isolates, recovered from Atlantic salmon in Scotland, produced caseinase and gelatinase. Both of these enzymes were inhibited by PMSF (a serine protease inhibitor) and to a lesser extent by EDTA (a divalent metal ion chelator). This finding indicated that both enzyme activities could be attributed to a single serine protease, which was dependent upon divalent cations for activity. In contrast, an achromogenic isolate which was obtained from Iceland did not produce detectable quantities of haemolysin or gelatinase. It is noteworthy that the caseinase from this isolate differed from that of the Scottish strains, insofar as it resembled a metalloprotease. Hastings and Ellis (1985) noted that this enzyme appeared to be unique to fish pathogens. It is not known, however, if other achromogenic strains of *Aer. salmonicida* share similar properties regarding their ECPs. Nevertheless, it is relevant to note that yet another strain, recovered from the USA, differed from the Scottish isolates insofar as it lacked both caseinase and gelatinase activity in the ECP. Moreover, its haemolysin production was notably lower. Hence, it seems that there is a marked variation in the nature of the precise components of the ECPs from different strains of the pathogen. Thus, there may be some variation in the modes of pathogenesis.

Yet another factor with a potential role in virulence and pathogenicity is the leucocytolytic component of the ECP. Although this component was recognized in the

1930s, many years passed before detailed study ensued. Thus, Klontz *et al.* (1966) reported leucopenia in rainbow trout, following injection of either viable cells or a saline soluble extract of the culture. On the basis of these results it was hypothesized that a leucocytolytic compound was responsible for the limited leucocyte activity observed in infected fish. This group did not comment, however, on the biochemistry of the leucocytolytic compound. Nevertheless, this aspect was examined by Fuller *et al.* (1977), who deduced that the compound was a glycoprotein, which was distinct from the endotoxin, i.e. LPS, as previously studied by Ross (1966), Anderson (1975) and Paterson and Fryer (1974). The leucocytolytic factor was present in the supernatants from broth cultures; moreover, virulent strains produced more than the avirulent counterparts. Furthermore, the glycoprotein was cytolytic for leucocytes *in vivo*, and produced a pronounced leucopenia when injected intravenously into adult rainbow trout. In addition, the factor appeared to enhance pathogenicity, presumably by increasing susceptibility of the host. This opinion was reached after experiments in which small coho salmon were inoculated with the leucocytolytic compound in combination with live *Aer. salmonicida* (approximately a quantity sufficient to achieve a LD_{50}). The result was that 36/40 fish succumbed, in contrast to the death of only 14/40 animals injected with just the pathogen. So, the conclusion was reached that the glycoprotein constituted a virulence factor of *Aer. salmonicida*. However, the results of Cipriano *et al.* (1981) were not in accord with the findings of Fuller's team. Essentially, Cipriano and colleagues deduced that there was no correlation between virulence and the leucocytolytic properties. Therefore according to Cipriano *et al.* (1981), leucocytolytic factors could not be considered as a principal virulence mechanism. Alternatively, the activity may not solely relate to the attack on leucocytes. Instead, Cipriano and colleagues suggested that the leucocytolytic factor, contained within fraction II of the ECP, contributed to virulence not only by way of its leucocytolytic properties, but also through its role as a generalized cytotoxin capable of generating pathological changes. Indeed, these investigators recorded that intramuscular injection of fraction II (produced by a virulent isolate) into brook trout caused haemorrhaging at the mouth, base of the fins and site of inoculation. Death occurred within 24 h. Such deleterious changes did not ensue in fish which received fraction II derived from an avirulent isolate.

It must be emphasized that the lack of a leucocytic response, apparent in the majority of salmonids with furunculosis, has not been substantiated in infections of coarse fish. In one example, a chronic leucocytosis was observed in goldfish (Mawdesley-Thomas, 1969).

Ellis *et al.* (1981) outlined a hypothesis for the role of ECP in pathology. Moreover, they highlighted some of the difficulties involved in reaching a complete understanding of the pathogenic process. Importantly, they emphasized that in many respects furunculosis is an inconsistent disease, insofar as a variety of lesions have been associated with invasion of the aetiological agent. Yet, virtually none of the symptoms may be considered as unique to the disease (Wolke, 1975). Consequently, it is hardly surprising that inconsistencies have resulted in conflicting opinions over the pathogenicity mechanisms. Ellis and co-workers formulated a tentative explanation for the lesions caused by *Aer. salmonicida*. Thus, they reported that nearly all of the lesions normally associated with the disease may be achieved by i.p. or i.m. injection of ECP. However, it would appear that artificially high doses are required to accomplish such lesions. Munro *et al.* (1980) suggested that the presence of an α-globulin in normal trout serum may have the ability to neutralize ECP activities. Indeed, other workers have confirmed such effects of fish serum on ECP. Rockey *et al.* (1989) published an article detailing the inhibition of haemolysin activity by salmonid serum. Sakai (1984) mentioned a decrease in, or absence of, mortality among rainbow trout which had received ECP first treated with large volumes of rainbow trout serum prior to injection. These results indicated involvement of complement in the detoxification of ECP. Continuing this theme, Grisley *et al.* (1984) reported the presence of an α-migrating protein (a possible homologue of mammalian α_2-macroglobulin) in normal rainbow trout serum. This protein apparently exerts a role in a non-specific defence function against microbial proteolytic toxins. Ellis and Grisley (1985) pursued the theme, concluding that normal trout serum inhibits ECP protease but neutralization is effected by different antiproteases and less efficiently than trypsin. They contended that the data, to some extent, explained the potency of ECP in causing disease. Ellis *et al.* (1981) assumed that in natural infections lesions would be produced after the ECP had exhausted any inhibiting factors, either locally or systemically. They thought that the various symptoms of furunculosis were explained by the colonization of different host tissues by the pathogen. It was concluded that the pathological effects resulting from infection by *Aer. salmonicida* were probably caused by the ECP released by the pathogen. Thus the leucocytolytic component might act against leucocytes, eventually resulting in leucopenia, and preventing the destruction of the bacterial colonies, thus allowing microbes to be transmitted to other organs via the circulatory system where they may initiate the development of more colonies. It was further submitted by these authors that lesions and mortalities are due to the collagenolytic activity of the ECP (this is one of the notable features of furunculosis), with haemorrhaging resulting in the vicinity of bacterial colonization. Generalized circulatory failure could ensure if the ECP subsequently entered the circulatory system.

Just when the role of ECP and proteases was becoming clarified, some elegant work with deletion mutants caused a fundamental re-think. It was obvious that to be sure of the role of a specific component, eliminate the genes from the bacteria and determine the effect on the host. Using this approach, Vipond *et al.* (1998) confirmed that mutants lacking GCAT or serine protease (AspA) were not less virulent than the parental cell following i.p. or cohabitation challenge of Atlantic salmon.

Finally, it is appropriate to recall the words of Munn *et al.* (1982), who commented that the interrelationships of ECP suggest that the pathogen exerts its toxic effects *in vivo* by means of multiple factors which interact synergistically.

Scavenging for iron
A current theme, which has prompted some excellent work, concerns the ability of *Aer. salmonicida* cells to successfully scavenge for ion in iron-limited conditions. These would be created in the host, and function as a defence mechanism against invasion by pathogens. Indeed, there is evidence that IROMP are produced *in vivo* (Ellis *et al.*, 1997). Thus, free iron would be bound to proteins, such as transferrin, resulting in iron-restricted conditions in the host. Initially, Chart and Trust (1983) demonstrated that typical strains of *Aer. salmonicida* were capable of sequestering iron. Then, Kay *et al.* (1985)

determined that the A-layer was implicated as a component of an iron-uptake mechanism. The conclusion was that the A-layer functioned as the initial stage of iron-uptake, being a binding site for porphyrins, i.e. haemin and protoporphyrin. The difference between typical and atypical isolates was reinforced by the conclusion that there was a fundamental difference in the mechanism of utilization of non-haem bound sources of iron. Hirst *et al.* (1991) and Hirst and Ellis (1996) described an inducible siderophore (these are soluble low-molecular-weight iron-chelators)-dependent iron-chelating system in typical strains and an identified siderophore-independent system in atypical *Aer. salmonicida*. Among *Aer. salmonicida* subsp. *salmonicida* (17 isolates from Scotland and Spain were examined), the siderophore is regarded as homogeneous (Fernandez *et al.*, 1998).

In summary, a variety of cell-associated and extracellular factors have been investigated in order to determine their role in virulence and pathogenicity of diseases of *Aer. salmonicida* aetiology. Unfortunately, the overview which emerges for the current understanding of pathogenicity mechanisms is confused. Much of the evidence, about the various factors suspected to be involved with virulence, is contradictory, or is based solely upon *in vitro* studies. Thus, although the presence of an A-layer is firmly believed to be a primary determinant of virulence, reports of avirulent isolates with A-layer (Udey, 1978) raises further questions, Conversely, ECP contain such a diverse array of different factors implicated with virulence and pathogenicity, that to pinpoint the function of each *in vivo* has proved difficult. Hence, a definitive assessment of the role of the various haemolysins, proteases and leucocidins in the natural disease process still eludes us. However, it is conceded that substantial progress has been made in the isolation, purification and biochemical characterization of the ECP. Moreover, strong evidence exists that the ECP are capable of eliciting a pathology reminiscent of the natural disease (Cipriano *et al.*, 1981; Ellis *et al.*, 1981). However, the interrelationships between the various sub-components remain unclear. Thus, Cipriano *et al.* (1981) believed that the leucocytolytic and proteolytic activities were dual expressions of a component, i.e. the chromatographic fraction II; a notion which requires more information for confirmation. This group opined that the generalized cytotoxicity for rainbow trout gonad cell lines by ECP was a better indicator of virulence. There was, however, agreement with the suggestion of Sakai (1977) that proteases constituted the most pathogenic element of the ECP. Yet, it is apt to recall the warnings of Sheeran *et al.* (1984), who emphasized that it is vital to establish the levels of the proteolytic enzymes in naturally infected fish tissues. Until this can be done, even conclusions drawn from *in vivo* experiments remain speculative. Results from some investigations have demonstrated that there is some degree of variation in the quantities of potential virulence factors produced by different isolates of *Aer. salmonicida*. Thus, there may be some variation in the precise mode of pathogenesis (Hastings and Ellis, 1985). Yet, Titball and Munn (1985b) did not find any appreciable differences in the levels of ECP between virulent and avirulent isolates. Nevertheless, these authors admitted that this did not exclude a role for the substances as aggressins, although it was paradoxical that a delayed release of ECP by A-layer[+] strains was observed. As a possible explanation, the disadvantage of late release of toxic material may be counterbalanced by the role of A-layer in conferring resistance to host defence mechanisms (Munn *et al.*, 1982). In yet another comment, Hackett *et al.* (1984) proposed that in peracute or acute forms of furunculosis, virulence is independent of the presence of protease and haemolysin. Accordingly, these workers suggested that death of the fish may result from organ dysfunction, due principally to massive growth of the pathogen. Alternatively, it was speculated that there may be involvement by an as yet unidentified component of the ECP.

Aeromonas sobria

Cultures were pathogenic to rainbow trout, with the LD_{50} reported as 2×10^5 cells. Dead animals revealed the presence of haemorrhagic septicaemia. In addition, thermolabile ECP were cytotoxic and lethal (30 μg protein/fish) to rainbow trout. Further work on *Aer. sobria* by Paniagua *et al.* (1990) highlighted the role of caseinase, haemolysins and cytotoxins in the pathogenic process.

F. ENTEROBACTERIACEAE REPRESENTATIVES

Citrobacter freundii

The pathogenicity of isolates was not confirmed in laboratory experiments by Sato *et al.* (1982). Nevertheless, Baya *et al.* (1990a) and Karunasagar *et al.* (1992) demonstrated pathogenicity following i.p. injection of rainbow trout and carp, respectively. Here, the LD_{50} was in the range of 10^5–10^6 cells. Injection of cell-free extracts did not result in mortalities (Karunasagar *et al.*, 1992).

Edwardsiella ictaluri

Mortalities of up to 50% have been recorded. In one comparative study, an injection of 1.5×10^3 cells of the pathogen was sufficient to cause 100% mortality among a group of channel catfish; tilapia only demonstrated slight susceptibility, whereas golden shiner, bighead carp and largemouth bass were completely resistant (Plumb and Sanchez, 1983). Survival of natural outbreaks has led to high humoral antibody levels and protection from fresh onslaught with *Edw. ictaluri* (Vinitnantharat and Plumb, 1993). The aetiological agent has been also associated with disease outbreaks in non-ictalurid fish, e.g. danio (*Danio devario*) (Waltman *et al.*, 1985).

Little is known about the pathogenicity mechanisms of *Edw. ictaluri*. Extracellular products have been associated with virulence (Stanley *et al.*, 1994). Saeed (1983) showed that cells are highly piliated, and inferred that the pili might be associated with virulence. By means of intragastric intubation and a comparatively high dose of 1×10^9 cells, *Edw. ictaluri* crossed the intestinal mucosa of channel catfish in 15 min (Baldwin and Newton, 1993). Using 1×10^6 cells/ml and an application directly into the olfactory organs of channel catfish, light and electron microscopy revealed damage after 1 h (Morrison and Plumb, 1993). Certainly, it has been firmly established that channel catfish are highly susceptible to the organism, with an injected dose of 1.5×10^3 cells capable of killing the host within 10 days at a water temperature of 26°C (Plumb and Sanchez, 1983). It does not appear that the organism produces abundant exo-enzymes, which would function as exotoxins in fish. It has been argued that both gut and nares are primary sites for the invasion of *Edw. ictaluri* in natural outbreaks of disease (Shotts *et al.*, 1986). Evidence pointed to the uptake of the pathogen into fish via a water-borne route, for which the outcome was a bacteraemia within 24 h (Wise *et al.*, 1997). By 72 h, the pathogen was recoverable from the blood. Then by 216 h, there was evidence of the pathogen clearing from the blood, with all survivors developing agglutinating antibodies to *Edw. ictaluri*. The infection process was accompanied by shedding of the pathogen into the water; a process contributing to transmission of the disease (Wise *et al.*, 1997). It does not appear that the level of dietary iron affected antibody production and thereby influenced the course of an infection (Sealey *et al.*, 1997).

Edwardsiella tarda

To date, the disease has been recorded in a diverse array of fish species, including chinook salmon (Amandi *et al.*, 1982), channel catfish (Meyer and Bullock, 1973), mullet (Kusuda *et al.*, 1976b), carp (Sae-Oui *et al.*, 1984), eels (Wakabayashi and Egusa, 1973), tilapia (Kubota *et al.*, 1981) and flounder (Nakatsugawa, 1983; Mekuchi *et al.*, 1995a). From laboratory-based experiments, pathogenicity has also been demonstrated in steelhead and rainbow trout (Amandi *et al.*, 1982), yellowtail (Nakatsugawa, 1983) and loach (Park *et al.*, 1983).

There has been no difficulty achieving experimental infections of fish with *Edw. tarda*. Thus, using channel catfish, Meyer and Bullock (1973) established an infection by using the i.p. route of administration. At a water temperature of 27°C, deaths in 80% of the population of 5–10 cm long fingerling fish followed within 10 days of injecting an artificially high dose of 8.0×10^7 cells. Similar experiments, with 8×10^6 and 8×10^5 cells resulted in only 40% cumulative mortalities within 10 days. However, these workers pointed to the host specificity of *Edw. tarda*, insofar as brown trout, held at a water temperature of 13°C, did not show any mortalities following injection with the pathogen. Subsequent investigations, however, showed that *Edw. tarda* could indeed infect salmonids. Thus Amandi *et al.* (1982) demonstrated that the LD_{50} for chinook salmon and steelhead trout, was 4.1×10^6 and 5.6×10^6 cells, respectively. It is worth emphasizing that ictalurids were determined to be more sensitive than salmonids, insofar as these workers determined the LD_{50} for channel catfish to be only 4.0×10^5 cells. At lower water temperatures, e.g. 12°C, the LD_{50} was in the region of one order of magnitude higher. Ironically, a water-borne challenge proved to be a failure, although this may reflect the inability of cells, grown in nutrient-rich media and therefore not typical of the natural physiological state, to survive in the aquatic environment. Nevertheless, Song *et al.* (1982) deduced that of 110 isolates, the lowest LD_{50} (by water-borne challenge) was 3.1×10^7 cells/ml. Clearly, this is not conducive to the notion of a particularly virulent pathogen.

Using immersion, oral, and i.p. and i.m. routes, Mekuchi *et al.* (1995a) succeeded in infecting and killing Japanese flounder (by all routes). The LD_{50} dose was calculated at 7.1×10^1/fish (via i.m. injection), 1.2×10^2/fish (via the i.p. route), 3.6×10^6/fish (by immersion) and 1.3×10^6/fish (orally) (Mekuchi *et al.*, 1995a).

The pathogenic mechanisms were investigated by Ullah and Arai (1983a,b), who reported that in contrast to *Edw. ictaluri*, pili were absent. Instead, cells were observed to be surrounded by a slime layer. This may help with the adhesion to host cells, and also protect the bacteria from host defences. Haemolysins and dermatotoxins, but not lipases or proteolytic enzymes, were produced *in vitro*, and it was postulated that these exo-enzymes may confer pathogenicity on *Edw. tarda* (Ullah and Arai, 1983a,b). Further-more, the pathogenic role of dermatotoxins was highlighted in additional experiments (Ullah and Arai, 1983b). This work concluded that two high-molecular-weight, heat-sensitive dermatotoxins were produced, which in rabbits (not fish!) were found to have separate functions. Thus, one toxin caused erythema within 3–8 h of intracutaneous injec-tion, whereas the second caused oedema followed by necrotic erythema in 5–7 days. Hopefully at some point, this work will be repeated in fish. Proteolytic toxins (molecular weight = 37 kDa), from ECP, have been purified from avirulent culture (the toxin is not present in avirulent isolates) and the LD_{50} equated to 1.6 g of toxin/g of fish (Suprapto *et al.*, 1996).

On iron-deficient medium, *Edw. tarda* produced siderophore (Kokubo *et al.*, 1990). Such components permit the pathogen to scavenge for iron in the blood of the host. Certainly, it appears that the ability of *Edw. tarda* to acquire iron is an important part of the infection process (Park, 1986; Iida and Wakabayashi, 1990), and it is relevant that an iron-regulated haemolysin gene has been reported (Hirono *et al.*, 1997a). The virulent strains are more resistant to the bacteriostasis of iron-chelating reagents than their avirulent counterparts.

Escherichia vulneris

There is scant information about the pathogenicity of *Esch. vulneris*. It would appear that infection was achieved in rainbow trout of 120 g in weight with death ensuing in 175 h (Aydin *et al.*, 1997). Details about dosages were not included in the original publication.

Hafnia alvei

The culture caused clinical disease in laboratory experiments with rainbow trout. Injection (subcutaneous) of rainbow trout, weight 150–200 g, and maintained at a water temperature of 4–6°C, resulted in clinical disease, with mortalities occurring between 3 and 10 days.

In the subsequent study by Teshima *et al.* (1992), it was reported that disease took 3 months to develop at 15°C following i.p. injection with 5×10^6 to 3×10^7 cells/ml. So, the inference was that the organism was not very aggressive.

Klebsiella pneumoniae

Cultures could induce fin and tail rot by immersion for 5 min in 10^7 bacteria/ml, but only after prior abrading the surface of the fins. Disease signs became evident after 3 days, with mortalities being recorded after a further 2 days. Within 5 days of injecting 10^4 cells/fish, there was some reddening of the muscle in animals injected intra-muscularly. Seven days after i.m. and i.p. injection, mortalities began, with disease signs including gastro-enteritis, liquefaction of the kidney, and the presence of ascitic fluid in the peritoneal cavity. All fish were dead within 12 days (Daskalov *et al.*, 1998).

Proteus rettgeri

Experimental infection of silver carp by i.m. injection of 5×10^2 bacteria, or by scarifica-tion of the fish surface and subsequent exposure to a broth culture, resulted in mortalities of 50% at a water temperature of 18–20°C. Experimentally infected fish showed lesions typical of the farmed stock (Bejerano *et al.*, 1979).

Salmonella choleraesuis **subsp.** *arizonae*
The pathogenicity of the single isolate was not confirmed (Kodama *et al.*, 1987).

Serratia liquefaciens
Injection of 10^3 cells killed Atlantic salmon within 72 h. Typically, there was pronounced muscle liquefaction within the vicinity of i.m. injections. ECP resulted in death within 48 h (McIntosh and Austin, 1990).

Serratia marcescens
Laboratory-based studies confirmed pathogenicity for striped bass ($LD_{50} = 1 \times 10^5$ cells) [LD_{50} in rainbow trout = 5×10^3 cells] with death occurring 1–3 or 1–7 days following administration of the cells via the i.m. or i.p. route, respectively (Baya *et al.*, 1992c). Experimentally infected fish displayed muscle necrosis and some signs of haemorrhagic septicaemia. Further work indicated a role for ECP, which possessed marked proteolytic and phospholipase activity. In fish, the ECP caused a cytotoxic response, with mortality occurring 24–48 h after administration. Here, the LD_{50} dose for rainbow trout and striped bass was 0.4 and 4.8 μg of protein/g of fish, respectively (Baya *et al.*, 1992c).

Serratia plymuthica
Laboratory experiments revealed that the LD_{50} dose for rainbow trout was 10^4–10^5 cells. Moreover, such infected fish displayed discoloration and abdominal swelling (Nieto *et al.*, 1990) and extensive surface lesions (Austin and Stobie, 1992b). Isolates were strongly hydrophobic (Rodriguez *et al.*, 1990); a property which may be involved in the adherence of the organism to surfaces.

Yersinia intermedia
It was considered that the organism was of endogenous origin, being a pathogen of cold-stressed (possibly immunocompromised) fish (Carson and Schmidtke, 1993). Pathogenicity experiments have not been carried out.

Yersinia ruckeri
In terms of pathogenicity to rainbow trout, Type 1 (Hagerman) is the most virulent, followed by Type 2 (O'Leary) and then Type 3 (Australian) (Bullock *et al.*, 1983). The LD_{50} dose has been established to be 3.0×10^5 cells/ml for Type 1, 1.0×10^7 cells/ml for Type 2 and an as yet undetermined dose for Type 3. The degree of pathogenicity of serovars IV and V needs clarification. It should be emphasized that exposure to sublethal concentrations of copper, i.e. 7 μg/l, for 96 h, rendered the fish more susceptible to infection by *Y. ruckeri* (Knittel, 1981). Moreover, the nature of the virulence factors is incompletely understood. The role of plasmids in virulence has been indicated insofar as the more pathogenic Type 1 isolates possess a large 40–50 mDa plasmid, and some contain a smaller 20–30 mDa plasmid. The larger plasmid is absent from Type 2 cultures (Cook and Gemski, 1982; De Grandis and Stevenson, 1985). Further work is necessary,

however, to resolve the precise role of this large plasmid in pathogenicity, especially as it is not thought to carry virulence factors (Guilvout *et al.*, 1988).

The O antigen, from the LPS of serogroup O1, has been determined to comprise a branched tetrasaccharide, with repeating units containing 2-acetamido-2,6-dideoxy-L-galactose, 2-acetamido-2-deoxy-D-glucose and 7-acetamido-3,5,7,9-tetradeoxy-5-(4-hydroxybutyramido)-D-glycero-L-galacto-nonulosonic acid (Beynon *et al.*, 1994).

ECP have been recovered, and demonstrated to have an LD_{50} of 2–9.12 μg of protein/g of fish. The ECP has been found to contain amylase, caseinase, gelatinase, haemolytic (salmon, sheep and trout erythrocytes) lipase and phospholipase activity. Also, aesculin has been attacked (Romalde and Toranzo, 1993). Adhesion activity, at least to the CHSE-214 cell culture, has been recorded (Romalde and Toranzo, 1993).

Evidence has been presented that *Y. ruckeri* may have a siderophore-mediated iron uptake system (Romalde *et al.*, 1991), in parallel to *Aer. salmonicida* and *V. anguillarum*. These workers revealed that >3 OMP were induced in iron limiting conditions. As a result of a commendable study, Davies (1991) reported on the OMP of 135 *Y. ruckeri* isolates. Several 36.5–40.5 kDa peptidoglycan associated proteins and a 36.5 or 38 kDa heat-modifiable protein were characterized. This 39.5 kDa peptidoglycan-associated protein was apparently not produced during logarithmic growth, but increased quantitatively in the stationary phase.

An example of the new approach in examining pathogenicity involved an inbred fish model, i.e. platyfish (*Xiphophorus maculatus*) and cultured fish cells. In particular, infection with *Y. ruckeri* involved bathing in suspensions containing 10^6 to 10^8 cells (Kawula *et al.*, 1996). The result, in terms of invasion of key tissues, may be summarized, as follows:

Cell type	Percentage invasion
Rainbow trout gonad	2.6
Rainbow trout kidney	2.3
Minnow epithelial cells	10.2

Y. ruckeri may be able to out-compete other micro-organisms, which may be an advantage for any potential pathogen, by the production of water soluble antimicrobial compounds (Michel and Faivre, 1987).

F. VIBRIONACEAE REPRESENTATIVES

Moritella marina
The LD_{50} dose to Atlantic salmon was 3.5×10^3 cells (Benediktsdóttir *et al.*, 1998).

Photobacterium damselae
The pathogen has been implicated with disease in sharks, turbot and yellowtail (Fujioka *et al.*, 1988; Sakata *et al.*, 1989; Fouz *et al.*, 1991). Laboratory infections of *C. punctipinnis* have been established by removing four to six scales from the flank, scarifying the dermis, and swabbing the wound with 10^7–10^8 viable cells of *Ph. damselae*. At

water temperatures of 16.0–16.5°C, the fish developed large ulcers in 3 days, with death following 24 h later. Similar data were recorded after experimental challenge of unscarified animals. However, in the initial study, fish from other families appeared to be unaffected by *Ph. damselae*, pointing to host specificity of the pathogen. Thus, representatives of Atherinidae, Clinidae, Cottidae, Embiotocidae, Girellidae and Gobiidae resisted experimental challenge. This is interesting because representatives of these families co-habited the reefs with blacksmith (Love *et al.*, 1981). However, Grimes *et al.* (1984) successfully infected dogfish by i.p. injection with *Ph. damselae*. Death ensued within 18 h at an unspecified water temperature. These authors reported that the organism was highly cytotoxic. A neurotoxic acetylcholinesterase has been described (Pérez *et al.*, 1998).

ECP have been implicated with disease (cytotoxic) processes, with the LD_{50} dose ranging from 0.02–0.43 μg of protein/g of fish with death occurring between 4 and 72 h after administration (Fouz *et al.*, 1993). The ECP were considered to have low proteolytic activity, without evidence of any caseinase, elastinase or gelatinase. In contrast, pronounced phospholipase and haemolytic activity was recorded for turbot (and human and sheep) erythrocytes. It was possible that LPS contributed to heat stability of the toxic fractions (Fouz *et al.*, 1993).

A siderophore-mediated iron sequestering system has now been described, and almost certainly contributes to the pathogenicity of the organism (Fouz *et al.*, 1994; 1997).

Photobacterium damselae subsp. *piscicida*

Experimental infection may be achieved by i.m. injection, oral uptake or immersion. The fate of the pathogen has been examined by FAT (Kawahara *et al.*, 1989). Thus following i.m. injection, the pathogen became located initially in the kidney and spleen, before spreading to the gills, heart, intestine and pyloric caeca. Following oral uptake, the pathogen appeared in the stomach, before spreading to the internal organs. After immersion, *Ph. damselae* subsp. *piscicida* located in the gills, and then spread widely to the heart, kidney, liver, pyloric caeca and spleen (Kawahara *et al.*, 1989). Within the tissues of infected fish, *Ph. damselae* subsp. *piscicida* was seen to accumulate in the macrophages (Nelson *et al.*, 1989) perhaps after an initial cell adherence stage (Magariños *et al.*, 1996a,b), which appears to involve capsular material (Magariños *et al.*, 1996b). However, a separate study reported that cells of the pathogen were killed by macrophages *in vitro* in 3–5 h (Skarmeta *et al.*, 1995).

Resistance to the pathogen may well reflect the size of the fish and the efficiency of the phagocytes (Nova *et al.*, 1995). There is a distinct role for the capsule to protect against phagocytosis (Arijo *et al.*, 1998). Comparing five capsulated, virulent and one non-capsulated, avirulent culture, Arijo *et al.* (1998) recorded significant differences in phagocytosis with the former resisting being engulfed and killed by the macrophages. In a separate development, it was considered that mucus from turbot—thought to contain a glycoprotein—inhibited *Ph. damselae* subsp. *piscicida*, but less so mucus from sea bass and sea bream (Magariños *et al.*, 1995). Perhaps, such observations explain the comparative sensitivity of some fish species, e.g. sea bass and sea bream, to the pathogen.

A siderophore-mediated iron-sequestering mechanism has been found in *Ph. damselae* subsp. *piscicida*, with IROMP of 105, 118 and 145 kDa in size (Magariños *et al.*,

1994b). Some variability has been detected insofar as isolates expressed a 75 kDa IROMP. Also, others have reported different sizes for the novel proteins associated with iron limitation. Thus, European isolates were considered to express four novel proteins of 63 kDa and three at $\geqslant 200$ kDa, whereas Japanese isolates did not form any different proteins (Bakopoulos *et al.*, 1997b).

Shewanella putrefaciens
Following i.p. injection, fish (average weight = 50 g) developed clinical disease, with 80% mortalities within 48 h. The organism was recovered from the kidney, liver and spleen of dead fish. Infection was not achieved following immersion in a dense suspension of the organism (Saeed *et al.*, 1987).

Vibrio alginolyticus
According to Lee (1995), the organism produced ECP, which was lethal at 0.52 μg/g of fish. This ECP contained a 44 kDa toxic protease, for which the minimum lethal dose was 0.17 μg/g of fish.

Vibrio anguillarum
The exact mode of infection is unclear, but undoubtedly involves colonization of (attachment to) the host, and thence penetration of the tissues. It has been postulated that chemotactic motility is necessary for invasion of the host. A 40.1 kDa flagellin A protein (encoded by the *flaA* gene) is essential for virulence (Milton *et al.*, 1996). Thus, loss of flagella by transposon mutagenesis led to a 500-fold reduction in virulence following an immersion challenge (O'Toole *et al.*, 1996). Flagellum production and virulence by the water-borne but not i.p. route was correlated with RpoN (O'Toole *et al.*, 1997). As a cautionary note, it is possible that other changes to the bacterial cells may have occurred with the loss of flagella. Ransom (1978) postulated that infection probably begins with colonization of the posterior gastro-intestinal tract and rectum. This conclusion resulted from the observations that *V. anguillarum* was seen initially in these sites. Horne and Baxendale (1983) reported adhesion of *V. anguillarum* to intestinal sections derived from rainbow trout. All regions of the intestine were colonized (approximately 10^3 cells/cm^2), with maximum attachment occurring within 100 min. Orally administered *V. anguillarum* survived in the stomach of juvenile turbot for several hours, persisted in the intestine, and proliferated in faeces (Olsson *et al.*, 1998). This view has been reinforced by a study, which concluded that >50% of the spleens of turbot contained cells of *V. anguillarum* after infection via the oral and rectal routes (Olsson *et al.*, 1996). The skin appears to become colonized within 12 h of immersion in a virulent culture (Kanno *et al.*, 1990). Then, invasion of the liver, spleen, muscle, gills and intestine follows (Muroga and De La Cruz, 1987). Resistance to the potential debilitating effect of fish serum (Trust *et al.*, 1981) may hasten the invasion processes. Some degree of host specificity has been indicated, insofar as strains from rainbow trout were poorly pathogenic to saithe, and vice versa (Egidius and Andersen, 1978). This raises the question concerning the size of inoculum necessary to achieve clinical disease. Levine *et al.* (1972) reported lesions at the site of infection in winter flounder after exposure to only 640 cells. These were administered by intradermal injection. Much larger inocula resulted in sizeable

mortalities. For example, Evelyn (1971b) determined that *Oncorhynchus keta* and *O. nerka* died within 48 h of receiving, by i.p. injection, 0.1 ml containing 10^7 viable cells of *V. anguillarum*. In a much more spectacular demonstration of virulence, Sawyer *et al.* (1979) established 80–100% mortality in a population of Atlantic salmon following exposure to 1–2.5×10^5 cells/ml as a bath for 1 h. In this demonstration, the fish were maintained at a water temperature of 10–15°C.

Turbot larvae have been successfully challenged with *V. anguillarum* orally via live feed (Grisez *et al.*, 1996). Using 10^3 *Artemia* nauplii/ml and 10^9 *V. anguillarum* cells/ml, the recipient fish died within 4 h (Grisez *et al.*, 1996).

The precise nature of the virulence mechanism of *V. anguillarum* has prompted some excellent work. Perhaps the highlight of these studies was the discovery that *bona fide* strains of serogroup O1 of the pathogen contained a virulence plasmid, which was associated with an iron-uptake system expressed under iron-limited conditions (Crosa *et al.*, 1977; 1980; Crosa, 1980; Wolf and Crosa, 1986; Chen *et al.*, 1996). This plasmid, designated pJM1 and of 67 kb, is always present in virulent isolates (Pedersen *et al.*, 1996; 1997b) and may be included on a transposon-like structure (Tolmasky and Crosa, 1995), but absent from those of low virulence. Conversely, the pJM1 plasmid has been found in some avirulent isolates (Pedersen *et al.*, 1997b). Yet, virulence may be attenuated by curing this plasmid (Crosa *et al.*, 1980) or by deleting three plasmid encoded gene products (Singer *et al.*, 1991). The role for pJM1 concerns specifying an iron-sequestering mechanism, i.e. the low-molecular-weight siderophore anguibactin for which the precursor is chromosome-mediated 2,3-dihydroxybenzoic acid (Chen *et al.*, 1994), and specific iron-transport proteins, of which the angR protein (this is regulated by the regulatory gene *angR* [Salinas and Crosa, 1995], which has been reported as similar to bacteriophage P22; Farrell *et al.*, 1990) acts as a positive regulator of anguibactin biosynthesis and the transcription of the iron-transport genes *FatA* and *FatB* (Actis *et al.*, 1995; Chen *et al.*, 1996). Also, *V. anguillarum* has a plasmid-encoded histamine decarboxylase gene *angH*, which is essential for the biosynthesis of anguibactin (Barancin *et al.*, 1998). The overall effect is that the system enables the bacterial cell to compete for available iron in the fish tissues. Two OMP have been designated as OM2 (molecular weight = 86 kDa) and OM3 (molecular weight = 79 kDa). The siderophore and OM2 are coded by plasmid pJM1, whereas OM3 is a function of chromosomal involvement. The basic mechanism involves diffusion of the siderophore into the environment, and the formation of iron complexes which attach to OM2, presumably leading to transport of the iron into bacterial cells (Crosa and Hodges, 1981; Crosa *et al.*, 1983; Tolmasky and Crosa, 1984; Actis *et al.*, 1985; Mackie and Birkbeck, 1992). Thus, invading bacteria may multiply in the host by scavenging successfully for the iron which is bound by high-affinity iron-binding proteins, such as transferrin, lactoferrin and ferritin. These are present in the serum, secretions and tissues, respectively (Bullen *et al.*, 1978). Toranzo *et al.* (1983a) complicated the issue by publishing data which showed that virulent strains, obtained from striped bass, did not contain plasmids. Yet, all the isolates grew in iron-limiting conditions, during which new OMP and a siderophore were found. Chromosomal DNA sequences, which hybridized with pJM1, were present. Thus, it seems likely that the plasmid DNA had become integrated into the bacterial chromosome.

Subsequent work proved that a separate iron-uptake system was contained on the chromosome. This differed from the plasmid-mediated system, insofar as the diffusible siderophore is not utilized as an external siderophore, and different OMP are synthesized (Lemos *et al.*, 1988; Conchas *et al.*, 1991; Mackie and Birkbeck, 1992). Siderophores of the phenolate class, possibly related to enterobactin, have been found in New Zealand isolates (Pybus *et al.*, 1994). Interestingly, iron-uptake mechanisms have been reported in non-pathogenic cultures (Lemos *et al.*, 1991), casting some doubt on the precise relevance of the mechanism to pathogenicity.

But what about the other serogroups? The virulence plasmid, pJM1, has not been found in representatives of any other serogroup (Austin *et al.*, 1995a). Yet, we regard that other serogroups, especially serogroup O2, which produces a 50 kDa porin (Davey *et al.*, 1998) may be more aggressive to fish than serogroup O1.

The presence of common antigens between *V. anguillarum* and other taxa would explain the cross-protection observed with fish vaccines (e.g. Nakai *et al.*, 1989b). What is the nature of these cross-protecting antigens? From capsular antigens (Rasmussen and Larsen, 1987), attention focused on the OMP (Chart and Trust, 1984)—a porin of ~40 kDa molecular weight, which is now regarded as a common antigen (Simón *et al.*, 1996). This porin, OM1, was examined by immunoblotting, ELISA and dot blot; antiserum to OM1 of serogroup O1 cross-reacted with *Vibrio* spp., but not with other genera—except *Plesiomonas shigelloides* (Simón *et al.*, 1998).

Debate has centred on the possible interaction of exotoxins and/or endotoxins (Bullock and Conroy, 1971; Abe, 1972; Umbreit and Ordal, 1972; Grischkowsky, 1973; Inamura *et al.*, 1984; De La Cruz and Muroga, 1989). Umbreit and Ordal (1972) reported mortalities in goldfish following injection of filter-sterilized supernatant derived from 24 h broth cultures. Thus, the debilitating effect of bacterial ECP was suggested (but not proven!) at this stage; but later held responsible for virulence [a neurotoxic acetylcholinesterase has been isolated from the ECP; Pérez *et al.*, 1998] (Lamas *et al.*, 1994a,b). This experiment was repeated, and resulted in >70% mortalities among a group of goldfish (Umbreit and Tripp, 1975). Unfortunately, the 3-year interval between publications did not achieve any significant improvement in knowledge. A similar basic theme was used by Abe (1972), who injected 'endotoxins' into chinook salmon and recorded the presence of haemorrhaging lesions at the point of injection. A significant development stems from the work of Wolke (1975) and Roberts (1976), who suggested that 'haemolytic toxins' might be responsible for the anaemic response in infected fish. Subsequent efforts by Munn (1978; 1980) demonstrated conclusively that haemolysins were involved. These were described as thermolabile enzymes (activity was lost by heating to 50°C for 10 min) with optimum pH of 7.2–7.4, which were inactivated by gangliosides. The molecular weight of one of these enzymes was estimated as 191 kDa (Munn, 1980). In the ensuing experiments, haemolytic activity was first detectable in cultures (filtrates) after 19 h incubation at 20°C. Production peaked at 39 h, and then declined. One explanation for this result is that haemolysin production only occurs during the stationary phase of growth. Another possibility is that production of the enzyme occurs intracellularly, with release into the environment taking place at a later time, perhaps during autolysis of the bacterial cells. Munn postulated that inactive haemolysins may also be secreted by the bacteria. These enzymes could then be re-activated by as-yet-unknown phenomena, at a later period. More recently, proteases of 36 kDa molecular weight have also been implicated with virulence (Inamura *et al.*, 1984; 1985; Kodama *et al.*, 1984; Kanemori *et al.*, 1987). In particular, a zinc metalloprotease has been associated with invasion processes (Norquist *et al.*, 1991). Crude ECP has led to the development of an inflammatory response, including leucopenia, in rainbow trout (Lamas *et al.*, 1994a,b). No doubt the debate will continued unabated for some considerable period, but the ultimate result should be a better understanding of the pathogenicity mechanisms. Basically at present, the scenario involves uptake and penetration of the host tissues, scavenging for iron as a result of a plasmid/chromosomal mediated trait, and damage to the fish by means of haemolysins and proteases.

In a rather elegant set of experiments, Harbell *et al.* (1979) catalogued the precise changes in the blood following infection with the pathogen. As anaemia is one of the signs of vibriosis, it seems hardly surprising that the haemoglobin level decreases concomitant with an increase in the erythrocyte osmotic fragility, and a decline in the number of leucocytes. Reductions were also recorded in osmolarity, and in the amounts of plasma protein, albumin, chloride, sodium and alkaline phosphatase. In contrast, levels of plasma glucose, lactate dehydrogenase, and glutamic oxaloacetic transaminase increased.

Vibrio cholerae (non-O1)

The evidence suggests that *V. cholerae* is a highly virulent fish pathogen, insofar as ayu and eels may be infected following immersion in only 1.26×10^4 cells/ml and 1.26×10^2 cells/ml, respectively. Yamanoi's team noted that with ayu, mortalities began in 2 to 7 days at water temperatures of 21 and 26°C, but no deaths occurred if the water temperature was at 16°C. In comparison, an eel population suffered 10% mortalities within 5 days at a water temperature of 21°C, and 30% deaths in 3–7 days at 26°C. Clearly, this information suggests that *V.cholerae* is likely to be troublesome only in higher water temperatures.

Vibrio harveyi

Infected (i.p. injection) dogfish (*Squalus acanthias*) died within 18 h. Lemon sharks were more resistant to infection, although internal damage followed the injection of 5×10^7 cells. Yii *et al.* (1997) reported an LD_{50} dose for grouper of 2.53×10^7 colony forming units/g of body weight, with disease signs reminiscent of the natural infection, i.e. swollen intestine full of a yellowish fluid. Although the pathogenic mechanism remains to be elucidated, it is relevant to note that the organism is slightly cytotoxic. Also, siderophores are produced (Owens *et al.*, 1996).

Vibrio ordalii

The virulence plasmid, pJM1, has not been detected in *V. ordalii* (Crosa, 1980). However, a 30 kb plasmid, designated pMJ101, has been found in all isolates of *V. ordalii* (Bidinost *et al.*, 1994). Moreover, haemolysins and proteases have not been found (Kodama *et al.*, 1984).

Vibrio pelagius

Infectivity experiments with rainbow trout (10 g) and turbot (5 g in size) confirmed virulence, and an LD_{50} of 1.9×10^5 cells/fish and 9.5×10^4 cells/fish, respectively (Angulo *et al.*, 1992).

Vibrio salmonicida

Intraperitoneal injection and immersion of Atlantic salmon and rainbow trout with broth cultures led to clinical disease. Atlantic salmon were more susceptible than rainbow trout (Egidius *et al.*, 1986; Hjeltnes *et al.*, 1987). The LD_{50} dose ranges from 4×10^6–1×10^8 cells/fish (Wiik *et al.*, 1989). In addition, *V. salmonicida* has caused mortalities in cod (Jørgensen *et al.*, 1989). The plasmids, which are regarded as being present in all strains, do not appear to be related to virulence (Valla *et al.*, 1992). Using isolated macrophages from Atlantic salmon and rainbow trout with immunofluorescence techniques, the pathogen has been observed to be internalized (Brattgjerd *et al.*, 1995).

What about the risk of disease after transferring salmon from freshwater to seawater? Eggset *et al.* (1997) concluded that the susceptibility of Atlantic salmon to Hitra disease in seawater possibly reflected the overall quality of the smolts.

Vibrio splendidus

The bacterium was pathogenic to rainbow trout and turbot, with an LD_{50} of 2.2×10^4 cells and 1.2×10^4 cells, respectively (Angulo *et al.*, 1994).

Vibrio trachuri

Comparatively high doses of the organism were required to cause disease in Japanese horse mackerel. Thus using 36.8 g fish at a water temperature of 26°C, 1.1×10^8 cells/fish caused 100% mortalities within 24 h of i.p. injection. A dose of 1.1×10^7 cells/ml led to 50% mortalities within 4 days. By immersion in 3.6×10^7 cells/ml for 2 min, 100% mortalities ensued within 3 days (Iwamoto *et al.*, 1995). The disease signs mimicked those on naturally infected fish, i.e. erratic swimming and melanosis. Incidentally, the organism failed to infect red sea bream.

Vibrio vulnificus

Intramuscular injection of eels with large numbers of bacterial cells, i.e. 4.85×10^8, resulted in 80% mortality in the population within 7 days, at a water temperature of 25°C. This confirms the pathogenicity, albeit weak, of *V. vulnificus*. The pathogenicity mechanisms, contained in ECP, which are lethal to fish (Lee *et al.*, 1997), included haemolysins, lipases, phospholipases and proteases (Amaro *et al.*, 1997). Interestingly, cultures from diseased European eel produced opaque, translucent colonies, and the cells possessed a capsule, which was not essential for the development of disease (Biosca *et al.*, 1993).

G. *CYTOPHAGA–FLAVOBACTERIUM–FLEXIBACTER* GROUP

Little is known about the pathogenicity of the cytophagas, flavobacteria or flexibacters. Kimura *et al.* (1978a) established that fatal infections by *Fla. branchiophilum* were only induced following water-borne challenge in salmonids weighing $\leqslant 1.1$ g each. Pathogenicity experiments with juvenile rainbow trout demonstrated that the organism occurred abundantly in the gills within 18–24 h after exposure to a dilute bacterial suspension, i.e. 10–20 ml of a 48 h broth culture in 2 l of fresh water. The gill lamellae became very swollen, but the reasons for subsequent fatalities remain unknown (Wakabayashi *et al.*, 1980). Initial attachment of the pathogen to host cells may be by means of pili (Heo *et al.*, 1990). In laboratory-based experiments, black sea bream were more susceptible to *Fle. maritimus* than red sea bream following i.m. injection of 0.02 ml of culture per fish, or infection by bathing or direct application of cultures to the tail or mouth. Mortalities, of up to 10%, occurred in 3 days (Wakabayashi *et al.*, 1984). ECP and subcellular components have been implicated with pathogenesis. Haemolysin (26.5 μg/fish) and ECP (25.5 μg/fish) killed black sea bream. Pathological signs included the presence of ascitic fluid, enlarged spleen, petechial haemorrhages in the visceral fat and intestine, and suppurating (= pus filled) liver. Also, mortalities ensued following i.p. injection of crude LPS with protease (but not when administered separately).

There is a marked variation in the virulence of isolates of *Fla. columnare*. Pacha and Ordal (1963) classified cultures into four grades of virulence, from high to low. Using a highly virulent culture, it was possible to achieve infection of chinook salmon and sockeye salmon following a 2-min dip in a diluted broth culture. Possibly, the organism entered the host through damaged areas of skin. Fujihara *et al.* (1971) achieved 100% mortality in chinook salmon following exposure for 25 min in a suspension containing 2.5×10^5 cells/ml. Deaths followed within 96 h and 8 h at water temperatures of 10 and 22°C, respectively. However, the pathogenic mechanism was unclear.

Fla. johnsoniae has infected barramundi and goldfish via water-borne challenge (Soltani *et al.*, 1994).

Exposure of disinfected eye ova of rainbow trout with an extremely high dose of 10^{10} cells of *Fla. psychrophilum*/ml in PBS for 30 and 60 min at 10°C led to the development of clinical signs of RTFS (Rangdale *et al.*, 1997b).

In fact, the precise nature of the pathogenic mechanism remains to be elucidated. Extracellular proteases have been implicated with *Fla. psychrophilum* (Pacha, 1968) and *Fla. columnare* (Newton *et al.*, 1997). Using cell-free extracts (culture supernatants) of *Fla. columnare*, Pacha (1961) obtained muscle damage following injection with the material. However, the result could not be substantiated by detailed *in vitro* chemical analyses of the supernatant. Later, two proteases of 53 and 58 kDa were isolated from *Fla. columnare* (Newton *et al.*, 1997), and we assume that these are responsible for tissue damage. *Fla. hydatis*, although not a proven fish pathogen, produces extracellular, thermostable, glucose-repressible collagenases, which could be involved in pathogenicity (Strohl, Gibb and Tait, personal communication).

Intraperitoneal injection of cell-free culture supernatants of *Chrys. scophthalmum* resulted in clinical disease in turbot, with swollen oedematous gill lamellae, and haemorrhaging in the stomach, gastro-intestinal tract, kidney and liver (Mudarris and Austin, 1989).

Iron may influence the pathogenicity of *Fla. columnare*, insofar as the presence of 0.35–1.4 mg of iron/100 g fish reduced the survival time following experimental challenge with the pathogen from 20 days to 1 day (Kuo *et al.*, 1981). In contrast, transferrin exerted a negligible effect.

H. PSEUDOMONADACEAE REPRESENTATIVES

Pseudomonas anguilliseptica

Eel, challenged by i.p. injection with 3-day-old broth cultures, eventually displayed the same symptoms attributed to natural outbreaks of the disease. Thus, they became gradually inactive, and developed petechial haemorrhages prior to death, which usually occurred in 6 to 10 days (Wakabayashi and Egusa, 1972). This is a short period for death to ensue. Of course, this depends upon the number of cells in the initial challenge. We consider that this is indicative of the effect of exotoxins, probably exoenzymes, although Dear (1985) could not obtain mortalities following injection of European eels with ECP. Of course, *Ps. anguilliseptica* is not metabolically very active, but most isolates attack proteins (gelatin) and lipids (Tween 80). Therefore, it is suggested that the pathogenicity mechanism involves proteases and lipases. There is no evidence for the presence of an extracellular layer in virulent isolates as occurs in *Aer. salmonicida*. There is some evidence of species-based susceptibility to Sekiten-byo, with Japanese eels seemingly more prone to the disease than European eels.

It is speculative whether or not this infers that the organism may be more common in and around European eels (or may have originated with this species). A comparative observation is that brown trout are more susceptible than rainbow trout to furunculosis caused by *Aer. salmonicida*. Certainly, *Ps. anguilliseptica* is capable of infecting a greater range of species than represented by the genus *Anguilla*. Thus, experimental infections have been achieved in ayu, bluegill, carp, goldfish and loach (Muroga *et al.*, 1975). The organism is only of low pathogenicity to rainbow trout (Lönnström *et al.*, 1994).

The presence of sublethal concentrations of copper (100–250 μg/l) in water exacerbates the disease (Mushiake *et al.*, 1984). Evidence points to a reduction in lymphocytes and granulocytes, which leads to lowered phagocytosis (Mushiake *et al.*, 1985).

Pseudomonas chlororaphis

The isolates from Amago trout and the neotype culture of *Ps. chlororaphis* were pathogenic to carp, eels and trout, following challenge by i.m. injection. Total mortalities occurred within 48 h at a water temperature of 22°C, with disease symptoms paralleling those on the naturally infected fish. However, the pathogenicity mechanism is unknown (Hatai *et al.*, 1975).

Pseudomonas fluorescens

Following invasion of the fish, extracellular proteases are probably responsible for the ensuing damage (Li and Fleming, 1967; Li and Jordan, 1968). Sakai *et al.* (1989a) reported the LD_{50} for rainbow trout as 4.2×10^5 cell at 18°C and 1.1×10^5 cells at 12°C.

Pseudomonas pseudoalcaligenes

Injection of 10^5 cells by i.p. or i.m. injection into rainbow trout (average weight = 12 g), held at 15°C, resulted in total mortalities within 7 days. Moribund fish revealed the presence of haemorrhaging (internal and around the vent) and ascitic fluid in the peritoneal cavity (Austin and Stobie, 1992b).

I. HALOMONADACEAE REPRESENTATIVE

Halomonas cupida
Pathogenicity was confirmed in laboratory-based experiments involved water-borne challenge with 10^3–10^5 cells/ml. Approximately 75% mortalities were recorded within 4 days Kusuda *et al.* (1986).

J. MORAXELLACEAE REPRESENTATIVES

Acinetobacter sp.
Nothing is known about the pathogenic mechanisms of this organism. Due to the comparative inactivity of cultures, it seems unlikely that exotoxins are implicated. This leaves a possible role for endotoxins. Nevertheless, it has been established that pure cultures will reproduce the disease condition. Thus, peptone water cultures, administered by i.m. injection into Atlantic salmon fingerlings, each of 15 g average weight, resulted in total mortalities of the population within 72 h at a water temperature of 12°C (Roald and Hastein, 1980).

Moraxella sp.
Laboratory-based experiments revealed that the organism was pathogenic to rainbow trout and striped bass, with an LD_{50} dose of 10^5–10^6 cells (Baya *et al.*, 1990b).

K. SMALL, MORPHOLOGICALLY SIMPLE BACTERIA

Piscirickettsia salmonis
Both coho salmon and Atlantic salmon were infected (mortalities approached 100%) by i.p. injection of cell lines with the rickettsia (Garcés *et al.*, 1991). The nature of the antigens to which animals respond has been addressed by Kuzuk *et al.* (1996), who used rickettsias purified from CSE cells by differential and Percoll density gradient centrifugation and rabbit antiserum. The conclusion was that the rabbit antiserum reacted with four protein and two carbohydrate (core region of the LPS) surface expressed antigens of 65, 60, 54, 51, 16 and ~11 kDa. However, a complication arises as a result of a subsequent study by Barnes *et al.* (1998), who concluded that the major antigens were of 56, 30 and 20 kDa.

L. MISCELLANEOUS PATHOGENS

Aquaspirillum sp.
Slight lesions were reported to occur after 24 h of infection in experimental fish (Lio-Po *et al.*, 1998). This limited pathology casts doubt on the role of the organism as a fish pathogen. Perhaps synergism with *Aer. hydrophila*, *Pseudomonas* sp. and *Streptococcus*,

may occur. Alternatively, *Aquaspirillum* may be an opportunistic invader or saprophyte living on diseased tissue.

Janthinobacterium lividum
Injection of 5×10^2 cells/fish i.m. and i.p. resulted in 100% mortalities within 14 days. Generally, infected rainbow trout fry and fingerlings became lethargic within 2 days. Moribund and newly dead fish displayed pale (almost white) gills, elongated spleen, pale liver, swollen watery kidney, internal haemorrhaging, pronounced gastro-enteritis, and slight amounts of ascitic fluid in the peritoneal cavity (Austin *et al.*, 1992b).

Unknown Gram-negative rod
According to Palmer *et al.* (1994) infectivity was achieved by injecting cells into Atlantic salmon, which became moribund after 6–10 days, and died within 30 days. Moribund fish displayed haemorrhagic areas in the jaw, cranium and at the base of the fins.

10

Control

It is worth remembering the age-old adage that 'prevention is better than cure', and certainly it is possible to devote more attention to preventing the occurrence of disease in fish. This is especially true for farmed fish, which tend to be at the mercy of all the extremes which their owners are capable of devising. Principally in the industrialized nations, farmed fish are subjected to questionable water quality and high stocking regimes. These are among the known prerequisites for the onset of disease cycles. Yet, the owners are among the first to seek help if anything adverse happens to the valuable stock. Fish may be reared under ideal conditions, in which case, the stock are inevitably in excellent condition without signs of disease. Such sites, for example located in Venezuela and the former Yugoslavia, are usually supplied by fast-flowing, clear river water. Careful feeding regimes are adopted, and the stocking levels are comparatively low. The latter point would make the enterprise unacceptable in the more industrialized nations of western Europe. Therefore, much attention has been devoted to control measures. These have been categorized in Table 10.1. Although most emphasis has been placed on aquaculture, some effort has gone towards considering disease in wild fish stocks.

A. WILD FISH STOCKS

It is questionable what, if anything, can be done to control disease in wild fish stocks. Perhaps, the first step should be to determine the precise extent of disease among wild fish populations. Surveys have been carried out with a view to assessing the incidence of 'abnormalities' in marine fish. Indeed, some workers have attempted to correlate the incidence of disease with pollution (see Chapter 11). It is suggested that attention be focused on archive material, collected at or before the turn of the century. Such material is housed in some museums, e.g. in Liverpool, UK, and access is usually granted to interested individuals. A detailed study would soon demonstrate whether or not 'abnormalities' in fish are a new or old phenomenon. Surely, this information could then be correlated with the changes in pollution of the aquatic environment. It is our contention that dense populations of fish have always maintained a given level of diseased individuals, regardless of whether the populations are shoals in the sea or aquacultural

Table 10.1. Methods of controlling bacterial fish diseases

Classification of fish stocks	Disease control measures
Wild	Control of pollutants (water quality)
Farmed	1. Adequate husbandry practices 2. Use of genetically resistant fish strains 3. Adequate diet or, where appropriate, use of dietary supplements 4. Use of vaccines 5. Use of nonspecific immunostimulants 6. Use of antimicrobial compounds 7. Water treatments 8. Preventing the movement of infected stock 9. Probiotics/biological control

stocks. Therefore, it is possible that reducing pollution will not noticeably alter the health index of wild fish. Nevertheless, by using a circuitous route, it may be possible to ensure that wild fish stocks are not likely to be exposed to pathogens, and therefore be at less risk of disease. Theoretically, this could be achieved by controlling outbreaks of disease in farmed fish and thereby reducing the possibility of pathogens escaping into the environment. It should be emphasized, however, that there is a dearth of information which suggests that disease may be transferred from farmed to wild fish stocks. At worst, there is a perceived problem, and this could easily escalate into adverse propaganda for the aquaculture industry. It is essential that consideration should be urgently given to control measures which will reduce any possible risk of pathogens escaping into the natural environment.

B. FARMED FISH

There are nine basic approaches which need to be adopted in order to control bacterial disease in farmed fish (Table 10.1). These will be explained separately below.

B.1 Husbandry

To reiterate a previous point, it is a common problem that under severe economic pressures, the aquaculturist is tempted to produce the maximum yield of fish in a finite volume of water. Sympathy must be addressed towards the fish farmer, especially in the UK, insofar as the prices of the principal products, i.e. Atlantic salmon and rainbow trout, have not kept pace with inflation. Smaller profit margins (if any!) have resulted, which may be offset only by increased production levels. Quite simply, within the intensive cultivation systems, the fish may be 'stressed' (an abstract term of which there is no clear meaning) beyond the limit commensurate with the production of healthy specimens. This may be complicated by bad management practices, in which aeration and water flow are insufficient, overfeeding occurs, and hygiene declines below the threshold at which disease is more likely to ensue. It may need only one individual to act as a reservoir of infection to the rest of the stock. Unsatisfactory occurrences, which are readily controlled, include:

- the accumulation of organic matter, namely faecal material and uneaten fish food, within the fish holding facilities;
- the presence of dead fish for prolonged periods (bad sanitation);
- the accumulation of a biofouling community, i.e. algae and slime, in the fish tanks;
- the depletion of the oxygen content of the water with a concomitant increase in nitrogen levels, especially as ammonium salts;
- the lack of proper disinfection for items entering the fish holding facilities. Reference is made here to nets, protective footwear, and size grading machinery.

Good basic hygiene (water quality) and farm husbandry practices may successfully alleviate many of the problems attributed to vibriosis. In the case of eels, Kocylowski (1963) recommended transfer to cold, well-aerated water, i.e. to conditions less conducive to the pathogen. Perhaps, it is appropriate to cite the old adage that 'cleanliness is next to godliness'. Control of Sekiten-byo is possible by means of raising the water temperature in the fish holding areas to 26–27°C. By keeping the water at this temperature for 2 weeks, followed by reducing it to approximately 21°C, there was no further outbreak of the disease for 5 months (Wakabayashi and Egusa, 1972). On a parallel theme, as columnaris is most severe at high water temperatures, it has been suggested that control may be exercised by keeping the water as cool as possible. However, it is recognized that, in the fish farm environment, this method of control is likely to be virtually impossible.

B.2 Genetically resistant stock

This is a topic worthy of greater attention, insofar as there are numerous observations which point to the value of genetically resistant strains for reducing the problems of disease. As a word of caution, however, comparative studies need to be carefully controlled so that meaningful results are obtained. In any comparison, the age, size and relative condition of the animals need to be standardized. Nevertheless, there has been prolonged interest in breeding disease-resistant fish. The starting point was the work of Embody and Hayford (1925) who increased resistance in brook trout to furunculosis by selective breeding. Subsequently, Wolf (1954) reported the start of an investigation aimed at developing ulcer disease and furunculosis-resistant strains of brook trout and brown trout. Five years later, Snieszko *et al.* (1959) concluded that disease was, indeed, genetically determined. Ehlinger (1964; 1977) echoed this opinion by determining resistance to furunculosis in the progeny of brook trout. Thereafter, a substantial leap forward in knowledge occurred following a publication by Cipriano (1982), who reported varying degrees of resistance to furunculosis among 11 different strains of rainbow trout, and correlated this with the serum neutralization titre. Cipriano determined that the McConnaughy strain was the most susceptible, with 83% of the animals dying within 14 days of challenge with 1.2×10^9 cells administered in a 1-min bath. The serum neutralization titre was 1:80 against one of the extracellular fractions of *Aer. salmonicida*. In contrast, there was no mortality among the Wytheville strain, which demonstrated a serum neutralization titre of 1:2560. Cipriano (1983) concluded that serum from rainbow trout (which are naturally resistant to furunculosis) could protect passively immunized brook trout from challenge with a virulent culture. In contrast, the administration of serum from susceptible Atlantic salmon was unsuccessful in conferring resistance upon brook trout. The protective effect of rainbow trout serum was believed to be attributed to the neutralization of toxic components produced by the pathogen. Recent unpublished data have pointed to the ability of certain strains of rainbow trout to tolerate the rigours of furunculosis. Genetic variation in susceptibility of Atlantic salmon has been examined in the study of one-year-old fish (Gjedrem and Gjoen, 1995).

Do resistant fish strains exist against other diseases? An affirmative response must be given as work has pointed to some strains of fish which are resistant to BKD (Suzumoto *et al.*, 1977; Winter *et al.*, 1979; Withler and Evelyn, 1990). Studying comparative resistance to BKD in three juvenile coho salmon and steelhead trout strains (transferrin genotypes AA, AC and CC), it was found that the AA genotype was the most susceptible to BKD, whereas the CC genotype was the most resistant. Withler and Evelyn (1990) found a variation in resistance to BKD in two strains of coho salmon from British Columbia, Canada. In particular, survival was greater and the time to death was longer in juvenile animals from the Kitimat River strain than from the Robertson Creek strain.

There is contradictory evidence about the role of genetically resistant fish strains at conferring resistance to vibriosis. Winter *et al.* (1980) determined that there was no variation in resistance to vibriosis among different transferrin geotypes of coho salmon and steelhead trout. We support this observation, with regards to rainbow trout. However, in contrast, Pratschner (1978) reported the presence of differential resistance to vibriosis between transferrin genotypes of coho salmon. Similarly, Gjedrem and Aulstad (1974) noted significant variation in resistance—also to *V. salmonicida*—among strains of Atlantic salmon.

It is obvious that the breeding of disease-resistant fish may be a valuable addition in the armoury of disease control in aquaculture. However, in fish farming where more than one disease is prevalent, it is not necessarily the case that a fish strain which is resistant to one disease would be similarly tolerant of others. Nevertheless, we consider that disease-resistant strains of fish have potential for areas in which diseases are enzootic. Further effort is clearly required to bring the concept to fruition.

B.3 Adequate diets/dietary supplements

An area of comparatively recent interest is that of dietary influence on fish health. The precise nutritional value of commercial feeds is largely unknown. Could essential nutrient be lacking, or other compounds be present in dangerous excess? The answers are largely unknown, although it has been established that some dietary supplements may be beneficial for maintaining the health of fish. For example, Ketola (1983) highlighted a requirement for arginine and lysine by rainbow trout fry, with fin erosion resulting from a deficiency of lysine. Eya and Lovell (1998) reported the beneficial effect of dietary phosphorus at enhancing resistance of channel catfish to infection by *Edw. ictaluri*. Paterson *et al.* (1981) discussed the importance of nutrition in the manifestation of BKD in Atlantic salmon. These workers noted that infected fish had lower serum levels of vitamin A, zinc and iron than uninfected animals. Subsequent experimentation showed that the level of BKD could be reduced by feeding with high levels of trace elements, notably cobalt, copper, iodine, iron, fluorine and manganese, and reducing the quantity of calcium. In further experiments, Lall *et al.* (1985) concluded that high levels of iodine and fluorine, each dosed at 45 mg/kg of food, reduced the occurrence of natural infections of BKD to 3% and 5%, respectively, as compared to 95% and 38% infection in Atlantic salmon fed with commercial diets. Earlier, Woodall and Laroche (1964) demonstrated a reduction in BKD infections by feeding chinook salmon with high levels of iodine (i.e. 10.1 μg/g). This theme was continued by Bell *et al.* (1984), who investigated the effects of sodium-L-ascorbate, zinc, iron and manganese as dietary supplements on the manifestation of BKD. They noted that survival time was inversely related to dietary ascorbate levels when the food was otherwise low in zinc and manganese.

Vitamins, viz. A, C and E, in diets are of value for controlling infections by a range of bacterial pathogens (Navarre and Halver, 1989; Hardie *et al.*, 1990), including *Edw. tarda* (Durve and Lovell, 1982) and *Y. ruckeri* (Vigneulle and Gérard, 1986). For example, Vigneulle and Gérard (1986) reported that 48 000 IU/kg, 8650 mg/kg and 500 mg/kg of vitamin A, C and E for 5 days respectively enhanced resistance to ERM. Clearly, field trials are warranted.

Incorporated into a purified basal medium (Table 10.2), vitamin C (dosed at 150 mg/kg of food) dramatically increased the resistance of catfish to *Edw. tarda* infections. This was carried out at a water temperature of 23°C, but curiously this observation was not confirmed at higher temperatures, e.g. 33°C, when less resistance to infection was noted (Durve and Lovell, 1982). Control groups of fish held at 23°C and fed with diets devoid of vitamin C, all died within 96 h of infection with 10^3 cells administered via the i.p. route. With only 30 mg of vitamin C/kg of diet, 85% mortality resulted in the recipient fish after challenge. These mortalities were reduced to 60% and 20% following administration of 60 mg and 150 mg of vitamin C/kg of diet, respectively. However, it is relevant to enquire whether or not the effective dose should be higher. Studies with channel catfish have shown than only 30 mg of vitamin C/kg of diet was sufficient to prevent vitamin deficiencies, which manifest themselves as scoliosis and lordosis. Doubling the dose to 60 mg of vitamin C/kg of diet enabled the maximum rate of wound repair in channel catfish (Lim and Lovell, 1978). As might be expected, there is some variation in the precise levels of vitamin C needed for nutritional requirement among various species of fish. Among the salmonids, rainbow trout needed 100 mg of vitamin C/kg of diet for normal growth, but a tenfold increase enabled maximum wound repair to occur (Halver *et al.*, 1969). The precise effect of vitamin C on retarding bacterial infections remains largely unknown, although several mechanisms, have been postulated. Perhaps, leucocytolytic or phagocytic activity is stimulated, or synthesis and release of glucocorticoids enhanced (Durve and Lovell, 1982). The potential benefit of vitamin B_6 as a nutritional supplement for Atlantic salmon has been examined, albeit without success (Albrektsen *et al.*, 1995). Thus, fish of 14 g weight were fed with diets supplemented with 0–160 mg of vitamin B_6/kg of feed for 20 weeks. However, challenge with *Aer. salmonicida* revealed that increased dietary levels of vitamin B_6 did not increase resistance to furunculosis.

Table 10.2. Composition of the purified basal medium to which different concentrations of vitamin C at 0–150 mg/kg were added[a]

Ingredients	Percentage composition
Carboxymethyl-cellulose	3.0
Cellulose	10.0
Cod liver oil (contains 850 IU of vitamin A and 85 IU of vitamin D/g)	3.0
Dextrin	33.05
Gelatin	9.4
Mineral mix of Williams and Briggs (1963) supplemented with cobalt chloride (1 mg/kg of diet), aluminium potassium sulphate (0.7 mg/kg of diet) and sodium selinite (0.08 mg/kg of diet)	4.0
Soya bean oil	4.0
Vitamin-free casein	32.6
Vitamin mix[b] (minus vitamin C)	0.95

[a] From Durve and Lovell (1982).

[b] This contains thiamin (10 mg/kg of diet), riboflavin (20 mg/kg of diet), pyridoxine (10 mg/kg of diet), folic acid (5 mg/kg of diet), calcium pantothenate (40 mg/kg of diet), choline chloride (3000 mg/kg of diet), niacin (150 mg/kg of diet), vitamin B_{12} (0.6 mg/kg of diet), retinyl acetate (5 000 000 IU/kg of diet), α-tocopherol (50 mg/kg of diet, cholecalciferol (1 000 000 ICU/g) (6 mg/kg of diet), menadione sodium bisulphite (80 mg/kg of diet), inositol (400 mg/kg of diet), biotin (2 mg/kg of diet) and ethoxyquin (200 mg/kg of diet).

Glucans appear to enhance the non-specific resistance to disease, including ERM, Hitra disease and vibriosis, probably by immunostimulation (Robertsen *et al.*, 1990). Moreover, spray-dried, heterotrophically grown preparations of the unicellular alga *Tetraselmis suecica* have been credited with antimicrobial activity and possibly immunostimulatory activity when used as dietary supplements (Austin *et al.*, 1992a). In contrast, the presence of heavy metals, principally copper, has been implicated as an initiating factor of vibriosis in eels (Rødsaether *et al.*, 1977). Heavy metals may be exposed to the fish in diets (such as via fish meal), or by way of environmental pollution. It seems likely that the emergence and spread of some diseases, such as RTFS, may be aided by or result from the use of inadequately prepared or stored diets. It is not difficult to see that a poorly nourished fish would be more prone to disease, but greater research effort is needed if this area is to be better understood.

B.4 Vaccines

The rationale for the development of fish vaccines parallels that of other aspects of veterinary and human medicine, i.e. a Utopian desire to rid fish stocks of disease coupled with a healthy regard for profit. In practical terms, the aquaculturist needs to control specific diseases which may be financially crippling in terms of high mortalities. From the opposite viewpoint, the vaccine manufacturer needs substantial (perhaps even multi-national) markets in order to ensure profitability of the products. A complicating factor concerned cost of the vaccines to the user. Generally, fish farmers who produce fish for human consumption demand inexpensive, easy-to-use, reliable products; whereas the vaccine supplier needs to charge high fees, which are sufficient to recoup developing and licensing costs, pay current expenses and invest for the future. This difference in opinion between user and supplier may lead to difficulty. Moreover, with a comparatively small aquaculture industry, private vaccine manufacturers are likely to invest resources only in developing vaccines against diseases which are prevalent in many countries, rather than those restricted to small geographical areas or representing novel and emerging conditions. This attitude undermines the whole basis of prophylaxis. No easy solution is envisaged unless research costs are supported by public monies or even from the aquaculture industry itself, as already happens in Scotland with the salmon growers.

Historically, the first serious attempt to develop a bacterial fish vaccine may be traced to the work of Duff (1942), who used chloroform-inactivated cells to protect cutthroat trout (*Salmo clarki*) against furunculosis. Since then, vaccines have been formulated against approximately half of the total number of bacterial fish pathogens. From these endeavours, vaccines to protect against edwardsiellosis, ERM, furunculosis, Hitra disease and vibriosis have reached large-scale commercial production. This is hardly encouraging for a primary prophylactic tool. It is noteworthy that the simplistic approach of using formalin-inactivated whole cells, which works well with edwardsiellosis, ERM, Hitra disease and vibriosis, has met with conflicting results with furunculosis. However, more sophisticated approaches, such as involving genetic engineering techniques, offer hope for the future.

Composition of bacterial fish vaccines
The composition of bacterial fish vaccines may be categorized as follows:

* Chemically or heat-inactivated whole cells. These vaccines may be mono- or polyvalent.
* Inactivated soluble cell extracts, i.e. toxoids.
* Cell lysates.
* Attenuated live vaccines, possibly genetically engineered cells. These would be unacceptable to some regulatory authorities because of the perceived risk that the vaccine strain may revert to a pathogenic mode.
* Purified sub-cellular components, e.g. OMP and LPS. These vaccines require a detailed understanding of microbial chemistry, aspects of which are deficient for many of the bacterial fish pathogens.
* Serum, for passive immunization. This is largely of academic interest only, insofar as it is difficult to envisage use of the technique in the fish farm environment. A possible exception is for brood stock or pet fish.
* Mixtures of the components, detailed above.

It is difficult to identify any particular type of preparation which excels in terms of protection. Generally, the simplest approach of using inactivated whole cells has received greatest attention. This technique has been successful with a wide assortment of pathogens, including *Aer. hydrophila*, *Edw. ictaluri*, *Fla. columnare*, *Ph. damselae* subsp. *piscicida*, *V. anguillarum*, *V. ordalii*, *V. salmonicida* and *Y. ruckeri*. Indeed with these pathogens (except *V. anguillarum* and *V. ordalii*), whole cell vaccines gave superior results to other more complex forms of vaccines (Austin, 1984b). However, even the best vaccines do not completely prevent the occurrence of disease, necessitating the use of costly drugs to combat low levels of infection. Clearly, more research is needed, particularly, in determining the precise nature of the protective antibody and of the important antigens. With this information, it may be possible to synthesize the antigens or use genetic engineering techniques to create vaccine strains suitable for inactivation in straightforward ways.

Methods of vaccine inactivation
Attention has focused on seven methods for inactivating bacterial cells for incorporation into fish vaccines (Austin, 1984b). These are the use of chemicals, namely 3% (v/v) chloroform, 0.2–0.5% (v/v) formalin and 0.5–3.0% (v/v) phenol, heat (e.g. 56°C or 100°C for 30 or 60 min), sonication, and lysis with sodium hydroxide at pH 9.5 or with SDS. Commercially, most interest has centred around use of formalin, which has given encouraging results with *Aer. hydrophila*, *Edw. ictaluri*, *Ph. damselae* subsp. *piscicida*, *Ps. anguilliseptica*, *V. anguillarum*, *V. ordalii* and *V. salmonicida*. However, it is unfortunate that only a few studies have been carried out to compare different inactivated preparations.

Methods of administering vaccines to fish
A number of methods of administering vaccines to fish have been tried with varying degrees of success (see Austin, 1984b), and include:

- Injection, with or without the presence of adjuvant, such as FCA/FIA. This technique is slow, and will inevitably require prior anaesthesia of the animals. Injection is only feasible for valuable fish, brood stock or pet fish. Fortunately, mass injection techniques are available.
- Oral uptake, via food. This should be the method of choice insofar as fish could be fed and vaccinated simultaneously. However, there may be problems with the degradation of the vaccine in the gastro-intestinal tract, although this is being overcome by new oralizing compounds.
- Immersion in a solution/suspension of the vaccine. This is quick (i.e. taking 30–120 s to perform) and easy, permitting large numbers of fish to be readily vaccinated. However, there could be problems regarding disposal of the spent vaccine. Thus, it is debatable whether or not disposal should take place in the fish farm effluent.
- Bathing in a very dilute preparation of the vaccine for prolonged periods, i.e. several hours. This is obviously very economic in the use of vaccine. It is feasible that the technique could be carried out during routine periods of confinement, such as during transportation of the stock between sites. However, with immersion, careful thought needs to be given to the question of disposal.
- Spraying or showering the vaccine onto fish. This can be automated, such that fish are vaccinated on conveyor belts during routine grading.
- Hyperosmotic infiltration. This involves a brief immersion (30–60 s) in a strong salt solution, i.e. 3–8% (w/v) sodium chloride, followed by dipping for 30–60 s into the

vaccine. This method is very stressful to fish, and its use has been consequently reduced.

- Anal/oral intubation. In particular, anal intubation offers possibilities for bypassing the deleterious effects of the stomach and intestine. The technique is, however, cumbersome and requires further development.

It is often difficult to determine which is the most effective method of vaccine application. The method of choice often reflects the whims of the user as much as scientific reasoning. The available evidence suggests that oral administration fares least well, although it offers some potential, especially as booster doses, for *Aer. hydrophila, Aer. salmonicida, Fla. columnare, Ph. damselae* subsp. *piscicida, V. anguillarum, V. ordalii* and *Y. ruckeri* vaccines; injection is better than oral uptake in terms of the resultant humoral antibody titre and protection (Austin, 1984b). However, with *Edw. ictaluri, Edw. tarda, V. anguillarum* and *V. ordalii* preparations, immersion has been demonstrated as superior to injection. Similarly with the 'vibrio' vaccines, the shower method exceeds injection in terms of resulting protection. Only with a *Y. ruckeri* product has injection been determined to be better than immersion vaccination (Austin, 1984b). It is apparent that further detailed comparative work is required to emphasize the most appropriate methods for vaccinating fish. However, it must be accepted that vaccines will be the prime prophylactic measure of the future.

Vaccine development programmes

GRAM-POSITIVE BACTERIA—THE 'LACTIC ACID' BACTERIA

- ***Lactococcus garvieae* and the streptococci**
A formalized suspension of β-haemolytic *Streptococcus* was successful when applied to rainbow trout by immersion or by injection with or without FCA (Sakai *et al.*, 1987; 1989d). Seemingly, a RPS of 70% was achieved, which was superior to the results of Iida *et al.* (1982). Yet, only low titres of agglutinating antibody occurred in fish vaccinated by injection. Conversely, antibodies were not detected in trout which were vaccinated by immersion (Sakai *et al.*, 1987; 1989d).

A toxoid enriched whole cell vaccine, administered to turbot by i.p. injection and immersion, gave long-term protection, with RPS of 89–100% and 67–86% recorded for 45 g and 150 g fish, respectively (Toranzo *et al.*, 1995).

Formalin-killed cells of *Str. difficilis* and a culture extract containing 50% protein conjugated to alum, administered intraperitoneally, protected tilapia against challenge with a virulent strain (Eldar *et al.*, 1995c). Protection was correlated by the presence of humoral antibodies. Of relevance for vaccine development, Western blots indicated that only a few proteins were actually protective (Eldar *et al.*, 1995c).

AEROBIC GRAM-POSITIVE RODS AND COCCI

- ***Mycobacterium* spp.**
Although there are no vaccines available against fish pathogenic mycobacteria, it is recognized that there is a cell-mediated response in fish, i.e. rainbow trout (Bartos and

Sommer, 1981). Immunization with *Myc. salmoniphilum* mixed with Freund's adjuvant resulted in delayed hypersensitivity reactions. This would indicate a possibility of eliciting protection in fish against some of the fish pathogenic mycobacteria. Therefore, there is potential for the development of vaccines.

• *Nocardia* spp.
The maxim must be that prevention is better than cure. Vaccines have been investigated, but with limited success.

• *Renibacterium salmoninarum*
There is evidence that under some conditions, renibacteria elicit an immune response in fish (Sanders *et al.*, 1978; Young and Chapman, 1978; Bruno, 1987; Jansson and Ljungberg, 1998). However, it is apparent that early exposure to the p57 antigen can lead to long-term immunosuppression (Brown *et al.*, 1996). Conversely, removal of the p57 antigen from the surface of renibacterial cells has led to enhanced immunogenicity (Wood and Kaattari, 1996). Administration of experimental vaccines prepared in FCA resulted in the development of humoral antibody (Evelyn, 1971c; Baudin-Laurençin *et al.*, 1977). Evelyn (1971c) detected antibodies in immature sockeye salmon at least 16 months after an i.p. injection with a heat-killed suspension in adjuvant. A second injection after 13 months resulted in a sharp increase in antibody titre from 1:2560 (after the first injection) to 1:10 247. The protective ability of vaccines is, however, questionable (Sakai *et al.*, 1989e; 1993). Sakai *et al.* (1993b) compared formalized (RPS = 10–23.8%), heat-killed, pH lysed (RPS = 35–36%) and UV killed (RPS = 25%) cells of *Ren. salmoninarum* and streptococci, and concluded that protection of rainbow trout did not develop. Also, Baudin-Laurençin *et al.* (1977) found no protective effect after injection of coho salmon with cells contained in FCA. Paterson *et al.* (1981), using a similar vaccine in Atlantic salmon, reported high agglutination titres and a reduced incidence of BKD lesions after one year, but FAT revealed the same number of bacteria in both vaccinated and unvaccinated (control) fish (Paterson *et al.*, 1981). Although McCarthy *et al.* (1984) reported optimistically that their vaccine worked in fish, close scrutiny of the data suggests success comparable to that of Paterson *et al.* (1981). McCarthy and co-workers used a number of vaccine formulations without adjuvants, including a formalized (0.3% v/v formaldehyde) suspension of cells grown in KDM2, a lysed cell suspension (this was lysed at pH 9.5 by the addition of 10 N sodium hydroxide for 1 h, after which the pH was re-adjusted to 7.2 with 10 N hydrochloric acid), and 50% concentrates of the vaccine. Juvenile rainbow trout were vaccinated by i.p. injection, hyperosmotic infiltration, and by 2 min immersion. Vaccinated fish were maintained for 6 weeks at 11°C and then challenged by i.p. injection with living cells of the homologous organism. Best success occurred with the lysed preparation, administered by i.p. injection, although failure greeted attempts to vaccinate fish by immersion or hyperosmotic infiltration. When ≥80% of the unvaccinated controls were infected, ≤10% of the vaccinated fish were affected. This seems encouraging until it is realized that the workers measured the presence of infection by the presence of macroscopic lesions and the occurrence of Gram-positive bacteria in the anterior part of the kidney. The occurrence of carriers could not be assessed, because the Gram staining method is not the most

sensitive technique for ascertaining the presence of renibacteria. Therefore, at present, an effective vaccine for the control of BKD is not available.

AEROMONADACEAE REPRESENTATIVES

• *Aeromonas hydrophila*

Some attention has been devoted to developing vaccines, although commercial products are still not available. Simple preparations of inactivated whole cells, which may be administered by immersion, injection or via the oral route, appear to work quite well (Schachte, 1978; Acuigrup, 1980b; Lamers and de Haas, 1983; Ruangpan *et al.*, 1986). In this connection, Schachte (1978) recorded that the most convincing immune response, measured in terms of antibody titre, was achieved after using injection techniques. Using formalized whole cells applied by i.p. injection, Ruangpan *et al.* (1986) recorded complete protection in Nile tilapia within only two weeks. Some protection, i.e. 53–61%, occurred only one week after vaccination. The next most promising method was immersion vaccination, and thence oral methods of administration. Concerning the method of vaccine inactivation, Lamers and de Haas (1983) deduced that heat-inactivated vaccines (60°C/1 h) gave superior results to formalized products (0.3% formalin). However, it was apparent that concentration of the vaccine, in terms of the numbers of cells, was very important in eliciting an immune response. Thus, using carp as the experimental animal, Lamers and de Haas (1983) concluded that 10^7–10^9 cells generated a distinct agglutinating response whereas 10^5 cells did not. Moreover, secondary doses of vaccine were shown to be beneficial. Nevertheless, single doses of a formalin-inactivated vaccine (containing 10^7–10^9 cells), administered via i.m. injection, were capable of eliciting an immune response which was maintained for 360 days. This demonstrates that fish have immunological memory (Lamers *et al.*, 1985a). Continuing the work, Lamers *et al.* (1985b) vaccinated carp by bathing. Although a single immersion did not result in significant serum antibody levels, secondary vaccination after 1, 3 or 8 months gave rise to a dramatic immune response. In particular, the highest response resulted from using booster doses at 3 months. However, at 12 months, there was no response. Subcellular components, particularly LPS, offer promise as components of vaccines. Indeed, evidence has been presented that LPS induces cell-mediated protection (= regulates T-cell like macrophage system) in carp (Baba *et al.*, 1988). Loghothetis and Austin (1996) echoed this view about the immunogenicity of LPS, but also emphasized that rainbow trout responded to exopolysaccharide. In short, there is every possibility that vaccines against *Aer. hydrophila* should work.

• *Aeromonas salmonicida*

The development of an effective vaccine against the rigours of *Aer. salmonicida* infections remains one of the great challenges to researchers. Interest in vaccine development may be traced to the pioneering work of Duff (1942), who produced an orally administered, chloroform-inactivated whole-cell preparation. It is enigmatic that his reported success has not been surpassed, and indeed often not equalled, by subsequent workers. Unfortunately, efforts concerning vaccine development languished as chemotherapy became established as the principal means of disease control. Eventually, however, aforementioned resistance problems with chemotherapeutants led researchers to recognize the need for alternative control measures, and, thus, a resurgence of interest in vaccines ensued. However, it now appears that *Aer. salmonicida* is an inefficient antigen, in terms of its overall capability of stimulating a protective immune response (Tatner, 1989). There is some controversy over the effectiveness of formulations based on ECP. Some studies indicate that they may well be immunosuppressive (Sövényi *et al.*, 1990), whereas others describe their benefit in terms of immunogenicity (Kawahara *et al.*, 1990). Notwithstanding, modern molecular techniques, principally the PCR, have demonstrated that vaccine antigens do get taken up into the body of fish, namely the head kidney and spleen (Høie *et al.*, 1996).

Some of the problems associated with vaccine development have been summarized below. Essentially, the problems reflect economics, i.e. the perceived need for low-cost products on the part of the fish farmer, versus the desire for substantial profit margins on the part of the vaccine manufacturer/suppliers. Scientific problems exist due to an incomplete understanding of the biology of *Aer. salmonicida*. Specifically, progress has been hindered by the uncertainty surrounding the nature of the antigenic components of *Aer. salmonicida*, the effect of strain differences [n.b.: Gudmundsdóttir and Gudmundsdóttir (1997) while examining the cross-protection of vaccines against typical and atypical isolates of *Aer. salmonicida*, concluded that the best protection resulted with autogenous products], and the lack of a consistent and reliable challenge method, although the latter has been improved by the development of effective co-habitation and bath methods (e.g. Bricknell, 1995; Nordmo *et al.*, 1998). Also, there is evidence of temporary immunosuppression following the administration of some vaccines (Inglis *et al.*, 1996). One solution to this problem has been the use of antibiotics, namely amoxycillin dosed at 0.1 ml containing 150 mg/fish, which are administered by injection with the vaccine (Inglis *et al.*, 1996).

The precise composition of the vaccine is of critical importance. To date, scientists have evaluated inactivated whole cells (including those based on IROMP, inactivated L-forms, soluble extracts, attenuated live cells (such as those lacking A-layer and O-antigen; Thornton *et al.*, 1994), inactivated cells supplemented with toxoids and/or purified sub-cellular components, immune serum (for passive immunization) and polyvalent preparations, usually including inactivated whole cells of *Aer. salmonicida* and *Vibrio* spp. (e.g. Hoel *et al.*, 1997). Most of the early formulations yielded poor or equivocal results (Table 10.3). The notable exceptions are passive immunization and the use of attenuated live vaccines (Cipriano and Starliper, 1982; Ellis *et al.*, 1988a,b; Vaughan *et al.*, 1993). The latter was particularly effective in Atlantic salmon, in which experimental use resulted in 12.5% mortalities in the vaccinated group compared to 87.5% mortality among control fish, after challenge with a virulent culture of *Aer. salmonicida*. A live aromatic-dependent *Aer. salmonicida* vaccine, *aroA* was administered intraperitioneally at 2×10^6 to 2×10^9 live bacteria/fish, resulting in a 253-fold increase in LD_{50} (Vaughan *et al.*, 1993). This live vaccine stimulated T-cells rather than B-cell responses in rainbow trout (Marsden *et al.*, 1996a). But how long did this live vaccine remain in fish tissues? The evidence revealed that following i.p. injection, the live vaccine became widely distributed throughout fish (in this case rainbow trout) tissues, with clearance taking 7–9 d at 16°C. Lower temperatures led to more prolonged retention of the bacterial cells within the vaccinated fish (Marsden *et al.*, 1996b).

Table 10.3. Development of vaccines for *A. salmonicida*

A. salmonicida strain used for vaccine and/or challenge	Nature of vaccine	Method of administration	Nature of challenge	Type of fish used	Water temp. (°C)	Ability of vaccine to protect fish	Ability of vaccine to induce antibody response	Reference
Whole cell virulent	chloroform inactivated	oral	immersion/i.p. /cohabitation	cutthroat trout (1–2 years)	19	+	+ (average 1:80)	Duff (1942)
NG	heat activated	oral	i.p.	brook trout	—	–	–	Snieszko and Friddle (1949)
NG	heat activated	oral	natural	as above	—	+	–	as above
ex-brook trout (vaccine)	formalized	i.p.	i.p.	brook and brook trout (0 to 2+ years)	11	–	+ (1:160)	Krantz *et al.* (1964b)
ex-brook trout (vaccine)	formalized + adjuvant	i.p.	i.p.	as above	11	+	+ (1:10 520)	Krantz *et al.* (1964b)
ATCC 14174 + 6 strains from hatcheries	formalized	oral	water-borne	coho salmon (0 to 1+ years)	13	–	–	Spence *et al.* (1965)
as above	formalized + FCA	i.p.	NG	rainbow trout (2 to 4+ years)	12	ND	+ (1:5120)	Spence *et al.* (1965)
AS 67	formalized + FCA	i.p.	ND	coho salmon (1+ years)	13	ND	+ (1:20 480)	Cisar and Fryer (1974)
ex-salmon	formalized + FCA	i.p.	i.p.	coho salmon (juveniles)	12	+	+ (1:40 960)	Paterson and Fryer (1974a)
SS-70 (virulent)	formalized	oral	natural	coho salmon (juveniles)	NG	–	–	Udey and Fryer (1978)

Table 10.3. (*Continued*)

A. salmonicida strain used for vaccine and/or challenge	Nature of vaccine	Method of administration	Nature of challenge	Type of fish used	Water temp. (°C)	Ability of vaccine to protect fish	Ability of vaccine to induce antibody response	Reference
SS-70 (virulent)	formalized + $Al(OH)_3$	oral	natural	coho salmon (juveniles)	NG	–	–	Udey and Fryer (1978)
S-70 (virulent)	formalized + FCA	i.p.	natural	coho salmon (juveniles)	NG	+	+ (NG)	Udey and Fryer (1978)
36/75	formalized	oral	i.m.	rainbow trout	15	–	–	Michel (1979)
36/75	formalized	i.p.	i.m.	rainbow trout	15	–	+ (1:80 000)	Michel (1979)
FD-2-75	formalized + FCA	i.p.	ND	Atlantic salmon (1+ years)	12–15	ND	+ (1:0–1:640)	Weber and Zwicker (1979)
A47R	formalized + FCA	i.p.	natural	Atlantic salmon (1+ years)	ND	–	+ (1:32)	Palmer and Smith (1980)
A47R	as above	h.i.	natural	as above	ND	–	+ (1:16)	as above
A	formalized	oral	natural	brown trout (0+ years)	ambient	+	+ (1:40)	Smith *et al.* (1980)
virulent	formalized	oral	natural	brown trout (0+ years)	ambient	+	+ (NG)	Austin and Rodgers (1981)
virulent	unwashed cells, formalized	i.p.	water-borne	brook trout (0+ years)	12.5	–	+ (1:205)	Cipriano (1982a)

Table 10.3. (*Continued*)

A. *salmonicida* strain used for vaccine and/or challenge	Nature of vaccine	Method of administration	Nature of challenge	Type of fish used	Water temp. (°C)	Ability of vaccine to protect fish	Ability of vaccine to induce antibody response	Reference
virulent	washed cells, formalized	i.p.	water-borne	brook trout (0+ years)	12.5	–	+ (1:14)	Cipriano (1982a)
avirulent	attenuated, live	immersion/dip	water-borne	brook trout, Atlantic salmon	12.5	+	+ (NG)	Cipriano and Starliper (1982)
virulent	formalized + FCA + Al(OH)$_3$	i.p.	immersion, injection, or co-habitation	salmon	10–12	+	ND	McCarthy *et al.* (1983)
avirulent	as above	i.p.	as above	salmon	10–12	–	ND	as above
virulent	formalized	immersion	immersion	chinook and coho salmon	8–18	+	ND	Johnson and Amend (1984)
virulent	formalized	i.p.	i.p.	coho salmon (0+ years)	13–15	+	+ (1:5120)	Olivier *et al.* (1985a)
avirulent	formalized	i.p.	i.p.	as above	13–15	+/–	+ (1:2560)	as above
virulent (MT004; = A-layer⁻)	formalized + FCA or FIA	i.p.	water-borne	Atlantic salmon, rainbow trout	NG	–	ND	Adams *et al.* (1988)
virulent	attenuated live cells of *V. anguillarum*	immersion	i.p.	rainbow trout	18	+	ND	Norqvist *et al.* (1989)

Table 10.3. (*Continued*)

A. salmonicida strain used for vaccine and/or challenge	Nature of vaccine	Method of administration	Nature of challenge	Type of fish used	Water temp. (°C)	Ability of vaccine to protect fish	Ability of vaccine to induce antibody response	Reference
virulent (TG 36-75)	culture supernatant + purified antigen	i.p.	i.m.	rainbow trout	15	–	+ (1:~844)	Michel *et al.* (1990)
MT 423	formalized + mineral oil adjuvant	i.p.	i.p.	Atlantic salmon	9–15	+	+	Inglis *et al.* (1997)
S24-92, V341-95	formalized autogenous + mineral oil adjuvant	i.p.	i.m.	Atlantic salmon fingerlings	10	+	ND	Gudmundsdóttir and Gudmundsdóttir (1997)
Disrupted cells/sub-cellular components								
virulent	water-soluble extract, toxoided with alum	oral	natural	coho salmon (juveniles)	ambient	–	ND	Klontz and Anderson (1970)
virulent (AS-Sil 67 to AS SS 70)	LPS endotoxin	i.p.	ND	coho salmon (juveniles)	7–18	ND	+ (1:10 000)	Paterson and Fryer (1974b)
NG	toxoid	oral/i.p.	natural	coho salmon (juveniles)	ambient	+/–	+ (1:16–1:2048)	Udey and Fryer (1978)

Table 10.3. (*Continued*)

A. salmonicida strain used for vaccine and/or challenge	Nature of vaccine	Method of administration	Nature of challenge	Type of fish used	Water temp. (°C)	Ability of vaccine to protect fish	Ability of vaccine to induce antibody response	Reference
B	disrupted with SDS	h.i.	natural	brown trout	ambient	+/−	+ (1:40)	Smith *et al.* (1980)
B	disrupted with ultra-sonication	h.i.	natural	brown trout	ambient	+/−	ND	Smith *et al.* (1980)
virulent	toxoid, formalized	oral	natural	brown trout (0+ years)	ambient	−	−	Austin and Rodgers (1981)
virulent	ECP, precipi-tated $(NH_4)_2SO_4$ + ethanol	i.p.	water-borne (juveniles)	brook trout	12.5	+	+ (1:122)	Cipriano (1982a)
virulent/aviru-lent	lysed broth cultures + FCA + $Al(OH)_3$	i.p.	immersion	salmon	10–18	+	ND	McCarthy *et al.* (1983)
virulent	toxoid, for-malized and with chloro-form	oral	natural	rainbow trout	ambient	−	ND	Rodgers and Austin (1985)
virulent	ECP, precipi-tated with $(NH_4)_2SO_4$	i.p.	i.p.	coho salmon (juveniles	13–15	+/−	+ (1:1280)	Olivier *et al.* (1985a)
avirulent	as above	i.p.	i.p.	as above	13–15	−	+ (1:1280)	as above

Table 10.3. (*Continued*)

A. salmonicida strain used for vaccine and/or challenge	Nature of vaccine	Method of administration	Nature of challenge	Type of fish used	Water temp. (°C)	Ability of vaccine to protect fish	Ability of vaccine to induce antibody response	Reference
avirulent	protease	i.m.	i.m.	Atlantic salmon (0+ years)	NG	+	NG	Shieh (1985)
virulent (MT004, = A-layer⁻)	ECP, toxoided	immersion	water-borne	Atlantic salmon, rainbow trout	NG	+/−	ND	Adams *et al.* (1988)
265-87, M108-91, S24-92	formalized ECP + FIA	i.p.	i.m.	Atlantic salmon fingerlings	10	+	+ (1:102 400)	Gudmunds-dóttir *et al.* (1997)
various	heated or formalized + adjuvant	i.p.	immersion (MT16)	rainbow trout	13	NG	NG	Lutwyche *et al.* (1995)
mixed	polyvalent formalized + adjuvant	i.p.	waterborne	Atlantic salmon	7–9	+	+	Hoel *et al.* (1997)
VI 88/09/03175	formalized + adjuvant	i.p.	co-habitation	Atlantic salmon	11	+	+ (1:20)	Hoel *et al.* (1998)
avirulent	formalized	i.p.	immersion (MT 26)	rainbow trout	13	+	+ (1:128)	Thornton *et al.* (1994)
virulent	monovalent + trivalent + adjuvant	i.p., oral + immersion	co-habitation	Atlantic salmon pre-smolts	10	+	NG	Midtlyng *et al.* (1996)

Table 10.3. (*Continued*)

A. salmonicida strain used for vaccine and/or challenge	Nature of vaccine	Method of administration	Nature of challenge	Type of fish used	Water temp. (°C)	Ability of vaccine to protect fish	Ability of vaccine to induce antibody response	Reference
NG	formalized + mineral oil adjuvant	i.p.	natural	Atlantic salmon pre-smolts	6–9	+	NG	Midtlyng *et al.* (1996)
virulent (Linne, LL, S24, 256/91) avirulent (AS 14)	formalized L-forms	immersion	i.p.	Atlantic salmon + rainbow trout	NG	+ (varying)	+ (0–1:256)	McIntosh and Austin (1993)
Compound vaccines/approaches								
avirulent	whole cells chloroform inactivated + supernatant	immersion	natural	brown trout	NG	+	ND	Cipriano (1983)
virulent	polyvalent whole cells, formalized + toxoid, formalin + chloroform inactivated	oral	natural	rainbow trout	ambient	–	ND	Rodgers and Austin (1985)

Table 10.3. (*Continued*)

A. salmonicida strain used for vaccine and/or challenge	Nature of vaccine	Method of administration	Nature of challenge	Type of fish used	Water temp. (°C)	Ability of vaccine to protect fish	Ability of vaccine to induce antibody response	Reference
virulent	whole cells, formalized with A-layer + toxoid, formalin, chloroform + lysine inacti-vated	oral	natural	rainbow trout (0+ years)	ambient	+	ND	Rodgers and Austin (1985)
virulent	whole cells, formalized + A-layer	oral	natural	rainbow trout (0+ years)	ambient	+/−	ND	Rodgers and Austin (1985)
virulent, 3SA	whole cells, formalized + toxoid + LPS +/− liposomes	immersion	natural	rainbow trout fry	ambient	+	ND	Rodgers (1990)
derived from 644RB (Brivax II)	formalized or sonicated	i.p.	—	rainbow trout	16	+	+ (1:162 755)	Marsden *et al.* (1996a)
various virulent	cloned aromatic dependent mutant	i.p.	i.m.	rainbow trout, At-lantic salmon	5–14	+	NG	Vaughan *et al.* (1993)
virulent	formalized + therapy with oxolinic acid	i.p.	natural	Atlantic salmon	NG	+	NG	Ford *et al.* (1998)

Table 10.3. (*Continued*)

A. *salmonicida* strain used for vaccine and/or challenge	Nature of vaccine	Method of administration	Nature of challenge	Type of fish used	Water temp. (°C)	Ability of vaccine to protect fish	Ability of vaccine to induce antibody response	Reference
Passive immunization								
virulent	immune serum	i.p.	scarification/ water-borne	coho salmon (0+ years)	NG	ND	NG	Spence *et al.* (1965)
virulent	immune serum —rainbow trout	i.p.	water-borne	brook trout	12.5	+	ND	Cipriano (1983)
virulent	immune serum —Atlantic salmon	i.p.	water-borne	brook trout	12.5	–	ND	Cipriano (1983)
R, virulent strain	immune serum —rainbow trout	i. p.	water-borne	sockeye salmon	10–18	+	+ (1:2048)	McCarthy *et al.* (1983)
R, virulent -boiled	as above	i.p.	water-borne	as above	10–18	–	+ (1:2048)	as above
virulent	as above	i.p.	water-borne	as above	10–18	–	+ (1:512)	as above
virulent	immune serum —rabbit	i.p.	i.p.	coho salmon	NG	+	ND	Olivier *et al.* (1985a)
virulent (MT028, MT048)	immune serum —rabbit	i.p.	i.p.	rainbow trout	11–14	+	ND	Ellis *et al.* (1988a)

NG = not given.
ND = not done.
h.i. = hyperosmotic infiltration.

The novel approach of using IROMP has met with success, but inactivated L-forms have been demonstrated to offer better protection (McIntosh and Austin, 1993). Avirulent cells, with altered A-layer, have also been proposed as candidates for live vaccines (Thornton *et al.*, 1991). However, a complication to the various developmental studies comes from the fascinating work of Norqvist *et al.* (1989), who used live attenuated cells of a different bacterial taxon, namely *V. salmonicida*, and reported their effectiveness at controlling infections by *Aer. salmonicida*.

A detailed study revealed that a 28 kDa outer membrane pore forming protein (= porin) from *Aer. salmonicida* led to the development of protective immunity in rainbow trout (Lutwyche *et al.*, 1995).

The commercial interest in polyvalent vaccines has resulted in several products, which are regularly used in Europe and elsewhere. The benefit of this approach to controlling furunculosis may be illustrated by the observation that *Vibrio* antigens, particularly *V. salmonicida*, appear to enhance the humoral immune response to *Aer. salmonicida* (Hoel *et al.*, 1997). Moreover, vaccination with *V. salmonicida* antigens led to protection against *Aer. salmonicida* following challenge by co-habitation (Hoel *et al.*, 1998a). This approach could well overcome the perceived problem that *Aer. salmonicida* is a weak antigen (Tatner, 1989). Also, this cross-protection may explain the often superior protection afforded by polyvalent vaccines (Hoel *et al.*, 1998a).

Concerning the use of rough and smooth strains for vaccine preparation, discrepancies are apparent among results obtained by different groups of investigators. Michel (1979) reported that there was no difference in the effectiveness of vaccines prepared with either rough or smooth cultures, when administered orally or via i.p. injection to rainbow trout. In fact, neither type of vaccine was protective. Yet, circulating antibodies were present in fish which received the vaccines via injection. Cipriano (1982a), examining the effectiveness of vaccines prepared from virulent and avirulent cultures, determined an equal level of protection from passive immunization of brook trout. Similar agglutinin titres, i.e. 1:512, were found in both groups of vaccinated fish. He concluded, therefore, that protective immunogens were common to both virulent and avirulent cultures. In contrast, McCarthy *et al.* (1983) reported that, in general, only rough variants conferred protective immunity. A parallel result emanated from the work of Olivier *et al.* (1985a), who ascertained that avirulent cells were less effective immunogens that their virulent counterparts. Both of these groups regarded the A-layer protein as the antigen which probably conferred a protective response by the fish. In another development, Hastings and Ellis (1988) recorded that rainbow trout responded to A-protein and LPS O-antigen and some of the components of the ECP (including proteases; Ellis *et al.*, 1988b). So, it is not surprising that Shieh (1985) demonstrated protection with protease fractions.

Researchers should consider the interesting work of Olivier *et al.* (1985b), who noted protection to *Aer. salmonicida* in coho salmon after i.p. injection of formalized cells as well as after injection with FCA. Undoubtedly, the use of adjuvant stimulated non-specific immunity, probably involving macrophage activity. From this work and the later study of Norqvist *et al.* (1989), it is necessary to question the need for incorporation of *Aer. salmonicida* cells or their cellular components into furunculosis vaccines.

Injection techniques appear to be the most efficacious, whereas the oral route is least promising (Midtlyng *et al.*, 1996). The use of adjuvants, especially mineral oil, in injectable vaccines is clearly beneficial (Midtlyng, 1996). Indeed, Midtlyng (1996) determined from a field study in Norway that i.p. administration of furunculosis vaccine in a mineral oil adjuvant gave the best protection in Atlantic salmon. Apart from the obvious benefits of FCA (Olivier *et al.*, 1985b), the use of β-1,3 glucan (Vita-Stim-Taito), lentinan and formalin-killed cells of *Ren. salmoninarum* have enhanced the effectiveness of vaccines based on formalized *Aer. salmonicida* cells (Nikl *et al.*, 1991). Possibly with oral uptake, degradation of the vaccine in the gastro-intestinal tract may occur. In this case, there may be potential for the use of micro-encapsulation techniques to avoid such pitfalls.

Immersion techniques have generated much useful data. Rodgers (1990) reported the benefits of using inactivated whole cells, toxoided ECP and LPS for the protection of juvenile salmonids. Moreover, the vaccinated animals grew better than the controls. Work has indicated that the duration of the immersion vaccination process does not affect the uptake of the vaccine, providing that the antigens are not in low concentrations (Tatner, 1987). Therefore, there appears to be some promise for the widely used immersion vaccination technique with furunculosis vaccines.

Some of the difficulties with ascertaining the efficacy of vaccines have been ascribed to methods of experimental challenge. Indeed, it is not unusual for vaccines to appear to work in laboratory conditions but to fail dismally in field trials. Under such circumstances, it is questionable whether or not meaningful challenge techniques have been used. For example, the precise dosage of cells to be employed remains undetermined. Apparently, there is substantial variation in virulence among strains. In addition, the most effective means of administering the challenges remains to be elucidated. In this respect, Michel (1980) and Cipriano (1982b) suggested standardized methods of challenge. However, the effectiveness of these techniques awaits clarification.

It is readily admitted that much effort has been expended on the development of furunculosis vaccines. Yet, after 40 years the quest continues. Most studies, to date, have measured effectiveness in terms of the humoral antibody response (e.g. Michel *et al.*, 1990). Unfortunately, there is now some doubt as to whether the presence of humoral agglutinins actually correlates with protection. Maybe, it would be preferable to emphasize other aspects of fish immunology, such as cell-mediated immunity, a notion which has been suggested by McCarthy and Roberts (1980).

Ford *et al.* (1998) treated sea run salmon broodstock with oxolinic acid and vaccinated with a formalized whole cell vaccine in an attempt to reduce the impact of furunculosis. Encouraging results were obtained insofar as of 2552 fish captured from the rivers Connecticut and Merrimack and treated in 1986–92, only 362 died of which 65 (18%) were diagnosed with furunculosis. In comparison, 206 fish served as untreated controls, with just over half, i.e. 109, dying, of which 63 (= 58%) had furunculosis.

ENTEROBACTERIACEAE REPRESENTATIVES

• *Edwardsiella ictaluri*

Studies have been carried out to demonstrate the feasibility of developing a vaccine against *Edw. ictaluri* (Plumb, 1984). Fortunately, the organism is highly immunogenic, with agglutination titres of 1:10 000 found in the serum of channel catfish after receiving only single injections of *Edw. ictaluri* cells mixed in Freund's adjuvant (cited in Plumb, 1984). Furthermore, Saeed (1983) and Saeed and Plumb (1986), using a LPS extract, demonstrated protection following i.p. injection. In these experiments, 0.2 mg of LPS injected into channel catfish (individual weight = 60 g) induced agglutination titres of >1:500, which was sufficient to confer ⩾80% survival of the population. This compared with <30% survival of the unvaccinated controls. Antigenicity of the LPS extract was enhanced in FCA, with protection conferred by single or multiple injections. Similarly, it should be emphasized that an inactivated whole-cell vaccine administered with FCA also showed promise. Immersion and oral boosting of channel catfish fry with commercial vaccine was successful at controlling mortalities (Plumb and Vinitnantharat, 1993). With this study, mortalities were as follows:

controls	96.7% mortalities
immersion vaccinated group	6.7% mortalities
immersion vaccinated + oral boosted group	3.3% mortalities

In a comparison of techniques (immersion, immersion and oral in combination, and injection), injection led to the highest antibody titre after 10 weeks in channel catfish fry (Thune *et al.*, 1997). However, a booster by immersion after 25 weeks gave a significant increase in titre. A modern approach reflected the development of an adenine-auxotrophic strain, the virulence of which was attenuated (Laurence *et al.*, 1997). Following injection, fish were protected against challenge with a virulent culture.

• *Edwardsiella tarda*

Prophylaxis by vaccination has shown some signs of success (e.g. Gutierrez and Miyazaki, 1994). Eels responded to the administration of heat- or formalin-killed cells (preferably by injection) by producing humoral antibodies with titres of up to 1:4096 (Song and Kou, 1981). Song *et al.* (1982) vaccinated eels by immersion for periods of 20 s to 3 min in suspensions containing 10^5 to 10^8 bacterial cells/ml. Following the i.m. injection of 10 mg amounts of formalized cells/100 g of Japanese eels and a booster after 7 days, an oral challenge resulted in 60–87.5% mortalities in the vaccinates, compared to 80–100% deaths among controls (Gutierrez and Miyazaki, 1994). Rather better protection resulted from use of 1 mg of LPS/100 g body weight of Japanese eels. With this vaccine, 40–57% mortalities were recorded compared to 80–90% of the controls (Gutierrez and Miyazaki, 1994). Unfortunately, Mekuchi *et al.* (1995b) did not find any clear sign of protection in Japanese flounder, that had been vaccinated with formalized cells via the i.m., oral or immersion routes. Other work has doubted the importance of cell-mediated immunity in the protection of fish against disease (Miyazaki and Egusa, 1976). This is perhaps surprising in view of the current opinions concerning fish immunology.

Intramuscular injection of eels and red sea bream with LPS resulted in protection from challenge with a virulent culture of *Edw. tarda* (Salati *et al.*, 1987a,b). Moreover, there was a demonstrable humoral immune response (titre = 1:2048) and phagocytosis by T-lymphocytes. Phagocytic activity in the eels peaked three weeks after vaccination (Kusuda and Taira, 1990). Indeed, the evidence showed that LPS was much more successful as an immunogen than vaccination with a formalized culture (Salati *et al.*, 1987a,b).

• *Serratia liquefaciens*
Whole cell formalized vaccines and toxoid preparations were effective for prophylaxis in laboratory-based experiments with Atlantic salmon (McIntosh and Austin, 1990).

• *Yersinia ruckeri*
The development of vaccines is a story of complete success, with commercially available products now being marketed for use in aquaculture. Ironically, it is still unknown how these vaccines are taken up by the fish, or for that matter, the exact mechanism of their action.

The initial attempts to produce vaccines for ERM may be traced back to the work of Ross and Klontz (1965). These workers used a phenol-inactivated vaccine, which was administered orally, via the food. Success was most encouraging, insofar as 90% of the vaccinated fish survived subsequent infection with *Y. ruckeri*. A comparison of different methods of bacterial inactivation convinced Anderson and Ross (1972) that 3% chloroform was better than sonication, 1% formalin, or 0.5% or 3% phenol. Passive immunization confirmed that the fish produced humoral antibodies to chloroform inactivated cells (Busch, 1978). Commercial interest in *Y. ruckeri* vaccines grew with the involvement of the now defunct Tavolek Company. Amend *et al.* (1983), examining factors affecting the potency of preparations, reported that potency was not affected by pH values of 6.5 to 7.7, or by cultivation for up to 96 h in TSB at room temperature. This team concluded that inactivation, whether by formalin or chloroform, did not matter. However, there was good evidence that protection was enhanced by culturing the cells for 48 h at pH 7.2, lysing them at pH 9.8 for 1–2 hours, and then adding 0.3% (w/v) formalin.

Application of these vaccine formulations may be by the oral route, i.e. on food (Klontz, 1963; Ross and Klontz, 1965; Anderson and Nelson, 1974), by injection (Anderson and Nelson, 1974; Cossarini-Dunier, 1986), by immersion, shower or spray (Johnson and Amend, 1983a,b), or by anal intubation (Johnson and Amend, 1983b). Problems have been recorded with the oral method, insofar as protection is short-lived. Thus in a comparison of injection and oral methods of uptake with a chloroform inactivated vaccine, Anderson and Nelson (1974) did not find any antibody in fish fed with a vaccine for 7 days; whereas a low titre of 1:16 and 1:32 resulted in trout which were injected. Moreover, injected fish were protected for 12 weeks compared to only 6 weeks in the group which received the oral preparation. Similarly, a comparison of injection, immersion, shower and spray methods showed that injection offered the best protection against artificial challenge with *Y. ruckeri* (Johnson and Amend, 1983a). In a comparison of the efficacy of injection, oral uptake and anal intubation, Vigneulle (1990) favoured the first-mention in terms of protection. Interestingly, antibodies were found in the serum of rainbow trout vaccinated by injection and anal intubation, but not by the oral route. Unfortunately, it must be emphasized that injection is only feasible for large and/or valuable fish and not for the millions of fry/fingerlings which abound on the typical fish farm. However, ERM vaccines have been used successfully on fish farms when administered by bathing (Tebbit *et al.*, 1981). In one investigation, 22 959 239 rainbow trout were vaccinated by a 90-s dip in a commercial vaccine preparation. The results were very encouraging with significantly reduced losses attributable to ERM. In this particular trial, the vaccine effected an 84% reduction in mortalities due to ERM (Tebbit *et al.*, 1981). Of additional benefit, there was a concomitant decrease in the use of medication by 77%, and an increase in food conversion of 13.7%. Analogous findings have been reported by Amend and Eschenour (1980) and Newman and Majnarich (1982).

Addressing the question of the interval necessary for the onset and duration of immunity to develop, Johnson *et al.* (1982a) reported that 5-s immersion in a vaccine suspension was sufficient to induce protection within 5 days at 18°C, or 10 days at 10°C. The minimum size of salmonids necessary for maximal protection was estimated to be in the range of 1.0–2.5 g. In fact, these authors concluded that protection was correlated with size of the fish, and not their age. Thus, with 1.0, 2.0 and 4.0 g fish, immunity lasted for approximately 4, 6 and 12 months, respectively (Johnson *et al.*, 1983b). There was some variation in results between species, with coho salmon and sockeye salmon retaining immunity for longer than pink salmon. Lamers and Muiswinkel (1984) concluded that a secondary immune response occurred as long as 7 months after primary contact with the antigen, indicating the presence of a fairly long-lived memory. Cossarini-Dunier (1986) found protection lasted for 445 days after intraperitoneal injection of a formalized culture which was suspended in saline or oil adjuvant. Thus, after challenge, 88.5% of the controls died but only a few vaccinates.

CYTOPHAGA–FLAVOBACTERIUM–FLEXIBACTER GROUP

Vaccines have been developed only for *Fla. columnare*, although a commercial product is not available. Fujihara and Nakatani (1971) experimented with heat-killed cells, which were administered via food to juvenile coho salmon. The fish responded with the production of antibody (titre = 1:5120). Schachte and Mora (1973) concurred with the general view by demonstrating agglutinating antibody in channel catfish. Survivors of infection experiments resisted re-infection, suggesting the presence of a protective immune response (Fujihara *et al.*, 1971). Formalized cells of *Fla. columnare* were administered to eel by immersion and injection, resulting in an immune response (in the skin) two weeks later, and survival of 60% and 20%, respectively (Mano *et al.*, 1996). Interestingly, 14 days following vaccination by immersion and injection of eel with formalin-killed cells of *Fla. columnare* agglutinating antibody could not be detected in the serum or mucus. Instead, there was inhibition of bacterial adhesion in the skin of the immersion vaccinated eel (Mano *et al.*, 1996).

PSEUDOMONADACEAE REPRESENTATIVES

• *Pseudomonas anguilliseptica*
Attempts have been made to develop formalin-inactivated vaccines. It is encouraging that fish are capable of eliciting an immune response against *Ps. anguilliseptica*, insofar as experimentally vaccinated eels developed agglutinating antibody within 2 weeks at water temperatures of between 15 and 28°C. The maximum titre recorded was 1:256, which was reached during the 7-week period that an immune response could be detected. Although injection in Freund's adjuvant produced the highest immune response in terms of production of agglutinating antibody (titre = 1:4096), all the commonly used vaccination techniques protected the recipient fish against experimental challenge with

virulent cells (Nakai and Muroga, 1979). From field trials with batches of eels, each comprising 2000 animals, Nakai *et al.* (1982) confirmed the efficacy of injectable heat-killed vaccine.

VIBRIONACEAE REPRESENTATIVES

• *Photobacterium damselae* subsp. *piscicida*

Much effort has been expended on vaccine development, with recent research highlighting major antigenic proteins of 7 kDa and 45 kDa (Hirono *et al.*, 1997b). Programmes have included the use of passive immunization (Fukuda and Kusuda, 1981a), which is of dubious practical value, the more conventional approach of using formalin-inactivated whole-cell preparations (Kusuda and Fukuda, 1980; Fukuda and Kusuda, 1981b) and the more modern approaches of genetic engineering. It was demonstrated that administration of a formalin-inactivated preparation in Freund's complete adjuvant by i.p. injection induced agglutinating antibodies in yellowtail. Thus titres of 1:256–1:2048 were achieved 5 weeks after vaccination (Kusuda and Fukuda, 1980). Further work, using a variety of vaccines and application methods, demonstrated conclusively that fish could be protected against subsequent infection by *Ph. damselae* subsp. *piscicida*, although this has been refuted by some workers (e.g. Hamaguchi and Kusuda, 1989). Toxoid enriched whole cells applied by immersion led to a low antibody response and an RPS of 37–41% in sea bream (Magariños *et al.*, 1994c). An improved RPS of >60% after 35 days resulted from use of an LPS mixed chloroform-killed whole-cell vaccine (Kawakami *et al.*, 1997). Using formalin-inactivated cells with or without FCA and a range of application methods, namely i.p. injection, 5–7-s spray, hyperosmotic infiltration and oral uptake via food, Fukuda and Kusuda (1981b) reported encouraging results within 21 days following artificial challenge with *Ph. damselae* subsp. *piscicida*. The best results, conferring 100% protection to the fish, were obtained by use of i.p. injection or by spraying. The titre of agglutinating antibodies was measured at between 1:4 and 1:128. A subsequent study by these authors has pointed to the value of vaccinating with sub-cellular components, notably bacterial LPS (Fukuda and Kusuda, 1982).

A ribosomal vaccine has been evaluated following administration by i.p. injection into yellowtail. Certainly, the initial evidence pointed to success with ribosomal antigen P (Kusuda *et al.*, 1988a; Ninomiya *et al.*, 1989). In a further development, this group experimented with a potassium thiocyanate extract and acetic acid treated 'naked cells' obtained from a virulent culture (Muraoka *et al.*, 1991). Yellowtail were vaccinated twice i.p., at one-week intervals with the extract—with or without the naked cells—and were challenged two weeks after the second injection. Results indicated partial success for the extract when used alone. However, the extract used in conjunction with naked cells led to good protection (RPS = 36.5%). Yet, the corresponding antibody levels were low, suggesting to the researchers that humoral antibodies did not play an important role in protection (Muraoka *et al.*, 1991).

• *Shewanella putrefaciens*

A formalin-killed suspension showed promise at controlling mortalities when applied (twice) by i.p. injection (Saeed *et al.*, 1987). Thus, two injections resulted in 40% less mortalities than the unvaccinated controls. Vaccination by immersion was unsuccessful.

- *Vibrio anguillarum*

V. anguillarum has been one of the few successful candidates for vaccine development. Commercial formalin-inactivated vaccines are available, which have gained widespread use in mariculture Ironically, the reasons for the success of these products are largely obscure. Moreover, there is some evidence to suggest that vaccinated fish generally fair much better, i.e. they exhibit better all-round health and growth characteristics, than the unvaccinated counterparts. Moreover, immunostimulants, e.g. levamisole, further enhance protection (Kajita *et al.*, 1990).

The immunogenicity appears to be a reflection on the presence of heat-stable (to 100–121°C) LPS in the cell wall (Salati *et al.*, 1989b; Kawai and Kusuda, 1995), which may be released in the culture supernatant (Chart and Trust, 1984; Evelyn, 1984), and OMP (Boesen *et al.*, 1997). It has been postulated that a probable mechanism of protection concerns the inhibition of bacterial attachment by unknown factors in the skin mucus (Kawai and Kusuda, 1995). That supernatants are also among the most immunogenic parts of *V. anguillarum* vaccines was verified following the anal uptake of different vaccine fractions in carp and rainbow trout (Joosten *et al.*, 1996). The large molecular weight LPS, i.e. 100 kDa (Evelyn and Ketcheson, 1980), are considered to confer protection to the recipient host. Moreover, the compounds are able to withstand severe extraction methods. Also, Chart and Trust (1984) isolated, from the outer membrane, two minor proteins with molecular weights of 49–51 kDa, which were potent antigens. A weakly antigenic protein, with a molecular weight of ~40 kDa was also present. Perhaps, these are heat-labile, and explain the reasons for the greater protection achieved with formalin-inactivated vaccines compared to heat-killed products (Kusuda *et al.*, 1978c; Itami and Kusuda, 1980). The potential of LPS as an immunogen was clearly demonstrated by Salati *et al.* (1989b). These workers injected i.p. crude LPS (0.05–0.5 mg) into ayu. Following challenge, mortalities among the vaccinates and controls were 0% and 86.7%, respectively. Similarly, O-antigen preparations induced an immune response following injection in a wide range of fish species, including ayu, carp, Japanese eel, Japanese flounder, rainbow trout and red sea bream (Nakamura *et al.*, 1990). Development of live attenuated vaccines have been tried, with some success (Norquist *et al.*, 1989; 1994). A field trial with an attenuated live *V. anguillarum* vaccine [VAN1000] involved bathing 10-g rainbow trout in a dose equivalent to 1×10^6 cells/ml for 60 min at 9°C in brackish water. Following a natural challenge, 68% of the unvaccinated controls succumbed compared to 14% of the vaccinates (Norquist *et al.*, 1994). Interestingly, these workers considered that the live vaccine protected against both furunculosis and vibriosis. However, there may be problems with regulatory authorities regarding licensing for fisheries use.

To date, most of the vaccine development programmes have concentrated on bivalent products, containing cells of *V. anguillarum* and *V. ordalii* (e.g. Nakai *et al.*, 1989b). At various times, these have been applied to fish by injection (of dubious practicality for masses of fish), on food (oral administration), by bathing/immersion, by spraying, and by anal and oral intubation. The evidence has shown that oral application, perhaps the most convenient method, fares least successfully. Indeed, comparative vaccine trials have produced a wealth of information. For example, Baudin-Laurençin and Tangtrongpiros

(1980) reported cumulative percentage mortalities among experimental groups of fish, as follows:

unvaccinated controls	33.8%
oral-vaccinated fish	31.7%
immersion-vaccinated group	2.1%
group vaccinated by injection	1.4%

Similar findings, although generally more favourable for orally administered vaccines, were published by Amend and Johnson (1981) and Horne *et al.* (1982). Thus, Amend and Johnson (1981) revealed the following mortalities in vaccinated salmonids:

unvaccinated controls	52%
oral-vaccinated fish	27%
immersion-vaccinated group	4%
spray-vaccinated fish	1%
group vaccinated by injection	0%

This compares with the work of Horne *et al.* (1982), who reported mortalities of:

unvaccinated controls	100%
oral-vaccinated fish	94%
immersion-vaccinated group	53%
group vaccinated by injection	7%

In a detailed examination of the effects of oral administration of formalin-inactivated vaccines in chinook salmon, Fryer *et al.* (1978) noted that maximal protection followed the feeding of 2 mg of dried vaccine/g of food for 15 days at temperatures even as low as 3.9°C. An important corollary was the observation that longer feeding regimes did not result in enhanced protection. This should be considered if prolonged durations of vaccination, via the oral route, are advocated. The reason for the apparently discouraging results with oral vaccination regimes may reflect the breakdown of vaccine inside the digestive tract (Johnson and Amend, 1983). To resolve this problem, Johnson and Amend (1983) incorporated a vaccine into gelatin, and applied it orally and anally in attempts to overcome digestion in the stomach and intestine. Encouraging results were obtained, in which mortalities following challenge were:

unvaccinated controls	97%
vaccine (minus gelatin) applied orally	35%
vaccine (with gelatin) applied orally	69%
vaccine (minus gelatin) applied anally	37%
vaccine (with gelatin) applied anally	7%

Similar encouraging data were published by Dec *et al.* (1990). These workers used a commercial vaccine (produced by Rhône-Merieux), which was administered orally to turbot and sea bass. Following challenge 28 days later, the following mortalities were reported:

oral-vaccinated sea bass	11.3%
unvaccinated sea bream	40.9%
oral-vaccinated turbot	19.2%
unvaccinated turbot	65.4%

Incorporation of vaccine with natural food, i.e. plankton, has shown promise with ayu (Kawai *et al.*, 1989). Thus in one set of experiments, 7.6% of the vaccinates died, compare to 35.8% of the controls.

Bypassing the potential deleterious effects of the stomach and upper regions of the gastro-intestinal tract enables effective vaccination to proceed. This suggests that micro-encapsulation techniques may be important for the development of successful oral vaccines. In this respect, the use of alginate microparticles has given promising results with an orally administered *V. anguillarum* vaccine (Joosten *et al.*, 1997). An interesting point is the implication that the posterior region of the gastro-intestinal tract is involved with the correct functioning of oral vaccines. This region has also been determined to be one of the initial sites of attachment of the pathogen. Therefore, it may be inferred that the best protection stems from methods paralleling those of the natural infection cycle.

In contrast to oral methods, injection has proved to be excellent as a means of vaccinating fish against vibriosis, with the development of high levels of immunity (Antipa, 1976; Antipa and Amend, 1977; Sawyer and Strout, 1977; Harrell, 1978; Evelyn and Ketcheson, 1980). Unfortunately, the technique is slow, and seems feasible only for large and/or valuable fish. Nevertheless, several types of preparations, including heat-killed and formalized vaccines, have been evaluated by injection. In addition, passive immunization (by injection) has demonstrated the transfer of immunity between fish. In one comparison, it was clearly demonstrated that heat-killed preparations were more successful than products treated with formalin, when administered by injection. Reference is made to the work of Antipa (1976), who injected chinook salmon with vaccines and, following challenge with the pathogen, reported cumulative mortalities of:

unvaccinated controls	85.4%
formalized vaccine	37.8%
heat-killed vaccine	22.3%

Sonicated heat-killed vaccines, administered in adjuvant, also stimulate elevated levels of antibody in the skin and mucus (Harrell *et al.*, 1976; Evelyn, 1984). At least these studies indicate the presence of heat-stable antigen, which features significantly in the establishment of protective immunity.

Immersion techniques are most suited for the vaccination of animals in the fish farm environment. Formerly, considerable attention was focused on hyperosmotic infiltration, involving use of a strong salt solution prior to immersion in a vaccine suspension (Croy and Amend, 1977; Aoki and Kitao, 1978; Nakajima and Chikahata, 1979; Antipa *et al.*, 1980; Giorgetti *et al.*, 1981). However, it is now appreciated that the technique is extremely stressful to fish (Busch *et al.*, 1978), and the level of protection achieved is only comparable to the much simpler direct immersion method (Antipa *et al.*, 1980), which is consequently favoured. Indeed, many articles have been published about the benefit of immersion vaccination (Hastein *et al.*, 1980; Song *et al.*, 1980; Amend and

Johnson, 1981; Giorgetti *et al.*, 1981; Horne *et al.*, 1982; Johnson *et al.*, 1982a,b; Kawai and Kusuda, 1995) and the longer, i.e. 2 h, 'bath' technique (Egidius and Andersen, 1979).

A further refinement involves use of low-pressure sprays, which are easy to use, and apparently economic in the quantity of vaccine administered (Gould *et al.*, 1978). The success was illustrated by 0% mortalities in a group of fish spray-vaccinated compared to 80% mortalities among unvaccinated controls after challenge (Gould *et al.*, 1978).

All of the aforementioned methods enable fish to develop an immune response to the pathogen. This aspect has been discussed comprehensively, as regards chinook salmon, by Fryer *et al.* (1972). It is thought that the maximum agglutination titre is in the region of 1:8192, depending on the fish species used (Groberg, 1982). The development of immunity is clearly a function of water temperature, and generally humoral antibodies are formed more rapidly at high rather than low temperatures. For example, in coho salmon, humoral antibodies appeared in 25 days and 10 days at water temperatures of 6°C and 18°C, respectively (Groberg, 1982). The poor relative performance of orally administered vaccines has been partially attributed to an inability of the fish to develop humoral antibodies (Fryer *et al.*, 1978; Gould *et al.*, 1978; Kusuda *et al.*, 1978c; Groberg, 1982). However, the role of these antibodies in protection against disease is unclear.

- *Vibrio ordalii*

The methods discussed for *V. anguillarum* apply. Likewise with *V. anguillarum*, the immunogenicity of LPS has been demonstrated (Velji *et al.*, 1990; 1991; 1992).

- *Vibrio salmonicida*

There has been success with formalized vaccines for the prophylaxis of Hitra disease. Immersion of Atlantic salmon in these vaccines resulted in protection, even after 6 months (Holm and Jørgensen, 1987). It has emerged that *V. salmonicida* vaccines exert adjuvant activities on T-dependent and T-independent antigens in salmonids, namely rainbow trout. Essentially, vaccine preparations enhance antibody responses. Thus, the inclusion of inactivated *V. salmonicida* antigens in vaccine preparations may have an overall beneficial effect on the recipient fish (Hoel *et al.*, 1998b).

At least one vaccine has been commercialized in a polyvalent form.

SMALL, MORPHOLOGICALLY SIMPLE BACTERIA

- *Piscirickettsia salmonis*

Attempts have been made to develop a vaccine. In one study, formalized cells ($10^{6.7}$ $TCID_{50}$/ml) administered i.p. led to the development of good protection in a field trial with coho salmon (Smith *et al.*, 1995).

B.5 Non-specific immunostimulants

A potential success story concerns the use of immunostimulatory compounds in fish. Such compounds, which have inevitably been applied by i.p. injection, include Baypamun, dimerized lysozyme, β-1,3 glucans, killed cells of mycobacteria, laminaran, sulphated laminaran, LPS, oligosaccharides and synthetic peptides (Dalmo and Seljelid, 1995; Yoshida *et al.*, 1995; Ortega *et al.*, 1996; Siwicki *et al.*, 1998). Initially, Olivier *et al.* (1985) observed that administration of killed cells of mycobacteria enhanced resistance in coho salmon to various bacteria. Then, Kitao and Yoshida (1986) found that synthetic peptides could enhance resistance of rainbow trout to *Aer. salmonicida*. The use of bovine lactoferrin, dosed orally at 100 mg/kg for 3 days enhanced the resistance of rainbow trout to subsequent challenge by streptococci and *V. anguillarum* (Sakai *et al.*, 1993a). Administration of Baypamun to rainbow trout led to a reduction in symptoms and mortalities attributed to furunculosis (Ortega *et al.*, 1996). Dimerized lysozyme, which is regarded as less toxic than the monomer, was injected into rainbow trout at a dose of 10 or 100 μg/kg, and stimulated cellular and humoral mechanisms giving protection against furunculosis (Siwicki *et al.*, 1998). One and three injections of lysozyme led to 45% and 25% mortalities following challenge with *Aer. salmonicida*. This compares to 85% mortality among the untreated controls (Siwicki *et al.*, 1998). Some immunomodulatory compounds, e.g. laminaran, accumulate in the kidney and spleen of Atlantic salmon (Dalmo *et al.*, 1995).

The greatest interest has been towards the potential for β-1,3 glucans. Certainly a rapidly growing literature points to the success of glucans in preventing disease (Yano *et al.*, 1989; Raa *et al.*, 1990; Robertson *et al.*, 1990; Nikl *et al.*, 1991; Mitsuyama *et al.*, 1992; Chen and Ainsworth, 1992). For example, Yano *et al.* (1989) published data that showed β-1,3 glucans, when applied by i.p. injection at 2–10 mg/kg of fish, enhanced resistance to infection by *Edw. tarda*. This effect was measured by heightened phagocytic activity. Use of β-1,3-glucan and chitosan for 30-min immersion in 100 μg/ml or as single i.p. injections with 100 μg led to protection in brook trout against *Aer. salmonicida* from 1–3 days after administration (Anderson and Siwicki, 1994). Generally, injection was superior to immersion (Anderson and Siwicki, 1994). Others have found oral administration to be superior to immersion (Nikl *et al.*, 1993). Nikl *et al.* (1991) reported success at preventing infection by *Ren. salmoninarum*. Similarly, Matsuyama *et al.* (1992) used the glucans schizophyllan and scleroglucan to protect against streptococci. Thus, 2–10 mg of glucans/kg of fish when administered by i.p. injection, enhanced resistance of yellowtail to streptococcicosis. In particular, there was an elevation of serum complement and lysozyme, and an increase in phagocytic activity of pronephros cells. Initially, success only appeared to result from injection of the glucans into fish. Yet, claims have now been made that application via food also meets with success (Onarheim, 1992). Also, resistance to streptococccicocis and vibriosis has been enhanced following the oral administration of peptidoglycan from *Bifidobacterium* (Itami *et al.*, 1996) and *Cl. butyricum* (Sakai *et al.*, 1995), respectively.

Peptidoglycan, derived from *Bifidobacterium thermophilus*, was administered in feed (fed at 3% of body weight daily) at 0.2 and 2 mg/kg to rainbow trout of 0.12 g average weight for 56 days (Matsuo and Miyazono, 1993). These doses were the equivalent of 6 or 60 μg of peptidoglycan/kg body weight of fish/day. Sub-groups of the fish were challenged on day 26 and 56 by immersion in *V. anguillarum*, with mortalities monitored over a 21-day period. At the halfway point of the feeding trial, survival following challenge with *V. anguillarum* was markedly higher than among the controls. Yet at day 56, there was not any apparent difference in survival between the experimental groups and controls. So, it would appear that the benefits of this approach were short-lived, and in the long term were not beneficial (Matsuo and Miyazono, 1993).

Vitamin E and iron sulphate, dosed at 2500 mg/kg and 60 mg/kg respectively, have been reported to be beneficial in enhancing the immune response of channel catfish, especially by improved phagocytosis, to *Edw. ictaluri* (Wise *et al.*, 1993; Lim *et al.*, 1996; Sealey *et al.*, 1997). Certainly, this aspect of research looks promising, and it is envisaged that other immunostimulatory compounds will be identified in the future.

B.6 Antimicrobial compounds

Use of antimicrobial compounds in fisheries is a highly emotive issue in which the possibility of tissue residues and the development of bacterial resistance feature prominently in any list of complaints. It is astounding that so many compounds (these have been reviewed by Snieszko, 1978a,b; Herwig, 1979; Austin, 1984a) have found use in aquaculture. The complete list reads like an inventory from any well-equipped pharmacy. Antibiotics, many of which are important in human medicine, appear side by side with compounds used almost exclusively in fisheries. In many instances, the introduction of a compound into fisheries use has followed closely after the initial use in human medicine. Perhaps in retrospect, it is surprising that there has not been any significant furore from the medical profession about, what could be perceived as, misuse of pharmaceutical compounds. Unfortunately, any backlash may come in the foreseeable future; therefore, it is in the interest of aquaculture that antimicrobial compounds should be carefully used. This theme will be discussed later.

The use of antimicrobial compounds in fisheries essentially started with the work of Gutsell (1946), who recognized the potential of sulphonamides for combating furunculosis. Indeed, it may be argued that the effectiveness of sulphonamides led to a temporary decline of interest in vaccine development. In fact, the eventual emergence of antibiotic-resistant strains of fish pathogenic bacteria led to renewed interest in vaccines. However, during the years following the Second World War, sulphonamides appeared to be the mystical saviour of fish farming. Important developments included the work of Rucker *et al.* (1951), who identified sulphadiazine as an effective chemotherapeutant for BKD. This claim was subsequently refuted by Austin (1985). The next substantial improvement with sulphonamides resulted from potentiation, i.e. the use of mixtures of trimethoprim and sulphonamide. These have proved to be extremely useful for the treatment of furunculosis. Indeed, formulations are currently licensed for fisheries use in Great Britain.

Following the introduction of sulphonamides, the range of antimicrobial compounds in aquaculture rapidly expanded to encompass chloramphenicol (Wold, 1955), oxytetracycline (Snieszko and Griffin, 1951), kanamycin (Conroy, 1961), nifurprazine (Shiraki *et al.*, 1970), oxolinic acid (Endo *et al.*, 1973), sodium nifurstyrenate (Kashiwagi *et al.*, 1977), flumequine (Michel *et al.*, 1980) and Baytril (Bragg and Todd, 1988). Unfortunately, detailed comparative studies of the various antimicrobial compounds are rare; consequently it is often difficult to assess the value of one drug (= any medicinal compound; Sykes, 1976) over another. Nevertheless, a pattern has emerged which points to the benefits of quinolines for controlling diseases caused by a wide range of Gram-negative bacteria. Currently, there is extensive use of oxolinic acid and flumequine in Europe. Newer quinolones offer hope for the future, although some, as yet unpublished evidence, points to possible problems with this class of molecules.

Whatever the range of compounds available, their effectiveness is a function of the method of administration to fish (and in the way in which it is carried out). We have listed seven basic approaches to the administration of antimicrobial compounds to fish (Table 10.4). These are the oral route via medicated food and bioencapsulation, bath, dip and flush treatments, injection and topical application. With the oral method, drugs are mixed with food, and then fed to the fish. Usually, the treatment regime leads to the administration of a unit weight of drug to a standard weight of fish per day for a predetermined period. Examples of commonly used antimicrobial compounds have been included in Table 10.5. Fortunately, medicated food appears to be quite stable (McCracken and Fidgeon, 1977). Moreover, this method is advantageous insofar as the quantities of compound fed to the fish are carefully controlled, and if sensible feeding regimes are adopted, only minimal quantities would reach the waterways. Three provisos exist, namely that:

- the fish are capable of feeding,
- the drug is palatable,
- the drug is capable of absorption intact through the gut.

Table 10.4. Methods for application of antimicrobial compounds to fish

Method of application	Comments
Oral route (on food)	need palatable components; minimal risk of environmental pollution
Bioencapsulation	need palatable compounds; minimal risk of environmental pollution
Bath	need for fairly lengthy exposure to compound, which must be soluble or capable of being adequately dispersed; problem of disposal of spent drug
Dip	brief immersion in compound, which must be soluble or capable of being adequately dispersed; problem of disposal of dilute compound
Flush	compound added to fish-holding facility for brief exposure to fish; must be soluble or capable of being adequately dispersed; poses problem of environmental pollution
Injection	feasible for only large and/or valuable fish; usually requires prior anaesthesia; slow; negligible risk of environmental pollution
Topical application	feasible for treatment of ulcers on valuable/pet fish

A more recent approach has involved bioencapsulation, principally of quinolones (Duis *et al.*, 1995). This theme was expanded with some excellent work which examined the potential for *Artemia* nauplii to serve as carriers to sulphamethoxazole and trimethoprim for the chemotherapy of diseased marine fish fry (Touraki *et al.*, 1996). Both these

Table 10.5. Methods of administering commonly used antimicrobial compounds to fish

Antimicrobial compound	Diseases controlled	Method(s) of administration
Acriflavine, neutral	columnaris	5–10 mg/l in water for several hours to several days
Amoxycillin	furunculosis, gill disease	60–80 mg/kg body weight of fish/day for 10 days
Benzalkonium chloride	fin rot, gill disease	1–2 mg/l of water for 1 h, 100 mg/l of water for 2 min
Chloramine B or T	fin rot, gill disease, mycobacteriosis	18–20 mg/l of water at pH 7.5–8, treat for 2–3 days
Chloramphenicol	CE, columnaris, ERM, fin rot, furunculosis, haemorrhagic septicaemia, pasteurellosis, ulcer disease, vibriosis	(a) 50–70 mg/kg of food/ day for 5–10 days (b) 10–50 mg/l of water, as a bath
Difloxacin	furunculosis	5 mg/kg body weight/d for 5 or 10 days
Doxycycline	streptococcicosis	20 mg/kg body weight of fish/day for an unspecified duration
Enrofloxacin (= Baytril)	BKD, furunculosis	10 or 20 mg/kg body weight/day for 10 days
Erythromycin	BKD, streptococcicosis	25–100 mg/kg of fish/day for 4–21 days 20 mg of erythromycin/kg of broodstock as an injec-tion
Florfenicol,	furunculosis	10 mg/kg body weight of fish/day for 10 days
Flumequine	furunculosis, ERM, vibriosis	6 mg/kg of fish/day for 6 days
Formalin	fin/tail rot	2–5 mg/l for 15–60 min
Fosfomycin	pasteurellosis	40 mg/kg body weight of fish/day for 5 days
Furanace	coldwater disease, columnaris, fin rot, gill disease, haemorrhagic septicaemia, vibriosis	(a) 2–4 mg/kg of fish/day for 3–5 days (b) 0.5–1 mg/l of water for 5–10 min, as a bath

Table 10.5. (*Continued*)

Antimicrobial compound	Diseases controlled	Method(s) of administration
Furazolidone	CE, furunculosis, vibriosis	25–75 mg/kg of fish/day for up to 20 days
Iodophors	acinetobacter disease, BKD, flavobacteriosis, furunculosis, haemorrhagic septicaemia, mycobacteriosis	50–200 mg of available iodine/l of water for 10–15 min
Josamycin	streptococcicosis	30 mg/kg body weight of fish/day for 3 days (or dosed at 20 mg/kg body weight of fish/day for 5 days)
Kanamycin	fin rot, haemorrhagic septicaemia, mycobacteriosis, vibriosis	50 mg/kg of fish/day for 7 days
Malachite green	columnaris, fin rot, gill disease	1–3 mg/l of water for up to 1 h
Methylene blue	salmonid blood spot	1–3 mg/l of water for 3–5 days
Nifurprazine hydrochloride	furunculosis, vibriosis	(a) 10 mg/kg of food, fed for 3–6 days (b) 0.01–0.1 mg/l of water, as a bath for an undefined period
Nitrofurantoin	vibriosis	50 mg/l of water, as a bath for 1 h
Oxolinic acid	columnaris, ERM, furunculosis, haemorrhagic septicaemia, vibriosis	(a) 10 mg/kg of fish/day for 10 days (b) 1 mg/l of water, as a bath for 24 h (recommended for columnaris)
Oxytetracycline	acinetobacter disease, CE, coldwater disease, columnaris, edwardsiellosis, emphysematous putrefactive disease, ERM, enteric septicaemia, fin rot, furunculosis, gill disease, haemorrhagic septicaemia,	50–75 mg/kg of fish/day for 10 days (doses of 300 mg/kg of fish/day for indefinite periods are used to treat RTFS)

Table 10.5. (*Continued*)

Antimicrobial compound	Diseases controlled	Method(s) of administration
	redpest, salmonid blood spot, saltwater columnaris, streptococcicosis, ulcer disease, RTFS	
Penicillin G	BKD	75–100 mg/kg of fish/day for 10–21 days
Polymyxin B nonapeptide	furunculosis	8–20 mg/kg of fish/day for 1–2 days
Potentiated sulphonamide	ERM, furunculosis, haemorrhagic septicaemia, vibriosis	30 mg/kg of fish/day for 10 days
Sodium nifurstyrenate	streptococcicosis	50 mg/kg of fish/day for 3-5 days
Streptomycin	haemorrhagic septicaemia, mycobacteriosis	50–75 mg/kg of fish/day for 5–10 days
Sulphonamides (sulphisoxazole, sulphamerazine, sulphamethazine)	BKD, coldwater disease, columnaris, ERM, furunculosis, haemorrhagic septicaemia, mycobacteriosis, nocardiosis, vibriosis	100–200 mg/kg of fish/day for 10–20 days
Tetracycline	CE, columnaris, furunculosis, streptococcicosis	75–100 mg/kg of fish/day for 10–14 days
Tiamulin	ERM	5 mg/kg of fish/day for 14 days

compounds accumulated in the nauplii, with maximal levels recorded after 8 h. In a trial with sea bass larvae challenged with *V. anguillarum*, an improvement in survival followed use of the medicated nauplii (Touraki *et al.*, 1996). Whether or not the fish will feed is largely a function of the nature and severity of the disease. Often in advanced cases of disease the fish will not feed. Therefore, it is vitally important that treatment begins as soon as possible after diagnosis has been established. The aquaculturist will need to seek specialist advice as soon as any abnormal behaviour or unhealthy condition is noted. This means that good management practices need to be routinely adopted.

Palatability of fisheries antimicrobial compounds receives only scant attention. Whereas it is accepted that little can be done to improve the palatability of the active ingredient, effort could be directed towards improving binders and bulking agents, which are commonly contained in proprietary mixes. Perhaps, consideration could be given to using chemical attractants.

Application by the water-borne route becomes necessary if the fish refuse to eat, and, therefore, would be unlikely to consume any medicated food. With these methods, the fish are exposed to solutions/suspensions of the drug for a predetermined period. This may be only briefly, i.e. a few seconds duration ('dip'), or for many minutes to several hours ('bath'). It is essential that the compounds are soluble or, if insoluble, are dispersed evenly in the water by means of surfactants or other dispersants (Austin *et al.*, 1981). Also, seawater cations may well antagonize antimicrobial compounds in seawater (Barnes *et al.*, 1995). One major drawback, however, concerns the disposal of the spent compound. Ideally, it should not be released into the aquatic environment, particularly if there are any abstraction points for potable water supply systems in the vicinity. Neglect of this point could lead to legal repercussions.

Flush treatments, e.g. with malachite green, also involve the addition of drugs, albeit at high concentrations, to the water in stock-holding areas. After addition, the drug is flushed through the system by normal water flow. Flushing inevitably results in only a brief exposure to the inhibitory compound; therefore, quick-acting agents are absolutely necessary. As before, the major problem is adequate disposal of the spent drug.

Injection of drug solutions is feasible for valuable stock, such as brood fish and ornamental/pet fish. However, the technique is slow and will undoubtedly require prior anaesthesia of the animals.

The topical application of antimicrobial compounds is worthy of consideration for valuable and/or pet fish. In the case of ulcers, we recommend that the animal should be gently removed from the water, and the antimicrobial compound (preferably as a powder) applied to the lesion, which is then sealed with a waterproof covering, e.g. with dental paste. The lesions tend to heal quickly, with only limited evidence of scarring.

Whatever the chosen method of application, drugs may be used for prevention, i.e. prophylaxis, or treatment, i.e. chemotherapy, of fish diseases. Certainly, it is comforting to note that there are treatments available for the majority of the bacterial fish pathogens. Providing that drugs are used prudently and correctly, they will continue to offer relief from the rigours of disease for the foreseeable future.

Chemotherapy development programmes

ANAEROBES

• *Eubacterium tarantellae*

Udey *et al.* (1977) reported that isolates were sensitive to chloramphenicol, erythromycin, novobiocin, penicillin and tetracycline when examined by *in vitro* methods. It is possible that one or more of these compounds may be useful for chemoprophylaxis, but the value of any of these antimicrobial agents for chemotherapy is dubious. However, it seems likely that once the pathogen has entered brain tissue, antimicrobial compounds would probably not be able to reach the site of infection.

GRAM-POSITIVE BACTERIA—THE 'LACTIC ACID' BACTERIA

• *Carnobacterium piscicola* **and the lactobacilli**

In vitro methods have shown that isolates are sensitive to ampicillin, cephaloridine, chloramphenicol, furazolidone and tetracycline, but not to erythromycin, novobiocin, streptomycin, sulphamerazine or sulphamethoxazole (Michel *et al.*, 1986b; B. Austin, unpublished data). Unfortunately, these antimicrobial compounds were not evaluated in experiments with fish.

• *Enterococcus*-like bacteria

The benefits of ionophores, namely lasalocid, monensin, narasin and salinomycin, were reported for an *Enterococcus*-like pathogen of rainbow trout in Australia. Following a comparison of 40 isolates, it was noted that the MIC of the ionophores was markedly less than erythromycin (0.1–0.8 μg/ml). Thus, the MIC for lasalocid, monensin, narasin and salinomycin were 0.8 μg/ml, 0.4–1.5 μg/ml, 0.2–0.4 μg/ml and 0.4–0.8 μg/ml, respectively. Perhaps, there are future opportunities for the use of some of these compounds in aquaculture.

• *Lactococcus garvieae* and the streptococci

Both *Str. difficilis* and *Str. shiloi* (= *Str. iniae*) revealed identical antimicrobial susceptibility patterns, with susceptibility to ampicillin, cefuroxime, cephalothin, chloramphenicol, ciprofloxacin, erythromycin, fusidic acid, methicillin, mezlocillin, nitrofurantoin, penicillin, potentiated sulphonamide, tetracycline and vancomycin, but resistance to amikacin, colistin, gentamicin and nalidixic acid. Erythromycin, dosed at 25 mg/kg body weight of fish/day for 4 to 7 days, controlled streptococcicosis in yellowtail (Kitao, 1982b), and worked better than oxytetracycline or ampicillin (Shiomitsu *et al.*, 1980). Doxycycline, at 20 mg/kg body weight of fish/day for an unspecified duration (Nakamura, 1982), and josamycin, dosed at 30 mg/kg body weight of fish/day for 3 days (or dosed at 20 mg/kg body weight of fish/day for 5 days) (Kusuda and Takemaru, 1987; Takemaru and Kusuda, 1988) have also been advocated. It is of particular interest that a novel fisheries therapeutant, namely sodium nifurstyrenate dosed at 50 mg/kg body weight of fish/day for 3–5 days (Kashiwagi *et al.*, 1977a,b), has found use for streptococcicosis. This drug appears to be particularly effective, and should not have problems with plasmid-mediated resistance.

• *Streptococcus iniae*

Success has been reported with the fluoroquinolone compound, enrofloxacin, dosed at 5 or 10 mg/kg body weight of fish/day for 10 days (Stoffregen *et al.*, 1996). Mortalities declined rapidly upon instigating treatment, such that by the end of treatment, a total of 11% and 17% of the treated fish had died compared to 55% and 40% of the untreated controls, respectively.

• *Streptococcus milleri*

The two isolates were sensitive to tetracycline, which might be of value for chemotherapy (Austin and Robertson, 1993).

• *Streptococcus parauberis*

It was reported that isolates were resistant to flumequine, oxolinic acid and streptomycin, moderately susceptible to oxytetracycline, tetracycline and sulphamethoxazole-trimethoprim, and highly sensitive to ampicillin, chloramphenicol, erythromycin, nitrofurantoin and penicillin (Doménech *et al.*, 1996).

AEROBIC GRAM-POSITIVE RODS AND COCCI

- ***Bacillus* sp.**
It was reported that *Bacillus* sp. was sensitive to tetracycline but not penicillin (Oladosu *et al.* (1994).

- ***Bacillus mycoides***
Sensitivity was reported to erythromycin, nalidixic acid, nitrofurazone, novobiocin and oxytetracycline but not to Romet (= orthometoprim-sulphadimethoxine) (Goodwin *et al.*, 1994).

- ***Corynebacterium aquaticum***
The organism was sensitive to ampicillin, erythromycin, oxytetracycline and potentiated sulphonamide, one or more of which may be useful for chemotherapy (Baya *et al.*, 1992b).

- ***Micrococcus luteus***
Sensitivity was recorded to chloramphenical, streptomycin, potentiated sulphonamides and tetracycline. These compounds may be useful at controlling the progress of infections in rainbow trout fry (Austin and Stobie, 1992a).

- ***Mycobacterium* spp.**
Some workers advocate that clinically diseased fish should be destroyed, by incineration or burying in quicklime, because of the necessity for prolonged use of chemotherapeutic agents, and the potential hazard to human health (Dulin, 1979; Van Duijn, 1981). Other scientists have described treatments with chloramine B or T, cycloserine, doxycycline, erythromycin, ethambutol, ethionamide, isoniazid, kanamycin, minocycline, penicillin, rifampicin, streptomycin, sulphonamides and tetracycline. Of course, the most economical treatment is with chloramine B or T at 10 mg/l of tank water for an exposure period of 24 h, after which the water should be changed (Van Duijn, 1981). Erythromycin, rifampicin and streptomycin appear to be highly effective against some isolates (Kawakami and Kusuda, 1989; 1990).

- ***Nocardia* spp.**
As with mycobacteria, it has been argued that infected fish should be destroyed so as to prevent any human health hazard. However, some success may result with chemotherapy, in particular with sulphonamides, e.g. sulphisoxazole at 2 mg/g of food (Van Duijn, 1981). An improvement in the fish becomes apparent within 10 days of the commencement of treatment, but it is advisable to continue chemotherapy for 21 days. This method may be suitable for the treatment of pet fish.

- ***Planococcus* sp.**
The organism was sensitive to carbenicillin, erythromycin, penicillin G and tetracycline, which may be effective for chemotherapy (Austin *et al.*, 1988; Austin and Stobie, 1992a).

- ***Renibacterium salmoninarum***
Chemotherapy offers some promise of success (Bandín *et al.*, 1991b). Although BKD has become regarded as one of the most difficult bacterial fish diseases to treat (Bullock *et al.*, 1975; Fryer and Sanders, 1981), some success at chemotherapy has been reported

with erythromycin (Wolf and Dunbar, 1959), sulphonamides (Rucker *et al.*, 1951), chloramphenicol (Rucker *et al.*, 1953; Wood and Wallis, 1955; Millan, 1977), penicillin (Decew, 1972), clindamycin, kitasamycin and spiramycin (Austin, 1985) and enrofloxacin [Baytril] (Hsu *et al.*, 1994). A MIC of 0.25–0.5 μg of enrofloxacin/ml was calculated (Hsu *et al.*, 1994). Furthermore, some beneficial effects have been indicated from trials using a dose of 20 mg of enrofloxacin/kg body weight/day for 10 days when there was a reduction in mortalities compared to controls. Over two trials, the deaths in the treated groups and the controls were 43% and 72%, and 93% and 100%, respectively (Hsu *et al.*, 1994). In addition, cephradine, lincomycin and rifampicin were found to be effective for prophylaxis of BKD, although they were of no use for therapeutic purposes (Austin, 1985). Undoubtedly, many of the problems with control measures revolve around the intracellular nature of the organism (Young and Chapman, 1978). Quite simply, many of the drugs probably do not reach the actual foci of infection. Nevertheless, experiments with liposomes, which targets drugs to given organs, proved to be disastrous, insofar as BKD was exacerbated (Austin, 1985). Perhaps, the value of micro-encapsulation techniques should be assessed.

The pioneering work with drugs for the control of BKD was undertaken by Rucker *et al.* (1951). They reported a decrease in the level of mortalities following administration of sulphadiazine, via the oral route, at 250 mg of drug/kg body weight of fish/day for 15 days. This was confirmed by Earp *et al.* (1953) and Allison (1958). However, the drug failed to eliminate the pathogen from the fish. Subsequently, Wolf and Dunbar (1959) in a comparison of 34 compounds concluded that erythromycin, dosed at 100 mg of drug/kg body weight of fish/day for 21 days, gave the best result. The value of erythromycin at this concentration was confirmed by Austin (1985), although it was suggested that treatment need only be continued for 10 days. Erythromycin has also been reported to prevent vertical transmission of renibacteria (Evelyn *et al.*, 1986; Brown *et al.*, 1990). Also, an injection of 20 mg of erythromycin/kg of broodstock fish is useful in preventing vertical transmission (Lee and Evelyn, 1994).

• *Staphylococcus epidermidis*
This was not discussed in the original paper, but Wang *et al.* (1996) reported success with erythromycin, dosed at 20 mg/kg body weight of fish/day for 10 days.

AEROMONADACEAE REPRESENTATIVES

• *Aeromonas hydrophila*
Chemotherapy of *Aer. hydrophila* infections corresponds closely to that of *Aer. salmonicida*. For example, the relevance of oxytetracycline has been well documented (Meyer, 1964). Unfortunately, plasmid-mediated resistance by means of 20–30 mDa plasmids is similarly widespread in fish farms, e.g. eel ponds (Aoki, 1988), and may negate the potential benefit of some antimicrobial compounds (Aoki and Egusa, 1971; Toranzo *et al.*, 1983a). It is alarming that R plasmids with common sequence DNA structures have now been found in several unrelated species of fish pathogens, including *Aer. hydrophila*, *Aer. salmonicida* and *Edw. tarda* (Aoki, 1988). Resistance in *Aer. hydrophila* has been recorded to a wide range of antimicrobial compounds, including ampicillin, chloramphenicol, erythromycin, nitrofurantoin, novobiocin, streptomycin, sulphonamides and

tetracycline (Aoki, 1988; De Paola *et al.*, 1988). Indeed, it has been estimated that as many as 38% of the *Aer. hydrophila* isolates from diseased catfish are resistant to oxytetracycline (De Paola *et al.*, 1988). For the future, new antimicrobial compounds, such as enrofloxacin, offer promise. This compound is antibacterial even at low dosages, i.e. with a MIC reported as 0.002 μg/ml (Bragg and Todd, 1988).

- *Aeromonas salmonicida*

Over the past few decades, a variety of inhibitory agents have been applied with varying degrees of success to the treatment of furunculosis. Early studies established that sulphonamides, notably sulphamerazine, were successful in controlling furunculosis when administered orally with food at dose of 22 g of drug/100 kg of fish/day (Gutsell, 1946; Snieszko, 1958a). Among the antibiotics, Snieszko (1958a) showed the usefulness of chloramphenicol and oxytetracycline, when dosed at 5–7 g/100 kg of fish/day. Furazolidone was also briefly mentioned as having promise. More recently, polymyxin B nonapeptide has been found to inhibit *Aer. salmonicida*, probably by disrupting the A-layer (McCashion and Lynch, 1987).

Curiously, the use of some compounds follows a geographical pattern, e.g. in France flumequine (a quinolone) is favoured (Michel *et al.*, 1980), whereas in England and Japan, oxolinic acid has been used extensively (see Endo *et al.*, 1973; Austin *et al.*, 1983b). It is debatable as to whether flumequine or oxolinic acid has been more successful at combating furunculosis (Barnes *et al.*, 1991a). Additionally in England, potentiated sulphonamides have been used widely (McCarthy *et al.*, 1974). Within the UK, four compounds, namely amoxycillin, oxolinic acid, oxytetracycline and potentiated sulphonamide, are currently licensed for fisheries used in the treatment of furunculosis. For other diseases caused by *Aer. salmonicida*, i.e. CE and goldfish ulcer disease, less information is available. It appears, however, that potentiated sulphonamide and oxytetracycline are generally effective against the pathogen, regardless of the disease manifestation and assuming that treatment begins at an early stage in the disease cycle (Gayer *et al.*, 1980). Unfortunately despite its comparatively recent arrival in the armoury of fisheries chemotherapeutants, resistance to amoxycillin has been documented in Scotland (Barnes *et al.*, 1994). For the future, florfenicol, dosed at 10 mg/kg body weight of fish/day for 10 days, offers promise, insofar as it has already been used with some success against furunculosis in Norway (Nordmo *et al.*, 1994; Samuelsen *et al.*, 1998).

Since the initial work with oxolinic acid (Endo *et al.*, 1973; Austin *et al.*, 1983b), it is apparent that substantial quantities have been used in aquaculture within many countries. Consequently, it is hardly surprising that resistant strains of *Aer. salmonicida* have emerged (Hastings and McKay, 1987; Tsoumas *et al.*, 1989; Barnes *et al.*, 1990a). Yet, the widespread usefulness of the compound at controlling furunculosis has prompted a search for other related compounds. The fruits of this research may be illustrated by the apparent success of 4-quinolones/fluoroquinolones at inhibiting the pathogen (Barnes *et al.*, 1990b,c; 1991b; Bowser and House, 1990; Lewin and Hastings, 1990; Martinsen *et al.*, 1991; Inglis and Richards, 1991; Stoffregen *et al.*, 1993; Elston *et al.*, 1995). Thus, enrofloxacin and sarofloxacin have been found to be more effective than oxolinic acid, in terms of MIC at inactivating *Aer. salmonicida*. The effectiveness of enrofloxacin at 10 mg/kg body weight of fish/day for 10 days has been attested by field trials with lake trout (*Salvelinus namaycush*) (Hsu *et al.*, 1995). In addition, enrofloxacin has been effective in controlling atypical *Aer. salmonicida* in tom cod, insofar as a single injection with 5 mg of enrofloxacin/kg of fish stopped furunculosis (Williams *et al.*, 1997). Difloxacin, dosed for 5 or 10 days at 5 mg/kg body weight, and 10 days at 1.25 or 2.5 mg/kg body weight resulted in significantly lowered mortalities compared to controls following i.m. injection with a virulent culture of *Aer. salmonicida*. Of relevance, there was little difference in the results between 5- and 10-day treatment regimes with 5 mg/kg body weight (Elston *et al.*, 1995).

With the realization that *Aer. salmonicida* occurs on the external surfaces, i.e. gills and mucus, attempts have been made at disinfection. Cipriano *et al.* (1996c,d) evaluated chloramine T, dosed at 15 mg/l for 60 min on three consecutive days, but the infection was not controlled. Success occurred with 77 mg of oxytetracycline/kg body weight of fish/kg for 10 days with mortalities stopping within 4 days of starting treatment (Cipriano *et al.*, 1996c).

There can be no dispute that chemotherapeutic agents are (and will continue to be) invaluable for preventing heavy mortalities during outbreaks of furunculosis. Nonetheless, there are substantial reasons for avoiding total reliance upon such compounds. For instance, the development of resistance by the pathogens to some widely used drugs is cause for serious concern (Wood *et al.*, 1986). Thus, it is disquieting that plasmids carrying antibiotic resistance factors (R factors) have been isolated from *Aer. salmonicida* strains (Aoki *et al.*, 1971). Indeed, a strain resistant to sulphathiazole and tetracycline was recovered as early as 1959. Moreover, Snieszko and Bullock (1957) reported the occurrence of cultures which were resistant to sulphonamides, although at the time the mechanism of resistance was not known. Now, it seems that resistance may reflect alterations in the outer membrane of *Aer. salmonicida* (Barnes *et al.*, 1992). Aoki *et al.* (1983) examined 175 isolates, which had been isolated from cultured and wild salmonids in Japan, for susceptibility to a wide range of antimicrobial agents. They noted that 96% of the isolates from cultured fish were resistant to at least one, and up to six, of the drugs, particularly nalidixic acid and nitrofuran derivatives. In addition, transferrable R plasmids, coding for resistance to chloramphenicol, streptomycin and sulphonamides, and non-transferable plasmids conveying resistance against tetracycline, were found in several strains. It was concluded from these results that drug resistant strains of *Aer. salmonicida* had increased in direct proportion to the enhanced use of antimicrobial compounds in fish culture. This was particularly evident in view of the observation that few isolates recovered from wild salmonids exhibited drug resistance. Toranzo *et al.* (1983a,b) characterized the plasmids, determining that the organism frequently possessed more than one plasmid. In particular, five strains possessed six plasmids of varying molecular weights. However, these workers did not observe any correlation between loss of plasmids, or changes in plasmid mobilities, and the loss of resistance to sulphadiazine, in the case of one strain which was studied in detail. Nevertheless, the molecular genetic studies of Mitoma *et al.* (1984) identified gene sequences in R plasmids coding for either chloramphenicol or tetracycline resistance. Hedges *et al.* (1985), using aeromonad isolates obtained from France, Ireland, Japan and the UK, determined that plasmids from *Aer. hydrophila* and *Aer. salmonicida* were similar. These workers reported that some plasmids were transmissible to *Esch. coli,* whereas others were unstable in this recipient organism. The R plasmids of *Aer. salmonicida* were considered to confer upon the pathogen the potential to withstand the onslaught of a wide variety of inhibitory agents, thus diminishing the effectiveness of chemotherapy. It may be hoped that the development of new compounds, particularly synthetic or semi-synthetic molecules, and the strict rotation in the use of currently available drugs may assist with the problem of resistance in the pathogen. Obviously, a constant awareness of the problem must be maintained.

Another aspect of chemotherapy concerns the presence and retention of the compounds in fish tissues. In fact, difficulties of this nature were recognized as early as 1951 with a report by Snieszko and Friddle who expressed concern with tissue levels of sulphamerazine in trout. McCarthy and Roberts (1980) pointed out that in some countries, such as the USA, there existed legislation restricting the number of anti-microbial compounds which may be used on fish destined for human consumption. In many countries, drugs may only be obtained on veterinary prescription. The caveat to the use of chemotherapeutants is that a suitable period of time must lapse following the conclusion of treatment, before the fish may be sold for human consumption. This should allow for the purging from the fish of all traces of the active compound and the metabolites. It is worth remembering the opinions of Snieszko (1958a), who cautioned that drug therapy should only be considered as a stop-gap measure until the sources of infection by *Aer. salmonicida* could be eliminated, or disease-resistant strains of fish introduced.

ENTEROBACTERIACEAE REPRESENTATIVES

• *Citrobacter freundii*
According to Sato *et al.* (1982) and Baya *et al.* (1990a), isolates were resistant to chloramphenicol, potentiated sulphonamide and tetracycline. Indeed, chemotherapy was unsuccessful at controlling the disease in the Japanese aquarium.

• *Edwardsiella ictaluri*
Clinical cases of disease have been greatly reduced, although not completely eliminated, using oxytetracycline, dosed at 2.5 g of drug/45.4 kg body weight of fish/day for 4–5 days (Hawke, 1979). Certainly, isolates have been found to be susceptible to a wide range of inhibitory compounds, including cefaperazone, cinoxacin, kanamycin, moxalactam, neomycin, nitrofurantoin, oxolinic acid, streptomycin, ticarcillin and trimethoprim (Waltman and Shotts, 1986a). Yet, plasmid-mediated resistance to antimicrobial compounds (including tetracycline) has been detected (Waltman *et al.*, 1989). Therefore, problems with chemotherapy may be envisaged in the future.

• *Edwardsiella tarda*
It was reported that isolates were highly susceptible to cinoxacin, nitrofurantoin, oxolinic acid, kanamycin, moxalactam, trimethoprim, piperacillin, potentiated sulphonamide, neomycin, mezlocillin and streptomycin, but not to colistin, cloxallin, clindamycin, bacitracin, erythromycin, lincomycin, methicillin, penicillin G, novobiocin or spectinomycin (Waltman and Shotts, 1986b). Chemotherapy by means of oxytetracycline, dosed at 55 mg of drug/kg body weight of fish/day for 10 days, has been recommended (Meyer and Bullock, 1973). Assuming that infected fish consume the medicated

diet, mortalities apparently dwindle away within 48–72 h of initiating treatment. However, recovery is slow, and the survivors may exhibit scar tissue. A complication with chemotherapy concerns R plasmids which have been demonstrated in cultures of *Edw. tarda* isolated from eels (Aoki *et al.*, 1977). Conceivably, this may cause problems for chemotherapy in the future.

- ### *Enterobacter agglomerans*

The pathogen was sensitive to ampicillin, chloramphenicol, streptomycin and tetracycline but not to novobiocin or penicillin (Hansen *et al.*, 1990). Based on these data, it would be prudent to evaluate tetracycline as a chemotherapeutant on fish farms, if the disease recurs.

- ### *Salmonella choleraesuis* subsp. *arizonae*

Although control regimes were not adopted, the isolate was sensitive to chloramphenicol, fradiomycin, gentamicin, kanamycin, nalidixic acid, oxytetracycline, streptomycin and tetracycline, but resistant to erythromycin, spiramycin and sulphadimethoxine (Kodama *et al.*, 1987).

- ### *Serratia liquefaciens*

There was contradictory evidence regarding the value of chemotherapy with oxytetracycline. Nevertheless, the disease could be controlled with oxolinic acid (McIntosh and Austin, 1990).

- ### *Serratia marcescens*

Isolates were sensitive to flumequine, oxolinic acid and potentiated sulphonamide (Baya *et al.*, 1992c).

- ### *Serratia plymuthica*

Isolates were sensitive to chloramphenicol, flumequine, oxolinic acid, oxytetracycline, potentiated sulphonamide and streptomycin, but not to nitrofurantoin or sulphadiazine (Nieto *et al.*, 1990; Austin nd Stobie, 1992b). Presumably, effective chemotherapy could be achieved with one or more of these compounds.

- ### *Yersinia ruckeri*

Control of clinical cases of ERM may be mediated by means of antimicrobial compounds, including sulphamerazine and oxytetracycline (Rucker, 1966), methylene blue and oxytetracycline (Llewellyn, 1980), potentiated sulphonamides (Bullock *et al.*, 1983), tiamulin (Bosse and Post, 1983) and oxolinic acid (Rodgers and Austin, 1982). Early success was obtained by administering medicated diet containing sulphamerazine (200 mg/kg body weight of fish/day for 3 days; Rucker, 1966; Klontz and Huddleston, 1976), followed by oxytetracycline (50 mg/kg body weight of fish/day for 3 days). A parallel therapy worked against salmonid blood spot (Llewellyn, 1980). This involved treatment with methylene blue dosed at 1 g of dye/kg of food for 5 days), followed by oxytetracycline (66 mg/kg body weight of fish/day for 10 days) and then a repeat dose of methylene blue. Llewellyn (1980) reported that in hatchery conditions, the disease cleared up in 10 to 14 days.

Two groups of workers demonstrated success with potentiated sulphonamides. Bullock *et al.* (1983) described the beneficial effects of a mixture of sulphadimethoxime and ormetroprim, dosed at 50 mg/kg body weight of fish/day for 5 days, whereas Bosse and Post (1983) discussed the usefulness of a combination of sulphadiazine and trimethoprim at the very low dose of 1 mg/kg body weight of fish/day for 14 days. This latter group also emphasized the benefit of using tiamulin at 5 mg/kg body weight of fish/day for 14 days. Finally, on the basis of laboratory experiments, Rodgers and Austin (1982) reported the effectiveness of oxolinic acid (10 mg/kg body weight of fish/day for 10 days). Unfortunately, the presence of R plasmids may reduce the effect of some antibiotics (De Grandis and Stevenson, 1985).

CYTOPHAGA–FLAVOBACTERIUM–FLEXIBACTER GROUP

• *Chryseobacterium scophthalmum*

Furazolidone, used by i.p. injection at 50 mg/kg body weight of fish, or by immersion (50 mg/l for 30 min daily for 10 days) was effective at controlling mortalities caused by the organism (Mudarris and Austin, 1989).

• *Flavobacterium* spp.

Apparently, some level of control was exercised with oxytetracycline (Acuigrup, 1980a). *In vitro* experiments have also pointed to the value of oxytetracycline and other drugs commonly used in aquaculture (Farkas, 1985).

• *Flavobacterium columnare*

Farkas and Oláh (1980) suggested the use of a salt (sodium chloride) bath for controlling infections. We agree with this suggestion, insofar as a 30-s dip in 8% (w/v) sodium chloride cleared up an infection in rainbow trout fingerlings within a few days. Other remedies, which have met with varying degrees of success, include:

— arsenic, cadmium, copper, lead and selenium mixture, dosed at 1–3 μg/l for 1 day (MacFarlane *et al.*, 1986);
— copper sulphate, used at a dilution of 1:2000, for a 1–2-min dip (Snieszko, 1958b);
— malachite green, used at a dilution of 1:15000, for a 1–30-s dip (Snieszko, 1958b);
— pyridylmercuric acetate, used at 2 mg/l for 1 h (Snieszko, 1958b);
— diquat, used at 1–2 mg/l for 30–60 min (McCarthy, 1975c);
— quaternary ammonium compounds, used at 2 mg/l for 1 h (McCarthy, 1975c);
— oxytetracycline, used at 50–100 mg/kg body weight of fish/day for 10 days (Snieszko, 1964; Ferguson, 1977). For external infections, oxytetracycline or chlortetracycline may also be used as a bath, i.e. 26–60 mg/l for 1 h (Snieszko and Hoffman, 1963; Wood, 1968);
— oxolinic acid, used at 10 mg/kg of body weight/day for 10 days (Soltani *et al.*, 1995);
— chloramphenicol, used at 5–10 mg/l, has been suggested for aquarium fish (Snieszko, 1958b);
— sulphadiazine, sulphamerazine or sulphamethiazone, used at 220 mg/kg body weight of fish/day for 10 days (Snieszko, 1954; Wolf and Snieszko, 1963). Sulphamerazine has been used successfully to treat rainbow trout, but not chinook salmon (Johnson and Brice, 1952).

- *Flavobacterium hydatis*

Antibiogrammes confirmed sensitivity to chloramphenicol (30 μg), erythromycin (15 μg), nalidixic acid (30 μg), streptomycin (10 μg), sulphathiazole (1 mg) and tetracycline (30 μg), but not to lincomycin (2 μg), methicillin (5 μg), penicillin (10 IU) or sulphadiazine (300 μg) (Strohl and Tait, 1978). From the uniform sensitivity to nalidixic acid, sulpha drugs and tetracycline, these could be considered for chemotherapeutic use, should the need arise.

- *Flavobacterium johnsoniae*

Acriflavine and oxolinic acid were considered to be effective treatments (Carson *et al.*, 1993).

- *Flavobacterium psychrophilum*

Chemotherapy with oxytetracycline (Winton *et al.*, 1983), sulphonamides (Amend *et al.*, 1965) and furanace (Holt *et al.*, 1975) has been advocated. In particular with infected fry, furanace dosed at 0.5 μg/ml for 1 h on every third day has been useful. From a comparison of 48 isolates from RTFS, sensitivity was recorded to doxycycline, enrofloxacin, florfenicol and sarafloxacin (Rangdale *et al.*, 1997a).

- *Flexibacter maritimus*

From an examination of 75 isolates, it was determined that *Fle. maritimus* was highly susceptible to ampicillin, erythromycin, josamycin, nifurpirinol, penicillin G and sodium nifurstyrenate, moderately sensitive to chloramphenicol, doxycycline, oleandomycin, oxytetracycline, sulphamonomethoxine and thiamphenicol, weakly sensitive to nalidixic acid, oxolinic acid, spiramycin and sulphisoxazole, and resistant to colistin and streptomycin (Baxa *et al.*, 1988b). In *in vivo* experiments, the efficacy of sodium nifurstyrenate was confirmed for chemotherapy. Thus, immersion (0.5 μg/μl for 1 h) or oral administration (30 mg/kg body weight of fish/day for 4 days) reduced mortalities in yellowtail to 60.5%, compared to 100% among the untreated controls (Baxa *et al.*, 1988b). Trimethoprim and amoxycillin (dosed at 80 mg/kg body weight of fish) have been recommended in Australia (Soltani *et al.*, 1995).

- *Sporocytophaga* sp.

Data have indicated that control may be exercised by use of pyridylmercuric acetate, ethylmercuric phosphate, oxytetracycline or chlortetracycline, at 1 mg/l for 1 h (Wood, 1968), or a 2-min dip in copper sulphate (repeated on 3 consecutive days) at 1:2000 (Anderson and Conroy, 1969).

PSEUDOMONADACEAE REPRESENTATIVES

- *Pseudomonas anguilliseptica*

Jo (1978) discussed the usefulness of nalidixic acid, oxolinic acid and piromidic acid for chemotherapy.

- *Pseudomonas fluorescens*

Bath treatments with benzalkonium chloride (1–2 mg/l of water/1 h), furanace (0.5–1 mg/l of water/5–10 min) or malachite green (1–5 mg/l of water/1 h) may help control early clinical cases of disease (Austin, 1984b). In one study, isolates showed susceptibility to kanamycin, nalidixic acid and tetracycline (Sakai *et al.*, 1989a). A second investigation reported widespread susceptibility to gentamicin, kanamycin and neomycin, less to amikacin and oxytetracycline, and total resistance to chloramphenicol, erythromycin, penicillin and potentiated sulphonamide (Markovic *et al.*, 1996).

• *Pseudomonas pseudoalcaligenes*
Antibiogrammes revealed sensitivity to oxytetracycline, oxolinic acid and potentiated sulphonamides (Austin and Stobie, 1992b).

VIBRIONACEAE REPRESENTATIVES

• *Photobacterium damselae* subsp. *piscicida*
Little is known about the value of chemotherapeutants. An *in vitro* study highlighted the value of ampicillin (Kusuda and Inoue, 1976), but field trials were not carried out. In a further study, Kusuda *et al.* (1988) reported marked sensitivity to ampicillin and oxolinic acid, moderate sensitivity to nalidixic acid and sodium nifurstyrenate, but resistance to chloramphenicol, chlortetracycline, oxytetracycline and tetracycline. Again, field evidence was not supplied. Sano *et al.* (1994) noted the value of fosfomycin (MIC = 1.56–3.13 μg/ml) at controlling laboratory infections. An effective dose was 40 mg of fosfomycin/kg body weight of fish/day for 5 days, albeit administered only one hour after infection. This dose reduced mortalities by *Ph. damselae* subsp. *piscicida* to 0% (Sano *et al.*, 1994). R plasmids have been identified among isolates, conferring resistance to chloramphenicol, kanamycin, sulphamonomethoxine and tetracycline (Aoki and Kitao, 1985) and florfenicol (Kim *et al.*, 1993). Although the isolates described by Aoki and Kitao (1985) were confined to one locality in Japan, there is the likelihood that the resistance will spread quickly to other sites.

• *Plesiomonas shigelloides*
A 10-day treatment regime with potentiated sulphonamide (sulphadiazine at 200 mg/kg of body weight of fish/day and trimethoprim at 50 mg/kg body weight of fish/day) was effective in reducing mortality levels (Cruz *et al.*, 1986).

• *Vibrio alginolyticus*
Colorni *et al.* (1981) achieved success with chloramphenicol, dosed at 50 mg of drug/kg body weight of fish/day for an unspecified period, and nitrofurantoin (50 mg/l of water/1 h), both of which alleviated mortalities. However, we would caution against the use of chloramphenicol in fisheries, in view of the report of the Swann Committee (Report, 1969). In essence, chloramphenicol should be restricted to use in human beings.

• *Vibrio anguillarum*
Antimicrobial compounds have proved to be very useful in controlling vibriosis. It is perhaps ironic that emphasis has been placed on using drugs as food additives, because vibriosis is typified by inappetence. Consequently, antimicrobial compounds need to be administered (by food) very early in the disease cycle, if success is to be achieved. Workers have indicated the value of many compounds, including chloramphenicol, furanace, nitrofurazone, oxolinic acid, oxytetracycline and sulphamerazine. As a general comment, we advise upon caution when contemplating the need for pharmaceuticals,

particularly antibiotics, because of the potential risk of resistance which may be attributed to plasmids, i.e. R factors (Aoki, 1988). Aoki *et al.* (1974) reported that 65/68 *V. anguillarum* isolates carried R factors, conveying resistance to chloramphenicol, streptomycin, sulphonamides and tetracycline. Therefore if R factors abound, it is unlikely that the common antibiotics will do much to retard the disease cycle.

- *Vibrio fischeri*

Chemotherapy was ineffective at reducing mortalities (Lamas *et al.*, 1990).

- *Vibrio harveyi*

Kraxberger-Beatty *et al.* (1990) reported success with Prefuran (Argent) dosed at 0.1 mg/l for an unspecified period. Saeed (1995) found success with oxytetracycline as a food additive. Yii *et al.* (1997) determined susceptibility to a wide range of inhibitory compounds, including chloramphenicol, doxycycline, nalidixic acid, oxolinic acid, oxytetracycline and sulphonamide, but not ampicillin or penicillin G.

- *Vibrio pelagius*

Treatment with oxytetracyline was effective at stopping mortalities (Angulo *et al.*, 1992). Also, it was considered that potentiated sulphonamides and flumequine would be successful.

- *Vibrio salmonicida*

Oxolinic acid controls mortalities in Atlantic salmon (B. Austin, unpublished data).

- *Vibrio splendidus*

Susceptibility was recorded to chloramphenicol, flumequine, nitrofurantoin, nifurpirinol, oxolinic acid and potentiated sulphonamide, but not to ampicillin, oxytetracycline or streptomycin. Treatment with oxolinic acid was partially successful. However, the presence of virus undoubtedly complicated the chemotherapeutic regime (Lupiani *et al.*, 1989). In a subsequent study, Angulo *et al.* (1994) reported success with flumequine as a feed additive.

- *Vibrio vulnificus*

Although this aspect has not been addressed, it seems likely that infections will respond to broad-spectrum antimicrobial compounds, such as flumequine, oxolinic acid, oxytetracycline and potentiated sulphonamides (Muroga *et al.*, 1976a,b; Nishibuchi and Muroga, 1977; 1980; Nishibuchi *et al.*, 1979; 1980).

MORAXELLAXEAE REPRESENTATIVES

- *Acinetobacter* **sp.**

Oxytetracycline, at a single dose of 100 mg/fish, appeared to be effective for chemotherapy when administered by intramuscular injection.

- *Moraxella* **sp.**

The pathogen was susceptible to chloramphenicol, nitrofurantoin, oxolinic acid, penicillin and tetracycline (Baya *et al.*, 1990b).

SMALL, MORPHOLOGICALLY SIMPLE BACTERIA

• *Piscirickettsia salmonis*
Sensitivity was recorded to clarithromycin, chloramphenicol, erythromycin, gentamicin, oxytetracycline, sarafloxacin, streptomycin and tetracycline, but not to penicillin or spectinomycin (Cvitanich *et al.*, 1991).

• *Rickettsia*-like organisms
Treatment with oxytetracycline was reported to be successful (Chern and Chao, 1994).

MISCELLANEOUS PATHOGENS

• *Janthinobacterium lividum*
Cultures were sensitive to furazolidone, oxolinic acid, oxytetracyline and potentiated sulphonamides (Austin *et al.*, 1992b). Therefore, it is surmised that one or more of these compounds would be useful in chemotherapy.

• **Unknown Gram-negative rod**
According to Palmer *et al.* (1994), the pathogen was susceptible to amoxycillin and penicillin G, but resistant to cotrimoxazole and oxytetracycline.

B.7 Disinfection/water treatments
Apart from the use of antibiotics and related compounds, the application of other chemicals to water as disinfectants is effective for disease control. Such chemicals include benzalkonium chloride, chloramine B and T, chlorine, formalin, iodophores, malachite green and methylene blue. For example, Sato *et al.* (1982) used chlorine to disinfect water, and thus control infection by *Cit. freundii*.

Another approach is to alter (increase or decrease, according to the season) the temperature of water within fish-holding facilities. We have noticed that this approach appears to be effective at reducing the incidence of diseases such as columnaris. Specifically:

• *Aeromonas salmonicida*
Adequate husbandry practices are necessary, such as maintenance of good water quality, disinfection of fish farm equipment and utensils especially when disease outbreaks occur, and routine disinfection policies for eggs upon arrival at receiving sites (Herman, 1972; McCarthy and Roberts, 1980).

• *Citrobacter freundii*
Outbreaks of the disease abated following adoption of water disinfection (with chlorine) practices (Sato *et al.*, 1982).

• *Clostridium botulinum*
As there is no effective chemotherapy for botulism in fish, disinfection of the ponds has been advocated. Success has resulted from moving the stock to clean areas, draining the contaminated ponds, and removing surface mud and detritus before disinfecting with slaked lime at $1.6 \, \text{kg/m}^2$. The disinfectant should be worked well into the layers at the bottom of the ponds, left for a period of not less than 7 days, and the pond then returned

to use (Huss *et al.*, 1974; Cann and Taylor, 1982). To remove clostridia from the gastro-intestinal tract of trout, it has been recommended that the fish should be starved for 5 days (Wenzel *et al.*, 1971). However, Huss *et al.* (1974) showed that whereas starvation does indeed reduce contamination with *Cl. botulinum*, the overall rate of success depends on other factors, including the nature of the water supply. It should be emphasized that clostridia may be normal inhabitants of the fish intestine.

• *Renibacterium salmoninarum*
Disinfection of egg surfaces has also been utilized to control BKD. Iodophors, at 25–100 mg/l for 5 min, have proved beneficial at reducing transmission of the disease (Amend and Pietsch, 1972; Ross and Smith, 1972; Bullock *et al.*, 1978b), although they will not eliminate the pathogen from inside eggs (Evelyn *et al.*, 1984). The use of erythromycin phosphate, at 1–2 mg/l for 30 min, has been advocated as an additive for water-hardening of eggs (Klontz, 1978). However, it is debatable whether or not it is wise to use antibiotics in this way.

Another approach has been to disinfect the water in fish farms. In particular, a level of only 0.05 mg of free chlorine/l was sufficient to inactivate cells of the pathogen in 18 s (Pascho *et al.*, 1995). With such rapid inactivation, there must surely be a use for the technique in hatcheries.

• *Staphylococcus aureus*
A bath of potassium permanganate (1 ml/l) for 5–10 min, together with treating the pond water with 250 mg/l of lime and 1 mg/l of potassium permanganate ever fourth day, was considered effective at halting mortalities except with advanced cases of the disease (Shah and Tyagi, 1986).

B.8 Preventing the movement of infected stock
Some diseases, e.g. BKD, ERM and furunculosis, are suspected to be spread through the movement of infected stock. Therefore, it is sensible to apply movement restrictions to diseased stock, as a means of disease control. This may prevent the spread of disease to both farmed and wild fish. Of course, the issue of movement restrictions is highly emotive among fish farmers. However, the procedure may be beneficial to the industry when viewed as a whole. The concept of movement restrictions usually involves legislative machinery, of which the Diseases of Fish Act (1983) in Great Britain is a prime example. To work effectively, there is a requirement for both the efficient monitoring of all stock at risk of disease, and the dissemination of the information to all interested parties. However, we believe that in any allegedly democratic society where such measures are adopted, there should be adequate compensation to the fish farmer for loss of revenue.

B.9 Probiotics/biological control
There is increasing evidence that members of the natural aquatic microflora are effective at inhibiting fish pathogens, inevitably by competitive exclusion which may involve the production of antibiotics or low-molecular-weight inhibitors. Dopazo *et al.* (1988) discovered the presence in the marine environment of antibiotic-producing bacteria, which inhibited a range of bacterial fish pathogens, including *Aer. hydrophila*. These

inhibitors produced low-molecular-weight (<10 kDa) anionic, thermolabile antibiotics. Subsequently, Chowdhury and Wakabayashi (1989), Austin and Billaud (1990) and Westerdahl *et al.* (1991) reported the presence of microbial inhibitors of *Fla. columnare*, *Ser. liquefaciens* and *V. anguillarum*. Smith and Davey (1993) identified a fluorescent pseudomonad which antagonized *Aer. salmonicida*. The progress of infection of *Fla. columnare* was mediated by *Aer. hydrophila* and *Cit. freundii* (Chowdhury and Wakabayashi, 1989). Sugita *et al.* (1997) isolated a *Vibrio*, coined strain NM10, from ponyfish (*Leiognathus nuchalis*) intestines, and determined antagonism of *Ph. damselae* subsp. *piscicida* by a heat-labile proteinaceous compound of 5 kDa. Putative *Aeromonas* and *Vibrio*, from halibut, have been found to inhibit the growth of fish pathogenic *Vibrio* (Bergh, 1995). Moreover, an isolate of *V. alginolyticus*, previously used as probiotic in Ecuadorian shrimp hatcheries, has been effective at controlling diseases caused by *Aer. salmonicida*, *V. anguillarum* and *V. ordalii* (Austin *et al.*, 1995b) The relative incidence of microbial antagonists is indicated from a study of >400 bacterial isolates from the gastro-intestinal tract and surface of turbot and fish food and water, in which 28% (mostly from the intestinal mucus) were inhibitory to *V. anguillarum* (Westerdahl *et al.*, 1991). Consequently, it is apparent that the native aquatic microflora may be important as a natural means of disease control. In addition, there is a report of the benefits for disease control of using the biopesticide, *Bacillus thuringiensis* (Meshram *et al.*, 1998).

11

Conclusions

It should be abundantly clear that only a few aspects of the biology of bacterial fish pathogens have been studied in comparative detail. In particular, it has been fashionable to describe the pathology, serology and control measures (by vaccines and antimicrobial compounds). However, even these areas are incomplete. The outlook for the future is nevertheless encouraging, and it seems likely that there will be increasing interest in the study of fish diseases. To some extent, this will be justified by the anticipated expansion of aquaculture and the expected increase in outbreaks of disease. Similar attention is likely to be devoted to the problems of wild fish populations, particularly with regard to the highly emotive issue of pollution and, thus, pollution-related diseases. It seems likely that future developments will encompass the following:

The recognition of emerging conditions

Any systematic study of fish populations will probably reveal the presence of novel disease conditions. Of course, whether or not such conditions will be truly emerging or merely ignored in the past is the subject of conjecture. Moreover, new pathogens may appear to take the place of some of the existing problems. During the preparation of this book, we have noted the spread of winter ulcer disease, from its initial focal point in Iceland, to Norway, and then to Scotland, possibly via Shetland. What will be next?

Taxonomy and diagnosis

As the rules governing bacterial taxonomy have been considerably improved, it is anticipated that the descriptions of novel pathogens will be much better than published in some of the earlier reports. However, there is an annoying tendency of naming new taxa after studying only a few (perhaps even one) isolates. We argue that this practice is not reflective of good science. Nevertheless with overall improvements in bacterial taxonomy, the direct benefit will be reflected in more reliable identification of fresh isolates, i.e. identification should lose some of its subjectivity. It would be nice to think that diagnosticians may even use systems designed for fish pathogens, rather than relying on kits developed for medical microbiology. There are some signs that new and exciting developments in microbiology will lead to better diagnostic methods. For example, the advent of highly specific monoclonal antibodies revolutionized serology. Already, the development of reliable diagnostic kits, such as those based on ELISA, have become a commercial reality. Other sophisticated methods, such as pyrolysis techniques, may offer excellent opportunities for the detection of pathogens and the identification of pure cultures (see Gutteridge and Norris, 1979). Perhaps, future developments will lead to molecular techniques that can be used by diagnosticians outside specialized laboratories, and preferably in field conditions.

Isolation and selective isolation of pathogens
Since 1980, there have been many improvements with isolation procedures for some fish pathogens. For example, selective media for *Ren. salmoninarum* (Austin *et al.*, 1983a) and *Y. ruckeri* (Waltman and Shotts, 1984) have been formulated. This impetus should continue, particularly for some of the difficult-to-isolate pathogens, such as *Myco-bacterium*.

Ecology (epizootiology)
Overall, the ecology of fish pathogens has been neglected. Consequently, it is often unclear whether a pathogen is a representative of the natural aquatic microflora or is restricted to fish. Nevertheless, a wealth of information has been gathered on a few organisms, including *Aer. salmonicida*, although it could be argued that the true reservoir of that pathogen remains to be discovered. True, there is an interest in ecological matters, but progress seems to reflect changing attitudes—sometimes ecology is fashionable, but not so at other times. We argue that an improvement in epizootiology will be invaluable in the design of disease control measures, particularly those involving the legislative machinery for movement-restriction policies.

Pathogenicity mechanisms
There has been commendable progress with some pathogens, notably *Aer. salmonicida* and *V. anguillarum*. However, the research input needs to continue with attention being diverted to other pathogens, such as mycobacteria. Information about pathogenicity mechanisms is especially relevant for vaccine development programmes.

Control measures
Although conventional vaccine development programmes seem likely to continue, further progress on difficult candidates, such as *Rickettsia* and *Ren. salmoninarum*, will necessitate use of more sophisticated techniques, such as are offered by genetic engineering. However, until improved vaccines become available, antimicrobial compounds will remain the primary means of disease control. Improvements in chemotherapy are envisaged, including tighter regulations on the use of antimicrobial compounds in fisheries, and the development of new products. With tighter control on their use, the risk of acquiring resistance to drugs may be reduced, thereby prolonging the effective life of the product. Moreover, the effect of antimicrobial compounds on the natural environment will undoubtedly receive increasing attention. Attention should also be focused on stock improvement programmes, which may highlight genetically determined disease-resistant fish strains. The recent interest in probiotics, nutritional supplements and non-specific immunostimulants is gathering momentum, and commercial products are a reality.

The effects of pollution
There is increasing concern about the possible role of pollution in disease, particularly of wild fish stocks (see also Pippy and Hare, 1969; Mahoney *et al.*, 1973; Robohm *et al.*, 1979). As this is a politically emotive issue, there is likely to be an increase in monies available and, thus, a stimulation of research interest. At present, there is considerable confusion over the precise role of pollution and fish health (Fig. 11.1; Bucke, 1991; 1997). Nevertheless, there are good data that long-term exposure to pollutants has adversely affected the health of fish, especially in the North Sea and Great Lakes. However, mortalities among fish populations do not necessarily imply disease. Furthermore, disease may develop long after the pollutant has been effectively removed from the aquatic environment. Much of the work attempting to correlate fish disease with aquatic pollution has resulted from surveys, many of which have been carried out in the North Sea (e.g. Dethlefsen and Watermann, 1980, Dethlefsen *et al.*, 1987; McVicar *et al.*, 1988; Vethaak and ap. Rheinallt, 1992). Briefly, fish are caught with nets, and the relative incidence of disease determined. One conclusion from these surveys is that larger numbers of diseased fish occur generally in the polluted compared to clean/unpolluted locations. However, the distinction between polluted and clean sites is imprecise. Therefore, there would be some uncertainty as to what comprises a truly polluted or clean site. Moreover, it is uncertain from surveys how long fish might have been in a polluted environment prior to capture. Thus, the effects of fish migration on the incidence of disease needs to be considered (Vethaak *et al.*, 1992; Bucke *et al.*, 1992; Jacquez *et al.*, 1994).

Pollution has been associated with some bacterial diseases, namely fin and tail rot (Vethaak 1992; Vethaak *et al.*, 1996), gill disease/hyperplasia (Kirk and Lewis, 1993) and skin disease/ulceration (Vethaak 1992; Vethaak and Jol, 1996). The trigger has been attributed to contaminated diets (Landsberg, 1995), heavy metals (e.g. Rødsaether *et al.*, 1997), hydrocarbons (e.g. Khan, 1987), nitrogenous compounds, i.e. ammonia (Kirk and Lewis, 1993) and nitrites (Hanson and Grizzle, 1985), pesticides (e.g. Voigt, 1994), sewage (e.g. Austin and Stobie, 1992b) and unspecified pollutants (e.g. Vethaak and Jol, 1996). The reasons for the association between pollution and disease need to be better researched. However, proof of correlation between the occurrence of specific pollutants and disease has seldom been documented. Surveys, which have pointed to a correlation between pollution and disease, have generally not considered the nature or concentration of the pollutant(s).

An association has been made between fish diseases and unknown components of sewage dumping (Siddall *et al.*, 1994). For example in a survey of 16 sites in the Dutch Wadden Sea, a higher incidence of skin ulcers and fin rot was noted in fish caught near fresh water drainage sluices than elsewhere (Vethaak, 1992). Pollution by domestic sewage, i.e. leakage from a septic tank, was attributed to a new skin disease, which was characterized by the presence of extensive skin lesions and muscle necrosis, in rainbow trout (otherwise infected with ERM for which there might also be a link with sewage sludge; Dudley *et al.*, 1980) in Scotland during 1992 (Austin and Stobie, 1992b). Interestingly, the skin lesions—but not ERM—declined substantially after the leaking septic tank was repaired.

There is accumulating evidence that contaminated waters lead to immunosuppression.

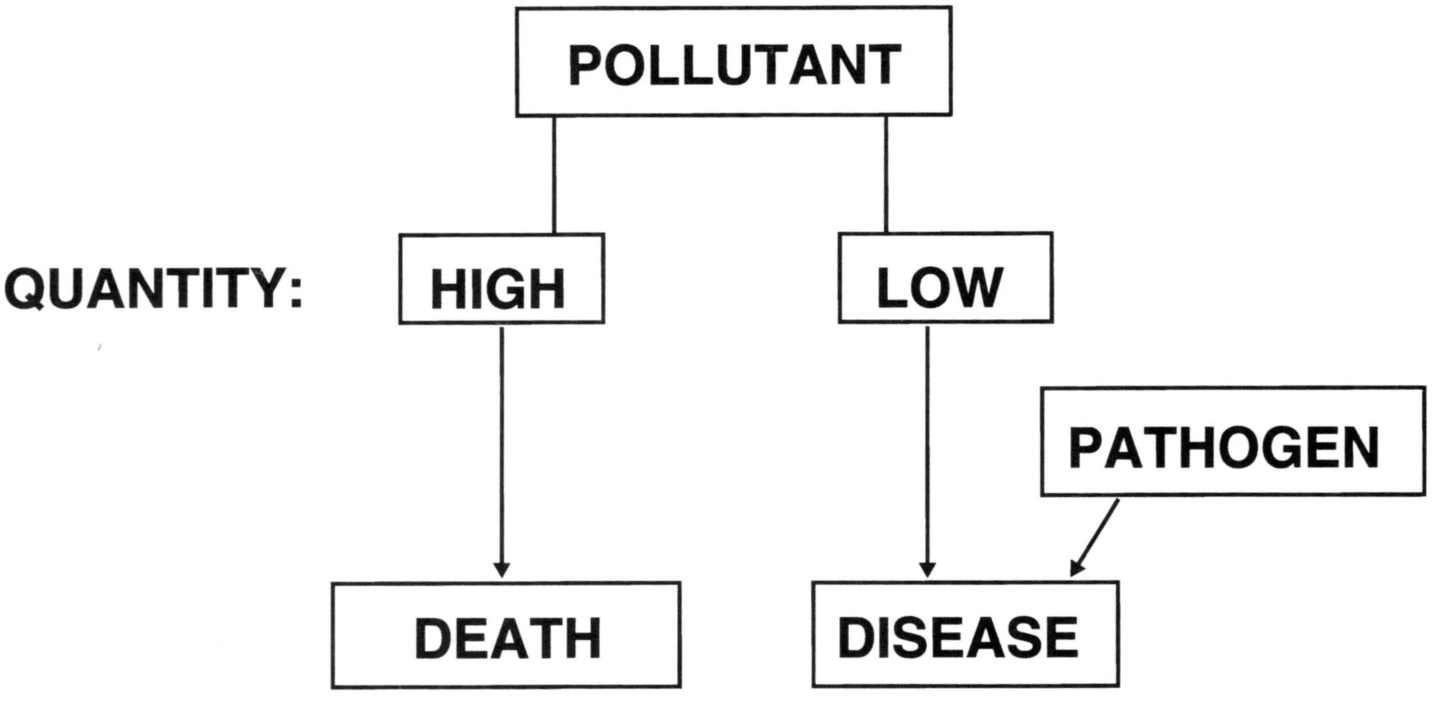

Fig. 11.1. The presence of high levels of pollutants may lead to disease. However, at lower levels, there is evidence that pollutants may weaken fish rendering them more susceptible to the occurrence of disease caused by bacteria (after Austin, 1999).

One recent example describes the increased susceptibility of chinook salmon from a contaminated estuary to *V. anguillarum* (Arkoosh *et al.*, 1998). Undoubtedly, the future will bring further examples.

Zoonoses

There is a growing awareness that some bacterial fish pathogens may also infect human beings. Although the number of cases is mercifully small, some have attracted media attention. Generally, official statistics are absent, and, therefore, only an educated guess can be taken as to the types and severity of risks which may exist for humans. Some publications, e.g. Austin (1981), Grimes (1986) and Austin and Austin (1989) have begun to address the problem. The range of fish-associated bacteria (pathogens and commensals), which may infect humans include:

- *Aer. hydrophila* (causing diarrhoea and septicaemias)
- *Campylobacter jejuni* (gastro-enteritis)
- *Cl. botulinum* type E (botulism)
- *Edw. tarda* (diarrhoea)
- *Erysipelothrix rhusiopathiae* (fish rose)
- *Leptospira interrogans* (leptospirosis—Weil's disease)
- *Myc. fortuitum* (mycobacteriosis; fish tank granuloma)
- *Myc. marinum* (mycobacteriosis; fish tank granuloma)
- *Plesiomonas shigelloides* (gastro-enteritis)
- *Ps. aeruginosa* (wound infections)
- *Ps. fluorescens* (wound infections)
- *Salmonella* spp. (food poisoning)
- *Str. iniae* ('mad fish disease')
- *V. parahaemolyticus* (food poisoning)
- *V. vulnificus* (wound infections)

The source of some of these organisms may well be the polluted waters in which the fish are to be found. A comparative few originate from diseased fish. Therefore, the transfer to humans will probably reflect the handling of diseased specimens. Uptake into humans may be via cuts or grazes, or, less likely, via the digestive tract.

Aeromonas hydrophila

Aer. hydrophila has been implicated as an opportunistic and occasionally as a primary pathogen of humans. In one example, a fish farm worker developed a bacterial septicaemia, which was attributed to *Aer. hydrophila*, and presumed to have resulted from exposure in the fish farm (Austin, 1981).

Edwardsiella tarda

Enterics have occasionally spread from fish to humans, and caused gastro-enteritis/diarrhoea. In one documented case, *Edw. tarda* from tropical ornamental fish, *Pterophyllum scalare*, caused diarrhoea in an infant (Vandepitte *et al.*, 1983).

Leptospira interrogans

During 1980 and 1981, three cases of leptospirosis attributed to serogroup icterohaemorrhagiae, of which one was a fatality, occurred in fish farm workers in the UK. Workers on one of the sites, which harboured rats, possessed antibodies to the pathogen, indicating a risk of infection (Robertson *et al.*, 1991). Previously, only one other case of leptospirosis in a fish farm worker had been reported in the UK, i.e. in 1968 (Gill *et al.*, 1983). Elsewhere, the disease had been found among prawn farmers in Hawaii (Anderson *et al.*, 1982). Subsequently, an examination of blood from >200 fish farm workers from 82 fish farms in the UK revealed the presence of antibodies to serogroup icterohaemorrhagiae in three of the samples (Gill *et al.*, 1983). Consequently, it would appear that there is minimal risk to leptospirosis among fish farm workers.

Mycobacterium marinum

Myc. marinum, and occasionally *Myc. fortuitum*, may be transmitted from diseased fish to humans (especially aquarists), where the organisms cause localised superficial (papulopustular) lesions on the hands and arms (referred to as fish tank granuloma; Bruckner-Tuderman and Blank, 1985; Gray *et al.*, 1990; Boyce, 1997). Although not fatal, the condition may need months or occasionally years of chemotherapy (Black and Eykyn, 1977; Austin, 1981). Within the UK, there appears to be a steady low number of infections each year (e.g. Ryan and Bryant, 1997; Boyce, 1997). Typically, patients who are usually in middle age, have handled ornamental fish or the tanks/tank water prior to infection (Ryan and Bryant, 1997).

Streptococcus iniae

During 1995–96, *Str. iniae* became associated with a cluster of four cases, totalling nine patients, of invasive disease (including endocarditis and cellulitis of the hand), coined by the press as 'mad fish disease' because of the neurological symptoms in fish, in females of Asian origin in the Toronto area of Canada (Centers for Disease Control, 1996; Weinstein *et al.*, 1997). Anecdotal evidence suggested that the women purchased clinically disease tilapia and following handling/preparation of the whole fish became infected. By PFGE, all the isolates from the infected women and some from diseased fish were indistinguishable. The conclusion was reached that the fish pathogen *Str. iniae* could cause human disease. As streptococcal fish infections inevitably produce clearly visible lesions, it is curious why individuals would purchase something which did not look in prime condition. Also, it is relevant to enquire why a fish farmer would sell diseased stock for human consumption.

Vibrio vulnificus

In Amsterdam, The Netherlands, a 63-year-old man was hospitalized because of severe pain in the right arm (Veenstra *et al.*, 1992). The arm became progressively swollen, and was painful. Also, it was noted that there were small wounds on the hands. In hospital, the patient's condition deteriorated with skin and muscle necrosis, necessitating surgery to remove the diseased tissue, before eventual recovery. Interestingly, the day before the onset of symptoms, he had cleaned eels purchased from a local market. Bacteriology revealed the presence of *V. vulnificus* from blood. Indeed, this isolate was considered to resemble one recovered previously from eels in a Dutch eel farm (Veenstra *et al.*, 1992). It was presumed that the man became infected from the eels through open wounds on his hand. Interestingly, we determined that the isolate caused disease in laboratory-infectivity experiments with Atlantic salmon.

Clearly, there should not be complacency in the handling of diseased fish, otherwise the infection may well spread to the handlers. This is certainly an area deserving of further consideration.

Bibliography

Abe, P.M. (1972) Certain chemical and immunological properties of the endotoxin from *Vibrio anguillarum*, M.S. thesis, Oregon State University, Corvallis.

Actis, L.A., Potter, S.A. and Crosa, J.W. (1985) Iron-regulated outer membrane protein OM2 of *Vibrio anguillarum* is encoded by virulence plasmid pJM1. *Journal of Bacteriology* **161**, 736–742.

Actis, L.A., Tolmasky, M.E., Crosa, L.M. and Crosa, J.H. (1995) Characterization of the expression of FATB, an iron transport protein encoded by the pJM1 virulence plasmid. *Molecular Microbiology* **17**, 197–204.

Acuigrup (1980a) Flavobacteriosis in coho salmon (*Oncorhynchus kisutch*). In: Ahne, W. (ed.), *Fish Diseases, Third COPRAQ-Session*, Berlin, Springer-Verlag, p. 212–217.

Acuigrup (1980b) Trial vaccination of rainbow trout against *Aeromonas liquefaciens*. In: Ahne, W. (ed.), *Fish Diseases, Third COPRAQ-Session*. Berlin, Springer-Verlag, p. 206–211.

Adams, A., Auchinachie, N., Bundy, A., Tatner, M.F. and Horne, M.T. (1988) The potency of adjuvanted injected vaccines in rainbow trout (*Salmo gairdneri* Richardson) and bath vaccines in Atlantic salmon (*Salmo salar* L.) against furunculosis. *Aquaculture* **69**, 15–26.

Adams, A., Thompson, K.D., Morris, D., Farias, C. and Chen, S.C. (1995) Development and use of monoclonal antibody probes for immunohistochemistry, ELISA and IFAT to detect bacterial and parasitic fish pathogens. *Fish and Shellfish Immunology* **5**, 537–547.

Ahne, W. (1981) Serological techniques currently used in fish virology. *Developments in Biological Standardization* **49**, 3–27.

Ahne, W., Popp, W. and Hoffmann, R. (1982) *Pseudomonas fluorescens* as a pathogen of tench (*Tinca tinca*). *Bulletin of the European Association of Fish Pathologists* **4**, 56–57.

Ainsworth, A.J., Capley, G., Waterstreet, P. and Munson, D. (1986) Use of monoclonal antibodies in the indirect fluorescent antibody technique (IFA) for the diagnosis of *Edwardsiella ictaluri*. *Journal of Fish Diseases* **9**, 439–444.

Ajmal, M. and Hobbs, B.C. (1967) Species of *Corynebacterium* and *Pasteurella* isolated from diseased salmon, trout and rudd. *Nature (London)* **215**, 142–143.

Akazaka, H. (1968) Bacterial disease of marine fishes. *Bulletin of the Japanese Society of Scientific Fisheries* **34**, 271–272.

Albrektsen, S., Snadnes, K., Glette, J. and Waagbo, R. (1995) The influence of dietary vitamin B_6 on tissue vitamin B_6 contents and immunity in Atlantic salmon, *Salmo salar* L. *Aquaculture Research* **26**, 331–339.

Al-Harbi, A.H. (1994) First isolation of *Streptococcus* sp. from hybrid tilapia (*Oreochromis niloticus* × *O. aureus*) in Saudi Arabia. *Aquaculture* **128**, 195–201.

Alexander, D.M. (1913) A review of piscine tubercles, with a description of an acid-fast bacillus found in the cod. *Report of the Lancashire Sea Fish Laboratory* **21**, 43–49.

Ali, A., Carnahan, A.M., Altwegg, M., Lüthy-Hotenstein, J. and Joseph, S.W. (1996) *Aeromonas bestiarum* sp. nov. (formerly genomospecies DNA group 2 *A. hydrophila*) a new species isolated from non-human sources. *Medical Microbiology Letters* **5**, 156–165.

Allan, B.J. and Stevenson, R.M.W. (1981) Extracellular virulence factors of *Aeromonas hydrophila* in fish infections. *Canadian Journal of Microbiology* **27**, 1114–1122.

Allen, D.A. (1982) Bacteria associated with freshwater fish farming, with emphasis on the fish pathogen, *Aeromonas salmonicida*. Ph.D. thesis, University of Maryland, USA.

Allen, D.A., Austin, B. and Colwell, R.R. (1983a) *Aeromonas media*, a new species isolated from river water. *International Journal of Systematic Bacteriology* **33**, 599–604.

Allen, D.A., Austin, B. and Colwell, R.R. (1983b) Numerical taxonomy of bacterial isolates associated with a freshwater fishery. *Journal of General Microbiology* **129**, 2043–2062.

Allen-Austin, D., Austin, B. and Colwell, R.R. (1984) Survival of *Aeromonas salmonicida* in river water. *FEMS Microbiology Letters* **21**, 143–146.

Allison, L.N. (1958) Multiple sulfa therapy of kidney disease among brook trout. *Progressive Fish Culturist* **20**, 66–68.

Almendras, F.E., Fuentealba, I.C., Jones, S.R.M., Markham, F. and Spangler, E. (1997) Experimental infection and horizontal transmission of *Piscirickettsia salmonis* in freshwater-raised Atlantic salmon, *Salmo salar* L. *Journal of Fish Diseases* **20**, 409–418.

Alsina, M. and Blanch, A.R. (1994a) A set of keys for biochemical identification of environmental *Vibrio* species. *Journal of Applied Bacteriology* **76**, 79–85.

Alsina, M. and Blanch, A.R. (1994b) Improvement and update of a set of keys for biochemical identification of *Vibrio* species. *Journal of Applied Bacteriology* **77**, 719–721.

Alsina, M., Martínez-Picardo, J., Jofre, J. and Blanch, A.R. (1994) A medium for the presumptive identification of *Vibrio anguillarum*. *Applied and Environmental Microbiology* **60**, 1681–1683.

Amandi, A., Hiu, S.F., Rohovec, J.S. and Fryer, J.L. (1982) Isolation and characterization of *Edwardsiella tarda* from chinook salmon (*Oncorhynchus tshawytscha*). *Applied and Environmental Microbiology* **43**, 1380–1384.

Amaro, C., Biosca, E.G., Esteve, C., Fouz, B. and Toranzo, A.E. (1992) Comparative study of phenotypic and virulence properties in *Vibrio vulnificus* biotypes 1 and 2 obtained from a European eel farm experiencing mortalities. *Diseases of Aquatic Organisms* **13**, 29–35.

Amaro, C., Biosca, E.G., Fouz, B., Alcaide, E. and Esteve, C. (1995) Evidence that water transmit *Vibrio vulnificus* biotype 2 infections to eels. *Applied and Environmental Microbiology* **61**, 1133–1137.

Amaro, C., Fouz, B., Biosca, E.G., Marco-Noales, E. and Collado, R. (1997) The lipopolysaccharide O side chain of *Vibrio vulnificus* serogroup E is the virulence determinant for eels. *Infection and Immunity* **65**, 2475–2479.

Amend, D.F. (1970) Myxobacterial infections of salmonids: prevention and treatment. In: Snieszko, S.F. (ed.), *A. Symposium on Diseases of Fishes and Shellfishes*. Washington D.C., American Fisheries Society Special Publication No. 5, p. 258–265.

Amend, D.F., Fryer, J.L. and Pilcher, K.S. (1965) Production trials utilizing sulfonamide drugs for the control of 'cold-water' disease in juvenile coho salmon. *Research Briefs, Fish Commission of Oregon* **11**, 14–17.

Amend, D.F. and Pietsch, J.P. (1972) Virucidal activity of two iodophors to salmonid viruses. *Journal of the Fisheries Research Board of Canada* **29**, 61–65.

Amend, D.F. and Eschenour, R.W. (1980) Development and use of commercial fish vaccines. *Salmonid* (March/April), 8–12.

Amend, D.F. and Johnson, K.A. (1981) Current status and future needs of *Vibrio anguillarum* bacterins. *Developments in Biological Standardization* **49**, 403–417.

Amend, D.F., Johnson, K.A., Croy, T.R. and McCarthy, D.H. (1983) Some factors affecting the potency of *Yersinia ruckeri* bacterins. *Journal of Fish Diseases* **6**, 337–344.

Amlacher, E. (1961) *Taschenbuch der Fischkrankheiten*. Jena, Gustav Fisher Verlag, 286p.

Anacker, R.L. and Ordal, E.J. (1959) Studies on the myxobacterium *Chondrococcus columnaris*. 1. Serological typing. *Journal of Bacteriology* **78**, 25–32.

Anderson, B.S., Brock, J.A., Higa, H.H., Gooch, J.M., Wienbenga, N.H., Palumbo, N.E., Perri, S. and Sato, V.T. (1982) *Epidemiology of Leptospirosis on Aquaculture Farms in Hawaii*. State of Hawaii, Department of Health Publication.

Anderson, D.P. (1973) Investigations of the lipopolysaccharide fractions from *Aeromonas salmonicida* smooth and rough forms. In: *Symposium on the Major Communicable Fish Diseases in Europe and their Control*. Rome, United Nations Food and Agricultural Organization, p. 175–179.

Anderson, D.P. and Ross, A.J. (1972) Comparative study of Hagerman redmouth disease oral bacterins. *Progressive Fish Culturist* **34**, 226–228.

Anderson, D.P. and Nelson, J.R. (1974) Comparison of protection in rainbow trout (*Salmo gairdneri*) inoculated with and fed Hagerman redmouth bacterins. *Journal of the Fisheries Research of Canada* **31**, 214–216.

Anderson, D.P. and Siwicki, A.K. (1994) Duration of protection against *Aeromonas salmonicida* in brook trout immunostimulated with glucan or chitosan by injection or immersion. *Progressive Fish Culturist* **56**, 258–261.

Anderson, J.I.W. and Conroy, D.A. (1969) Myxobacteria as fish pathogens. *Journal of Applied Bacteriology* **32**, 30–39.

Anderson, J.W. and Conroy, D.A. (1970) *Vibrio* diseases in fishes. In: Snieszko, S.F.

(ed.), *A Symposium on Diseases of Fishes and Shellfishes*, Washington, D.C., American Fisheries Society, Special Publication No. 5, p. 266–272.

Angka, S.L., Lam, T.J. and Sin, Y.M. (1995) Some virulence characteristics of *Aeromonas hydrophila* in walking catfish (*Clarias gariepinus*). *Aquaculture* **130**, 103–112.

Angulo, L., Lopez, J.E., Lema, C. and Vicente, J.A. (1992) *Vibrio pelagius* associated with mortalities in farmed turbot, *Scophthalmus maximus*. *Thalassas* **10**, 129–133.

Angulo, L., Lopez, J.E., Vicente, J.A. and Saborido, A.M. (1994) Haemorrhagic areas in the mouth of farmed turbot, *Scophthalmus maximus* (L.). *Journal of Fish Diseases* **17**, 163–169.

Antipa, R. (1976) Field testing of injected *Vibrio anguillarum* bacterins in pen-reared Pacific salmon. *Journal of the Fisheries Research Board of Canada* **33**, 1291–1296.

Antipa, R. and Amend, D.P. (1977) Immunization of Pacific salmon: comparison of intraperitoneal injection and hyperosmotic infiltration of *Vibrio anguillarum* and *Aeromonas salmonicida* bacterins. *Journal of the Fisheries Research Board of Canada* **34**, 203–208.

Antipa, R., Gould, R. and Amend, D.P. (1980) *Vibrio anguillarum* vaccination of sockeye salmon (*Oncorhynchus nerka*) by direct immersion and hyperosmotic immersion. *Journal of Fish Diseases* **3**, 161–165.

Aoki, T. (1988) Drug-resistant plasmids from fish pathogens. *Microbiological Sciences* **5**, 219–223.

Aoki, T. and Egusa, S. (1971) Drug sensitivity of *Aeromonas liquefaciens* isolated from freshwater fishes. *Bulletin of the Japanese Society of Scientific Fisheries* **37**, 176–185.

Aoki, T., Egusa, S., Kimura, T. and Watanabe, T. (1971) Detection of R factors in naturally occurring *Aeromonas salmonicida* strains. *Applied Microbiology* **22**, 716–717.

Aoki, T., Egusa, S. and Arai, T. (1974) Detection of R factor in naturally occurring *Vibrio anguillarum* strains. *Antimicrobial Agents and Chemotherapy* **6**, 534–538.

Aoki, T., Arai, T. and Egusa, S. (1977) Detection of R plasmids in naturally occurring fish-pathogenic bacteria, *Edwardsiella tarda*. *Microbiologia et Immunologia (Scandinavia)* **21**, 77–83.

Aoki, T. and Kitao, T. (1978) Vibriosis in ayu. *Fish Pathology* **13**, 19–24.

Aoki, T., Kitao, T., Itabashi, T., Wada, Y. and Sakai, M. (1981) Proteins and lipopolysaccharides in the membrane of *Vibrio anguillarum*. *Developments in Biological Standardization* **49**, 225–232.

Aoki, T., Kitao, T., Iemura, N., Mitoma, Y. and Nomura, T. (1983) The susceptibility of *Aeromonas salmonicida* strains isolated in cultured and wild salmonids to various chemotherapeutants. *Bulletin of the Japanese Society of Scientific Fisheries* **49**, 17–22.

Aoki, T. and Kitao, T. (1985) Detection of transferable R plasmids in strains of the fish-pathogenic bacterium *Pasteurella piscicida*. *Journal of Fish Diseases* **8**, 345–350.

Aoki, T., Ikeda, D., Katagiri, T. and Hirono, I. (1997) Rapid detection of the fish-pathogenic bacterium *Pasteurella piscicida* by polymerase chain reaction targeting nucleotide sequences of the species–species plasmid pZP1. *Fish Pathology* **32**, 143–151.

Arakawa, C.K. and Fryer, J.L. (1984) Isolation and characterization of a new subspecies of *Mycobacterium chelonei* infections for salmonid fish. *Helgoländer Meeresuntersuchengen* **37**, 329–342.

Arakawa, C.K., Sanders, J.E. and Fryer, J.L. (1987) Production of monoclonal antibodies against *Renibacterium salmoninarum*. *Journal of Fish Diseases* **10**, 249–253.

Arakawa, C.K. and Fryer, J.L. (1984) Isolation and characterization of a new subspecies of *Mycobacterium chelonei* infections for salmonid fish. *Helgoländer Meeresuntersuchungen* **37**, 329–342.

Argenton, F., De Mas, S., Malocco, C., Dalla Valle, L., Giorgetti, G. and Colombo, L. (1996) Use of random DNA amplification to generate specific molecular probes for hybridization tests and PCR-based diagnosis of *Yersinia ruckeri*. *Diseases of Aquatic Organisms* **24**, 121–127.

Arias, C.R., Verdonck, L., Swings, J., Garay, E. and Aznar, R. (1997a) Intraspecific differentiation of *Vibrio vulnificus* biotypes by amplified fragment length polymorphism and ribotyping. *Applied and Environmental Microbiology* **63**, 2600–2606.

Arias, C.R., Verdonck, L., Swings, J., Aznar, R. and Garay, E. (1997b) A polyphasic approach to study the intraspecific diversity amongst *Vibrio vulnificus* isolates. *Systematic and Applied Microbiology* **20**, 622–633.

Arijo, S., Borrego, J.J., Zorilla, I., Balebona, M.C. and Moriñigo, M.A. (1998) Role of the capsule of *Photobacterium damselae* subsp. *piscicida* in protection against phagocytosis and killing by gilt-head seabream (*Sparus aurata*, L) macrophages. *Fish and Shellfish Immunology* **8**, 63–72.

Arkoosh, M.R., Casillas, E., Huffman, P., Clemons, E., Evered, J., Stein, J.E. and Varanasi, U. (1998) Increased susceptibility of juvenile chinook salmon from a contaminated estuary to *Vibrio anguillarum*. *Transactions of the American Fisheries Society* **127**, 360–374.

Arkwright, J.A. (1912) An epidemic disease affecting salmon and trout in England during the summer of 1911. *Journal of Hygiene, Cambridge* **12**, 391–413.

Armstrong, R.D., Martin, S.W., Evelyn, T.P.T., Hicks, B., Dorward, W.J. and Ferguson, H.W. (1989) A field evaluation of the indirect fluorescent antibody-based broodstock screening test used to evaluate the vertical transmission of *Renibacterium salmoninarum* in chinook salmon (*Oncorhynchus tshawytscha*). *Canadian Journal of Veterinary Research* **53**, 385–389.

Aronson, J.D. (1926) Spontaneous tuberculosis in salt water fish. *Journal of Infectious Diseases* **39**, 315–320.

Ascencio, F., Ljungh, Å. and Wadstrom, T. (1991) Comparative study of extracellular matrix protein binding to *Aeromonas hydrophila* isolated from diseased fish and human infection. *Microbios* **65**, 135–146.

Ashburner, L.D. (1978) Management and diseases of hatchery fish. *Proceedings of the Fauna Course for Veterinarians, Taronga Zoo, Mosman, Australia* **36**, 387–449.

Atsuta, S., Yoshimoto, J., Sakai, M. and Kobayashi, M. (1990) Streptococcicosis occurred in pen-cultured coho salmon *Oncorhynchus kistuch*. *Suisanzoshoku* **38**, 215–219.

Austin, B. (1981) Human diseases associated with fish culture. *Proceedings of the Institute of Fisheries Management*, p.170–175.

Austin, B. (1982a) Taxonomy of bacteria isolated from a coastal, marine fish-rearing unit. *Journal of Applied Bacteriology* **53**, 253–268.

Austin, B. (1982b) Close up of a killer. *Fish Farmer* **5**, 30–31.

Austin, B. (1984a) The control of bacterial fish diseases by antimicrobial compounds. In: Woodbine, M. (ed.) *Antibiotics and Agriculture, Benefits and Malefits.* Sevenoaks, Butterworths, p.255–268.

Austin, B. (1984b) The future of bacterial fish vaccines. *Vaccine* **2**, 249–254.

Austin, B. (1985) Evaluation of antimicrobial compounds for the control of bacterial kidney disease in rainbow trout, *Salmo gairdneri* Richardson. *Journal of Fish Diseases* **8**, 209–220.

Austin, B. (1986) Ecology of *Renibacterium salmoninarum,* the causal agent of bacterial kidney disease in salmonids. *Proceedings of the 4th International Symposium on Microbial Ecology, Ljublana* p.650–654.

Austin, B. (1988) *Marine Microbiology.* Cambridge, Cambridge University Press.

Austin, B. (1993) Recovery of 'atypical' isolates of *Aeromonas salmonicida,* which grow at 37°C, from ulcerated non-salmonids in England. *Journal of Fish Diseases* **16**, 165–168.

Austin, B. (1999) The effect of pollution on fish health. *Journal of Applied Microbiology, Symposium Supplement* (in press).

Austin, B., Allen, D.A., Zachary, A., Belas, M.R. and Colwell, R.R. (1979) Ecology and taxonomy of bacteria attaching to wood surfaces in a tropical harbor. *Canadian Journal of Microbiology* **25**, 447–461.

Austin, B. and Rodgers, C.J. (1980) Diversity among stains causing bacterial kidney disease in salmonid fish. *Current Microbiology* **3**, 231–235.

Austin, B. and Rodgers, C.J. (1981) Preliminary observations on *Aeromonas salmonicida* vaccines. *Development in Biological Standardization* **49**, 387–393.

Austin, B., Morgan, D.A. and Alderman, D.J. (1981) Comparison of antimicrobial agents for the control of vibriosis in marine fish. *Aquaculture* **26**, 1–12.

Austin, B., Green, M. and Rodgers, C.J. (1982) Morphological diversity among strains of *Yersinia ruckeri. Aquaculture* **27**, 73–78.

Austin, B., Embley, T.M. and Goodfellow, M. (1983a) Selective isolation of *Renibacterium salmoninarum. FEMS Microbiology Letters* **17**, 111–114.

Austin, B., Rayment, J. and Alderman, D.J. (1983b) Control of furunculosis by oxolinic acid. *Aquaculture* **31**, 101–108.

Austin, B. and Rayment, J. (1985) Epizootiology of *Renibacterium salmoninarum,* the causal agent of bacterial kidney disease in salmonid fish. *Journal of Fish Diseases* **8**, 505–509.

Austin, B., Bucke, D., Feist, S. and Rayment, J. (1985). A false positive reaction in the indirect fluorescent antibody test for *Renibacterium salmoninarum* with a 'coryneform' organism. *Bulletin of the European Association of Fish Pathologists* **5**, 8–9.

Austin, B., Bishop, I., Gray, C., Watt, B. and Dawes, J. (1986). Monoclonal antibody-based enzyme-linked immunosorbent assays for the rapid diagnosis of clinical cases of enteric redmouth and furunculosis in fish farms. *Journal of Fish Diseases* **9**, 469–474.

Austin, B., McIntosh, D. and Murray, K.R. (1988) Infection of Atlantic salmon (*Salmo salar*) with *Planococcus. Bulletin of the European Association of Fish Pathologists* **8**, 21–22.

Austin, B. and Austin, D.A. (1989) 1. General introduction. In: Austin, B. and Austin, D.A. (eds), *Methods for the Microbiological Examination of Fish and Shellfish*, pp.1–24. Chichester, Ellis Horwood.

Austin, B. and Billaud, A.-C. (1990) Inhibition of the fish pathogen, *Serratia liquefaciens*, by an antibiotic-producing isolate of *Planococcus* recovered from sea water. *Journal of Fish Diseases* **13**, 553–556.

Austin, B. and Stobie, M. (1992a) Recovery of *Micrococcus luteus* and presumptive *Planococcus* from moribund fish during outbreaks of rainbow trout (*Oncorhynchus mykiss* Walbaum) fry syndrome (RTFS) in England. *Journal of Fish Diseases* **15**, 203–206.

Austin, B. and Stobie, M. (1992b) Recovery of *Serratia plymuthica* and presumptive *Pseudomonas pseudoalcaligenes* from skin lesions in rainbow trout, *Oncorhynchus mykiss* (Walbaum), otherwise infected with enteric redmouth. *Journal of Fish Diseases* **15**, 541–543.

Austin, B., Baudet, E. and Stobie, M. (1992a) Inhibition of bacterial fish pathogens by *Tetraselmis suecica. Journal of Fish Diseases* **15**, 55–61.

Austin, B., Gonzalez, C.J., Stobie, M., Curry, J.I. and McLoughlin, M.F. (1992b) Recovery of *Janthinobacterium lividum* from diseased rainbow trout, *Oncorhynchus mykiss* (Walbaum), in Northern Ireland and Scotland. *Journal of Fish Diseases* **15**, 357–359.

Austin, B. and Robertson, P.A.W. (1993) Recovery of *Streptococcus milleri* from ulcerated koi carp (*Cyprinus carpio* L.) in the U.K. *Bulletin of the European Association of Fish Pathologists* **13**, 207–209.

Austin, B., Stobie, M., Robertson, P.A.W., Glass, H.G., Stark, J.R. and Mudarris, M. (1993) *Vibrio alginolyticus*: the cause of gill disease leading to progressive low-level mortalities among juvenile turbot, *Scophthalmus maximus* L., in a Scottish aquarium. *Journal of Fish Diseases* **16**, 277–280.

Austin, B., Alsina, M., Austin, D.A., Blanch, A.R., Grimont, F., Grimont, P.A.D., Jofre, J., Koblavi, S., Larsen, J.L., Pedersen, K., Tiainen, T., Verdonck, L. and Swings, J. (1995a) Identification and typing of *Vibrio anguillarum* a comparison of different methods *Systematic and Applied Microbiology* **18**, 285–302.

Austin, B., Stuckey, L.F., Robertson, P.A.W., Effendi, I. and Griffith, D.R.W. (1995b) A probiotic strain of *Vibrio alginolyticus* effective in reducing diseases caused by *Aeromonas salmonicida, Vibrio anguillarum* and *Vibrio ordalii. Journal of Fish Diseases* **18**, 93–96.

Austin, B., Austin, D.A., Falconer, V.M., Pedersen, K., Larsen, J.L., Swings, J. and Verdonck, L. (1996) Dissociation of *Vibrio anguillarum* and *V. ordalii* cultures into two or three distinct colony types. *Bulletin of the European Association of Fish Pathologists* **16**, 101–103.

Austin, B., Austin, D.A., Blanch, A.R., Cerdà, M., Grimont, F., Grimont, P.A.D., Jofre, J., Koblavi, S., Larsen, J.L., Pedersen, K., Tiainen, T., Verdonck, L. and Swings, J. (1997) A comparison of methods for the typing of fish-pathogenic *Vibrio* spp. *Systematic and Applied Microbiology* **20**, 89–101.

Austin, B., Austin, D.A., Dalsgaard, I., Gudmundsdóttir, B.K., Høie, S., Thornton, J.M., Larsen, J.L., O'Hici, B. and Powell, R. (1998) Characterization of atypical *Aeromonas salmonicida* by different methods. *Systematic and Applied Microbiology* **21**, 50–64.

Austin, D.A., McIntosh, D. and Austin, B. (1989) Taxonomy of fish associated *Aeromonas* spp., with the description of *Aeromonas salmonicida* subsp. *smithia* subsp. nov. *Systematic and Applied Microbiology* **11**, 277–290.

Aydin, S., Çelebi, S. and Akyurt, I. (1997) Clinical, haematological and pathological investigations of *Escherichia vulneris* in rainbow trout (*Oncorhynchus mykiss*). *Fish Pathology* **32**, 29–34.

Baba, T., Imamura, J., Izawa, K. and Ikeda, K. (1988) Cell-mediated protection in carp, *Cyprinus carpio* L., against *Aeromonas hydrophila*. *Journal of Fish Diseases* **11**, 171–178.

Bach, R., Wenzel, S., Müller-Prasuhn, G. and Gläsker, M. (1971) Farmed trout as a carrier of *Clostridium botulinum* and a cause of botulism. III. Evidence of *Clostridium botulinum* type E on a fish farm with processing station and in fresh and smoked trout from different sources. *Archives für Lebensmittelhygiene* **22**, 107.

Bachman, S., Ferguson, H.W., Prescott, J.F. and Wilcock, B.P. (1990) Progressive panophthalmitis in chinook salmon, *Oncorhynchus tshawytscha* (Walbaum): a case report. *Journal of Fish Diseases* **13**, 345–353.

Bader, J.A., Shoemaker, C.A., Klesius, P.H., Connolly, M.A. and Barbaree, J.M. (1998) Genomic subtyping of *Edwardsiella ictaluri* isolated from diseased channel catfish by arbitrary primed polymerase chain reaction. *Journal of Aquatic Animal Health* **10**, 22–27.

Bagge, J. and Bagge, O. (1956) *Vibrio anguillarum* som arsag til ulcussygdom hos torsk (*Gadus callaris* Linné). *Nordisk Veterinarmedicin* **8**, 481–492.

Baird-Parker, A.C. (1963) A classification of micrococci and staphylococci based on physiological and biochemical tests. *Journal of General Microbiology* **30**, 409–427.

Baird-Parker, A.C. (1965) A classification of staphylococci and micrococci from worldwide sources. *Journal of General Microbiology* **39**, 363–387.

Baker, K.J., Knittel, M.D. and Fryer, J.L. (1983) Susceptibility of chinook salmon, *Oncorhynchus tshawytsha* (Walbaum), and rainbow trout, *Salmo gairdneri* Richardson, to infection with *Vibrio anguillarum* following sublethal copper exposure. *Journal of Fish Diseases* **6**, 267–275.

Bakopoulos, V., Volpatti, D., Papapanagiotou, E., Richards, R., Galleoti, M. and Adams, A. (1997a) Development of an ELISA to detect *Pasteurella piscicida* in culture and in 'spiked' fish tissue. *Aquaculture* **156**, 359–366.

Bakopoulos, V., Adams, A. and Richards, R.H. (1997b) The effect of iron limitation growth conditions on the cell and extracellular components of the fish pathogen *Pasteurella piscicida*. *Journal of Fish Diseases* **20**, 297–305.

Baldacci, E., Farina, G. and Locci, R. (1966) Emendation of the genus *Streptoverticillium* Baldacci (1958) and revision of some species. *Gionale Microbiologica* **14**, 153–171.

Baldwin, T.J. and Newton, J.C. (1993) Pathogenesis of enteric septicemia of channel catfish, caused by *Edwardsiella ictaluri*: bacteriologic and light and electron microscopic findings. *Journal of Aquatic Animal Health* **5**, 189–198.

Balebona, M.C., Zorrilla, I., Moriñgo, M.A. and Borrego, J.J. (1998) Survey of bacterial pathogens affecting farmed gilt-head sea bream (*Sparus aurata* L.) in southwestern Spain from 1990 to 1996. *Aquaculture* **166**, 19–35.

Balfry, S.K., Albright, L.J. and Evelyn, T.P.T. (1996) Horizontal transfer of *Renibacterium salmoninarum* among farmed salmonids via the faecal-oral route. *Diseases of Aquatic Organisms* **25**, 63–69.

Balfry, S.K., Shariff, M. and Iwama, G.K. (1997) Strain differences in non-specific immunity of tilapia *Oreochromis niloticus* following challenge with *Vibrio parahaemolyticus*. *Diseases of Aquatic Organisms* **30**, 77–80.

Bandín, I., Santos, Y., Barja, J.L. and Toranzo, A.E. (1989) Influence of the growth conditions on the hydrophobicity of *Renibacterium salmoninarum* evaluated by different methods. *FEMS Microbiology Letters* **60**, 71–78.

Bandín, I., Santos, Y., Bruno, D.W., Raynard, R.S., Toranzo, A.E. and Barja, J.L. (1991a) Lack of biological activities in the extracellular products of *Renibacterium salmoninarum*. *Canadian Journal of Fisheries and Aquatic Sciences* **48**, 421–425.

Bandín, I., Santos, Y., Toranzo, A.E. and Barja, J.L. (1991b) MICs and MBCs of chemotherapeutic agents against *Renibacterium salmoninarum*. *Antimicrobial Agents and Chemotherapy* **35**, 1011–1013.

Bandín, I., Santos, Y., Magariños, B., Barja, J.L. and Toranzo, A.E. (1992) The detection of two antigenic groups among *Renibacterium salmoninarum* isolates. *FEMS Microbiology Letters* **94**, 105–110.

Bandín, I., Santos, Y., Barja, J.L. and Toranzo, A.E. (1993) Detection of a common antigen among *Renibacterium salmoninarum*, *Corynebacterium aquaticum* and *Carnobacterium piscicola* by the Western Blot Technique. *Journal of Aquatic Animal Health* **5**, 172–176.

Bandín, I., Heinen, P., Brown, L.L. and Toranzo, A.E. (1996) Comparison of different ELISA kits for detecting *Renibacterium salmoninarum*. *Bulletin of the European Association of Fish Pathologists* **16**, 19–22.

Banner, C.R., Long, J.J., Fryer, J.L. and Rohovec, J.S. (1986) Occurrence of salmonid fish infected with *Renibacterium salmoninarum* in the Pacific Ocean. *Journal of Fish Diseases* **9**, 273–275.

Banner, C.R., Rohovec, J.S. and Fryer, J.L. (1991) A new value for mol percent guanine + cytosine of DNA for the salmonid pathogen *Renibacterium salmoninarum*. *FEMS Microbiology Letters* **79**, 57–60.

Barancin, C.E., Smoot, J.C., Findlay, R.H. and Actis, L.A. (1998) Plasmid-mediated histamine biosynthesis in the bacterial fish pathogens *Vibrio anguillarum*. *Plasmid* **39**, 235–244.

Barham, W.T., Schoonbee, H. and Smit, G.L. (1979) The occurrence of *Aeromonas* and *Streptococcus* in rainbow trout (*Salmo gairdneri*). *Journal of Fish Biology* **15**, 457–460.

Barnes, A.C., Lewin, C.S., Hastings, T.S. and Amyes, S.G.B. (1990a) Cross resistance between oxytetracycline and oxolinic acid in *Aeromonas salmonicida* associated with alterations in outer membrane proteins, *FEMS Microbiology Letters* **72**, 337–340.

Barnes, A.C., Lewin, C.S., Hastings, T.S. and Amyes, S.G.B. (1990b) In vitro activities of 4-quinolones against the fish pathogen *Aeromonas salmonicida. Antimicrobial Agents and Chemotherapy* **34**, 1819–1820.

Barnes, A.C., Lewin, C.S., Hastings, T.S. and Amyes, S.G.B. (1991a) *In vitro* susceptibility of the fish pathogen *Aeromonas salmonicida* to flumequine. *Antimicrobial Agents and Chemotherapy* **35**, 2634–2635.

Barnes, A.C., Amyes, S.G.B., Hastings, T.S. and Lewin, C.S. (1991b) Fluoroquinolones display rapid bactericidal activity and low mutation frequencies against *Aeromonas salmonicida. Journal of Fish Diseases* **14**, 661–667.

Barnes, A.C., Lewin, C.S., Hastings, T.S. and Amyes, S.G.B. (1992) Alterations in outer membrane proteins identified in a clinical isolate of *Aeromonas salmonicida* subsp. *salmonicida. Journal of Fish Diseases* **15**, 279–282.

Barnes, A.C., Hastings, T.S. and Amyes, S.G. (1994) Amoxycillin resistance in Scottish isolates of *Aeromonas salmonicida. Journal of Fish Diseases* **17**, 357–363.

Barnes, A.C., Hastings, T.S. and Amyes, S.G.B. (1995) Aquaculture antibacterials are antagonized by seawater cations. *Journal of Fish Diseases* **18**, 463–465.

Barnes, M.N., Landolt, M.L., Powell, D.B. and Winton, J.R. (1998) Purification of *Piscirickettsia salmonis* and partial characterization of antigens. *Diseases of Aquatic Organisms* **33**, 33–41.

Barry, T., Powell, R. and Gannon, F. (1990) A general method to generate DNA probes for microorganisms. *Biotechnology* **8**, 233–236.

Bartlett, K.H. and Trust, T.J. (1976) Isolation of salmonellae and other potential pathogens from the freshwater aquarium snail, *Ampullaria. Applied and Environmental Microbiology* **31**, 635–639.

Bartos, J.M. and Sommer, C.V. (1981) *In vitro* cell mediated immune response to *M. tuberculosis* and *M. salmoniphilum* in rainbow trout (*Salmo gairdneri*). *Developmental and Comparative Immunology* **5**, 75–83.

Bast, L., Daly, J.G., DeGrandis, S.A. and Stevenson, R.M.W. (1988) Evaluation of profiles of *Aeromonas salmonicida* as epidemiological markers of furunculosis infections in fish. *Journal of Fish Diseases* **11**, 133–145.

Bataillon, E., Dubard and Terre, L. (1897) Un nouveau type de tuberculose. *Comptes rendus des Seances de la Société Biologie* **49**, 446–449.

Baudin-Laurençin, F., Vigneulle, M. and Mevel, M. (1977) Premières observations sur la Corynebacterioses des salmonides en Bretagne. *Bulletin de l'Office International des Epizooties* **87**, 505–507.

Baudin-Laurençin, F. and Tangtrongpiros, J. (1980) Some results of vaccination against vibriosis in Brittany. In: Ahne, W. (ed.), *Fish Diseases, Third COPRAQ-Session*. Berlin, Springer-Verlag, p.60–68.

Baudin-Laurençin, F., Castric, J.-C., Vigneulle, M. and Tixerant, G. (1989) La myxobacteríose viscérale de la truite arc-en-ciel *Salmo gairdneri* R.: une forme nouvelle de la maladie de l'eau froide à *Cytophaga psychrophila. Bulletin d'Academie de Véterinaire de France* **62**, 147–157.

Bauer, O.N., Musselius, V.A. and Strelkov, Yu. A. (1973) *Diseases of Pond Fishes*. Jerusalem, Keter Press, p.39–40.

Baumann, L., Baumann, P., Mandel, M. and Allen, R.D. (1972) Taxonomy of aerobic marine eubacteria. *Journal of Bacteriology* **110**, 402–429.

Baumann, P., Bang, S.S. and Baumann, L. (1978) Phenotypic characterization of *Beneckea anguillara* biotypes I and II. *Current Microbiology* **1**, 85–88.

Baumann, L., Bowditch, R.D. and Baumann, P. (1983) Description of *Deleya* gen. nov. created to accommodate the marine species *Alcaligenes aestus, A. pacificus, A. cupidus, A. venustus* and *Pseudomonas marina. International Journal of Systematic Bacteriology* **33**, 793–802.

Baumann, P., Furniss, A.L. and Lee, J.V. (1984) Genus I, *Vibrio* Pacini 1854, 411[AL]. In: Krieg, N.R. and Holt, J.G. (eds), *Bergey's Manual of Systematic Bacteriology*, Vol. 1. Baltimore, Williams & Wilkins, p.518–538.

Baxa, D.V., Kawai, K. and Kusuda, R. (1988a) Detection of *Flexibacter maritimus* by fluorescent antibody technique in experimentally infected black sea bream fry. *Fish Pathology* **23**, 29–32.

Baxa, D.V., Kawai, K. and Kusuda, R. (1988b) Chemotherapy of *Flexibacter maritimus* infection. *Reports of the USA Marine Biological Institute, Kochi University* **10**, 9–14.

Baya, A.M., Lupiani, B., Hetrick, F.M. and Toranzo, A.E. (1990a) Increasing importance of *Citrobacter freundii* as a fish pathogen. *Fish Health Section/American Fisheries Society Newsletter* **18**, 4.

Baya, A., Toranzo, A.E., Núñez, S., Barja, J.L. and Hetrick, F.M . (1990b) Association of a *Moraxella* sp. and a reo-like virus with mortalities of striped bass, *Morone saxatilis. Pathology in Marine Science* 91–99.

Baya, A.M., Lupiani, B. Hetrick, F.M., Roberson, B.S., Lukacovic, R., May, E. and Poukish, C. (1990c) Association of *Streptococcus* sp. with fish mortalities in the Chesapeake Bay and its tributaries. *Journal of Fish Diseases* **13**, 251–253.

Baya, A.M., Toranzo, A.E., Lupiani, B., Li, T., Roberson, B.S. and Hetrick, F.M. (1991) Biochemical and serological characterization of *Carnobacterium* spp. isolated from farmed and natural populations of striped bass and catfish. *Applied and Environmental Microbiology* **57**, 3114–3120.

Baya, A.M., Li, T., Lupiani, B. and Hetrick, F. (1992a) *Bacillus cereus*, a pathogen for striped bass. Eastern Fish Health and American Fisheries Society Fish Health Section Workshop, 16–19 June 1992, Auburn University, Auburn, p.67.

Baya, A.M., Lupiani, B., Bandín, I., Hetrick, F.M., Figueras, A., Carnahan, A., May, E.M. and Toranzo, A.E. (1992b) Phenotypic and pathobiological properties of *Corynebacterium aquaticum* isolated from diseased striped bass. *Diseases of Aquatic Organisms* **14**, 115–126.

Baya, A.M., Toranzo, A.E., Lupiani, B., Santos, Y. and Hetrick, F.M. (1992c) *Serratia marcescens*: a potential pathogen for fish. *Journal of Fish Diseases* **15**, 15–26.

Becker, C.D. and Fujihawa, M.P. (1978) *The Bacterial Pathogen Flexibacter columnaris and its Epizootiology among Columbia River Fish.* Monograph No. 2, Washington D.C., American Fisheries Society.

Bein, S.J. (1954) A study of certain chromogenic bacteria isolated from 'red-tide' water with a description of a new species. *Bulletin of Marine Science in the Gulf and Caribbean* **4**, 110–119.

Bejerano, Y., Srig, S., Horne, M.T. and Roberts, R.J. (1979) Mass mortalities in silver carp *Hypophthalmichthys molitrix* (Valenciennes) associated with bacterial infection following handling. *Journal of Fish Diseases* **2**, 49–56.

Belding, D.L. (1927) Fish disease epidemics. *Transactions of the American Fisheries Society* **57**, 217–221.

Belding, D.L. and Merrill, B. (1935) A preliminary report upon a hatchery disease of the *Salmonidae. Transactions of the American Fisheries Society* **65**, 76–84.

Bell, G.R. (1961) Two epidemics of apparent kidney disease in cultured pink salmon (*Oncorhynchus gorbuscha*). *Journal of the Fisheries Research Board of Canada* **18**, 559–562.

Bell, G.R., Higgs, D.A. and Traxler, G.S. (1984) The effect of dietary ascorbate, zinc, and manganese on the development of experimentally induced bacterial kidney disease in sockeye salmon (*Oncorhynchus nerka*). *Aquaculture* **36**, 293–311.

Belland, R.J. and Trust, T.J. (1985) Synthesis, export, and assembly of *Aeromonas salmonicida* A-layer analyzed by transposon mutagenesis. *Journal of Bacteriology* **163**, 877–881.

Belland, R.J. and Trust, T.J. (1988) DNA:DNA reassociation analysis of *Aeromonas salmonicida. Journal of General Microbiology* **134**, 307–315.

Belland, R.J. and Trust, T.J. (1989) *Aeromonas salmonicida* plasmids: plasmid-directed synthesis of proteins *in vitro* and in *Escherichia coli* minicells. *Journal of General Microbiology* **135**, 513–524.

Benediktsdóttir, E., Helgason, S. and Gudmundsdóttir, S. (1991) Incubation time for the cultivation of *Renibacterium salmoninarum* from Atlantic salmon, *Salmo salar* L., broodfish. *Journal of Fish Diseases* **14**, 97–102.

Benediktsdóttir, E., Helgason, S. and Sigurjónsdóttir, H. (1998) *Vibrio* spp. isolated from salmonids with shallow skin lesions and reared at low temperature. *Journal of Fish Diseases* **21**, 19–28.

Bercovier, H. and Mollaret, H.H. (1984) Genus XIV. *Yersinia* van Loighem 1944, 15[AL]. In: Krieg, N.R. and Holt, J.G. (eds), *Bergey's Manual of Systematic Bacteriology*, Vol. I., Baltimore, Williams & Wilkins, p.489–506.

Berg, R.W. and Anderson, A.W. (1972) Salmonellae and *Edwardsiella tarda* in gull feces: a source of contamination in fish processing plants. *Applied Microbiology* **24**, 501–503.

Bergh, Ø. (1995) Bacteria associated with early life stages of halibut, *Hippoglossus hippoglossus* L., inhibit growth of a pathogenic *Vibrio* sp. *Journal of Fish Diseases* **18**, 31–40.

Bergh, Ø., Hjeltnes, B. and Skiftesvik, A.B. (1997) Experimental infection of turbot *Scophthalmus maximus* and halibut *Hippoglossus hippoglossus* yolk sac larvae with *Aeromonas salmonicida* subsp. *salmonicida. Diseases of Aquatic Organisms* **29**, 13–20.

Bergman, A.M. (1909) Die rote Beulenkrankheit des Aals. *Bericht aus der Königlichen Bayerischen Versuchsstation* **2**, 10–54.

Bernardet, J.-F. (1989) '*Flexibacter columnaris*': first description in France and comparison with bacterial strains from other origins. *Diseases of Aquatic Organisms* **6**, 37–44.

Bernadet, J.-F. and Grimont, P.A.D. (1989a) Deoxyribonucleic acid relatedness and phenotypic characterization of *Flexibacter columnaris* sp. nov., rom. rev., *Flexibacter psychrophilus* ssp. nov., rom. rev., and *Flexibacter maritimus* Wakabayashi, Hikida and Masumura 1986. *International Journal of Systematic Bacteriology* **39**, 346–354.

Bernardet, J.-F. and Kerouault, B. (1989b) Phenotypic and genomic studies of '*Cytophaga psychrophila*' isolated from diseased rainbow trout (*Oncorhynchus mykiss*) in France. *Applied and Environmental Microbiology* **55**, 1796–1800.

Bernardet, J.-F., Campbell, A.C. and Buswell, J.A. (1990) *Flexibacter maritimus* is the agent of 'black patch necrosis' in Dover sole in Scotland. *Diseases of Aquatic Organisms* **8**, 233–237.

Bernardet, J.-F., Kerouault, B. and Michel, C. (1994) Comparative study on *Flexibacter maritimus* strains isolated from farmed sea bass (*Dicentrarchus labrax*) in France. *Fish Pathology* **29**, 105–111.

Bernardet, J.-F., Segers, P., Vancanneyt, M., Berthe, F., Kersters, K. and Vandamme, P. (1996) Cutting a Gordian knot: emended classification and description of the family *Flavobacteriaceae*, and a proposal of *Flavobacterium hydatis* nom. nov. (Basonym, *Cytophaga aquatilis* Strohl and Tait 1978). *International Journal of Systematic Bacteriology* **46**, 128–148.

Bernheimer, A.W. and Avigad, L.S. (1974) Partial characterization of aerolysin, a lytic exotoxin from *Aeromonas hydrophila*. *Infection and Immunity* **9**, 1016–1021.

Bernoth, E.-M. (1990) Autoagglutination, growth on tryptone-soy-Coomassie-agar, outer membrane protein patterns and virulence of *Aeromonas salmonicida* strains. *Journal of Fish Diseases* **13**, 145–155.

Bernoth, E.-M., Ellis, A.E., Midtlyng, P.J., Olivier, G. and Smith, P. (1997) *Furunculosis Multidisciplinary Fish Disease Research*. San Diego, Academic Press.

Berthe, F.C.J., Michel, C. and Bernardet, J.-F. (1995) Identification of *Pseudomonas anguilliseptica* isolated from several fish species in France. *Diseases of Aquatic Organisms* **21**, 151–155.

Bertone, S., Gili, C., Moizo, A. and Calegari, L. (1996) *Vibrio carchariae* associated with a chronic skin ulcer on a shark, *Carcharhinus plumbeus* (Nardo). *Journal of Fish Diseases* **19**, 429–434.

Beynon, L.M., Richards, J.C. and Perry, M.B. (1994) The structure of the lipopolysaccharide O antigen from *Yersinia ruckeri* serotype O1. *Carbohydrate Research* **256**, 303–317.

Bidinost, C., Crosa, J.H. and Actis, L.A. (1994) Localization of the replication region of the pMJ101 plasmid from *Vibrio ordalii*. *Plasmid* **31**, 242–250.

Biosca, E.G., Amaro, C., Esteve, C., Alcaide, E. and Garay, E. (1991) First record of *Vibrio vulnificus* biotype 2 from diseased European eel, *Anguilla anguilla* L. *Journal of Fish Diseases* **14**, 103–109.

Biosca, E.G., Llorens, H., Garay, E. and Amaro, C. (1993) Presence of a capsule in *Vibrio vulnificus* biotype 2 and its relationship to virulence for eels. *Infection and Immunity* **61**, 1611–1618.

Biosca, E.G., Amaro, C., Larsen, J.L. and Pedersen, K. (1997a) Phenotypic and genotypic characterization of *Vibrio vulnificus*: proposal for the substitution of the subspecific taxon biotype for serovar. *Applied and Environmental Microbiology* **63**, 1460–1466.

Biosca, E.G., Marco-Noales, E., Amaro, C. and Alcaide, E. (1997b) An enzyme-linked immunosorbent assay for detection of *Vibrio vulnificus* biotype 2: development and field studies. *Applied and Environmental Microbiology* **63**, 537–542.

Black, M.M. and Eykyn, S.J. (1977) The successful treatment of tropical fish tank granuloma (*Mycobacterium marinum*) with co-trimoxazole. *British Journal of Dermatology* **97**, 689–692.

Blake, I. (1935) Some observations on other bacterial diseases of fish, with particular reference to the vascular and tissue reactions in these infections. In: Adamson, W. and Baxton, N. (eds), *Final Report of the Furunculosis Committee*. Edinburgh, H.M.S.O., p.1–13.

Blake, I. and Anderson, E.J.M. (1930) The identification of *Bacillus salmonicida* by the complement fixation test—a further contribution to the study of furunculosis of the Salmonidae. *Fisheries Board of Scotland, Salmon Fisheries No. 1*. Edinburgh, H.M.S.O.

Blake, I. and Clark, J.C. (1931) Observations on experimental infection of trout by *B. salmonicida*, with particular reference to 'carriers' of furunculosis and to certain factors influencing susceptibility. *Fisheries Board of Scotland, Salmon Fisheries* No. 7. Edinburgh, H.M.S.O., p.1–13.

Boesen, H.T., Pedersen, K., Koch, C. and Larsen, J.L. (1997) Immune response of rainbow trout (*Oncorhynchus mukiss*) to antigenic preparations from *Vibrio anguillarum* serogroup O1. *Fish and Shellfish Immunology* **7**, 543–553.

Bolinches, J., Romalde, J.L. and Toranzo, A.E. (1988) Evaluation of selective media for isolation and enumeration of vibrios from estuarine waters. *Journal of Microbiological Methods* **8**, 151–160.

Bolinches, J., Lemos, M.L., Fouz, B., Cambra, M., Larsen, J.L. and Toranzo, A.E. (1990) Serological relationships among *Vibrio anguillarum* strains. *Journal of Aquatic Animal Health* **2**, 21–29.

Bonaveri, G.F. (1761) quoted by Drouin de Bouville (1907).

Bonet, R., Simon-Pujol, M.D. and Congregardo, F. (1993) Effects of nutrients on exopolysaccharide production and surface properties of *Aeromonas salmonicida*. *Applied and Environmental Microbiology* **59**, 2437–2441.

Boomker, J., Imes, G.D., Cameron, C.M., Naudé, T.W. and Schoonbee, H.J. (1979) Trout mortalities as a result of *Streptococcus* infection. *Onderstepoort Journal of Veterinary Research* **46**, 71–77.

Bootsma, R. and Clerx, J.P.M. (1976) Columnaris disease of cultured carp *Cyprinus carpio* L. Characteristics of the causative agent. *Aquaculture* **7**, 371–384.

Bootsma, R., Fijan, N. and Blommaert, J. (1977) Isolation and identification of the causative agent of carp-erythrodermatitis. *Veterinarski Archiv* **47**, 291–302.

Borg, A.F. (1948) Studies on myxobacteria associated with diseases of salmonid fishes. Ph.D. thesis, University of Washington, Seattle.

Borg, A.F. (1960) Studies on myxobacteria associated with diseases in salmonid fishes. *Journal of Wildlife Diseases* **8**, 1–85.

Bosse, M.P. and Post, G. (1983) Tribrissen and tiamulin for control of enteric redmouth disease. *Journal of Fish Diseases* **6**, 27–32.

Bott, T.L., Johnson, J., Foster, E.M. and Sugiyama, H. (1968) Possible origin of the high incidence of *Clostridium botulinum* type E in an inland bay (Green Bay of Lake Michigan). *Journal of Bacteriology* **95**, 1542–1547.

Boulanger, Y., Lallier, R. and Cousineau, G. (1977) Isolation of enterotoxigenic *Aeromonas* from fish. *Canadian Journal of Microbiology* **23**, 1161–1164.

Bøvre, K. (1984) Genus II *Moraxella* Lwoff 1939, 173 emend. Henriksen and Bøvre 1968, 391[AL]. In: Krieg, N.R. and Holt, J.G. (eds), *Bergey' Manual of Systematic Bacteriology*, Vol. 1, Baltimore, Williams & Wilkins, p.296–303.

Bowser, P.R. (1973) Seasonal prevalence of *Chondrococcus columnaris* infection in black bullheads from Clear Lake, Iowa. *Journal of Wildlife Diseases* **9**, 115–119.

Bowser, P.R. and House, M. (1990) *In vitro* sensitivity of some fish pathogens to the quinolones nalidixic acid and oxolinic acid and the fluorophinolone enrofloxacin. *Bulletin of the European Association of Fish Pathologists* **10**, 48–49.

Boyce, S.H. (1997) Fish tank granuloma—an unusual cause of skin infection. *Journal of Accident and Emergency Medicine* **14**, 400.

Bragg, R.R. and Todd, J.M. (1988) *In vitro* sensitivity to Baytril of some Bacteria pathogenic to fish. *Bulletin of the European Association of Fish Pathologists* **8**, 5.

Branson, E.J. and Diaz-Munoz, D.N. (1991) Description of a new disease condition occurring in farmed coho salmon, *Oncorhynchus kisutch* (Walbaum), in South America. *Journal of Fish Diseases* **14**, 147–156.

Brattgjerd, S., Evensen, Ø., Speilberg, L. and Lauve, A. (1995) Internalization of *Vibrio salmonicida* in isolated macrophages from Atlantic salmon (*Salmo salar*) and rainbow trout (*Oncorhynchus mykiss*) evaluated by a paired immunofluorescence technique. *Fish and Shellfish Immunology* **5**, 121–135.

Bravo, S. (1994) Piscirickettsiosis in freshwater. *Bulletin of the European Association of Fish Pathologists* **14**, 137–138.

Brenner, D.J. (1984) Family 1. Enterobacteriaceae Rahn 1937, Nom. fam. cons. Opin. 15, Jud. Comm. 1958, 73; Ewing, Farmer and Breener 1980, 674; Judicial Commission 1981, 104. In: Krieg, N.R. and Holt, J.G. (eds), *Bergey's Manual of Systematic Bacteriology*, Vol. I., Baltimore, Williams & Wilkins, p.408–420.

Brenner, D.J., McWhorter, A.C., Knutson, J.K.L. and Steigerwalt, A.G. (1982) *Escherichia vulneris*—a new species of Enterobacteriaceae associated with human wounds. *Journal of Clinical Microbiology* **15**, 1133–1140.

Bricknell, I.R. (1995) A reliable method for the induction of experimental furunculosis. *Journal of Fish Diseases* **18**, 127–133.

Bridge, P.D. and Sneath, P.H.A. 91983) Numerical taxonomy of *Streptococcus*. *Journal of General Microbiology* **129**, 565–597.

Brisou, J., Tysset, C. and Vacher, B. (1964) Recherches sur les Pseudomonadaceae. Etude de deux souches de *Flavobacterium* isolées des poissons d'eau douce. *Annals Institut Pasteur* **74**, 633–638.

Brocklebank, J.R. (1998) Ulcerative dermatitis caused by *Aeromonas salmonicida* spp. *salmonicida* in farmed Atlantic salmon in British Columbia. *Revue Vetérinaire Canadienne* **39**, 110.

Broom, A.K. and Sneath, P.H.A. (1981) Numerical taxonomy of *Haemophilus*. *Journal of General Microbiology* **126**, 123–149.

Brown, L.L., Albright, L.J. and Evelyn, T.P.T. (1990) Control of vertical transmission of *Renibacterium salmoninarum* by injection of antibiotics into maturing female coho salmon *Oncorhynchus kisutch*. *Diseases of Aquatic Organisms* **9**, 127–131.

Brown, L.L., Iwana, G.K., Evelyn, T.P.T., Nelson, W.S. and Levine, R.P. (1994) Use of the polymerase chain reaction (PCR) to detect DNA from *Renibacterium salmoninarum* within individual salmonid eggs. *Diseases of Aquatic Organisms* **18**, 165–171.

Brown, L.L., Evelyn, T.P.T., Iwama, G.K., Nelson, W.S. and Levine, R.P. (1995) Bacterial species other than *Renibacterium salmoninarum* cross react with antisera against *Renibacterium salmoninarum* but are negative for the p57 gene of *Renibacterium salmoninarum* as detected by the polymerase chain reaction (PCR). *Diseases of Aquatic Organisms* **21**, 227–231.

Brown, L.L., Iwama, G.K. and Evelyn, T.P.T. (1996) The effect of early exposure of coho salmon (*Oncorhynchus kisutch*) eggs to the p57 protein of *Renibacterium salmoninarum* on the development of immunity to the pathogen. *Fish and Shellfish Immunology* **6**, 149–165.

Brown, L.L., Cox, W.T. and Levine, R.P. (1997) Evidence that the causal agent of bacterial cold-water disease *Flavobacterium psychrophilum* is transmitted within salmonid eggs. *Diseases of Aquatic Organisms* **29**, 213–218.

Bruckner-Tuderman, L. and Blank, A.A. (1985) Unusual cutaneous dissemination of a tropical fish tank granuloma. *Cutis* **36**, 405–406.

Bruno, D.W. (1987) Serum agglutinating titres against *Renibacterium salmoninarum*, the causative agent of bacterial kidney disease, in rainbow trout, *Salmo gairdneri* Richardson, and Atlantic salmon, *Salmo salar* L. *Journal of Fish Biology* **30**, 327–334.

Bruno, D.W. (1988) The relationship between auto-agglutination, cell surface hydrophobicity and virulence of the fish pathogen *Renibacterium salmoninarum*. *FEMS Microbiology Letters* **51**, 135–140.

Bruno, D.W., Hastings, T.S., Ellis, A.E. and Wootten, R. (1985) Outbreak of a cold-water vibriosis in Atlantic salmon in Scotland. *Bulletin of the European Association of Fish Pathologists* **5**, 62–63.

Bruno, D.W. and Munro, A.L.S. (1992) Detection of the causative agent of bacterial kidney disease. *Bulletin of the European Association of Fish Pathologists* **2**, 10–12.

Bruno, D.W., Griffiths, J., Michell, C.G., Wood, B.P., Fletcher, Z.J., Drobniewski, F.A. and Hastings, T.S. (1998) Pathology attributed to *Mycobacterium chelonae* infection among farmed and laboratory-infected Atlantic salmon *Salmo salar*. *Diseases of Aquatic Organisms* **33**, 101–109.

Bryant, T.N. and Smith, K.R. (1991) *Bacterial Identifier, Version M1.0*. Blackwell, Oxford.

Bruun, A.F. and Heiberg, B. (1935) Weitere Untersuchungen über die Rotseuche des Aales in den dänischen Gewäsern. *Zeitschrift für Fischerei und deren Hilfwissenschaften* **33**, 379–382.

Bucke, D. (1978) Bacterial kidney disease (BKD) of fish. *Fisheries Notice*. Lowestoft, M.A.F.F., **60**, 1–4.

Bucke, D. (1979) Investigation into the cause of an epizootic in perch (*Perca fluviatilis* L.). In: O'Hara, K. (ed.), *Proceedings of the First British Freshwater Fish Conference*. London, University of Liverpool, Janssen Services, p.45.

Bucke, D. (1991) Current approaches to the study of pollution-related diseases of fish. *Bulletin of the European Association of Fish Pathologists* **11**, 46–53.

Bucke, D. (1997) Facts and myths regarding pollution and fish health. *Bulletin of the European Association of Fish Pathologists* **17**, 191–196.

Bucke, D., Vethaak, A.D. and Lang, T. (1992) Quantitative assessment of melanomacrophage centres (MMCs) in dab *Limanda limanda* as indicators of pollution effects on the non-specific immune system. *Marine Progress Series* **91**, 193–196.

Buckley, J.T., Halasa, L.N. and MacIntyre, S. (1982) Purification and partial characterization of a bacterial phospholipid: cholesterol acyltransferase. *Journal of Biological Chemistry* **257**, 3320–3325.

Bullen, J.J., Roger, H.J. and Griffiths, E. (1978) Role of iron in bacterial infections. *Current Topics in Microbiology and Immunology* **80**, 1–35.

Bullock, G.L. (1965) Characteristics and pathogenicity of a capsulated *Pseudomonas* isolated from goldfish. *Applied Microbiology* **13**, 89–92.

Bullock, G.L. (1966) Precipitins and agglutinin reactions of aeromonads isolated from fish and other sources. *Bulletin de l'Office International des Epizooties* **65**, 805–824.

Bullock, G.L. (1981) Streptococcal infections of fishes. *United States Department of the Interior, Fish and Wildlife Service, Fish Disease Leaflet* **63**, 1–7.

Bullock, G.L. and Conroy, D.A. (1971) Vibrio diseases. In: Snieszko, S.F. and Axelrod, H.R. (eds), *Diseases of Fishes, 2A. Bacterial Diseases of Fishes*. Jersey City, New York, T.F.H. Publications, p.42–50.

Bullock, G.L. and McLaughlin, J.A.A. (1970) Advances in knowledge concerning bacteria pathogenic to fishes (1954–1968). In: Snieszko, S.F. (ed.), *A Symposium on Diseases of Fishes and Shellfishes*. Washington D.C., American Fisheries Society, Special Publication No. 5, p.231–242.

Bullock, G.L. and Snieszko, S.F. (1970) Fin rot, coldwater disease, and peduncle disease of salmonid fish. *United States Department of the Interior, Fish and Wildlife Service, Fish Disease Leaflet* **25**, 1–3.

Bullock, G.L., Conroy, D.A. and Snieszko, S.F. (1971) Bacterial diseases of fishes. In: Snieszko, S.F. and Axelrod, H.R. (eds), *Diseases of Fishes*, Book 2A. Neptune, New Jersey, T.F.H. Publications, 151p.

Bullock, G.L., Stuckey, H.M. and Chen, P.K. (1974) Corynebacterial kidney disease of salmonids: growth and serological studies on the causative bacterium. *Applied Microbiology* **28**, 811–814.

Bullock, G.L. and Snieszko, S.F. (1975) Hagerman redmouth, a disease of salmonids caused by a member of the Enterobacteriaceae. *United States Department of the Interior, Fish and Wildlife Service, Fish Disease Leaflet* **42**, 1–5.

Bullock, G.L. and Stuckey, H.M. (1975a) *Aeromonas salmonicida*: detection of asymptomatically infected trout. *Progressive Fish Culturist* **37**, 237–239.

Bullock, G.L. and Stuckey, H.M. (1975b) Fluorescent antibody identification and detection of the *Corynebacterium* causing kidney disease of salmonids. *Journal of the Fisheries Research Board of Canada* **32**, 2224–2227.

Bullock, G.L., Stuckey, H.M. and Wolf, K. (1975) Bacterial kidney disease of salmonid fishes. *Fish and Wildlife Service, Fish Diseases Leaflet 41*. Washington D.C., United States Department of the Interior.

Bullock, G.L., Stuckey, H.M. and Shotts, E.B. (1978a) Enteric redmouth bacterium: comparison of isolates from different geographical areas. *Journal of Fish Diseases* **1**, 351–354.

Bullock, G.L., Stuckey, H.M. and Mulcahy, D. (1978b) Corynebacterial kidney disease: egg transmission following iodophore disinfection. *Fish Health News* **76**, 51–52.

Bullock, G.L., Griffin, B.R. and Stuckey, H.M. (1980). Detection of *Corynebacterium salmoninus* by direct fluorescent antibody test. *Canadian Journal of Fisheries and Aquatic Sciences* **37**, 719–721.

Bullock, G.L., Maestrone, G., Starliper, C. and Schill, B. (1983) Potentiated sulfonamide therapy of enteric redmouth disease. *Canadian Journal of Fisheries and Aquatic Sciences* **40**, 101–102.

Bullock, G.L. and Anderson, D.P. (1984) Immunization against *Yersinia ruckeri*, cause of enteric redmouth disease. In: De Kinkelin, P. (ed.), *Symposium on Fish Vaccination; Theoretical Background and Practical Results on Immunization against Infectious Diseases*, Paris, Office International des Epizooties, p.151–166.

Bullock, G.L., Hsu, T.C. and Shotts, E.B. (1986) *Columnaris disease of fishes*. Fish disease leaflet No. 72, US Department of the Interior, Fish and Wildlife Service, Division of Fisheries and Wetlands Research, Washington, DC, 9pp.

Bullock, G.L., Cipriano, R.C. and Schill, W.B. (1997) Culture and serodiagnostic detection of *Aeromonas salmonicida* from covertly-infected rainbow trout given the stress-induced furunculosis test. *Biomedical Letters* **55**, 169–177.

Burke, J. and Rodgers, L. (1981) Identification of pathogenic bacteria associated with the occurrence of 'red spot' in sea mullet, *Mugil cephalus* L., in south-eastern Queensland. *Journal of Fish Diseases* **3**, 153–159.

Busch, R.A. (1973) The serological surveillance of salmonid populations for presumptive evidence of specific disease association. Ph.D. dissertation, University of Idaho.

Busch, R.A. (1978) Enteric redmouth disease (Hagerman) strain. *Marine Fisheries Review* **40**, 42–51.

Busch, R.A. (1981) The current status of diagnostic serology for the major bacterial disease of fish. *Developments in Biological Standardization* **49**, 85–96.

Busch, R.A. and Lingg, A. (1975) Establishment of an asymptomatic carrier state infection of enteric redmouth disease in rainbow trout (*Salmo gairdneri*). *Journal of the Fisheries Research Board Canada* **32**, 2429–2432.

Busch, R.A., Burmeister, N.E. and Scott, A.L. (1978) Field and laboratory evaluation of a commercial enteric redmouth disease vaccine for rainbow trout. *Proceedings of the Joint 3rd Biennial Fish Health Section and 9th Annual Midwest Fish Disease Workshop*, p.67.

Bustos, P.A., Calbuyahue, J., Montaña, J., Opazo, B., Entrala, P. and Solervicens, R. (1995) First isolation of *Flexibacter psychrophilus*, as causative agent of rainbow trout fry syndrome (RTFS), producing rainbow trout mortality in Chile. *Bulletin of the European Association of Fish Pathologists* **15**, 162–164.

Campbell, E.J.M., Scadding, J.G. and Roberts, M.S. (1979) The concept of disease. *British Medical Journal* **2**, 757–762.

Campbell, G. and MacKelvie, R.M. (1968) Infection of brook trout (*Salvelinus fontinalis*) by nocardiae. *Journal of the Fisheries Research Board of Canada* **25**, 423–425.

Campos-Pérez, J.J., Ellis, A.E. and Secombes, C.J. (1997) Investigation of factors influencing the ability of *Renibacterium salmoninarum* to stimulate rainbow trout macrophage respiratory burst activity. *Fish and Shellfish Immunology* **7**, 555–566.

Canestrini, G. (1893) La malatti dominate delle anguille. *Atti Institute Veneto Service* **7**, 809–814.

Candan, A., Küçüker, M.Ang and Karatas, S. (1995) Motile aeromonad septicaemia in *Salmo salar* cultured in the Black Sea in Turkey. *Bulletin of the European Association of Fish Pathologists* **15**, 195–196.

Cann, D.C., Wilson, B.B., Hobbs, G. and Shewan, J.M. (1965a) The growth and toxin production of *Clostridium botulinum* type E in certain vacuum packed fish. *Journal of Applied Bacteriology* **28**, 431–436.

Cann, D.C., Wilson, B.B., Hobbs, G., Shewan, J.M. and Johannsen, A. (1965b) The incidence of *Clostridium botulinum* type E in fish and bottom deposits in the North Sea and off the coast of Scandinavia. *Journal of Applied Bacteriology* **28**, 426–430.

Cann, D.C., Taylor, L.Y. and Hobbs, (1975) The incidence of *Cl. botulinum* in farmed trout raised in Great Britain. *Journal of Applied Bacteriology* **39**, 321–326.

Cann, D.C. and Taylor, L.Y. (1982) An outbreak of botulism in rainbow trout, *Salmo gairdneri* Richardson, farmed in Britain. *Journal of Fish Diseases* **5**, 393–399.

Cann, D.C. and Taylor, L.Y. (1984) An evaluation of residual contamination by *Clostridium botulinum* in a trout farm following an outbreak of botulism in the fish stock. *Journal of Fish Diseases* **7**, 391–396.

Carnahan, A.M. and Altwegg, M. (1996) 1. Taxonomy. In: *The Genus Aeromonas* (edited by B. Austin, M. Altwegg, P.J. Gosling and S. Joseph), pp. 1–38. Chichester, John Wiley.

Carlson, R.V. and Pacha, R.E. (1968) Procedure for the isolation and enumeration of myxobacteria from aquatic habitats. *Applied Microbiology* **16**, 795–796.

Carson, J. and Munday, B. (1990) Streptococcosis—an emerging disease in aquaculture. *Austrasia Aquaculture* **5**, 32–33.

Carson, J. and Schmidtke, L.M. (1993) Opportunistic infection by psychrotrophic bacteria of cold-comprised Atlantic salmon. *Bulletin of the European Association of Fish Pathologists* **13**, 49–52.

Carson, J. and Statham, P. (1993) The inhibition by ionophores *in vitro* of an *Enterococcus*-like pathogen of rainbow trout, *Oncorhynchus mykiss*. *Veterinary Microbiology* **36**, 253–259.

Carson, J., Schmidtke, L.M. and Munday, B.L. (1993) *Cytophaga johnsonae*: a putative skin pathogen of juvenile farmed barramundi, *Lates calcarifer* Bloch. *Journal of Fish Diseases* **16**, 209–218.

Cato, E.P., George, W.L. and Finegold, S.M. (1986) Genus *Clostridium* Prazmowski 1880, 23[AL]. In: Sneath, P.H.A., Mair, N.S., Sharpe, M.E. and Holt, J.G. (eds), *Bergey's Manual of Systematic Bacteriology*, Vol. 2. Baltimore, Williams & Wilkins, p.1141–1200.

Centers for Disease Control (1996) Invasive infection with *Streptococcus iniae*—Ontario, 1995–1996. *Morbidity and Mortality Weekly Report* **45**, 650–653.

Ceschia, G., Giorgetti, G., Giavenni, R. and Sarti, M. (1992) A new problem for Italian trout farms: streptococcosis in rainbow trout (*Oncorhynchus mykiss*). *Bulletin of the European Association of Fish Pathologists* **12**, 1–72.

Chakroun, C., Grimont, F., Urdaci, M.C. and Bernadet, J.-F. (1998) Fingerprinting of

Flavobacterium psychrophilum isolates by ribotyping and plasmid profiling. *Diseases of Aquatic Organisms* **33**, 167–177.

Chart, H. and Trust, T.J. (1983) Acquisition of iron by *Aeromonas salmonicida*. *Journal of Bacteriology* **156**, 758–764.

Chart, H. and Trust, T.J. (1984) Characterization of the surface antigens of the marine fish pathogens, *Vibrio anguillarum* and *Vibrio ordalii*. *Canadian Journal of Microbiology* **30**, 703–710.

Chen, C.-R.L., Chung, Y.-Y. and Kuo, G.-H. (1982) Studies on the pathogenicity of *Flexibacter columnaris*—1. Effect of dissolved oxygen and ammonia on the pathogenicity of *Flexibacter columnaris* to eel (*Anguilla japonica*). *CAPD Fisheries Series No. 8, Reports on Fish Disease Research* **4**, 57–61.

Chen, D. and Ainsworth, A.J. (1992) Glucan administration potentiates immune defence mechanisms of channel catfish, *Ictalurus punctatus* Rafinesque. *Journal of Fish Diseases* **15**, 295–304.

Chen, P.K., Bullock, G.L., Stuckey, H.M. and Bullock, A.C. (1974) Serological diagnosis of corynebacterial kidney disease of salmonids. *Journal of the Fisheries Research Board of Canada* **31**, 1939–1940.

Chen, S.-C. (1992) Study on the pathogenicity of *Nocardia asteroides* to the Formosa snakehead (*Channa maculata* (Lacepède)). *Journal of Fish Diseases* **15**, 47–53.

Chen, S.-C., Tung, M.-C., Chen, S.-P., Tsai, J.-F., Wang, P.-C., Chen, R.-S., Lin, S.-C. and Adams, A. (1994) Systematic granulomas caused by a rickettsia-like organism in Nile tilapia, *Oreochronius niloticus* (L), from southern Taiwan. *Journal of Fish Diseases* **17**, 591–599.

Chen, S.-C., Adams, A. and Richards, R.H. (1997) Extracellular products from *Mycobacterium* spp. in fish. *Journal of Fish Diseases* **20**, 19–25.

Chen, Q., Actis, L.A., Tolmasky, M.E. and Crosa, J.H. (1994) Chromosome-mediated 2,3-dihydroxybenzoic acid is a precursor in the biosynthesis of the plasmid-mediated siderophore anguibactin in *Vibrio anguillarum*. *Journal of Bacteriology* **176**, 4226–4234.

Chen, Q., Wertheimer, A.M., Tolmasky, M.E. and Crosa, J.H. (1996) The AngR protein and the siderophore anguibactin positively regulate the expression of iron-transport genes in *Vibrio anguillarum*. *Molecular Microbiology* **22**, 127–143.

Chern, R.S. and Chao, C.B. (1994) Outbreaks of disease caused by rickettsia-like organism in cultured tilapias in Taiwan. *Fish Pathology* **29**, 61–71.

Chester, F.D. (1897) A preliminary arrangement of the species of the genus *Bacterium*. Contributions of determinative bacteriology. Part 1. *Ninth Annual Report, Delaware College Agricultural Experimental Station*, 92p.

Chomarat, M., Guerin-Faublee, V., Kodjo, A., Breysse, F. and Flandrois, J.P. (1998) Molecular analysis of the fish pathogen *Aeromonas salmonicida* subsp. *salmonicida* by pulsed field gel electrophoresis. *Revue de Medecine Vetérinaire* **149**, 245–250.

Chowdbury, M.B.R. and Wakabayashi, H. (1989) Effects of competitive bacteria on the survival and infectivity of *Flexibacter columnaris*. *Fish Pathology* **24**, 9–15.

Christensen, P.J. (1977) The history, biology and taxonomy of the *Cytophaga* group. *Canadian Journal of Microbiology* **23**, 1599–1653.

Chun, S.-K. (1975) The pathogenicity of myxobacteria isolated from infected fish. *Bulletin of the National Fisheries University, Busan* **15**, 31–42.

Cipriano, R.C. (1982a) Immunization of brook trout (*Salvelinus fontinalis*) against *Aeromonas salmonicida*: immunogenicity of virulent and avirulent isolates and protective ability of different antigens. *Canadian Journal of Fisheries and Aquatic Sciences* **39**, 218–221.

Cipriano, R.C. (1982b) Furunculosis in brook trout: infection by contact exposure. *Progressive Fish Culturist* **44**, 12–14.

Cipriano, R.C. (1982c) Resistance of salmonids to *Aeromonas salmonicida*: relation between agglutinins and neutralizing activities. *Transactions of the American Fisheries Society* **112**, 95–99.

Cipriano, R.C. (1983) Resistance of salmonids to *Aeromonas salmonicida*: relation between agglutinins and neutralizing activities. *Transactions of the American Fisheries Society* **112**, 95–99.

Cipriano, R.C., Griffin, B.R. and Lidgerding, B.C. (1981) *Aeromonas salmonicida*: relationship between extracellular growth products and isolate virulence. *Canadian Journal of Fisheries and Aquatic Sciences* **38**, 1322–1326.

Cipriano, R.C. and Starliper, C.E. (1982) Immersion and injection vaccination of salmonids against furunculosis with an avirulent strain of *Aeromonas salmonicida*. *Progressive Fish Culturist* **44**, 167–169.

Cipriano, R.C., Ford, L.A., Teska, J.D. and Hale, L.E. (1992) Detection of *Aeromonas salmonicida* in the mucus of salmonid fishes. *Journal of Aquatic Animal Health* **4**, 114–118.

Cipriano, R.C., Bullock, G.L. and Noble, A. (1996a) Nature of *Aeromonas salmonicida* carriage on asymptomatic rainbow trout maintained in a culture system with recirculating water and fluidized sand biofilters. *Journal of Aquatic Animal Health* **8**, 47–51.

Cipriano, R.C., Ford, L.A., Teska, J.D., Schachte, J.H., Petrie, C., Novak, B.M. and Flint, D.E. (1996b) Use of non-lethal procedures to detect and monitor *Aeromonas salmonicida* in potentially endangered or threatened populations of migrating and post-spawning salmon. *Diseases of Aquatic Organisms* **27**, 233–236.

Cipriano, R.C., Ford, L.A., Starliper, C.E., Teska, J.D., Nelson, J.T. and Jensen, B.N. (1996c) Control of external *Aeromonas salmonicida*: topical disinfection of salmonids with chloramine T. *Journal of Aquatic Animal Health* **8**, 52–57.

Cipriano, R.C., Ford, L.A., Nelson, J.T. and Jenson, B.N. (1996d) Monitoring for early detection of *Aeromonas salmonicida* to enhance antibiotic therapy and control furunculosis in Atlantic salmon. *Progressive Fish Culturist* **58**, 203–208.

Cipriano, R.C., Ford, L.A., Smith, D.R., Schachte, J.H. and Petrie, C.J. (1997) Differences in detection of *Aeromonas salmonicida* in covertly infected salmonid fishes by the stress-inducible furunculosis test and culture-based assays. *Journal of Aquatic Animal Health* **9**, 108–113.

Cisar, J.O. and Fryer, J.L. (1974) Characterization of anti-*Aeromonas salmonicida* antibodies from coho salmon. *Infection and Immunity* **9**, 236–243.

Citarella, R.V. and Colwell, R.R. (1970) Polyphasic taxonomy of the genus *Vibrio*:

polynucleotide sequence relationships among selected *Vibrio* species. *Journal of Bacteriology* **104**, 434–442.

Clausen, H.J. and Duran-Reynals, F. (1937) Studies on the experimental infection of some reptiles, amphibia and fish with *Serratia anolium. American Journal of Pathology* **13**, 441–541.

Claveau, R. (1991) Néphrite granulomateuse à *Rhodococcus* spp dans un élevage de saumons de l'Atlantique (*Salmo salar*). *Le Médecin Vétérinaire du Québec* **21**, 160–161.

Collins, M.D., Farrow, J.A.E., Phillips, B.A., Ferusu, S. and Jones, D. (1987) Classification of *Lactobacillus divergens, Lactobacillus piscicola,* and some catalase-negative, asporogenous, rod-shaped bacteria from poultry in a new genus, *Carnobacterium. International Journal of Systematic Bacteriology* **37**, 310–316.

Collins, V.G. (1970) Recent studies on bacterial pathogens of freshwater fish. *Water Treatment and Examination* **19**, 3–31.

Colorni, A., Paperna, I. and Gordin, H. (1981) Bacterial infections in gilthead sea bream *Sparus aurata* cultured in Elat. *Aquaculture* **23**, 257–267.

Colwell, R.R. and Grimes, D.J. (1984) *Vibrio* diseases of marine fish populations. *Helgoländer Meeresuntersuchungen* **37**, 265–287.

Colwell, R.R., MacDonell, M.T. and De Ley, J. (1986) Proposal to recognize the family Aeromondaceae fam. nov. *International Journal of Systematic Bacteriology* **36**, 473–477.

Comps, M., Raymond, J.C. and Plassiart, G.N. (1996) Rickettsia-like organism infecting juvenile sea-bass *Dicentrarchus labrax. Bulletin of the European Association of Fish Pathologists* **16**, 30–33.

Conchas, R.F., Lemos, M.L., Barja, J.L. and Toranzo, A.E. (1991) Distribution of plasmid- and chromosome-mediated iron uptake systems in *Vibrio anguillarum* strains of different origins. *Applied and Environmental Microbiology* **57**, 2956–2962.

Cone, D.K. (1982) A *Lactobacillus* sp. from diseased female rainbow trout, *Salmo gairdneri* Richardson, in Newfoundland, Canada. *Journal of Fish Diseases* **5**, 479–485.

Conroy, D.A. (1961) Estudio in vitro de la Accion de la Kanamicina sobre bacterias patogenas para los peces. *Microbiología Españolo* **14**, 147–155.

Conroy, D.A. (1964) Notes on the incidence of piscine tuberculosis in Argentina. *Progressive Fish Culturist* **26**, 89–90.

Conroy, D.A. (1966) Observaciones sobre casos Espontáneous de tuberculosis ictica. *Microbiología Españolo* **19**, 93–113.

Conroy, D.A. (1986) Agents: Bacteria. In: Kinne, O. (ed.), *Diseases of Marine Animals,* Vol. IV, Helgoland, Biologische Anstalt.

Conroy, D.A. and Valdez, I.E. (1962) Un casos de tuberculosis on peces tropicales. *Revista Latinoamericana de Microbiología* **5**, 9–16.

Cooke, D.W. and Lofton, S.R. (1975) Pathogenicity studies with a *Streptococcus* sp. isolated from fishes in an Alabama-Florida fish kill. *Transactions of the American Fisheries Society* **104**, 286–288.

Cook, T.M. and Gemski, P. (1982) Studies of plasmids in the fish pathogen, *Yersinia ruckeri. Proceedings of the International Congress of Microbiologists* **13**, 97.

Cornick, J.W., Chudyk, R.V. and McDermott, L.A. (1969) Habitat and viability studies on *Aeromonas salmonicida*, causative agent of furunculosis. *Progressive Fish Culturist* **31**, 90–93.

Cossarini-Dunier, M. (1985) Indirect enzyme-linked immunosorbent assay (ELISA) to titrate rainbow trout serum antibodies against two pathogens: *Yersinia ruckeri* and Egtved virus. *Aquaculture* **49**, 197–208.

Cossarini-Dunier, M. (1986) Protection against enteric redmouth disease in rainbow trout, *Salmo gairdneri* Richardson, after vaccination with *Yersinia ruckeri* bacterin. *Journal of Fish Diseases* **9**, 27–33.

Cowan, S.T. (1974) *Cowan and Steel's Manual for the Identification of Medical Bacteria*, 2nd edn. Cambridge, Cambridge University Press.

Crosa, J.H. (1980) A plasmid associated with virulence in the marine fish pathogen *Vibrio anguillarum* specifies an iron-sequestering system. *Nature (London)* **283**, 566–568.

Crosa, J.H., Schiewe, M.H. and Falkow, S. (1977) Evidence for plasmid contribution to the virulence of the fish pathogen *Vibrio anguillarum. Infection and Immunity* **18**, 509–513.

Crosa, J.H. and Hodges, L.L. (1981) Outer membrane proteins induced under conditions of iron limitation in the marine fish pathogen *Vibrio anguillarum. Infection and Immunity* **31**, 223–227.

Crosa, J.H., Walter, M.A. and Potter, S.A. (1983) The genetics of plasmid-mediated virulence in the marine fish pathogen *Vibrio anguillarum*. In: Crosa, J.H. (ed.), *Bacteria and Viral Diseases of Fish, Molecular Studies*. Seattle, University of Washington, p.21–30.

Croy, T.R. and Amend, D.F. (1977) Immunization of sockeye salmon (*Oncorhynchus nerka*) against vibriosis using the hyperosmotic infiltration technique. *Aquaculture* **12**, 317–325.

Cruz, J.M., Saraiva, A., Eiras, J.C., Branco, R. and Sousa, J.C. (1986) An outbreak of *Plesiomonas shigelloides* in farmed rainbow trout, *Salmo gairdneri* Richardson, in Portugal. *Bulletin of the European Association of Fish Pathologists* **6**, 20–22.

Csaba, Gy., Körmendy, B . and Békési, L. (1981a) Observations on the causative agent of carp erthrodermatitis in Hungary. *The Proceedings of an International Seminar on Fish, Pathogens and Environment in European Polyculture, Szarvas, Hungary*, p. 95–110.

Csaba, Gy., Prigli, M., Békési, L., Kovács-Gayer, E., Bajmócy, E. and Fazekas, B. (1981b) Septicaemia in silver carp (*Hypophthalmichthys molitrix*, Val.) and bighead (*Aristichthys nobilis* Rich.) caused by *Pseudomonas fluorescens*. In: Oláh, J., Molnár, K. and Jeney, S. (eds), *Fish, Pathogens and Environment in European Polyculture*, Szarvas, F. Muller (Fisheries Research Institute), p. 111–123.

Cumberbatch, N., Gurwith, M.J., Langston, C., Sack, R.B. and Brunton, J.L. (1979) Cytotoxic enterotoxin produced by *Aeromonas hydrophila*: relationship of toxigenic isolates to diarrheal disease. *Infection and Immunity* **23**, 829–837.

Cvitanich, J.D. (1994) Improvements in the direct fluorescent antibody technique for the detection, identification, and quantification of *Renibacterium salmoninarum* in salmonid kidney smears. *Journal of Aquatic Animal Health* **6**, 1–12.

Cvitanich, J., Garate, O. and Smith, C.E. (1991) The isolation of a rickettsia-like organism causing disease and mortality in Chilean salmonids and its confirmation by Koch's postulates. *Journal of Fish Diseases* **14**, 121–145.

Dalmo, R.A., Ingebrigtsen, K., Bøgwald, J., Horsberg, T.E. and Seljelid, R. (1995) Accumulation of immunomodulatory laminaran (β (1,3)-D-glucan) in the spleen and kidney of Atlantic salmon, *Salmo salar* L. *Journal of Fish Diseases* **18**, 545–553.

Dalmo, R.A. and Seljelid, R. (1995) The immunomodulatory effect of LPS, laminaran and sulphated laminaran [β(1,3)-D-glucan] on Atlantic salmon, *Salmo salar* L., macrophages *in vitro*. *Journal of Fish Diseases* **18**, 175–185.

Dalsgaard, I., From, J. and Hølyck, V. (1984) First observations of *Yersinia ruckeri* in Denmark. *Bulletin of the European Association of Fish Pathologists* **4**, 10.

Dalsgaard, I., Nielsen, B. and Larsen, J.L. (1994) Characterization of *Aeromonas salmonicida* subsp. *salmonicida*: a comparative study of strains of different geographic origin. *Journal of Applied Bacteriology* **77**, 21–30.

Dalsgaard, I., Gudmundsdóttir, B.K., Helgason, S., Høie, S., Thoresen, O.F., Wichardt, U.-P. and Wiklund, T. (1998) Identification of atypical *Aeromonas salmonicida*: inter-laboratory evaluation and harmonization of methods. *Journal of Applied Microbiology* **84**, 999–1006.

Daly, J.G. and Stevenson, R.M.W. (1985a) Charcoal agar, a new growth medium for the fish disease bacterium *Renibacterium salmoninarum*. *Applied and Environmental Microbiology* **50**, 868–871.

Daly, J.G. and Stevenson, R.M.W. (1985b) Importance of culturing several organs to detect *Aeromonas salmonicida* in salmonid fish. *Transactions of the American Fisheries Society* **114**, 909–910.

Daly, J.G. and Stevenson, R.M.W. (1987) Hydrophobic and haemagglutinating properties of *Renibacterium salmoninarum*. *Journal of General Microbiology* **133**, 3575–3580.

Daly, J.G. and Stevenson, R.M.W. (1989) Agglutination of salmonid spermatozoa by *Renibacterium salmoninarum*. *Fish Health News*.

Daly, J.G. and Stevenson, R.M.W. (1990) Characterization of the *Renibacterium salmoninarum* haemagglutinins. *Journal of General Microbiology* **136**, 949–953.

Daly, J.G., Kew, A.K., Moore, A.R. and Olivier, G. (1996) The cell surface of *Aeromonas salmonicida* determines *in vitro* survival in cultivated brook trout (*Salvelinus fontinalis*) peritoneal macrophages. *Microbial Pathogenesis* **21**, 447–461.

Daskalov, H., Stobie, M. and Austin, B. (1998) *Klebsiella pneumoniae*: a pathogen of rainbow trout (*Oncorhynchus mykiss*, Walbaum). *Bulletin of the European Association of Fish Pathologists* **18**, 26–28.

David, H. (1927) Über eine durch cholera ähnliche Vibrionen hervorgerufene Fischseuche. *Zentralblatt für Bakteriologie und Parasitenkunde, Abteilung 1, Originale* **102**, 46–60.

Davey, M.L., Hancock, R.E.W. and Mutharia, L.M. (1998) Influence of culture conditions on expression of the 40-kilodalton porin protein of *Vibrio anguillarum* serotype O2. *Applied and Environmental Microbiology* **64**, 138–146.

Davies, J.A. (1969) Isolation and identification of clostridia from North Sea sediments. *Journal of Applied Bacteriology* **32**, 164–169.

Davies, R.L. (1990) O-serotyping of *Yersinia ruckeri* with special emphasis on European isolates. *Veterinary Microbiology* **22**, 299–307.

Davies, R.L. (1991) Outer membrane protein profiles of *Yersinia ruckeri*. *Veterinary Microbiology* **26**, 125–140.

Davis, H.S. (1922) A new bacterial disease of freshwater fishes. *U.S. Bureau of Fisheries Bulletin* **38**, 261–280.

Davis, H.S. (1929) Care and diseases of trout. *United States Department of Commerce, Bureau of Fisheries Document* **1061**, 155.

Davis, H.S. (1946) Care and diseases of trout. *Research Report, United States Fisheries and Wildlife Service* **12**, 98.

Dear, G. (1985) Studies on the biology, metabolism and pathogenicity of *Pseudomonas anguilliseptica*. Ph.D thesis, Heriot-Watt University.

Dec, C., Angelidus, P. and Baudin-Laurençin, F. (1990) Effects of oral vaccination against vibriosis in turbot, *Scophthalmus maximus* (L.), and sea bass, *Dicentrarchus labrax* (L.). *Journal of Fish Diseases* **13**, 369–376.

Decew, M.G. (1972) Antibiotic toxicity, efficacy and teratogenicity in adult spring chinook salmon (*Oncorhynchus tschawytscha*). *Journal of the Fisheries Research Board of Canada* **29**, 1513–1517.

Deere, D., Porter, J., Pickup, R.W. and Edwards, C. (1996a) Survival of cells and DNA of *Aeromonas salmonicida* released into aquatic microcosms. *Journal of Applied Bacteriology* **81**, 309–318.

Deere, D., Porter, J., Pickup, R. and Edwards, C. (1996b) Direct analysis of starved *Aeromonas salmonicida*. *Journal of Fish Diseases* **19**, 459–467.

De Grandis, S.A. and Stevenson, R.M.W. (1985) Antimicrobial susceptibility patterns and R plasmid-mediated resistance of the fish pathogen *Yersinia ruckeri*. *Antimicrobial Agents and Chemotherapy* **27**, 938–942.

De Grandis, S.A., Krell, P.J., Flett, D.E. and Stevenson, R.M.W. (1988) Deoxyribunocleic acid relatedness of serovars of *Yersinia ruckeri*, the enteric redmouth bacterium. *International Journal of Systematic Bacteriology* **38**, 49–55.

De La Cruz, M. and Muroga, K. (1989) The effects of *Vibrio anguillarum* extracellular products on Japanese eels. *Aquaculture* **80**, 201–210.

Del Corral, F., Shotts, E.B. and Brown, J. (1990) Adherence, haemagglutination and cell surface characteristics of motile aeromonads virulent for fish. *Journal of Fish Diseases* **13**, 255–268.

De Meuron, P.-A. and Peduzzi, R. (1979) Caractérisation de souches du genre *Aeromonas* isolées chez de poisons d'eau douce et quelques reptiles. *Zentralblatt für Veterinaermedizin Reihe* **B26**, 153–167.

Densmore, C.L., Smith, S.A. and Holladay, S.D. (1998) *In vitro* effects of the extracellular protein of *Renibacterium salmoninarum* of phagocyte function in brook trout (*Salvelinus fontinalis*). *Veterinary Immunology and Immunopathology* **62**, 349–357.

De Paolo, A., Flynn, P.A., McPhearson, R.M. and Levy, S.B. (1988) Phenotypic and genotypic characterization of tetracycline- and oxytetracycline-resistant *Aeromonas hydrophila* from cultured channel catfish (*Ictalurus punctatus*) and their environment. *Applied and Environmental Microbiology* **54**, 1861–1863.

Dethlefsen, V. and Watermann, B. (1980) Epidermal papilloma of North Sea dab, *Limanda limanda*: histology, epidemiology and relation to dumping from TiO_2 industry. *ICES Special Meeting on Diseases of Commercially Important Marine Fish and Shellfish* **8**, 1–30.

Dethlefsen, V. Watermann, B. and Hoppenheit, M. (1987) Diseases of North Sea dab (*Limanda limanda* L.) in relation to biological and chemical parameters. *Archive für Fishereiwissenschaft* **37**, 101–237.

Dobson, S.J. and Franzmann, P.D. (1996) Unification of the genera *Deleya* (Baumann *et al.*, (1983), *Halomonas* (Vreeland *et al.*, 1980) and *Halovibrio* (Robinson and Gibbon, 1952) into a single genus, *Halomonas*, and placement of the genus *Zymobacter* in the family *Halomonadaceae*. *International Journal of Systematic Bacteriology* **46**, 550–558.

Doetsch, R.N. (1981) 3. Determinative method of light microscopy. In: Gerhardt, P. (ed.), *Manual of Methods for General Bacteriology*, Washington, D.C., American Society of Microbiology, p. 21–23.

Doménech, A., Fernández-Garayzábal, J.F., Pascual, C., Garcia, J.A., Cutuli, M.T., Moreno, M.A., Collins, M.D. and Dominguez, L. (1996) Streptococcosis in cultured turbot, *Scophthalmus maximus* (L.), associated with *Streptococcus parauberis*. *Journal of Fish Diseases* **19**, 33–38.

Doménech, A., Fernández-Garayzábal, J.F., Lawson, P., García, J.A., Cutuli, M.T., Blanco, M., Gibello, A., Moreno, M.A., Collin, M.D. and Domínguez, L. (1997) Winter disease outbreak in sea-bream (*Sparus aurata*) associated with *Pseudomonas anguilliseptica* infection. *Aquaculture* **156**, 317–326.

Donlon, J., McGettigan, S., O'Brien, P. and O'Carra, P. (1983) Reappraisal of the nature of the pigment produced by *Aeromonas salmonicida*. *FEMS Microbiology Letters* **19**, 285–290.

Donta, S.T. and Haddow, A.D. (1978) Cytotoxic activity of *Aeromonas hydrophila*. *Infection and Immunity* **21**, 989–993.

Dooley, J.S.G., Lallier, R., Shaw, D.H. and Trust, T.J. (1985) Electrophoretic and immunochemical analyses of the lipopolysaccharides from various strains of *Aeromonas hydrophila*. *Journal of Bacteriology* **164**, 262–269.

Dooley, J.S.G., Lallier, R. and Trust, T.J. (1986) Surface antigens of virulent strains of *Aeromonas hydrophila*. *Veterinary Immunology and Immunopathology* **12**, 339–344.

Dooley, J.S.G. and Trust, T.J. (1988) Surface protein composition of *Aeromonas hydrophila* strains virulent for fish: identification of a surface array protein. *Journal of Bacteriology* **170**, 499–506.

Dooley, J.S.G., McCubbin, W.D., Kay, C.M. and Trust, T.J. (1988) Isolation and biochemical characterization of the S-layer protein from a pathogenic *Aeromonas hydrophila* strain. *Journal of Bacteriology* **170**, 2631–2638.

Dooley, J.S.G., Engelhardt, H., Baumeister, W., Kay, W.W. and Trust, T.J. (1989) Three-dimensional structure of an open form of the surface layer from the fish pathogen *Aeromonas salmonicida*. *Journal of Bacteriology* **171**, 190–197.

Dopazo, C.P., Lemos, M.L., Lodeiros, C., Bolinches, J., Barja, J.L. and Toranzo, A.E.

(1988) Inhibitory activity of antibiotic-producing marine bacteria against fish pathogens. *Journal of Applied Bacteriology* **65**, 97–101.

Drouin de Bouville, R. de (1907) Les maladies des poissons d'eau donce d'Europe. *Annales des Sciences Agronomique* **1**, 120–250.

Dubois-Darnaudpeys, A. (1977a) Epidemiologie de la furonculose des salmonides. II. Etude experimentale de divers facteurs microbiotiques de l'environment. *Bulletin Français de Pisciculture* **49**, 128–133.

Dubois-Darnaudpeys, A. (1977b) Epidemiologie de la furonculose des salmonides. III. Ecologie de *Aeromonas salmonicida* proposition d'un modele epidemiologique. *Bulletin Français de Pisciculture* **50**, 21–32.

Dudley, D.J., Guentzel, M.N., Ibarra, M.J., Moore, B.E. and Sagik, B.P. (1980) Enumeration of potentially pathogenic bacteria from sewage sludges. *Applied and Environmental Microbiology* **39**, 118–126.

Duff, D.C.B. (1937) Dissociation in *Bacillus salmonicida*, with special reference to the appearance of the G form of culture. *Journal of Bacteriology* **34**, 49–67.

Duff, D.C.B. (1939) Some serological relationships of the S, R and G phases of *Bacillus salmonicida*. *Journal of Bacteriology* **38**, 91–103.

Duff, D.C.B. (1942) The oral immunization of trout against *Bacterium salmonicida*. *Journal of Immunology* **44**, 87–94.

Duff, D.C.B. and Stewart, B. (1933) Studies on furunculosis of fish in British Columbia. *Contributions of Canadian Biology and Fisheries*, **8**, No. 8, Series A, General Number 35, 103.

Duguid, J.P. and Old, D.C. (1980) Adhesive properties of Enterobacteriaceae. In: Beachey, E.H. (ed.), *Receptors and Recognition*, Series B, vol. 6. London, Chapman & Hall, p.185–217.

Duis, K., Hammer, C., Beveridge, M.C.M., Inglis, V. and Braum, E. (1995) Delivery of quinolone antibacterials to turbot, *Scophthalmus maximus* (L.), via bioencapsulation: quantification and efficacy trial. *Journal of Fish Diseases* **18**, 229–238.

Dulin, M.P. (1979) A review of tuberculosis (mycobacteriosis) in fish. *Veterinary Medicine/Small Animal Clinician* **May 1979**, 735–737.

Dulin, M.P., Huddleston, T.R., Larson, R. and Klontz, G. (1976) Enteric redmouth disease. *Forest Wildlife Range Experimental Station, University of Idaho Bulletin* **8**, 1–15.

Duncan (1932) Cited by McCraw (1952).

Durve, V.S. and Lovell, T.T. (1982) Vitamin C and disease resistance in channel catfish (*Ictalurus punctatus*). *Canadian Journal of Fisheries and Aquatic Sciences* **39**, 948–951.

Earlix, D., Plumb, J.A. and Rogers, W.A. (1996) Isolation of *Edwardsiella ictaluri* from channel catfish by tissue homogenization, filtration and enzyme linked immunosorbent assay. *Diseases of Aquatic Organisms* **27**, 19–24.

Earp, B.J. (1950) Kidney disease in young salmon. M.S. thesis, University of Washington.

Earp, B.J., Ellis, C.H. and Ordal, E.J. (1953) Kidney disease in young salmon. *Washington, Department of Fisheries, Special Report* **1**, 1–74.

Eddy, B.P. (1960) Cephalotrichous, fermentative Gram-negative bacteria: the genus *Aeromonas*. *Journal of Applied Bacteriology* **23**, 216–248.

Eddy, B.P. (1962) Further studies on *Aeromonas*. I. Additional strains and supplementary biochemical tests. *Journal of Applied Bacteriology* **25**, 137–146.

Eddy, B.P. and Carpenter, K.P. (1964) Further studies on *Aeromonas*. II. Taxonomy of *Aeromonas* and C_{27} strains. *Journal of Applied Bacteriology* **27**, 96–100.

Effendi, I. and Austin, B. (1991) Survival of the fish pathogen *Aeromonas salmonicida* in seawater. *FEMS Microbiology Letters* **84**, 103–106.

Effendi, I. and Austin, B. (1993) A rapid method for the determination of viability of *Aeromonas salmonicida* in seawater. *Bulletin of the European Association of Fish Pathologists* **13**, 171–173.

Effendi, I. and Austin, B. (1994) Survival of the fish pathogen *Aeromonas salmonicida* in the marine environment. *Journal of Fish Diseases* **17**, 375–385.

Effendi, I. and Austin, B. (1995a) Dormant/unculturable cells of the fish pathogen *Aeromonas salmonicida*. *Microbial Ecology* **30**, 183–192.

Effendi, I. and Austin, B. (1995b) Uptake of *Aeromonas salmonicida* by Atlantic salmon (*Salmo salar* L.). *Bulletin of the European Association of Fish Pathologists* **15**, 115–118.

Eggset, G., Mortensen, A., Johansen, L.-H. and Sommer, A.-I. (1997) Susceptibility to furunculosis, cold water vibriosis, and infectious pancreatic necrosis (IPN) in post-smolt Atlantic salmon (*Salmo salar* L.) as a function of smolt status by seawater transfer. *Aquaculture* **158**, 179–191.

Egidius, E. and Andersen, K. (1978) Host-specific pathogenicity of strains of *Vibrio anguillarum* isolated from rainbow trout *Salmo gairdneri* Richardson and saithe *Pollachius virens* (L.). *Journal of Fish Diseases* **1**, 45–50.

Egidius, E. and Andersen, K. (1979) Bath immunization—a practical and non-stressing method of vaccinating farmed sea rainbow trout *Salmo gairdneri* Richardson against vibriosis. *Journal of Fish Diseases* **2**, 405–410.

Egidius, E., Anderson, K., Causen, E. and Raa, J. (1981) Cold water vibriosis or 'Hitra disease' in Norwegian salmonid farming. *Journal of Fish Diseases* **4**, 353–354.

Egidius, E., Wiik, R., Andersen, K., Hoff, K.A. and Hjeltnes, B. (1986) *Vibrio salmonicida* sp. nov., a new fish pathogen. *International Journal of Systematic Bacteriology* **36**, 518–520.

Egusa, S. (1983) Disease problems in Japanese yellowtail, *Seriola quinqueradiata*, culture: a review. In: Stewart, J.E. (ed.), *Diseases of Commercially Important Marine Fish and Shellfish*. Conseil International pour l'Exploration de la Mer, Copenhagen, p.10–18.

Ehlinger, N.F. (1964) Selective breeding of trout for resistance to furunculosis. *New York Fish and Game Journal* **11**, 78–90.

Ehlinger, N.F. (1977) Selective breeding of trout for resistance to furunculosis. *New York Fish and Game Journal* **24**, 25–36.

Eklund, M.W., Peterson, M.E., Poysky, F.T., Peck, L.W. and Conrad, J.F. (1982) Botulism in juvenile coho salmon (*Oncorhynchus kisutch*) in the United States. *Aquaculture* **27**, 1–11.

Eldar, A., Bejerano, Y. and Bercovier, H. (1994) *Streptococcus shiloi* and *Streptococcus difficile*: two new streptococcal species causing a meningoencephalitis in fish. *Current Microbiology* **28**, 139–143.

Eldar, A., Frelier, P.F., Assenta, L., Varner, P.W., Lawhon, S. and Bercovier, H. (1995a) *Streptococcus shiloi*, the name for an agent causing septicemia infection in fish, is a junior synonym of *Streptococcus iniae*. *International Journal of Systematic Bacteriology* **45**, 840–842.

Eldar, A., Bejerano, Y., Livoff, A., Horovitcz, A. and Bercovier, H. (1995b) Experimental streptococcal meningo-encephalitis in cultured fish. *Veterinary Microbiology* **43**, 33–40.

Eldar, A., Shapiro, O., Bejerano, Y. and Bercovier H. (1995c) Vaccination with whole-cell vaccine and bacterial protein extract protects tilapia against *Streptococcus difficile* meningoencephalitis. *Vaccine* **13**, 867–870.

Eldar, A., Ghittino, C., Asanta, L., Bozzetta, E., Goria, M., Prearo, M. and Bercovier, H. (1996) *Enterococcus seriolicida* is a junior synonym of *Lactococcus garvieae*, a causative agent of septicemia and meningoencephalitis in fish. *Current Microbiology* **32**, 85–88.

Elliott, D.G. and Shotts, E.B. (1980) Aetiology of an ulcerative disease in goldfish *Carassius auratus* (L.): microbiological examination of diseased fish from seven locations. *Journal of Fish Diseases* **3**, 133–143.

Elliott, D.G. and McKubben, C.L. (1997) Comparison of two fluorescent antibody techniques (FATs) for detection and quantification of *Renibacterium salmoninarum* in coelomic fluid of spawning chinook salmon *Oncorhynchus tshawytscha*. *Diseases of Aquatic Organisms* **30**, 37–43.

Ellis, A.E., Hastings, T.S. and Munro, A.L.S. (1981) The role of *Aeromonas salmonicida* extracellular products in the pathology of furunculosis. *Journal of Fish Diseases* **4**, 41–51.

Ellis, A.E., Dear, G. and Stewart, D.J. (1983) Histopathology of Sekiten-byo caused by *Pseudomonas anguilliseptica* in the European eel, *Anguilla anguilla* L., in Scotland. *Journal of Fish Diseases* **6**, 77–79.

Ellis, A.E. and Grisley, M.S. (1985) Serum antiproteases of salmonids: studies on the inhibition of trypsin and the proteolytic activity of *Aeromonas salmonicida* extracellular products. In: Ellis, A.E. (ed.), *Fish and Shellfish Pathology*. London, Academic Press, p.85–96.

Ellis, A.E., Burrows, A.S. and Stapleton, K.J. (1988a) Lack of relationship between virulence of *Aeromonas salmonicida* and the putative virulence factors: A-layer, extracellular proteases and extracellular haemolysins. *Journal of Fish Diseases* **11**, 309–323.

Ellis, A.E., Burrows, A.S., Hastings, T.S. and Stapleton, K.J. (1988b) Identification of *Aeromonas salmonicida* extracellular proteases as a protective antigen against furunculosis by passive immunization. *Aquaculture* **70**, 207–218.

Ellis, A.E. and Stapleton, K.J. (1988) Differential susceptibility of salmonid fishes to furunculosis correlates with differential serum enhancement of *Aeromonas salmonicida* extracellular protease activity. *Microbial Pathogenesis* **4**, 299–304.

Ellis, A.E., doVale, A., Bowden, T.J., Thompson, K. and Hastings, T.S. (1997) *In vivo* production of A-protein, lipopolysaccharide, iron-regulated outer membrane proteins and 70-kDa serine protease by *Aeromonas salmonicida* subsp. *salmonicida. FEMS Microbiology Letters* **149**, 157–163.

Ellis, R.W., Novotny, A.J. and Harrell, L.W. (1978) Case report of kidney disease in a wild chinook salmon, *Oncorhynchus tshawytscha* in the sea. *Journal of Wildlife Diseases* **14**, 120–123.

Elston, R., Drum, A.S. and Bunnell, P.R. (1995) Efficacy of orally administered difloxacin for the treatment of Atlantic salmon held in seawater. *Journal of Aquatic Animal Health* **7**, 22–28.

Embley, T.M. (1983) Aspects of the biology of *Renibacterium salmoninarum*. Ph.D. thesis, University of Newcastle upon Tyne.

Embley, T.M., Goodfellow, M. and Austin, B. (1982) A semi-defined growth medium for *Renibacterium salmoninarum. FEMS Microbiology Letters* **14**, 299–301.

Embley, T.M., Goodfellow, M., Minnikin, D.E. and Austin, B. (1983) Fatty acid, isoprenoid quinone and polar lipid composition in the classification of *Renibacterium salmoninarum. Journal of Applied Bacteriology* **55**, 31–37.

Embody, G.C. and Hayford, C.O. (1925) The advantage of rearing brook trout fingerlings from selected breeders. *Transactions of the American Fisheries Society* **55**, 135–142.

Emmerich, R. and Weibel, E. (1894) Über eine durch Bakterien erzengte Seuche unter den Forellen. *Archives für Hygiene und Bakteriologie* **21**, 1–21.

Endo, T., Ogishima, K., Hayasaki, H., Kaneko, S. and Ohshima, S. (1973) Application of oxolinic acid as a chemotherapeutic agent for treating infectious diseases in fish. I. Antibacterial activity, chemotherapeutic effect and pharmacokinetic effect of oxolinic acid in fish. *Bulletin of the Japanese Society of Scientific Fisheries* **3**, 165–171.

Enger, Ø. Husevåg, B. and Goksøyr, J. (1989) Presence of the fish pathogen *Vibrio salmonicida* in fish farm sediments. *Applied and Environmental Microbiology* **55**, 2815–2818.

Enger, Ø. Husevåg, B. and Goksøyr, J. (1991) Seasonal variation in presence of *Vibrio salmonicida* and total bacterial counts in Norwegian fish farm water. *Canadian Journal of Microbiology* **37**, 618–623.

Esteve, C. (1995) Numerical taxonomy of *Aeromonadaceae* and *Vibrionaceae* associated with reared fish and surrounding fresh and brackish water. *Systematic and Applied Microbiology* **18**, 391–402.

Esteve, C., Biosca, E.G. and Amaro, C. (1993) Virulence of *Aeromonas hydrophila* and some other bacteria isolated from European eels *Anguilla anguilla* reared in fresh water. *Diseases of Aquatic Organisms* **16**, 15–20.

Esteve, C., Amaro, C. and Toranzo, A.E. (1994) O-serotyping and surface components of *Aeromonas hydrophila* and *Aeromonas jandaei* pathogenic for eels. *FEM Microbiology Letters* **117**, 85–90.

Esteve, C., Gutierrez, M.C. and Ventosa, A. (1995a) DNA relatedness among *Aeromonas allosaccharophila* strains and DNA hybridization groups of the genus *Aeromonas. International Journal of Systematic Bacteriology* **45**, 390–391.

Esteve, C., Amaro, C., Garay, E., Santos, Y. and Toranzo, A.E. (1995b) Pathogenicity of live bacteria and extracellular products of motile *Aeromonas* isolated from eels. *Journal of Applied Bacteriology* **78**, 555–562.

Etchegaray, J.P., Matínez, M.A., Krauskopf, M. and León, G. (1991) Molecular cloning of *Renibacterium salmoninarum* DNA fragments. *FEMS Microbiology Letters* **79**, 61–64.

Eurell, T.E., Lewis, D.H. and Grumbles, L.C. (1978) Comparison of selected diagnostic tests for detection of motile *Aeromonas* septicaemia in fish. *American Journal of Veterinary Research* **39**, 1384–1386.

Euzéby, J.P. (1998) Taxonomic note: necessary correction of specific and subspecific epithets according to Rules 12c and 13b of the International Code of Nomenclature of Bacteria (1990 Revision). *International Journal of Systematic Bacteriology* **48**, 1073–1075.

Evelyn, T.P.T. (1971a) An aberrant strain of the bacterial fish pathogen *Aeromonas salmonicida* isolated from a marine host, the sablefish (*Anoplopoma fimbria*) and from two species of cultured Pacific salmon. *Journal of the Fisheries Research Board of Canada* **28**, 1629–1634.

Evelyn, T.P.T. (1971b) First records of vibriosis in Pacific salmon cultured in Canada, and taxonomic studies of the responsible bacterium, *Vibrio anguillarum*. *Journal of the Fisheries Research Board of Canada* **28**, 517–525.

Evelyn, T.P.T. (1971c) The agglutinin response in sockeye salmon vaccinated intraperitoneally with a heat-killed preparation of the bacterium responsible for salmonid kidney disease. *Journal of Wildlife Diseases* **7**, 328–335.

Evelyn, T.P.T. (1977) An improved growth medium for the kidney bacterium and some notes on using the medium. *Bulletin de l'Office International des Epizooties* **87**, 511–513.

Evelyn, T.P.T. (1978) Sensitivities of bacterial kidney disease detection methods with special remarks on the culture method. *Proceedings of the Joint 3rd Biennial Fish Health Section, American Fisheries Society and 9th Annual Midwest Fish Disease Workshops, Kansas City*, p.1–2.

Evelyn, T.P.T. (1984) Immunization against pathogenic vibrios. In: De Kinkelin, P. (ed.), *Symposium on Fish Vaccination*. Paris, Office, International des Epizooties, p.121–150.

Evelyn, T.P.T., Hoskins, G.E. and Bell, G.R. (1973) First record of bacterial kidney disease in an apparently wild salmonid in British Columbia. *Journal of the Fisheries Research Board of Canada* **30**, 1578–1587.

Evelyn, T.P.T. and Ketcheson, J.E. (1980) Laboratory and field observations on antivibriosis vaccines. In: Ahne, W. (ed.), *Fish Diseases, Third-COPRAQ Session*. Berlin, Springer-Verlag, p.45–54.

Evelyn, T.P.T., Ketcheson, J.E. and Prosperi-Porta, L. (1981) The clinical significance of immunofluorescence-based diagnosis of the bacterial kidney disease carrier. *Fish Pathology* **15**, 293–300.

Evelyn, T.P.T., Ketcheson, J.E. and Prosperi-Porta, L. (1984) Further evidence for the presence of *Renibacterium salmoninarum* in salmonid eggs and for the failure of povidine-iodine to reduce the intra-ovum infection in water-hardened eggs. *Journal of Fish Diseases* **7**, 173–182.

Evelyn, T.P.T., Prosperi-Porta, L. and Ketcheson, J.E. (1986a) Experimental intra-ovum infection of salmonid eggs with *Renibacterium salmoninarum* and vertical transmission of the pathogen with such eggs despite their treatment with erythromycin. *Diseases in Aquatic Organisms* **1**, 197–202.

Evelyn, T.P.T., Ketcheson, J.E. and Prosperi-Porta, L. (1986b) Use of erythromycin as a means of preventing vertical transmission of *Renibacterium salmoninarum. Diseases of Aquatic Organisms* **2**, 7–11.

Evelyn, T.P.T., Bell, G.R., Prosperi-Porta, L. and Ketcheson, J.E. (1989) A simple technique for accelerating the growth of the kidney disease bacterium *Renibacterium salmoninarum* on a commonly used culture medium (KDM2). *Diseases of Aquatic Organisms* **7**, 231–234.

Evelyn, T.P.T. and Prosperi-Porta, L. (1989) Inconsistent performance of KDM2, a culture medium for the kidney disease bacterium *Renibacterium salmoninarum*, due to variation in the composition of its peptone ingredient. *Diseases of Aquatic Ortganisms* **7**, 227–229.

Evelyn, T.P.T., Prospseri-Porta, L. and Ketcheson, J.E. (1990) Two new techniques for obtaining consistent results when growing *Renibacterium salmoninarum* on KDM2 medium. *Diseases of Aquatic Organisms* **9**, 209–212.

Evenberg, D. and Lugtenberg, B. (1982) Cell surface of the fish pathogenic bacterium *Aeromonas salmonicida*. II. Purification and characterization of a major cell envelope protein related to autoagglutination. *Biochemica et Biophysica Acta* **684**, 249–254.

Evenberg, D., van Boxtel, R., Lugtenberg, B., Frank, S., Blommaert, J. and Bootsma, R. (1982) Cell surface of the fish pathogenic bacterium *Aeromonas salmonicida*. 1. Relationship between autoagglutination and the presence of a major cell envelope protein. *Biochemica et Biophysica Acta* **684**, 241–248.

Evenden, A.J., Gilpin, M.L. and Munn, C.B. (1990) The cloning and expression of a gene encoding haemolytic activity from the fish pathogen *Renibacterium salmoninarum. FEMS Microbiology Letters* **71**, 31–34.

Evensen, Ø., Espelid, S. and Håstein, T. (1991) Immunohistochemical identification of *Vibrio salmonicida* in stored tissues of Atlantic salmon *Salmo salar* from the first known outbreaks of coldwater vibriosis ('Hitra disease'). *Diseases of Aquatic Organisms* **10**, 185–189.

Ewing, W.H., Hugh, R. and Johnson, J.G. (1961) Studies on the *Aeromonas* group. *United States Department of Health, Education and Welfare, Public Health Service, Communicable Disease Center.* 37p.

Ewing, W.H., MacWhorter, A.C., Escobar, M.R. and Lubin, A.H. (1965) *Edwardsiella*, a new genus of *Enterobacteriaceae* based on a new species, *E. tarda. International Bulletin of Bacteriological Nomenclature and Taxonomy* **15**, 33–38.

Ewing, W.H., Ross, A.J., Brenner, D.J. and Fanning, G.R. (1978) *Yersinia ruckeri* sp. nov., the redmouth (RM) bacterium. *International Journal of Systematic Bacteriology* **28**, 37–44.

Eya, J.C. and Lovell, R.T. (1998) Effects of dietary phosphorus on resistance of channel catfish to *Edwardsiella ictaluri* challenge. *Journal of Aquatic Animal Health* **10**, 28–34.

Ezura, Y., Tahima, K., Yoshimizu, M. and Kimura, T. (1980) Studies on the taxonomy and serology of causative organisms of fish vibriosis. *Fish Pathology* **14**, 167–179.

Fabre-Domerque, J.P. (1890) On a disease of fish. *Compte Rendu Memoranda du Sociéte Biologique, Paris* **42**, 127–129.

Facklam, R. and Moody, M. (1970) Presumptive identification of group D streptococci: the bile-esculin test. *Applied Microbiology* **20**, 245–250.

Farkas, J. (1985) Filamentous *Flavobacterium* sp. isolated from fish with gill diseases in cold water. *Aquaculture* **44**, 1–10.

Farkas, J. and Oláh, J. (1980) Characterization and antibiotic resistance of a 'gliding bacterium' isolated from sheatfish fry (*Silurus glanis* L.). *Aquacultura Hungarica (Szarvas)* **11**, 131–138.

Farmer, J.J. and McWhorter, A.L. (1984) Genus X *Edwardsiella* Ewing and McWhorter 1965, 37[AL]. In: Krieg, N.R. and Holt, J.G. (eds), *Bergey's Manual of Systematic Bacteriology*, Vol. I, Baltimore, Williams & Wilkins, p.486–491.

Farmer, J.J. and Kelly, M.T. (1991) Enterobacteriaceae. In: *Manual of Clinical Microbiology*, 5th edn. (ed. by Balows, A., Hausler, W.J., Herman, K.L., Isenberg, H.D. and Shadomy), pp. 360–383. American Society for Microbiology, Washington, D.C., USA.

Farrell, D., Mikesell, P., Actis, L.A. and Crosa, J.H. (1990) A regulatory gene, *angR*, of the iron uptake system of *Vibrio anguillarum*: similarity with phage P22 *cro* and regulation by iron. *Gene* **86**, 45–51.

Ferguson, H.W. (1977) Columnaris disease in rainbow trout (*Salmo gairdneri*) in Northern Ireland. *Veterinary Record* **101**, 55–56.

Ferguson, Y., Glover, L.A., McGillivray, D.M. and Prosser, J.I. (1995) Survival and activity of *lux*-marked *Aeromonas salmonicida* in seawater. *Applied and Environmental Microbiology* **61**, 3494–3498.

Fernandes, P.B., Hanson, C.W., Stamm, J.M., Chu, D.T.W., Bailer, R. and Vojtko, C. (1987) The frequency of *in vitro* resistance development to fluoroquinolones and the use of a murine pyelonephritis model to demonstrate selection of resistance *in vivo*. *Journal of Antimicrobial Chemotherapy* **19**, 446–449.

Fernandez, A.I.G., Fernández, A.F., Pérez, M.J., Nieto, T.P. and Ellis, A.E. (1998) Siderophore production by *Aeromonas salmonicida* subsp. *salmonicida*. Lack of strain specificity. *Diseases of Aquatic Organisms* **33**, 87–92.

Fiedler, F. and Draxl, R. (1986) Biochemical and immunochemical properties of the cell surface of *Renibacterium salmoninarum*. *Journal of Bacteriology* **168**, 799–804.

Field, J.B., Gee, L.L., Elvehjem, C.A. and Juday, C. (1944) The blood picture in furunculosis induced by *Bacterium salmonicida* in fish. *Archives of Biochemistry* **3**, 277–284.

Fijan, N.N. (1968) The survival in *Chondrococcus columnaris* in waters of different quality. *Bulletin de l'Office International des Epizooties* **69**, 1158–1166.

Fijan, N.N. (1969) Antibiotic additives for the isolation of *Chondrococcus columnaris* from fish. *Applied Microbiology* **17**, 333–334.

Fijan, N.N. (1972) Infectious dropsy of carp—a disease complex. *Proceedings of the Symposia of the Zoological Society of London* **30**, 39–57.

Fischel, D. and Enoch, H.P. (1892) Cited by Williamson (1929).

Fish, F.F. (1937) Furunculosis in wild trout. *Copeia* **37**, 40.

Fish, F.F. and Hanavan, M.G. (1948) A report upon the Grand Coulee Fish-Maintenance Project of 1939–1947. *United States Department of the Interior, Fish and Wildlife Service, Special Scientific Report* **55**, 1–63.

Ford, L.A. (1994) Detection of *Aeromonas salmonicida* from water using a filtration method. *Aquaculture* **122**, 1–7.

Ford, L.A., Barbash, P.A. and Cipriano, R.C. (1998) Control of furunculosis and enteric redmouth disease in sea run salmon broodstock in the Connecticut and Merrimack rivers. *Progressive Fish Culturist* **60**, 88–94.

Forel, F. (1868) Red disease of trout. *Bulletin de la Société Vandoise des Sciences Naturelles, Lausanne* **9**, 599–608.

Forsyth, R.B., Candido, E.P.M., Babich, S.L. and Iwama, G.K. (1997) Stress protein expression in coho salmon with bacterial kidney disease. *Journal of Aquatic Animal Health* **9**, 18–25.

Fouz, B.I., Larsen, J.L. and Toranzo, A.E. (1991) *Vibrio damsela* as a pathogenic agent causing mortalities in cultured turbot (*Scophthalmus maximus*). *Bulletin of the European Association of Fish Pathologists* **11**, 80–81.

Fouz, B., Larsen, J.L., Nielsen, B., Barja, J.L. and Toranzo, A.E. (1992) Characterization of *Vibrio damsela* strains isolated from turbot, *Scophthalmus maximus* in Spain. *Diseases of Aquatic Organisms* **12**, 155–166.

Fouz, B., Barja, J.L., Amaro, C., Rivas, C. and Toranzo, A.E. (1993) Toxicity of the extracellular products of *Vibrio damsela* isolated from diseased fish. *Current Microbiology* **27**, 341–347.

Fouz, B., Toranzo, A.E., Biosca, E.G., Mazoy, R. and Amaro, C. (1994) Role of iron in the pathogenicity of *Vibrio damsela* for fish and mammals. *FEMS Microbiology Letters* **121**, 181.

Fouz, B., Biosca, E.G. and Amaro, C. (1997) High affinity iron-uptake systems in *Vibrio damsela*: role in the acquisition of iron from transferrin. *Journal of Applied Microbiology* **82**, 157–167.

Frantsi, C., Flewelling, T.C. and Tidswell, K.G. (1975) Investigations on corynebacterial kidney disease and *Diplostomulum* sp. (eye-fluke) at Margaree Hatchery, 1972–73. *Technical Report Series*, MAR/T-75-9: 30pp. Resource Development Branch of the Fisheries and Marine Service, Canadian Department of the Interior.

Frerichs, G.N., Miller, S.D. and McManus, C. (1992) Atypical *Aeromonas salmonicida* isolated from health wrasse (*Ctenolabrus rupestris*). *Bulletin of the European Association of Fish Pathologists* **12**, 48–49.

Fryer, J.L., Nelson, J.S. and Garrison, R.L. (1972) Vibriosis in fish. In: Moore, R.W. (ed.), *Progress in Fishery and Food Science*. Seattle, University of Washington, Publications in Fisheries, p.129–133.

Fryer, J.L., Rohovec, J.S. and Garrison, R.L. (1978) Immunization of salmonids for control of vibriosis. *Marine Fisheries Review* **40**, 20–23.

Fryer, J.L. and Sanders, J.E. (1981) Bacterial kidney disease of salmonid fish. *Annual Review of Microbiology* **35**, 273–298.

Fryer, J.L., Hedrick, R.P., Park, J.W. and Hah, Y.C. (1988) Isolation of *Aeromonas salmonicida* from masu salmon in the Republic of Korea. *Journal of Wildlife Diseases* **24**, 364–365.

Fryer, J.L., Lannan, C.N., Garcés, L.H., Larenas, J.J. and Smith, P.A. (1990) Isolation of a rickettsiales-like organism from diseased coho salmon (*Oncorhynchus kisutch*) in Chile. *Fish Pathology* **25**, 107–114.

Fryer, J.L., Lannan, C.N., Giovannoni, S.J. and Wood, N.D. (1992) *Piscirickettsia salmonis* gen. nov., the causative agent of an epizootic disease in salmonid fishes. *International Journal of Systematic Bacteriology* **42**, 120–126.

Fuhrmann, H., Böhm, K.H. and Schlotfeldt, H.-J. (1983) An outbreak of enteric redmouth disease in West Germany. *Journal of Fish Diseases* **6**, 309–311.

Fujihara, M.P. and Nakatani, R.E. (1971) Antibody production and immune responses of rainbow trout and coho salmon to *Chondrococcus columnaris*. *Journal of the Fisheries Research Board of Canada* **28**, 1253–1258.

Fujihara, M.P., Olson, P.A. and Nakatani, R.E. (1971) Some factors in susceptibility of juvenile rainbow trout and chinook salmon to *Chondrococcus columnaris*. *Journal of the Fisheries Research Board of Canada* **28**, 1739–1743.

Fujioka, R.S., Greco, S.B., Cates, M.B. and Schroeder, J.P. (1988) *Vibrio damsela* from wounds in bottlenose dolphins, *Tursiops truncatus*. *Diseases of Aquatic Organisms* **4**, 1–8.

Fukuda, Y. and Kusuda, R. (1981a) Passive immunization of cultured yellowtail against pseudotuberculosis. *Fish Pathology* **16**, 85–89.

Fukuda, Y. and Kusuda, R. (1981b) Efficacy of vaccination for pseudotuberculosis in cultured yellowtail by various routes of administration. *Bulletin of the Japanese Society of Scientific Fisheries* **47**, 147–150.

Fukuda, Y. and Kusuda, R. (1982) Detection and characterization of precipitating antibody in the serum of immature yellowtail immunized with *Pasteurella piscicida* cells. *Fish Pathology* **17**, 125–127.

Fuller, D.W., Pilcher, K.S. and Fryer, J.L. (1977) A leucocytolytic factor isolated from cultures of *Aeromonas salmonicida*. *Journal of the Fisheries Research Board of Canada* **34**, 1118–1125.

Furones, M.D., Gilpin, M.L. and Munn, C.B. (1993) Culture media for the differentiation of isolates of *Yersinia ruckeri*, based on detection of a virulence factor. *Journal of Applied Bacteriology* **74**, 360–366.

Fyfe, L., Finley, A., Coleman, G. and Munro, A.L.S. (1986) A study of the pathological effect of isolated *Aeromonas salmonicida* extracellular protease on Atlantic salmon, *Salmo salar* L. *Journal of Fish Diseases* **9**, 403–409.

Fyfe, L., Coleman, G. and Munro, A.L.S. (1987a) Identification of major common extracellular proteins secreted by *Aeromonas salmonicida* strains isolated from diseased fish. *Applied and Environmental Microbiology* **53**, 722–726.

Fyfe, L., Coleman, G. and Munro, A.L.S. (1987b) A comparative study of the formation of extracellular proteins by *Aeromonas salmonicida* at two different temperatures. *Journal of Applied Bacteriology* **62**, 367–370.

Gaggero, A., Castro, H. and Sandino, A.M. (1995) First isolation of *Piscirickettsia salmonis* from coho salmon, *Oncorhynchus kisutch* (Walbaum), and rainbow trout, *Oncorhynchus mykiss* (Walbaum), during the freshwater stage of their life cycle. *Journal of Fish Diseases* **18**, 277–279.

Garcés, L.H., Larenas, J.J., Smith, P.A., Sandino, S., Lannan, C.N. and Fryer, J.L. (1991) Infectivity of a rickettsia isolated from coho salmon (*Oncorhynchus kisutch*). *Diseases of Aquatic Organisms* **11**, 93–97.

Garduño, R.A., Thornton, J.C. and Kay, W.W. (1993a) Fate of the fish pathogen

Aeromonas salmonicida in the peritoneal cavity of rainbow trout. *Canadian Journal of Microbiology* **39**, 1051–1058.

Garduño, R. H., Thornton, J.C. and Kay, W.W. (1993b) *Aeromonas salmonicida* grown *in vivo*. *Infection and Immunity* **61**, 3854–3862.

Garduño, R.A., Phipps, B.M. and Kay, W.W. (1994) Physiological consequences of the S-layer of *Aeromonas salmonicida* in relation to growth, temperature, and outer membrane permeation. *Canadian Journal of Microbiology* **40**, 622–629.

Garduño, R.A. and Kay, W.W. (1995) Capsulated cells of *Aeromonas salmonicida* grown *in vitro* have different functional properties than capsulated cells grown *in vivo*. *Canadian Journal of Microbiology* **41**, 941–945.

Garduño, R.A., Kuzyk, M.A. and Kay, W.W. (1997) Structural and physiological determinants of resistance of *Aeromonas salmonicida* to reactive radicals. *Canadian Journal of Microbiology* **43**, 1044–1053.

Garnjobst, L. (1945) *Cytophaga columnaris* (Davis) in pure culture: a myxobacterium pathogenic to fish. *Journal of Bacteriology* **49**, 113–128.

Gauthier, G., Lafay, B., Ruimy, R., Breittmayer, V., Nicolas, J.L., Gauthier, M. and Christen, R. (1995) Small-subunit rRNA sequences and whole DNA relatedness concur for the re-assignment of *Pasteurella piscicida* (Sniezko *et al.*) Janssen and Surgalla to the genus *Photobacterium* as *Photobacterium damsela* subsp. *piscicida* comb. nov. *International Journal of Systematic Bacteriology* **45**, 139–144.

Gayer, E.K., Bekesi, L. and Csaba, G. (1980) Some aspects of the histopathology of carp erythrodermatitis (CE). In: Ahne, W. (ed.), *Fish Diseases, Third COPRAQ-Session*. Berlin, Springer-Verlag, p. 127–136.

Geck, P. (1971) India ink immuno-reaction for the rapid detection of enteric pathogens. *Acta Microbiologica Academicae Scientiarum Hungaricae* **18**, 191–196.

Geldreich, E.E. (1977) Microbiology of water. *Journal of the Water Pollution Control Federation* **69**, 122–144.

Gelev, I. and Gelev, E. (1988) A new species of fish-pathogenic bacterium antigenically related to classical Brucellae. *Zentralblatt für Bakteriologie und Hygiene, Reihe A* **269**, 1–6.

Gelev, I., Gelev, E., Steigerwalt, A.G., Carter, G.P. and Brenner, D.J. (1990) Identification of the bacterium associated with haemorrhagic septicaemia in rainbow trout as *Hafnia alvei*. *Research in Microbiology, Institut Pasteur* **141**, 573–576.

Getchell, R.G., Rohovec, J.S. and Fryer, J.L. (1985) Comparison of *Renibacterium salmoninarum* isolates by antigenic analysis. *Fish Pathology* **20**, 149–159.

Ghittino, C. and Pearo, M. (1992) Segnalazione di streptococcosi nella trota iridea (*Oncorhynchus mykiss*) in Italia: nota preliminare. *Boll. Soc. It. Patol. Ittica* **8**, 4–9.

Ghittino, C. and Pearo, M. (1993) Comparazione di alcuni ceppi isolati da trote iridea affette da streptococcosi. *Boll. Soc. It. Patol. Ittica* **11**, 30–43.

Ghittino, P. (1967) Eziologia e lesioni anatomo-patologiche della malattia branchiale (MB) delle trotelline in Italie. *Rivista Italiane di Piscicultura e Ittiopatologia* **A11**, 24–29.

Ghittino, P. and Penna, P. (1968) Recherches microbiologiques sur la nocardios de la truite arc-en-ciel. *Bulletin de l'Office International des Epizooties* **69**, 1045–1056.

Ghittino, P. and Andruetto, S. (1977) Fish vibriosis in fresh and salt waters of Italy. *Bulletin de l'Office International des Epizooties* **87**, 483–485.

Gibb, A. and Austin, B. (1994) A stable L-form of the fish pathogen *Aeromonas salmonicida*. *Letters in Applied Microbiology* **19**, 473–476.

Gibb, A., McIntosh, D. and Austin, B. (1996) Characterization of a stable L-form of the fish pathogen *Aeromonas salmonicida*. *Letters in Applied Microbiology* **23**, 445–447.

Giles, J.S., Hariharan, H. and Heaney, S.B. (1995) The plasmid profiles of fish pathogenic isolates of *Aeromonas salmonicida, Vibrio anguillarum,* and *Vibrio ordalii* from the Atlantic and Pacific coasts of Canada. *Canadian Journal of Microbiology* **41**, 209–216.

Gill, O.N., Coghlan, J.D. and Calder, I.M. (1983) The risk of leptospirosis in United Kingdom fish farm workers. Results from a 1981 serological survey. *Journal of Hygiene, Cambridge* **94**, 81–86.

Gillespie, N.C. (1981) A numerical taxonomic study of *Pseudomonas*-like bacteria isolated from fish in southeastern Queensland and their association with spoilage. *Journal of Applied Bacteriology* **50**, 29–44.

Gilmour, A. (1977) Characteristics of marine vibrios isolated from fish farm tanks. *Aquaculture* **11**, 51–62.

Giorgetti, G., Tomasin, A.B. and Ceschia, G. (1981) First Italian anti-vibriosis vaccination experiments of freshwater farmed rainbow trout. *Developments in Biological Standardization* **49**, 455–459.

Gjedrem, T. and Aulstad, D. (1974) Selection experiments with salmonids. 1. Differences in resistance to vibrio disease of salmon parr (*Salmo salar*). *Aquaculture* **3**, 51–59.

Gjedrem, T. and Gjoen, H.M. (1995) Genetic variation in susceptibility of Atlantic salmon, *Salmo salar* L., to furunculosis, BKD and cold water vibriosis. *Aquaculture Research* **26**, 129–134.

Goerlich, R., Schlüsener, H.J., Lehmann, J. and Greuel, E. (1984) The application of monoclonal antibodies to diagnosis of *Aeromonas salmonicida* infections in fishes. *Bulletin of the European Association of Fish Pathologists* **4**, 66.

Gomez, S., Bernabe, A., Gomez, M.A., Navarro, J.A. and Sanchez, J. (1993) Fish mycobacteriosis: morphological and immunocytochemical aspects. *Journal of Fish Diseases* **16**, 137–141.

Goodfellow, M. (1971) Numerical taxonomy of some nocardioform bacteria. *Journal of General Microbiology* **69**, 33–80.

Goodfellow, M., Embley, T.M. and Austin, B. (1985) Numerical taxonomy and emended description of *Renibacterium salmoninarum*. *Journal of General Microbiology* **131**, 2739–2752.

Goodwin, A.E., Spencer Roy, Jr, J., Grizzle, J.M. and Terrell Goldsby Jr, M. (1994) *Bacillus mycoides*: a bacterial pathogen of channel catfish. *Diseases of Aquatic Organisms* **18**, 173–179.

Gordon, R.E. and Mihm, J.M. (1959) A comparison of four species of mycobacteria. *Journal of General Microbiology* **21**, 736–748.

Gould, R.W., O'Leary, P.J., Garrison, R.L., Rohovec, J.S. and Fryer, J.L. (1978) Spray vaccination: a method for the immunization of fish. *Fish Pathology* **13**, 63–68.

Graham, S., Jeffries, A.H. and Secombes, C.J. (1988) A novel assay to detect macrophage bacterial activity in fish: factors influencing the killing of *Aeromonas salmonicida*. *Journal of Fish Diseases* **11**, 389–396.

Gray, S.F., Smith, R.S., Reynolds, N.J. and Williams, E.W. (1990) Fish tank granuloma. *British Medical Journal* **300**, 1069–1070.

Grayson, T.H., Evenden, A.J., Gilpin, M.L. and Munn, C.B. (1995a) Production of a *Renibacterium salmoninarum* hemolysin fusion protein in *Escherichia coli* K12. *Diseases of Aquatic Organisms* **22**, 153–156.

Grayson, T.H., Bruno, D.W., Evenden, A.J., Gilpin, M.L. and Munn, C.B. (1995b) Iron acquisition by *Renibacterium salmoninarum*: contribution of iron reductase. *Diseases of Aquatic Organisms* **22**, 157–162.

Green, M. and Austin, B. (1982) The identification of *Yersinia ruckeri* and its relationship of other representatives of the Enterobacteriaceae. *Aquaculture* **34**, 185–192.

Griffin, P.J. (1953) The nature of bacteria pathogenic to fish. *Transactions of the American Fisheries Society* **83**, 241–253.

Griffin, P.J., Snieszko, S.F. and Friddle, S.B. (1953a) A more comprehensive description of *Bacterium salmonicida*. *Transactions of the American Fisheries Society* **82**, 129–138.

Griffin, P.J., Snieszko, S.F. and Friddle, S.B. (1953b) Pigment formation by *Bacterium salmonicida*. *Journal of Bacteriology* **65**, 652–659.

Griffiths, S.G., Olivier, G., Fildes, J. and Lynch, W.H. (1991) Comparison of western blot, direct fluorescent antibody and drop-plate culture methods for the detection of *Renibacterium salmoninarum* in Atlantic salmon (*Salmo salar* L.). *Aquaculture* **97**, 117–129.

Griffiths, S.G., Liska, K. and Lynch, W.H. (1996) Comparison of kidney tissue and ovarian fluid from broodstock Atlantic salmon for detection of *Renibacterium salmoninarum*, and use of SKDM broth culture with Western blotting to increase detection in ovarian fluid. *Diseases of Aquatic Organisms* **24**, 3–9.

Grimes, D.J. (1986) Human pathogens associated with elasmobranchs and other marine animals. In: Megusar, F. and Gantar, M. (eds), *Perspectives in Microbial Ecology*. Ljubljana, Slovene Society for Microbiology, pp. 417–421.

Grimes, D.J., Stemmler, J., Hada, H., May, E.B., Maneval, D., Hetrick, F.M., Jones, R.T., Stoskopf, M. and Colwell, R.R. (1984a) *Vibrio* species associated with mortality of sharks held in captivity. *Microbial Ecology* **10**, 271–282.

Grimes, D.J., Colwell, R.R., Stemmler, J., Hada, H., Maneval, D., Hetrick, F.M., May, E.B., Jones, R.T. and Stoskopf, M. (1984b). *Vibrio* species as agents of elasmobranch disease. *Helgoländer Meeresuntersuchungen* **37**, 309–315.

Grimes, D.J., Gruber, S.H. and May, E.B. (1985) Experimental infection of lemon sharks, *Negaprion brevirostris* (Poey), with *Vibrio* species. *Journal of Fish Diseases* **8**, 173–180.

Grimont, P.A.D. and Grimont, F. (1984) Genus VIII. *Serratia* Bizio 1823, 288[AL]. In: Krieg, N.R. and Holt, J.G. (eds), *Bergey's Manual of Systematic Bacteriology*, Vol. I., Baltimore, Williams and Wilkins, p.477–484.

Grischkowsky, R.S. (1973) Studies of the nature of Pacific oyster (*Crassostrea gigas* Thunberg) mortalities. 1. Implications of bacterial pathogenicity, and II. Pathogenicity testing of vibrios on chinook salmon (*Oncorhynchus tshawytscha* Walbaum) and Pacific oysters. Ph.D. dissertation, University of Washington, Seattle.

Grisez, L. and Ollevier, F. (1995) Comparative serology of the marine fish pathogen *Vibrio anguillarum*. *Applied and Environmental Microbiology* **61**, 4367–4373.

Grisez, L., Chair, M., Sorgeloos, P. and Ollevier, F. (1996) Mode of infection and spread of *Vibrio anguillarum* in turbot *Scophthalmus maximus* larvae after oral challenge through live feed. *Diseases of Aquatic Organisms* **26**, 181–187.

Grisley, M.S., Ellis, A.E., Hastings, T.S. and Munro, A.L.S. (1984) An alpha-migrating anti-protease in normal salmonid plasma and its relationship to the neutralization of *Aeromonas salmonicida* toxins. In: Acuigrup (ed.), *Fish Diseases, Fourth COPRAQ-Session*. Madrid, Editora ATP, p.77–82.

Groberg, W.J. (1982) Infection and the immune response induced by *Vibrio anguillarum* in juvenile coho salmon (*Oncorhynchus kisutch*). Ph.D. thesis, Oregon State University, Corvallis.

Groberg, W.J., McCoy, R.H., Pilcher, K.S. and Fryer, J.L. (1978) Relation of water temperature to infections of coho salmon (*Oncorhynchus kisutch*), chinook salmon (*O. tshawytscha*), and steelhead trout (*Salmo gairdneri*) with *Aeromonas salmonicida* and *A. hydrophila*. *Journal of the Fisheries Research Board of Canada* **35**, 1–7.

Gudmundsdóttir, B.K. (1996) Comparison of extracellular proteases by *Aeromonas salmonicida* strains, isolated from various fish species. *Journal of Applied Bacteriology* **80**, 105–113.

Gudmundsdóttir, B.K., Hastings, T.S. and Ellis, A.E. (1990) Isolation of a new toxic protease from a strain of *Aeromonas salmonicida* subspecies *achromogenes*. *Diseases of Aquatic Organisms* **9**, 199–208.

Gudmundsdóttir, S., Helgason, S. and Bendiktsdóttir, E. (1991) Comparison of the effectiveness of three different growth media for primary isolation of *Renibacterium salmoninarum* from Atlantic salmon, *Salmo salar* L., broodfish. *Journal of Fish Diseases* **14**, 89–96.

Gudmundsdóttir, S., Benediktsdóttir, E. and Helgason, S. (1993) Detection of *Renibacterium salmoninarum* in salmonid kidney samples: a comparison of results using double sandwich ELISA and isolation on selective medium. *Journal of Fish Diseases* **16**, 185–195.

Gudmundsdóttir, B.K. and Gudmundsdóttir, S. (1997) Evaluation of cross protection by vaccines against atypical and typical furunculosis in Atlantic salmon, *Salmo salar* L. *Journal of Fish Diseases* **20**, 343–350.

Gudmundsdóttir, B.K., Jónsdóttir, H., Stenhorsdóttir, B., Magnadóttir and Gudmundsdóttir, S. (1997) Survival and humoral antibody response of Atlantic salmon (*Salmo salar* L.) vaccinated against *Aeromonas salmonicida* ssp. *achromogenes*. *Journal of Fish Diseases* **20**, 351–360.

Guilvout, I., Quilici, M.L., Rabot, S., Lesel, R. and Mazigh, D. (1988) *Bam*HI restriction endonuclease analysis of *Yersinia ruckeri* plasmids and their relatedness to the genus *Yersinia* 42- to 47-megadalton plasmid. *Applied and Environmental Microbiology* **54**, 2594–2597.

Gunn, B.A., Singleton, F.L., Peele, E.R. and Colwell, R.R. (1982) A note on the isolation and enumeration of Gram-positive cocci from marine and estuarine waters. *Journal of Applied Bacteriology* **53**, 127–129.

Gunnlaugsdóttir, B. and Gudmundsdóttir, B.K. (1997) Pathogenicity of atypical *Aeromonas salmonicida* in Atlantic salmon compared with protease production. *Journal of Applied Microbiology* **83**, 542–551.

Gutenberger, S.K., Giovannoni, S.J., Field, K.G., Fryer, J.L. and Rohovec, J.S. (1991) A phylogenetic comparison of the 16S rRNA sequence of the fish pathogen, *Renibacterium salmoninarum*, to Gram-positive bacteria. *FEMS Microbiology Letters* **77**, 151–156.

Gutenberger, S.K., Duimstra, J.R., Rohovec, J.S. and Fryer, J.L. (1997) Intracellular survival of *Renibacterium salmoninarum* in trout mononuclear macrophages. *Diseases of Aquatic Organisms* **28**, 93–106.

Gutierrez, M.A. and Miyazaki, T. (1994) Response of Japanese eels to oral challenge with *Edwardsiella tarda* after vaccination with formalin-killed cells or lipopolysaccharide of the bacterium. *Journal of Aquatic Animal Health*, **6**, 110–117.

Gutsell, J. (1946) Sulfa drugs and the treatment of furunculosis in trout. *Science* **104**, 85–86.

Gutteridge, C.S. and Norris, J.R. (1979) The application of pyrolysis techniques to the identification of micro-organisms. *Journal of Applied Bacteriology* **47**, 5–43.

Hackett, J.L., Lynch, W.H., Paterson, W.D. and Coombs, D.H. (1984) Extracellular protease, extracellular hemolysis, and virulence in *Aeromonas salmonicida*. *Canadian Journal of Fisheries and Aquatic Sciences* **41**, 1354–1360.

Hahnel, G.B., Gould, R.W. and Boatman, E.S. (1983) Serological comparison of selected strains of *Aeromonas salmonicida* ssp. *salmonicida*. *Journal of Fish Diseases* **6**, 1–11.

Haley, R., Davis, S.P. and Hyde, J.M. (1967) Environmental stress and *Aeromonas liquefaciens* in American and threadfin shad mortalities. *Progressive Fish Culturist* **29**, 193.

Hall, J.D. (1963) An ecological study of the chestnut lamprey, *Ichthyomyzon castaneus* Girard, in the Manistee River, Michigan. Ph.D. thesis, University of Michigan, Ann Arbor, USA, 106 p.

Halver, J.E., Ashley, L.M. and Smith, R.E. (1969) Ascorbic acid requirements of coho salmon and rainbow trout. *Transactions of the American Fisheries Society* **98**, 762–772.

Hamaguchi, M. and Kusuda, R. (1989) Field testing of *Pasteurella piscicida* formalin killed bacterin against pseudotuberculosis in cultured yellowtail, *Seriola quinqueradiata. Bulletin of Marine Science and Fisheries, Kochi University*, **11**, 11–16.

Hamilton, A.J., Fallon, M.J.M., Alexander, J. and Canning, E.U. (1986) A modified enzyme linked immunosorbent assay (ELISA) for monitoring antibody production during experimental *Aeromonas salmonicida* infection in rainbow trout (*Salmo gairdneri*). *Developmental and Comparative Immunology* **10**, 443–448.

Hamilton, R.C., Kalnins, H., Ackland, N.R. and Ashburner, L.D. (1981) An extra layer in the surface layers of an atypical *Aeromonas salmonicida* isolated from Australian goldfish. *Journal of General Microbiology* **122**, 363–366.

Handlinger, J., Soltani, M. and Percival, S. (1997) The pathology of *Flexibacter maritimus* in aquaculture species in Tasmania, Australia. *Journal of Fish Diseases* **20**, 159–168.

Hänninen, M.-L., Ridell, J. and Hirvelä-Koski, V. (1995) RAPD analysis of *Aeromonas salmonicida* and *Aeromonas hydrophila*. *Journal of Applied Bacteriology* **79**, 181–185.

Hänninen, M.-L. and Hirvelä-Koski, V. (1997) Molecular and phenotypic methods for the characterization of atypical *Aeromonas salmonicida*. *Veterinary Microbiology* **56**, 147–158.

Hansen, C.B. and Lingg, A.J. (1976) Inert particle agglutination tests for detection of antibody to enteric redmouth bacterium. *Journal of the Fisheries Research Board of Canada* **33**, 2857–2860.

Hansen, G.H. and Olafsen, J.A. (1989) Bacterial colonization of cod (*Gadus morhua* L.) and halibut (*Hippoglossus hippoglossus*) eggs in marine aquaculture. *Applied and Environmental Microbiology* **55**, 1435–1446.

Hansen, G.H., Raa, J. and Olafsen, J.A. (1990) Isolation of *Enterobacter agglomerans* from dolphin fish, *Coryphaena hippurus* L. *Journal of Fish Diseases* **13**, 93–96.

Hansen, G.H., Bergh, Ø., Michaelsen, J. and Knappskog, D. (1992) *Flexibacter ovolyticus* sp. nov., a pathogen of eggs and larvae of Atlantic halibut, *Hippoglossus hippoglossus* L. *International Journal of Systematic Bacteriology* **42**, 451–458.

Hanson, L.A. and Grizzle, J.M. (1985) Nitrite-induced predisposition of channel catfish to bacterial diseases. *Progressive Fish Culturist* **47**, 98–101.

Hao, M.V. and Komagata, K. (1985) A new species of *Planococcus*, *P. kocurii* isolated from fish, frozen foods, and fish curing brine. *Journal of General and Applied Microbiology* **31**, 441–455.

Harbell, S.O., Hodgins, H.O. and Schiewe, M.H. (1979) Studies on the pathology of vibriosis in coho salmon (*Oncorhynchus kisutch*). *Journal of Fish Diseases* **2**, 527–535.

Hardie, L.J., Fletcher, T.C. and Secombes, C.J. (1989) The effect of vitamin E on the immune response of the Atlantic salmon (*Salmo salar* L.). *Aquaculture* **87**, 1–13.

Hardie, L.J., Ellis, A.E. and Secombes, C.J. (1996) *In vitro* activation of rainbow trout macrophages stimulates inhibition of *Renibacterium salmoninarum* growth concomitant with augmented generation of respiratory burst products. *Diseases of Aquatic Organisms* **25**, 175–183.

Hariharan, H., Qian, B., Despres, B., Kibenge, F.S., Heaney, S.B. and Rainnie, D.J. (1995) Development of a specific biotinylated DNA probe for the detection of *Renibacterium salmoninarum*. *Canadian Journal of Veterinary Research* **59**, 306–310.

Harrell, L.W. (1978) Vibriosis and current vaccination procedures in Puget Sound, Washington. *Marine Fisheries Review* **40**, 24–25.

Harrell, L.W., Etlinger, H.M. and Hodgins, H.O. (1976) Humoral factors important in resistance of salmonid fish to bacterial disease. II. Anti *Vibrio anguillarum* activity in mucus and observations on complement. *Aquaculture* **7**, 363–370.

Harrison, F.C. and Sadler, W. (1929) Discoloration of halibut. *Biological Board of Canada, Bulletin* **12**, 1–18.

Hashimoto, S., Muraoka, A. and Kusuda, R. (1989) Effects of carbohydrates, amino acids, vitamins, inorganic salts and peptones on the growth of *Pasteurella piscicida*. *Nippon Suisan Gakkaishi* **55**, 1791–1797.

Håstein, T. and Bullock, A.M. (1976) An acute septicaemic disease of brown trout (*Salmo trutta*) and Atlantic salmon (*Salmo salar*) caused by a *Pasteurella*-like organism. *Journal of Fish Biology* **8**, 23–26.

Håstein, T. and Smith, J.E. (1977) A study of *Vibrio anguillarum* from farms and wild fish using principal components analysis. *Journal of Fish Biology* **11**, 69–75.

Håstein, T., Hallingstad, F., Refsti, T. and Roald, S.O. (1980) Recent experience of field vaccination trials against vibriosis in rainbow trout (*Salmo gairdneri*). In: Ahne, W. (ed.), *Fish Diseases, Third COPRAQ-Session*. Berlin, Springer-Verlag, p.53–59.

Hastings, T.S. and Bruno, D.W. (1985) Enteric redmouth disease: survey in Scotland and evaluation of a new medium, Shotts-Waltman, for differentiating *Yersinia ruckeri*. *Bulletin of the European Association of Fish Pathologists* **5**, 32–35.

Hastings, T.S. and Ellis, A.E. (1985) Differences in the production of haemolytic and proteolytic activities by various isolates of *Aeromonas salmonicida* In: Ellis, A.E. (ed.), *Fish and Shellfish Pathology*. London, Academic Press, p.69–77.

Hastings, T.S. and McKay, A. (1987) Resistance of *Aeromonas salmonicida* to oxolinic acid. *Aquaculture* **61**, 165–171.

Hastings, T.S. and Ellis, A.E. (1988) The humoral immune response of rainbow trout, *Salmo gairdneri* Richardson, and rabbits to *Aeromonas salmonicida* extracellular products. *Journal of Fish Diseases* **11**, 147–160.

Hatai, K., Egusa, S., Nakajima, M. and Chikahata, H. (1975) *Pseudomonas chlororaphis* as a fish pathogen. *Bulletin of the Japanese Society of Scientific Fisheries* **41**, 1203.

Hatai, K., Lawhavinit, O., Toda, K. and Sugou, Y. (1993) Mycobacterium infection in pejerrey, *Odonthestes bonariensis* Cuvier & Valenciennes. *Journal of Fish Diseases* **16**, 397–402.

Hawke, J.P. (1979) A bacterium associated with disease of pond cultured channel catfish, *Ictalurus punctatus*. *Journal of the Fisheries Research Board of Canada* **36**, 1508–1512.

Hawke, J.P., McWhorter, A.C., Steigerwalt, A.G. and Brenner, D.J. (1981) *Edwardsiella ictaluri* sp. nov., the causative agent of enteric septicaemia of catfish. *International Journal of Systematic Bacteriology* **31**, 396–400.

Hazen, T.C., Raker, M.L., Esch, G.W. and Fliermans, C.B. (1978) Ultrastructure of red-sore lesions on largemouth bass (*Micropterus salmoides*); association of the ciliate *Epistylis* sp. and the bacterium *Aeromonas hydrophila. Journal of Protozoology* **25**, 351–355.

Hazen, T.C., Esch, G.W. and Raker, M.L. (1981) Agglutinating antibody to *Aeromonas hydrophila* in wild largemouth bass. *Transactions of the American Fisheries Society* **110**, 514–518.

Hazen, T.C., Esch, G.W., Dimock, R.V. and Mansfield, A. (1982) Chemotaxis of *Aeromonas hydrophila* to the surface mucus of fish. *Current Microbiology* **7**, 371–375.

Hedges, R.W., Smith, P. and Brazil, G. (1985) Resistance plasmids of aeromonads. *Journal of General Microbiology* **131**, 2091–2095.

Hellberg, H ., Moksness, E. and Høie, S. (1996) Infection with atypical *Aeromonas salmonicida* in farmed common wolffish, *Anarhichas lupus*. *Journal of Fish Diseases* **19**, 329–332.

Henley, M.W. and Lewis, D.L. (1976) Anaerobic bacteria associated with epizootics in grey mullet (*Mugil cephalus*) and redfish (*Sciaenops acellata*) along the Texas coast. *Journal of Wildlife Diseases* **12**, 448–453.

Hennigan, M., Vaughan, L.M., Foster, T.J., Smith, P. and Gannon, F. (1989) Characterization of *Aeromonas salmonicida* strains using DNA probe technology. *Canadian Journal of Fisheries and Aquatic Sciences* **46**, 877–879.

Heo, G.-J., Kawai, K. and Wakabayashi, H. (1990) Occurrence of *Flavobacterium branchiophila* associated with bacterial gill disease at a trout hatchery. *Fish Pathology* **25**, 99–105.

Herman, R.L. (1968) Fish furunculosis 1952–1966. *Transactions of the American Fisheries Society* **97**, 221–230.

Herman, R.L. (1972) A review of the prevention and treatment of furunculosis. *FI: EIFAC 72/SC II Symposium* **19**, 1–6.

Herwig, N. (1979) *Handbook of Drugs and Chemicals Used in the Treatment of Fish Diseases: A Manual of Fish Pharmacology and Material Medica.* Springfield, Illinois, Charles C. Thomas.

Hester, F.E. (1973) Fish health: a nationwide survey of problems and needs. *Progressive Fish Culturist* **35**, 11–18.

Hetrick, F.M., Hall, L.W., Wolski, S., Graves, W.C., Robertson, B.S. and Burton, D.T. (1984) Influence of chlorine on the susceptibility of striped bass (*Morone saxatilis*) to *Vibrio anguillarum*. *Canadian Journal of Fisheries and Aquatic Sciences* **9**, 1375–1380.

Hettiarachchi, D.C. and Cheong, C.H. (1994) Some characteristics of *Aeromonas hydrophila* and *Vibrio* species isolated from bacterial disease outbreaks in ornamental fish culture in Sri Lanka. *Journal of the National Science Council of Sri Lanka* **22**, 261–269.

Heuschmann-Brunner, G. (1965a) Nocardiose bei Fischen des Süsswassers und des Meeres. *Berliner und Munchener Tierärzliche Wochenschrift* **78**, 95–97.

Heuschmann-Brunner, G. (1965b) Ein Beitrag zur Erregerfrage der Infektiösen Bachwassersucht des Karpfens. *Allgemeine Fischerei Zeitung* **90**, 41–49.

Heuschmann-Brunner, G. (1978) Aeromonads of the 'hydrophila-punctata group' in freshwater fishes. *Archiv für Hydrobiologie* **83**, 99–125.

Hickman-Brenner, F.W., MacDonald, K.L., Steigerwalt, A.G., Fanning, G.R., Brenner, D.J. and Farmer, J.J. (1987) *Aeromonas veronii*, a new ornithine decarboxylase-positive species that may cause diarrhoea. *Journal of Clinical Microbiology* **25**, 900–906.

Hikada, T., Yanohara, Y. and Shibata, T. (1983) On the causative bacteria of head ulcer disease in cultured eels. *Memoirs of the Faculty of Fisheries, Kagoshima University* **32**, 147–165.

Hikida, M ., Wakabayashi, H., Egusa, S. and Masumura, K. (1979) *Flexibacter* sp., a gliding bacterium pathogenic to some marine fishes in Japan. *Bulletin of the Japanese Society of Scientific Fisheries* **45**, 421–428.

Hiney, M., Dawson, M.T., Heery, D.M., Smith, P.R., Gannon, F. and Powell, R. (1992) DNA probe for *Aeromonas salmonicida*. *Applied and Environmental Microbiology* **58**, 1039–1042.

Hiney, M.P., Kilmartin, J.J. and Smith, P.R. (1994) Detection of *Aeromonas salmonicida* in Atlantic salmon with asymptomatic furunculosis infections. *Diseases of Aquatic Organisms* **19**, 161–167.

Hiney, M.P. and Smith, P.R. (1998) Validation of polymerase chain reaction-based techniques for proxy detection of bacterial fish pathogens: framework, problems and possible solutions for environmental applications. *Aquaculture* **162**, 41–68.

Hirono, I., Tange, N. and Aoki, T. (1997a) Iron-regulated haemolysin gene from *Edwardsiella tarda*. *Molecular Microbiology* **24**, 851–856.

Hirono, I., Kato, M. and Aoki, T. (1997b) Identification of major antigenic proteins of *Pasteurella piscicida*. *Microbial Pathogenesis* **23**, 371–380.

Hirst, I.D., Hastings, T.S. and Ellis, A.E. (1991) Siderophore production by *Aeromonas salmonicida*. *Journal of General Microbiology* **137**, 1185–1192.

Hirst, I.D. and Ellis, A.E. (1996) Utilization of transferrin and salmon serum as sources of iron by typical and atypical strains of *Aeromonas salmonicida*. *Microbiology* **142**, 1543–1550.

Hirvelä-Koski, V., Koski, P. and Niiranen, H. (1994) Biochemical properties and drug resistance of *Aeromonas salmonicida* in Finland. *Diseases of Aquatic Organisms* **20**, 191–196.

Hispano, C., Nebra, Y. and Blanch, A.R. (1997) Isolation of *Vibrio harveyi* from an ocular lesion in the short sunfish (*Mola mola*). *Bulletin of the European Association of Fish Pathologists* **17**, 104–107.

Hjeltnes, B., Andersen, K., Ellingsen, H.-M. and Egidius, E. (1987) Experimental studies on the pathogenicity of a *Vibrio* sp. isolated from Atlantic salmon, *Salmo salar* L., suffering from Hitra disease. *Journal of Fish Diseases* **10**, 21–27.

Hjeltnes, B., Bergh, Ø., Wergeland, H. and Holm, J.C. (1995) Susceptibility of Atlantic cod *Gadus morhua*, halibut *Hippoglossus hippoglossus* and wrasse (Labridae) to *Aeromonas salmonicida* subsp. *salmonicida* and the possibility of transmission of furunculosis from farmed salmon *Salmo salar* to marine fish. *Diseases of Aquatic Organisms* **23**, 25–31.

Hoel, K. Salonius, K. and Lillehaug, A. (1997) *Vibrio* antigens of polyvalent vaccines enhance the humoral immune response to *Aeromonas salmonicida* antigens in Atlantic salmon (*Salmo salar* L.). *Fish and Shellfish Immunology* **7**, 71–80.

Hoel, K., Reitan, L.J. and Lillehaug, A. (1998a) Immunological cross reactions between *Aeromonas salmonicida* and *Vibrio salmonicida* in Atlantic salmon (*Salmo salar* L.) and rabbit. *Fish and Shellfish Immunology* **8**, 171–182.

Hoel, K., Holstad, G. and Lillehaug, A. (1998b) Adjuvant activities of a *Vibrio salmonicida* bacterin on T-dependent and T-independent antigens in rainbow trout (*Oncorhynchus mykiss*). *Fish and Shellfish Immunology* **8**, 287–293.

Hoff, K.A. (1989) Survival of *Vibrio anguillarum* and *Vibrio salmonicida* at different salinities. *Applied and Environmental Microbiology* **55**, 1775–1786.

Hoffman, R., Popp, W. and Van der Graaff, S. (1984) Atypical BKD predominantly causing ocular and skin lesions. *Bulletin of the European Association of Fish Pathologists* **4**, 7–9.

Hoffman, R.W., Bell, G.R., Pfeil-Putzien, C. and Ogawa, M. (1989) Detection of *Renibacterium salmoninarum* in tissue sections by different methods—a comparative study with special regard to the indirect immunohistochemical peroxidase technique. *Fish Pathology* **24**, 101–104.

Høi, L., Larsen, J.L., Dalsgaard, I. and Dalsgaard, A. (1998) Occurrence of *Vibrio vulnificus* biotypes in Danish marine environments. *Applied and Environmental Microbiology* **64**, 7–13.

Høie, S., Heum, M. and Thoresen, O.F. (1996) Detection of *Aeromonas salmonicida* by polymerase chain reaction in Atlantic salmon vaccinated against furunculosis. *Fish and Shellfish Immunology* **6**, 199–206.

Høie, S., Heum, M. and Thoresen, O.F. (1997) Evaluation of a polymerase chain reaction-based assay for the detection of *Aeromonas salmonicida* sp. *salmonicida* in Atlantic salmon *Salmo salar*. *Diseases of Aquatic Organisms* **30**, 27–35.

Holder-Franklin, M.A., Thorpe, A. and Cormier, C.J. (1981) Comparison of numerical taxonomy and DNA–DNA hybridization in diurnal studies of river bacteria. *Canadian Journal of Microbiology* **27**, 1165–1184.

Holm, K.O., Strøm, E., Stemsvåg, K., Raa, J. and Jørgensen, T. (1985) Characteristics of a *Vibrio* sp. associated with the 'Hitra disease' of Atlantic salmon in Norwegian fish farms. *Fish Pathology* **20**, 125–129.

Holm, K.O. and Jørgensen, T. (1987) A successful vaccination of Atlantic salmon, *Salmo salar* L., against 'Hitra disease' or coldwater vibriosis. *Journal of Fish Diseases* **10**, 85–90.

Holmes, B. (1992) Synonymy of *Flexibacter maritimus* Wakabayashi, Hikida, and Masumura 1986 and *Cytophaga marina* Reichenbach 1989. *International Journal of Systematic Bacteriology* **42**, 185.

Holmes, B., Owen, R.J. and McMeekin, T.A. (1984) Genus *Flavobacterium* Bergey, Harrison, Breed, Hammer and Huntoon 1923, 97[AL]. In: Krieg, N.R. and Holt, J.G. (eds), *Bergey's Manual of Systematic Bacteriology*, Vol. 1, Baltimore, Williams & Wilkins, p.353–361.

Holmes, B., Pinning, C.A. and Dawson, C.A. (1986) A probabilistic matrix for the identification of Gram-negative, aerobic, non-fermentative bacteria that grow on nutrient agar. *Journal of General Microbiology* **132**, 1827–1842.

Holt, R.A., Conrad, J.F. and Fryer, J.L. (1975) Furanace for control of *Cytophaga psychrophila* the causal agent of cold-water disease in coho salmon. *Progressive Fish Culturist* **37**, 137–139.

Holt, R.A., Amandi, A., Rohovec, J.S. and Fryer, J.L. (1989) Relation of water temperature to bacterial cold-water disease in coho salmon, chinook salmon and rainbow trout. *Journal of Aquatic Animal Health* **1**, 94–101.

Horiuchi, M., Sato, T., Takagi, H. and Tozuka, K. (1980) Studies on rapid diagnosis system of main bacterial diseases of pond cultured eels in Japan—1. Basic investigations on the diagnosis of edwardsiellosis by direct immunofluorescence. *Fish Pathology* **15**, 49–55.

Horne, J.H. (1928) Furunculosis in trout and the importance of carriers in the spread of disease. *Journal of Hygiene, Cambridge* **28**, 67–78.

Horne, M.T., Tatner, M., McDerment, S. and Agius, C. (1982) Vaccination of rainbow trout, *Salmo gairdneri* Richardson, at low temperatures and the long-term persistence of protection. *Journal of Fish Diseases* **5**, 343–345.

Horne, M.T. and Baxendale, A. (1983) The adhesion of *Vibrio anguillarum* to host tissues and its role in pathogenesis. *Journal of Fish Diseases* **6**, 461–471.

Horsley, R.W. (1973) The bacterial flora of the Atlantic salmon (*Salmo salar* L.) in relation to its environment. *Journal of Applied Bacteriology* **36**, 337–386.

Hoshina, T. (1956) An epidemic disease affecting rainbow trout in Japan. *Journal of Tokyo University, Fisheries* **42**, 15–66.

Hoshina, T. (1962) On a new bacterium *Paracolobactrum anguillimortiferum. Bulletin of the Japanese Society of Scientific Fisheries* **28**, 162–164.

Hoshina, T., Sano, T. and Morimoto, Y. (1958) A *Streptococcus* pathogenic to fish. *Journal of Tokyo University Fisheries* **44**, 57–58.

Howard, S.P. and Buckley, J.T. (1985) Activation of the hole-forming toxin aerolysin by extracellular processing. *Journal of Bacteriology* **163**, 336–340.

Hsu, H.-M., Bowser, P.R. and Schachte, J.H. (1991) Development and evaluation of a monoclonal antibody-based enzyme-linked immunosorbent assay for the diagnosis of *Renibacterium salmoninarum* infection. *Journal of Aquatic Animal Health* **3**, 168–175.

Hsu, H.-M., Wooster, G.A. and Bowser, P.R. (1994) Efficacy of enrofloxacin for the treatment of salmonids with bacterial kidney disease, caused by *Renibacterium salmoninarum. Journal of Aquatic Animal Health* **3**, 220–223.

Hsu, H.-M., Bowser, P.R., Schachte, J.H. Jr., Scarlett, J.M. and Babish, J.G. (1995) Winter field trials of enrofloxacin for the control of *Aeromonas salmonicida* infection in salmonids. *Journal of the World Aquaculture Society* **26**, 307–314.

Hsu, T.C., Waltman, W.D. and Shotts, E.B. (1981) Correlation of extracellular enzymatic activity and biochemical characteristics with regard to virulence of *Aeromonas hydrophila. Developments in Biological Standardization* **49**, 101–111.

Hsu, T., Shotts, E.B. and Waltman, W.D. (1983) A selective medium for the isolation of yellow pigmented bacteria associated with fish disease. *Newsletter for the Flavobacterium–Cytophaga Group* No. 3, 29–30.

Hsu, T.C., Shotts, E.B. and Waltman, W.D. (1983) Quantitation of biochemical and enzymatic characteristics with pathogenicity of *Aeromonas hydrophila* complexes in fish. *Proceedings of the Republic of China–Japan Symposium on Fish Diseases.*

Hugh, R. and Leifson, E. (1953) The taxonomic significance of fermentative versus oxidative metabolism of carbohydrate by various Gram-negative bacteria. *Journal of Bacteriology* **66**, 24–26.

Huh, G.-J. and Wakabayashi, H. (1989) Serological characteristics of *Flavobacterium branchiophila* isolated from gill diseases of freshwater fishes in Japan, USA and Hungary. *Journal of Aquatic Animal Health* **1**, 142–147.

Hui, S.F., Holt, R.A., Sriranganathan, N., Seidler, R.J. and Fryer, J.L. (1984) *Lactobacillus piscicola*, a new species from salmonid fish. *International Journal of Systematic Bacteriology* **34**, 393–400.

Huizinga, H.W., Esch, G.W. and Hazen, T.C. (1979) Histopathology of red-sore disease (*Aeromonas hydrophila*) in naturally and experimentally infected largemouth bass *Micropterus salmoides* (Lacépède). *Journal of Fish Diseases* **2**, 263–277.

Hunter, V.A., Knittel, M.D. and Fryer, J.L. (1980) Stress-induced transmission of *Yersinia ruckeri* infection from carriers to recipient steelhead trout *Salmo gairdneri* Richardson. *Journal of Fish Diseases* **3**, 467–472.

Huntly, P.J., Coleman, G. and Munro, A.L.S. (1992) The nature of the lethal effect on Atlantic salmon, *Salmo salar* L., of a lipopolysaccharide-free phospholipase activity isolated from the extracellular products of *Aeromonas salmonicida*. *Journal of Fish Diseases* **15**, 99–102.

Husevåg, B., Lunestad, B.T., Johannessen, P.J., Enger, Ø. and Samuelsen, O.B. (1991) Simultaneous occurrence of *Vibrio salmonicida* and antibiotic-resistant bacteria in sediments at abandoned aquaculture sites. *Journal of Fish Diseases* **14**, 631–640.

Huss, H.H. and Eskildsen, U. (1974) Botulism in farmed trout caused by *Clostridium botulinum* type E. *Nordic Journal of Veterinary Medicine* **26**, 733–738.

Huss, H.H., Pedersen, A. and Cann, D.C. (1974a) The incidence of *Clostridium botulinum* in Danish trout farms. 1. Distribution in fish and their environment. *Journal of Food Technology* **9**, 445–450.

Huss, H.H., Pedersen, A. and Cann, D.C. (1974b) The incidence of *Clostridium botulinum* in Danish trout farms. 2. Measures to reduce contamination of the fish. *Journal of Food Technology* **9**, 451–458.

Huys, G., Coopman, R., Janssen, P. and Kersters, K. (1996) High-resolution genotypic analysis of the genus *Aeromonas* by AFLP fingerprinting. *International Journal of Systematic Bacteriology* **46**, 572–580.

Ibrahim, A., Goebel, B.M., Liesack, W., Griffiths, M. and Stackebrandt, E. (1993) The phylogeny of the genus *Yersinia* based on 16S rDNA sequences. *FEMS Microbiology Letters* **114**, 173–178.

Iida, T., Wakabayashi, H. and Egusa, S. (1982) Vaccination for control of streptococcal disease in cultured yellowtail. *Fish Pathology* **16**, 201–206.

Iida, T., Furukawa, K., Sakai, M. and Wakabayashi, H. (1986) Non-haemolytic *Streptococcus* isolated from the brain of the vertebral deformed yellowtail. *Fish Pathology* **21**, 33–38.

Iida, T. and Wakabayashi, H. (1990) Relationship between iron acquisition ability and virulence of *Edwardsiella tarda*, the etiological agent of Paracolo disease in Japanese eel *Anguilla japonica*. In: R. Hirano and Hanyu, I. (eds), *The Second Asian Fisheries Forum*, Manila, Asian Fisheries Society, p. 667–670.

Iida, T. and Sorimachi, M. (1994) Cultural characteristics of the bacterium causing jaundice of yellowtail, *Seriola quinqueradiata*. *Fish Pathology* **29**, 25–28.

Iida, T., Sakata, C., Kawatsu, H. and Kukuda, Y. (1997) Atypical *Aeromonas salmonicida* infection in cultured marine fish. *Fish Pathology* **32**, 65–66.

Inamura, H., Muroga, K. and Nakai, T. (1984) Toxicity of extracellular products of *Vibrio anguillarum*. *Fish Pathology* **19**, 89–96.

Inglis, V. and Richards, R.H. (1991) The *in vitro* susceptibility of *Aeromonas salmonicida* and other fish-pathogenic bacteria to 29 antimicrobial agents. *Journal of Fish Diseases* **14**, 641–650.

Inglis, V., Robertson, D., Miller, K., Thompson, K.D. and Richards, R.H. (1996) Antibiotic protection against recrudescence of latent *Aeromonas salmonicida* during furunculosis vaccination. *Journal of Fish Diseases* **19**, 341–348.

Ishiguro, E.E., Kay, W.W., Ainsworth, T., Chamberlain, J.B., Austen, R.A., Buckley, J.T. and Trust, T.J. (1981) Loss of virulence during culture of *Aeromonas salmonicida* at high temperature. *Journal of Bacteriology* **148**, 333–340.

Ishimaru, K., Akagawa-Matsushita, M. and Muroga, K. (1996) *Vibrio ichthyoenteri* sp. nov., a pathogen of Japanese flounder (*Paralichthys olivaceus*). *International Journal of Systematic Bacteriology* **46**, 155–159.

Ishimaru, K. and Muroga, K. (1997) Taxonomical re-evaluation of two pathogenic *Vibrio* species isolated from milkfish and swimming crab. *Fish Pathology* **32**, 59–64.

Itami, T. and Kusuda, R. (1980) Studies on spray vaccination against vibriosis in cultured ayu. II. Duration of vaccination efficacy and effect of different vaccine preparation. *Bulletin of the Japanese Society of Scientific Fisheries* **46**, 699–703.

Itami, T., Kondo, M., Uozu, M., Suganuma, A., Abe, T., Nakagawa, A., Suzuki, N. and Takahashi, Y. (1996) Enhancement of resistance against *Enterococcus seriolicida* infection in yellowtail, *Seriola quinqueradiata* (Temminck & Schlegel), by oral administration of peptidoglycan derived from *Bifidobacterium thermophilum*. *Journal of Fish Diseases* **19**, 185–187.

Iverson, J.B. (1971) Strontium chloride B and E enrichment broth media for the isolation of *Edwardsiella*, *Salmonella* and *Arizona* species from Tiger snakes. *Journal of Hygiene, Cambridge* **69**, 323–330.

Iwamoto, Y., Suzuki, Y., Kurita, A., Watanabe, Y., Shimizu, T., Ohgami, H. and Yanagihara, Y. (1995) *Vibrio trachuri* sp. nov., a new species isolated from diseased Japanese horse mackerel. *Microbiology and Immunology* **39**, 831–837.

Izumi, S. and Wakabayashi, H. (1997) Use of PCR to detect *Cytophaga psychrophila* from apparently healthy juvenile ayu and coho salmon eggs. *Fish Pathology* **32**, 169–173.

Jacquez, G.M., Ziskowski, J. and Rolfe, F.J. (1994) Criteria for the evaluation of alternative environmental monitoring variables: Theory and an application using winter flounder (*Pleuronectes americanus*) and Dover sole (*Microstomus pacificus*). *Environmental Monitoring and Assessment* **30**, 275–290.

Janssen, W.A. and Surgalla, M.J. (1968) Morphology, physiology and serology of a *Pasteurella* species pathogenic for white perch (*Roccus americanus*). *Journal of Bacteriology* **96**, 1606–1610.

Jansson, E., Hongslo, T., Lindberg, R., Ljungberg, O. and Svensson, B.-M. (1991) Detection of *Renibacterium salmoninarum* and *Yersinia ruckeri* by the peroxidase–antiperoxidase immunohistochemical technique in melanin-containing cells of fish tissue. *Journal of Fish Diseases* **14**, 689–692.

Jansson, E., Hongslo, T., Höglund, J. and Ljungberg, O. (1996) Comparative evaluation of bacterial culture and two ELISA techniques for the detection of *Renibacterium salmoninarum* antigens in salmonid kidney tissues. *Diseases of Aquatic Organisms* **27**, 197–206.

Jansson, E. and Ljungberg, O. (1998) Detection of humoral antibodies to *Renibacterium salmoninarum* by immersion and in naturally infected populations. *Diseases of Aquatic Organisms* **33**, 93–99.

Jiwa, S.F.H. (1983) Enterotoxigenicity, haemagglutination and cell-surface hydrophobicity in *Aeromonas hydrophila, A. sobria* and *A. salmonicida. Veterinary Microbiology* **8**, 17–34.

Jo, Y. (1978) Therapeutic experiments on red spot disease. *Fish Pathology* **13**, 41–42.

Johnsen, B.O. and Jensen, A.J. (1994) The spread of furunculosis in salmonids in Norwegian rivers. *Journal of Fish Biology* **45**, 47–55.

Johnson, C.M., Tatner, M.F. and Horne, M.T. (1985) Comparison of the surface properties of seven strains of a fish pathogen, *Aeromonas salmonicida. Journal of Fish Pathology* **27**, 445–458.

Johnson, G.R., Wobeser, G. and Rouse, B.T. (1974) Indirect fluorescent antibody technique for detection of RM bacterium of rainbow trout (*Salmo gairdneri*). *Journal of the Fisheries Research Board of Canada* **31**, 1957–1959.

Johnson, G.S. (1977) Immunological studies on *Vibrio anguillarum. Aquaculture* **10**, 221–230.

Johnson, R., Colwell, R.R., Sakazak, R. and Tamura, K. (1975) Numerical taxonomy study of the *Enterobacteriaceae. International Journal of Systematic Bacteriology* **25**, 12–37.

Johnson, H. E. and Brice, R.F. (1952) Observations on columnaris in salmon and trout. *Progressive Fish Culturist* **14**, 104–109.

Johnson, K.A., Flynn, J.K. and Amend, D.F. (1982a) Onset of immunity in salmonid fry vaccinated by direct immersion in *Vibrio anguillarum* and *Yersinia ruckeri* bacterins. *Journal of Fish Diseases* **5**, 197–205.

Johnson, K.A., Flynn, J.K. and Amend, D.F. (1982b) Duration of immunity in salmonid fry vaccinated by direct immersion with *Yersinia ruckeri* and *Vibrio anguillarum* bacterins. *Journal of Fish Diseases* **5**, 207–213.

Johnson, K.A. and Amend, D.F. (1983a) Comparison of efficacy of several delivery methods using *Yersinia ruckeri* bacterin on rainbow trout, *Salmo gairdneri* Richardson. *Journal of Fish Diseases* **6**, 331–336.

Johnson, K.A. and Amend, D.F. (1983b) Efficacy of *Vibrio anguillarum* and *Yersinia ruckeri* bacterins applied by oral and anal intubation of salmonids. *Journal of Fish Diseases* **6**, 473–476.

Johnson, K.A. and Amend, D.F. (1984) Potential for immersion vaccination against *Aeromonas salmonicida. Journal of Fish Diseases* **7**, 101–105.

Jones, G.L., Hebert, G.A. and Cherry, W.B. (1978) Fluorescent antibody techniques and bacterial application. US Department of Health, Education and Welfare Publication Number (CDC) 78–83 64, 118p.

Jones, S.R.M., Markham, R.J.F., Groman, D.B. and Cusack, R.R. (1998) Virulence and antigenic characteristics of a cultured Rickettsiales-like organism isolated from farmed Atlantic salmon *salmo salar* in eastern Canada. *Diseases of Aquatic Organisms* **33**, 25–31.

Jónsdóttir, H., Malmquist, H.J., Snorrason, S.S., Gudbergsson, G. and Gudmundsdóttir,

S. (1998) Epidemiology of *Renibacterium salmoninarum* in wild Arctic charr and brown trout in Iceland. *Journal of Fish Biology* **53**, 322–339.

Joosten, P.H.M., Kruijer, W.J. and Rombout, J.H.W.M. (1996) Anal immunization of carp and rainbow trout with different fractions of *Vibrio anguillarum* bacterin. *Fish and Shellfish Immunology* **6**, 541–551.

Joosten, P.H.M., Tiermersma, E., Threels, A., Caumartin-Dhieux, C. and Rombout, J.H.W.M. (1997) Oral vaccination of fish against *Vibrio anguillarum* using alginate microparticles. *Fish and Shellfish Immunology* **7**, 471–485.

Jørgensen, T., Midling, K., Espelid, S., Nilsen, R. and Stensvåg, K. (1989) *Vibrio salmonicida*, a pathogen in salmonids, also causes mortality in net-pen captured cod (*Gadus morhua*). *Bulletin of the European Association of Fish Pathologists* **9**, 42–44.

Juni, E. (1984) Genus III *Acinetobacter* Brisou and Prevot 1954, 727[AL]. In: Krieg, N.R. and Holt, J.G. (eds), *Bergey's Manual of Systematic Bacteriology*, Vol. 1, Baltimore, Williams & Wilkins, p.303–307.

Kajita, Y., Sakai, M., Atsuta, S. and Kobayashi, M. (1990) The immunomodulatory effects of levamisole on rainbow trout, *Oncorhynchus mykiss*. *Fish Pathology* **25**, 93–98.

Kanai, K. and Takagi, Y. (1986) Alpha-haemolytic toxins of *Aeromonas hydrophila* produced *in vitro*. *Fish Pathology* **21**, 245–250.

Kaneko, T. and Colwell, R.R. (1974) Distribution of *Vibrio parahaemolyticus* and related organisms in the Atlantic Ocean off South Carolina and Georgia. *Applied Microbiology* **28**, 1009–1017.

Kanemori, Y., Nakai, T. and Muroga, K. (1987) The role of extracellular protease produced by *Vibrio anguillarum*. *Fish Pathology* **22**, 153–158.

Kanetsuna, F. and Bartoli, A. (1972) A simple chemical method to differentiate *Mycobacterium* from *Nocardia*. *Journal of General Microbiology* **70**, 209–212.

Kanno, T., Nakai, T. and Muroga, K. (1990) Scanning electron microscopy on the skin surface of ayu *Plecoglossus altivelis* infected with *Vibrio anguillarum*. *Diseases of Aquatic Organisms* **8**, 73–75.

Kaper, J.B., Sayler, G.S., Baldini, M.M. and Colwell, R.R. (1977) Ambient-temperature primary nonselective enrichment for isolation of *Salmonella* spp. from an estuarine environment. *Applied and Environmental Microbiology* **33**, 829–835.

Kaper, J.B., Seidler, R.J., Lockman, H. and Colwell, R.R. (1979) A medium for the presumptive identification of *Aeromonas hydrophila* and Enterobacteriaceae. *Applied and Environmental Microbiology* **38**, 1023–1026.

Kaper, J.B., Lockman, H., Remmers, E.F., Kristensen, K. and Colwell, R.R. (1983) Numerical taxonomy of vibrios isolated from estuarine environments. *International Journal of Systematic Bacteriology* **33**, 229–255.

Kapperud, G. (1981) Survey on the reservoirs of *Yersinia enterocolitica* and *Yersinia enterocolitica*-like bacteria in Scandinavia. *Acta Pathologica et Microbiologica, Scandinavia Section B* **89**, 29–35.

Kariya, T., Kubota, S., Nakamura, Y. and Kira, K. (1968) Nocardial infection in cultured yellowtail *Seriola quinqueradiata* and *S. purpurascens*. 1. Bacteriological study. *Fish Pathology* **3**, 16–23.

Karunasagar, I., Segar, K., Karunasagar, I., Ali, P.K.M.M. and Jeyasekaran, G. (1990)

Virulence of *Aeromonas hydrophila* strains from fish ponds and infected fishes. In: Chang, S.-T., Chan, K.-Y. and Norman, Y.S.W. (eds), *Recent Advances in Biotechnology and Applied Biology*.

Karunasagar, I., Karunasagar, I. and Pai, R. (1992) Systemic *Citrobacter freundii* infection in common carp, L., fingerlings. *Journal of Fish Diseases* **15**, 95–98.

Karlsson, K.A. (1962) An investigation of the *Aeromonas salmonicida* haemolysin. *Ninth Nordic Veterinary Congress, Copenhagen*.

Karlsson, K.A. (1964) Serological studies of *Aeromonas salmonicida*. *Zentralblatt für Bakteriologie, Parasitenkunde, Infektionskrankheiten und Hygiene* **194**, 73–80.

Kashiwagi, S., Sugimoto, N., Watanabe, K., Ohta, S. and Kusuda, R. (1977a) Chemotherapeutic studies on sodium nifurstyrenate against *Streptococcus* infection in cultured yellowtails—I. *In vitro* studies on sensitivity and bacteriocidal effect. *Fish Pathology* **12**, 11–14.

Kashiwagi, S., Sugimoto, N., Ohta, S. and Kusuda, R. (1977b) Chemotherapeutical studies on sodium nifurstyrenate against *Streptococcus* infection in cultured yellowtail—II. Effect of sodium nifurstyrenate against experimental streptococcal infection. *Fish Pathology* **12**, 157–162.

Kawahara, E. and Kusuda, R. (1987). Direct fluorescent antibody technique for diagnosis of bacterial disease in eel. *Nippon Suisan Gakkaishi* **53**, 395–399.

Kawahara, E., Kawai, K. and Kusuda, R. (1989) Invasion of *Pasteurella piscicida* in tissues of experimentally infected yellowtail *Seriola quinqueradiata*. *Nippon Suisan Gakkaishi* **55**, 499–501.

Kawahara, E., Oshima, S. and Nomura, S. (1990) Toxicity and immunogenicity of *Aeromonas salmonicida* extracellular products to salmonids. *Journal of Fish Diseases* **13**, 495–503.

Kawai, K., Yamamoto, S. and Kusuda, R. (1989) Plankton-mediated oral delivery of *Vibrio anguillarum* vaccine to juvenile ayu. *Nippon Suisan Gakkaishi* **55**, 35–40.

Kawai, K. and Kusuda, R. (1995) A review: *Listonella anguillarum* infection in ayu, *Plecoglossus altivelis*, and its prevention by vaccination. *Israeli Journal of Aquaculture Bamidgeh* **47**, 173–177.

Kawakami, K. and Kusuda, R. (1989) *In vitro* effect of some chemotherapeutics on the causative of *Mycobacterium* infection in yellowtail. *Nippon Suisan Gakkaishi* **55**, 2111–2114.

Kawakami, K. and Kusuda, R. (1990) Efficacy of rifampicin, streptomycin and erythromycin against experimental *Mycobacterium* infection in cultured yellowtail. *Nippon suisan Gakkaishi* **56**, 51–53.

Kawakami, H., Shinohara, N., Fukuda, Y., Yamashita, H., Kihara, H. and Sakai, M. (1997) The efficacy of lipopolysaccharide mixed chloroform-killed cell (LPS-CKC) bacterin of *Pasteurella piscicida* on yellowtail, *Seriola quinqueradiata*. *Aquaculture* **154**, 95–105.

Kawula, T.H., Lelivelt, M.J. and Orndorff, P.E. (1996) Using a new inbred fish model and cultured fish tissue cells to study *Aeromonas hydrophila* and *Yersinia ruckeri* pathogenesis. *Microbial Pathogenesis* **20**, 119–125.

Kay, W.W., Buckley, J.T., Ishiguro, E.E., Phipps, B.M., Monette, J.P.L. and Trust, T.J.

(1981) Purification and disposition of a surface protein associated with virulence of *Aeromonas salmonicida*. *Journal of Bacteriology* **147**, 1077–1084.

Kay, W.W., Phipps, B.M., Ishiguro, E.E. and Trust, T.J. (1985) Porphyrin binding by the surface array virulence protein of *Aeromonas salmonicida*. *Journal of Bacteriology* **164**, 1332–1336.

Kay, W.W. and Trust, T.J. (1991) Form and functions of the regular surface array (S-layer) of *Aeromonas salmonicida*. *Experientia* **47**, 412–414.

Kent, M.L. (1982) Characteristics and identification of *Pasteurella* and *Vibrio* species pathogenic to fishes using API-20E (Analytab Products) multitube test strips. *Canadian Journal of Fisheries and Aquatic Sciences* **39**, 1725–1729.

Kent, M.L., Dungan, C.F., Elston, R.A. and Holt, R.A. (1988) *Cytophaga* sp. (Cytophagales) infection in seawater pen-reared Atlantic salmon *Salmo salar*. *Diseases of Aquatic Organisms* **4**, 173–179.

Kent, M.L., Groff, J.M., Morrison, J.K., Yasutake, W.T. and Holt, R.A. (1989) Spiral swimming behavior due to cranial and vertebral lesions associated with *Cytophaga psychrophila* infections in salmonid fishes. *Diseases of Aquatic Organisms* **5**, 11–16.

Kersters, K. and De Ley, J. (1984) Genus *Alcaligenes* Castellani and Chalmers 1919, 936[AL]. In: Krieg, N.R. and Holt, J.G. (eds), *Bergey's Manual of Systematic Bacteriology*, Vol. I., Baltimore, Williams & Wilkins, p.361–373.

Ketola, H.G. (1983) Requirement for dietary lysine and arginine by fry of rainbow trout. *Journal of Animal Science* **56**, 101–107.

Khalil, A.H. and Mansour, E.H. (1997) Toxicity of crude extracellular products of *Aeromonas hydrophila* in tilapia, *Tilapia nilotica*. *Letters in Applied Microbiology* **25**, 269–273.

Khan, R.A. (1987) Crude oil and parasites of fish. *Parasitology Today* **3**, 99–100.

Khoo, L., Dennis, P.M. and Lewbat, G.A. (1995) Rickettsia-like organisms in the blue-eyed plecostomus, *Panaque suttoni* (Eigenman & Eigenmann). *Journal of Fish Diseases* **18**, 157–163.

Kiehn, E.D. and Pacha, R.E. (1969) Characterization and relatedness of marine vibrios pathogenic to fish: deoxyribonucleic acid homology and base composition. *Journal of Bacteriology* **100**, 1248–1255.

Kiiyukia, C., Nakajima, A., Nakai, T., Muroga, K., Kawakami, H. and Hashimoto, H. (1992) *Vibrio cholerae* non-01 isolated from ayu fish (*Plecoglossus altivelis*) in Japan. *Applied and Environmental Microbiology* **58**, 3078–3082.

Kilian, M. (1976) A taxonomic study of the genus *Haemophilus*, with the proposal of a new species. *Journal of General Microbiology* **93**, 9–62.

Kim, E.-H., Yoshida, T. and Aoki, T. (1993) Detection of R plasmid encoded with resistance to florfenicol in *Pasteurella piscicida*. *Fish Pathology* **28**, 165–170.

Kimura, T. (1969a) A new subspecies of *Aeromonas salmonicida* as an etiological agent of furunculosis on 'Sakuramasu' (*Oncorhynchus masou*) and pink salmon (*O. gorbuscha*) rearing for maturity. Part 1. On the serological properties. *Fish Pathology* **3**, 34–44.

Kimura, T. (1969b) A new subspecies of *Aeromonas salmonicida* as an etiological agent of furunculosis on 'Sakuramasu' (*Oncorhynchus masou*) and pink salmon (*O.*

gorbuscha) rearing for maturity. Part 2. On the morphological and physiological properties. *Fish Pathology* **3**, 45–52.

Kimura, T. (1970) Studies on a bacterial disease of adult 'Sakuramasu' (*Oncorhynchus masou*) and pink salmon (*O. gorbuscha*) reared for maturity. *Scientific Report of the Hokkaido Salmon Hatchery* **24**, 9–100.

Kimura, T., Wakabayashi, H. and Kudo, S. (1978a) Studies on bacterial gill disease in salmonids. 1. Selection of bacterium transmitting gill disease. *Fish Pathology* **12**, 233–242.

Kimura, T., Ezura, K., Tajima, K. and Yoshimizu, M. (1978b) Serological diagnosis of bacterial kidney disease (BKD); immunodiffusion test by heat stable antigen extracted from infected kidney. *Fish Pathology* **13**, 103–108.

Kimura, H. and Kusuda, R. (1979) Studies on the pathogenesis of streptococcal infection in cultured yellowtails *Seriola* spp.: effect of the cell free culture on experimental streptococcal infection. *Journal of Fish Disease* **2**, 501–510.

Kimura, T. and Yoshimizu, M. (1981) A coagglutination test with antibody-sensitized staphylococci for rapid and simple diagnosis of bacterial kidney disease (BKD). *Developments in Biological Standardization* **49**, 135–148.

Kimura, H. and Kusuda, R. (1982) Studies on the pathogenesis of streptococcal infection in cultured yellowtails *Seriola* spp.: effect of crude exotoxin fractions from cell free culture on experimental streptococcal infection. *Journal of Fish Diseases* **5**, 471–478.

Kimura, T. and Yoshimizu, M. (1983) Coagglutination test with antibody-sensitized staphylococci for rapid and simple serological diagnosis of fish furunculosis. *Fish Pathology* **17**, 259–262.

Kimura, T. and Yoshimizu, M. (1984) Coagglutination test with antibody-sensitized staphylococci for rapid serological identification of rough strains of *Aeromonas salmonicida*. *Bulletin of the Japanese Society of Scientific Fisheries* **50**, 439–442.

King, C.H. and Shotts, E.B. (1988) Enhancement of *Edwardsiella tarda* and *Aeromonas salmonicida* through ingestion by the ciliated protozoan *Tetrahymena pyriformis*. *FEMS Microbiology Letters* **51**, 95–100.

King, E.O., Ward, M.K. and Raney, D.E. (1954) Two simple media for the demonstration of pyocyanin and fluorescein. *Journal of Laboratory and Clinical Medicine* **44**, 301–307.

Kinne, O. (1980) *Diseases of Marine Animals*, Vol. 1. *General Aspects, Protozoa to Gastropoda*. Chichester, John Wiley.

Kirchhoff, H. and Rosengarten, R. (1984) Isolation of a motile mycoplasma from fish. *Journal of General Microbiology* **130**, 2439–2445.

Kirchhoff, H., Beyenne, P., Fischer, M., Flossdorf, J., Heitmann, J., Khattab, B., Lopatta, D., Rosengarten, R., Seidel, G. and Yousef, C. (1987) *Mycoplasma mobile* sp. nov., a new species from fish. *International Journal of Systematic Bacteriology* **37**, 192–97.

Kitao, T. (1982a) The methods for detection of *Streptococcus* sp. causative bacteria of streptococcal disease of cultured yellowtail (*Seriola quinqueradiata*). *Fish Pathology* **17**, 17–26.

Kitao, T. (1982b) Erythromycin—the application to streptococcal infection in yellowtails. *Fish Pathology* **17**, 77–85.

Kitao, T., Aoki, T. and Iwata, K. (1979) Epidemiological study on streptococcicosis of cultured yellowtail (*Seriola quinqueradiata*)—I. Distribution of *Streptococcus* sp. in sea water and muds around yellowtail farms. *Bulletin of the Japanese Society of Scientific Fisheries* **45**, 567–572.

Kitao, T., Aoki, T. and Sakoh, R. (1981) Epizootics caused by β-haemolytic *Streptococcus* species in cultured freshwater fish. *Fish Pathology* **15**, 301–307.

Kitao, T., Aoki, T., Fukudome, M., Kawano, K., Wada, Y. and Mizuno, Y. (1983) Serotyping of *Vibrio anguillarum* isolated from diseased freshwater fish in Japan. *Journal of Fish Diseases* **6**, 175–181.

Kitao, T., Yoshida, T., Aoki, T. and Fukudome, M. (1984). Atypical *Aeromonas salmonicida*, the causative agent of an ulcer disease of eel occurred in Kagoshima Prefecture. *Fish Pathology* **19**, 113–117.

Kitao, T. and Yoshida, Y. (1986) Effect of an immunopotentiator on *Aeromonas salmonicida* infection in rainbow trout (*Salmo gairdneri*). *Veterinary Immunology and Immunopathology* **12**, 287–296.

Klein, B.U., Kleingeld, D.W. and Böhm, K.H. (1993) First isolations of *Plesiomonas shigelloides* from samples of cultured fish in Germany. *Bulletin of the European Association of Fish Pathologists* **13**, 70–72.

Klein, B.U., Kleingeld, D.W. and Böhm, K.H. (1994) First isolation of a non-motile/Tween 80 negative *Yersinia ruckeri* strain in Germany. *Bulletin of the European Association of Fish Pathologists* **14**, 165-166.

Klesius, P.H. (1992) Carrier state of channel catfish infected with *Edwardsiella ictaluri*. *Journal of Aquatic Animal Health* **4**, 220–230.

Klesius, P. (1994) Transmission of *Edwardsiella ictaluri* from infected, dead to non-infected channel catfish. *Journal of Aquatic Animal Health* **6**, 1–12.

Klontz, G.W. (1963) Oral immunization of rainbow trout against redmouth. In: *Proceedings of the Northwest Fish Culture Conference*, Olympia, Washington, 5–6 December 1963, p.121.

Klontz, G.W. (1968) Oral immunization of coho salmon against furunculosis. In: *Progress in Sport Fishery Research. B.S.R.W. Resource Publication* **39**, 81–82.

Klontz, G. (1978) Prevention of bacterial kidney disease in adult and juvenile salmonids. *Proceedings of the Joint 3rd Biennial Fish Health Section and 9th Annual Midwest Fish Disease Workshops*, Kansas City, p.11–14.

Klontz, G.W., Yasutake, W.T. and Ross, A.J. (1966) Bacterial disease of the Salmonidae in the Western United States: pathogenesis of furunculosis in rainbow trout. *American Journal of Veterinary Research* **27**, 1455–1460.

Klontz, G.W. and Anderson, D.P. (1968) Fluorescent antibody studies of isolates of *Aeromonas salmonicida*. *Bulletin de l'Office International des Epizooties* **69**, 1149–1157.

Klontz, G.W. and Anderson, D.P. (1970) Oral immunization of salmonids: a review. In: Snieszko, S.F. (ed.), *Symposium on Diseases of Fishes and Shellfishes*. Washington D.C., Special Publication, No. 5, American Fisheries Society, p.16–20.

Klontz, G.W. and Wood, J.W. (1972) Observations on the epidemiology of furunculosis disease in juvenile coho salmon (*Oncorhynchus kisutch*). *FI:EIFAC 72/SC II Symposium* **27**, 1–8.

Klontz, G.W. and Huddleston, T.R. (1976) *Control of Enteric Redmouth Disease. A Twelve Month Activity Report on Contract 10690004 DEMO-1014.* Moscow, University of Idaho.

Kloos, W.E. and Schliefer, K.H. (1986) Genus IV. *Staphylococcus* Rosenbach 1884, 18[AL], Nom. Cons. Opin. 17 Jud. Comm. 1858, 153. In: Sneath, P.H.A., Mair, N.S., Sharpe, M.E. and Holt, J.G. (eds), *Bergey' Manual of Systematic Bacteriology*, Vol. 2, Baltimore, Williams & Wilkins, p.1013–1035.

Kluyver, A.J. and Van Niel, C.B. (1936) Prospects for a natural system of classification of bacteria. *Zentralblatt für Bakteriologie, Parasitenkunde, Infektionskrankheiten und Hygiene, Originale, 2 Abteilung* **94**, 369–403.

Knibb, W., Colorni, A., Ankaoua, M., Lindell, D., Daimant, A. and Gordin, H. (1993) Detection and identification of a pathogenic marine mycobacterium from the European seabass *Dicentrarchus labrax* using polymerase chain reaction and direct sequencing of 16S rRNA sequences. *Molecular Marine Biology and Biotechnology* **2**, 225–232.

Knittel, M.D. (1981) Susceptibility of steelhead trout *Salmo gairdneri* Richardson to redmouth infection by *Yersinia ruckeri* following exposure to copper. *Journal of Fish Diseases* **4**, 33–40.

Ko, Y.-M. and Heo, G.-J. (1997) Characteristics of *Flavobacterium branchiophilum* isolated from rainbow trout in Korea. *Fish Pathology* **32**, 97–102.

Kocur, M. (1986) Genus 1. *Micrococcus* Cohn 1872, 151[AL]. In: Sneath, P.H.A., Mair, N.S., Sharpe, M.E. and Holt, J.G. (eds), *Bergey's Manual of Systematic Bacteriology*, Vol. 2, Baltimore, Williams & Wilkins, p.1004–1008.

Kocylowski, B. (1963) Etat actuel dess maladies des poissons. Organisation de l'inspection des poissons et de leurs de consommation en Pologne. *Bulletin de l'Office International des Epizooties* **59**, 89–109.

Kodama, H., Moustafa, M., Ishiguro, S., Mikami, T. and Izawa, H. (1984) Extracellular virulence factors of fish *Vibrio*: relationships between toxic material, hemolysin, and proteolytic enzyme. *American Journal of Veterinary Research* **45**, 2203–2207.

Kodama, H., Nakanishi, Y., Yamamoto, F., Mikami, T., Izawa, H., Imagawa, T., Hashimoto, Y. and Kudo, N. (1987) *Salmonella arizonae* isolated from a pirarucu, *Arapaima gigas* Cuvier, with septicaemia. *Journal of Fish Diseases* **10**, 509–512.

Koshi, G. and Lalitha, M.K. (1976) *Edwardsiella tarda* in a variety of human infection. *Indian Journal of Medical Research* **64**, 1753–1759.

Koski, P., Hirvelä-Koski, V. and Bernardet, J.F. (1993) *Flexibacter columnaris* infection in Arctic char (*Salvelinus alpinus* (L.)), first isolation in Finland. *Bulletin of the European Association of Fish Pathologists* **13**, 66–69.

Kou, G.H. (1981) Some bacterial diseases of eel in Taiwan. *Proceedings of the Republic of China/U.S. Cooperative Science Seminar on Fish Diseases, National Science Council Series* **3**, 11–20.

Krantz, G.E., Reddecliff, J.M. and Heist, C.E. (1964a) Immune response of trout to *Aeromonas salmonicida*. Part 1. Development of agglutinating antibodies and protective immunity. *Progressive Fish Culturist* **26**, 3–10.

Krantz, G.E., Reddecliff, J.M. and Heist, C.E. (1964b) Immune response of trout to

Aeromonas salmonicida. Part 2. Evaluation of feeding techniques. *Progressive Fish Culturist* **26**, 65–69.

Kraxberger-Beatty, T., McGarey, D.J., Grier, H.J. and Lim, D.V. (1990) *Vibrio harveyi*, an opportunistic pathogen of common snook, *Centropomus undecimalis* (Block), held in captivity. *Journal of Fish Diseases* **13**, 557–560.

Krovacek, K., Faris, A., Ahne, W. and Månsson, I. (1987) Adhesion of *Aeromonas hydrophila* and *Vibrio anguillarum* to fish cells and to mucus-coated glass slides. *FEMS Microbiology Letters* **42**, 85–89.

Kubota, S.S. and Takakuwa, M. (1963) Studies on the disease of marine culture fishes. 1. General description and preliminary discussion of fish diseases at Mie Prefecture. *Journal of the Faculty of Fisheries of the Prefectural University of Mie* **6**, 107–124.

Kubota, S., Kariya, T., Nakanura, Y. and Kira, K. (1968) Nocardial infection in cultured yellowtails (*Seriola quinqueradiata* and *S. purpurascens*). II. Histological study. *Fish Pathology* **3**, 24–33.

Kubota, S.S. Kaige, N., Miyazaki, T. and Miyashita, T. (1981) Histopathological studies on edwardsiellosis of Tilapia—1. Natural infection. *Bulletin of the Faculty of Fisheries, Mie University* **9**, 155–165.

Kudo, S. and Kimura, N. (1983a) Transmission electron microscopic studies on bacterial gill disease in rainbow trout fingerlings. *Japanese Journal of Ichthyology* **30**, 247–260.

Kudo, S. and Kimura, N. (1983b) Scanning electron microscopic studies on bacterial gill disease in rainbow trout fingerlings. *Japanese Journal of Ichthyology* **30**, 393–403.

Kudo, T., Hatai, K. and Seino, A. (1988) *Nocardia seriolae* sp. nov. causing nocardiosis of cultured fish. *International Journal of Systematic Bacteriology* **38**, 173–178.

Kukubo, T., Iida, T. and Wakabayashi, H. (1990) Production of siderophore by *Edwardsiella tarda*. *Fish Pathology* **25**, 237–241.

Kuo, S.C., Chung, H.Y. and Kou, G.H. (1981) Studies on artificial infection of the gliding bacteria in cultured fishes. *Fish Pathology* **15**, 309–314.

Kusser, W. and Fiedler, F. (1983) Murein type and polysaccharide composition of cell walls from *Renibacterium salmoninarum*. *FEMS Microbiology Letters* **20**, 391–394.

Kusuda, R. and Yamaoka, M. (1972) Etiological studies on bacterial pseudotuberculosis in cultured yellowtail with *Pasteurella piscicida* as the causative agent—1. On the morphological and biochemical properties. *Bulletin of the Japanese Society of Scientific Fisheries* **38**, 1325–1332.

Kusuda, R., Taki, H. and Takeuchi, T. (1974) Studies on a nocardia infection of cultured yellowtail. II. Characteristics of *Nocardia kampachi* isolated from a gill-tuberculosis of yellowtail. *Bulletin of the Japanese Society of Scientific Fisheries* **40**, 369–373.

Kusuda, R. and Inoue, K. (1976) Studies on the application of ampicillin for pseudotuberculosis of cultured yellowtail—1. *In vitro* studies on sensitivity, development of drug-resistance and reversion of acquired drug-resistance characteristics of *Pasteurella piscicida*. *Bulletin of the Japanese Society of Scientific Fisheries* **42**, 969–973.

Kusuda, R., Kawai, T., Toyoshima, T. and Komatsu, I. (1976a) A new pathogenic bacterium belonging to the genus *Streptococcus*, isolated from an epizootic of cultured yellowtail. *Bulletin of the Japanese Society of Scientific Fisheries* **42**, 1345–1352.

Kusuda, R., Toyoshima, T., Iwamura, Y. and Sano, H. (1976b) *Edwardsiella tarda* from an epizootic of mullets (*Mugil cephalus*) in Okitsu Bay. *Bulletin of the Japanese Society of Scientific Fisheries* **42**, 271-275.

Kusuda, R. and Komatsu, I. (1978) A comparative study of fish pathogenic *Streptococcus* isolated from saltwater and freshwater fishes. *Bulletin of the Japanese Society of Scientific Fisheries* **44**, 1073–1078.

Kusuda, R., Kawai, K. and Masui, T. (1978a) Etiological studies on bacterial pseudotuberculosis in cultured yellowtail with *Pasteurella piscicida* as the causative agent—II. On the serological properties. *Fish Pathology* **13**, 79–83.

Kusuda, R., Komatsu, I. Kawai, K. (1978b) *Streptococcus* sp. isolated from an epizootic of cultured eel. *Bulletin of the Japanese Society of Scientific Fisheries* **44**, 295.

Kusuda, R., Kawai, K., Jo, Y., Akizuki, T., Fukunaga, M. and Kotake, N. (1978c) Efficacy of oral vaccination for vibriosis in cultured ayu. *Bulletin of the Japanese Society of Scientific Fisheries* **44**, 21–25.

Kusuda, R., Sako, H. and Kawai, K. (1979) Classification of vibrios isolated from diseased fishes—1. On the morphological, biological and biochemical properties. *Fish Pathology* **13**, 123–137.

Kusuda, R. and Fukuda, Y. (1980) Agglutinating antibody titers and serum protein changes of yellowtail after immunization with *Pasteurella piscicida* cells. *Bulletin of the Japanese Society of Scientific Fisheries* **46**, 801–807.

Kusuda, R. and Sugiyama, A. (1981) Studies on the characters of *Staphylococcus epidermidis* isolated from diseased fishes—1. On the morphological, biological and biochemical properties. *Fish Pathology* **16**, 15–24.

Kusuda, R., Kawai, K. and Shirakawa, T. (1982) Serological study of *Streptococcus* sp. pathogenic to cultured yellowtail. *Bulletin of the Japanese Society of Scientific Fisheries* **48**, 1731–1738.

Kusuda, R., Yokoyama, J. and Kawai, K. (1986) Bacteriological study on cause of mass mortalities in cultured black sea bream fry. *Bulletin of the Japanese Society of Scientific Fisheries* **52**, 1745–1751.

Kusuda, R. and Takemaru, I. (1987) Efficacy of josamycin against experimental streptococcal infection in cultured yellowtail. *Nippon Suisan Gakkaishi* **53**, 1519–1523.

Kusuda, R., Itaoka, M. and Kawai, K. (1988) Drug sensitivity of *Pasteurella piscicida* strains isolated from cultured yellowtail from 1984 to 1985. *Nippon Suisan Gakkaishi* **54**, 1521–1526.

Kusuda, R. and Taira, T. (1990) Change of biological activities of phagocytes from the eel immunized with *Edwardsiella tarda*. *Fish Pathology* **25**, 53–58.

Kusuda, R., Kawai, K., Salati, F., Banner, C.R. and Fryer, J.L. (1991) *Enterococcus seriolicida* sp. nov., a fish pathogen. *International Journal of Systematic Bacteriology* **41**, 406–409.

Kusuda, R., Dohata, N., Fukuda, Y. and Kawai, K. (1995) *Pseudomonas anguilliseptica* infection of striped Jack. *Fish Pathology* **30**, 121–122.

Kuzuk, M.A., Thornton, J.C. and Kay, W.W. (1996) Antigenic characterization of the salmonid pathogen *Piscirickettsia salmonis*. *Infection and Immunity* **64**, 5205–5210.

Kwon, M.G., Lee, J.-Y., Park, S., Iida, T., Hirono, I. and Aoki, T. (1997) RAPD analysis of atypical *Aeromonas salmonicida* isolated in Japan. *Fish Pathology* **32**, 109–115.

Laidler, L.A. (1980) Detection and identification of the bacterial kidney disease (BKD) organism by the indirect fluorescent antibody technique. *Journal of Fish Diseases* **3**, 67–69.

Lall, S.P., Paterson, W.D., Hines, J.A. and Adam, N.J. (1985) Control of bacterial kidney disease in Atlantic salmon, *Salmo salar* L., by dietary modification. *Journal of Fish Diseases* **8**, 113–124.

Lallier, R., Bernard, F. and Lalonde, G. (1984) Difference in the extracellular products of two strains of *Aeromonas hydrophila* virulent and weakly virulent for fish. *Canadian Journal of Microbiology* **30**, 900–904.

Lamas, J., Anadon, R., Devesa, S. and Toranzo, A.E. (1990) Visceral neoplasia and epidermal papillomas in cultured turbot *Scophthalmus maximus*. *Diseases of Aquatic Organisms* **8**, 179–187.

Lamas, J., Santos, Y., Bruno, D., Toranzo, A.E. and Anadón, R. (1994a) A comparison of pathological changes caused by *Vibrio anguillarum* and its extracellular products in rainbow trout (*Oncorhynchus mykiss*). *Fish Pathology* **29**, 79–89.

Lamas, J., Santos, Y., Bruno, D.W., Toranzo, A.E. and Anadón, R. (1994b) Non-specific cellular responses of rainbow trout to *Vibrio anguillarum* and its extracellular products (ECPs). *Journal of Fish Biology* **45**, 839–854.

Lamers, C.H.L. and de Haas, M.J.M. (1983) The development of immunological memory in carp (*Cyprinus carpio* L.) to a bacterial antigen. *Developmental and Comparative Immunology* **7**, 713–714.

Lamers, C.H.J. and Muiswinkel, W.B. (1984) Primary and secondary immune responses in carp (*Cyprinus carpio*) after administration of *Yersinia ruckeri* O-antigen. In: Acuigrup (ed.), *Fish Diseases*, Madrid, Editora ATP, p.119–127.

Lamers, C.H.J, de Haas, M.J.M. and van Muiswinkel, W.B. (1985a) Humoral response and memory formation in carp after injection of *Aeromonas hydrophila* bacterin. *Developmental and Comparative Immunology* **9**, 65–75.

Lamers, C.H.J., de Haas, M.J.M. and van Muiswinkel, W.B. (1985b) The reaction of the immune system of fish to vaccination: development of immunological memory in carp, *Cyprinus carpio* L., following direct immersion in *Aeromonas hydrophila* bacterin. *Journal of Fish Diseases* **8**, 253–262.

Lannan, C.N. and Fryer, J.L. (1991) Recommended methods for inspection of fish for the salmonid rickettsia. *Bulletin of the European Association of Fish Pathologists* **11**, 135–136.

Lannan, C.N., Ewing, S.A. and Fryer, J.L. (1991) A fluorescent antibody test for detection of the rickettsia causing disease in Chilean salmonids. *Journal of Aquatic Animal Health* **3**, 229–234.

Lannan, C.N. and Fryer, J.L. (1994) Extracellular survival of *Piscirickettsia salmonis*. *Journal of Fish Diseases* **17**, 545–548.

Lansdell, W., Dixon, B., Smithin, N. and Benjamin, L. (1993) Isolation of several *Mycobacterium* species from fish. *Journal of Aquatic Animal Health* **5**, 73–76.

Larenas, J., Astorga, C., Contreras, J., Garcés, H., Fryer, J.L. and Smith, P. (1996) Rapid detection of *Piscirickettsia salmonis* using microwave irradiation. *Fish Pathology* **31**, 231–232.

Larsen, J.L. (1982) *Vibrio anguillarum*: prevalence in three carbohydrate loaded marine recipients and a control. *Zentralblatt für Bakteriologie und Hygiene, 2, Abteilung Originale C* **3**, 519–530.

Larsen, J.L. and Jensen, N.J. (1977) An *Aeromonas* species implicated in ulcer-disease of the cod (*Gadus morhua*). *Nordisk Veterinaermedicin* **29**, 199–211.

Larsen, J.L. and Olsen, J.E. (1991) Occurrence of plasmids in Danish isolates of *Vibrio anguillarum* serovars O1 and O2 and association of plasmids with phenotypic characteristics. *Applied and Environmental Microbiology* **57**, 2158–2163.

Lawrence, M.L., Cooper, R.K. and Thune, R.L. (1997) Attenuation, persistence, and vaccine potential of an *Edwardsiella ictaluri purA* mutant. *Infection and Immunity* **65**, 4642–4651.

Leadbetter, E.R. (1974) Genus II. *Flexibacter* Soriano 1945, 92, Lewin 1969, 192 emend. mut. char. In: Buchanan, R.E. and Gibbons, N.E. (eds), *Bergey's Manual of Determinative Bacteriology*, 8th edn, Baltimore, Williams & Wilkins, p.105–107.

Leblanc, D., Mittal, K.R., Olivier, G. and Lallier, R. (1981) Serogrouping of motile *Aeromonas* species isolated from healthy and moribund fish. *Applied and Environmental Microbiology* **42**, 56–60.

Lee, E.G.-H. and Gordon, M.R. (1987) Immunofluorescence screening of *Renibacterium salmoninarum* in the tissues and eggs of farmed chinook salmon spawners. *Aquaculture* **65**, 7–14.

Lee, E.G.-H. and Evelyn, T.P.T. (1989) Effect of *Renibacterium salmoninarum* levels in the ovarian fluid of spawning chinook salmon on the prevalence of the pathogen in their eggs and progeny. *Diseases of Aquatic Organisms* **7**, 179–184.

Lee, E.G.-H. and Evelyn, T.P.T. (1994) Prevention of vertical transmission of the bacterial kidney disease agent *Renibacterium salmoninarum* by broodstock injection with erythromycin. *Diseases of Aquatic Organisms* **18**, 1–4.

Lee, J.V., Gibson, D.M. and Shewan, J.M. (1977) A numerical taxonomic study of some *Pseudomonas*-like marine bacteria. *Journal of General Microbiology* **98**, 439–451.

Lee, J.V., Shread, P., Furniss, A.L. and Bryant, T.N. (1981) Taxonomy and description of *Vibrio fluvialis* sp. nov. (Synonym Group F Vibrios, Group EF6). *Journal of Applied Bacteriology* **50**, 73–94.

Lee, J.V., Bashford, D.J., Donovan, T.J., Furniss, A.L. and West, P.A. (1982) The incidence of *Vibrio cholerae* in water, animals and birds in Kent, England. *Journal of Applied Bacteriology* **52**, 281–291.

Lee, K.-K. (1995) Pathogenesis studies on *Vibrio alginolyticus* in the grouper, *Epinephelus malabaricus*, Bloch et Schneider. *Microbial Pathogenesis* **19**, 39–48.

Lee, K.-K., Chiang, H.T., Yii, K.C., Su, W.M. and Liu, P.C. (1997) Effects of extracellular products of *Vibrio vulnificus* on *Acanthopagrus schlegeli* serum components *in vitro* and *in vivo*. *Microbios* **92**, 209–217.

Lee, S.Y., Yin, Z., Ge, R. and Sin, Y.M. (1997) Isolation and characterization of fish *Aeromonas hydrophila* adhesins important for *in vitro* epithelial cell invasion. *Journal of Fish Diseases* **20**, 169–175.

Lehmann, J., Mock, D., Stürenberg, F.-J. and Bernardet, J.-F. (1991) First isolation of *Cytophaga psychrophila* from a systemic disease in eel and cyprinids. *Diseases of Aquatic Organisms* **10**, 217–220.

Lehmann, K.B. and Neumann, R. (1896) *Atlas und Grundress der Bakteriologie*. Munich, J.F. Lehmann.

Leifson, E. (1963) Determination of carbohydrate metabolism of marine bacteria. *Journal of Bacteriology* **85**, 1183–1184.

Le Minor, L. (1984) Genus III. *Salmonella* Lignières 1900 389[AL]. In: Krieg, N.R. and Holt, J.G. (eds), *Bergey's Manual of Systematic Bacteriology*, Vol. I., Baltimore, Williams & Wilkins, p.427–458.

Lemos, M.L., Salinas, P., Toranzo, A.E., Barja, J.L. and Crosa, J.H. (1988) Chromosome-mediated iron uptake system in pathogenic strains of *Vibrio anguillarum*. *Journal of Bacteriology* **170**, 1920–1925.

Lemos, M.L., Mazoy, R., Conchas, R.F. and Toranzo, A.E. (1991) Presence of iron uptake mechanisms in environmental non-pathogenic strains of *Vibrio anguillarum*. *Bulletin of the European Association of Fish Pathologists* **11**, 150–152.

León, G., Maulén, N., Figueroa, J., Villanueva, J., Rodríguez, C., Vera, M.I. and Krauskopf, M. (1994a) A PCR based assay for the identification of the fish pathogen *Renibacterium salmoninarum*. *FEMS Microbiology Letters* **115**, 131–136.

León, G., Martinez, M.A., Etchegaray, J.P., Vera, M.I., Figueroa, J. and Krauskopf, M. (1994b) Specific DNA probes for the identification of the fish pathogen, *Renibacterium salmoninarum*. *World Journal of Microbiology & Biotechnology* **10**, 149–153.

Lesel, R., Lesel, M., Gavini, F. and Vuillaume, A. (1983) Outbreak of enteric redmouth disease in rainbow trout, *Salmo gairdneri* Richardson in France. *Journal of Fish Diseases* **6**, 385–387.

Leung, K.-Y. and Stevenson, R.M.W. (1988) Characteristics and distribution of extracellular proteases from *Aeromonas hydrophila*. *Journal of General Microbiology* **134**, 151–160.

Levine, M.A., Wolke, R.E. and Cabelli, V.J. (1972) *Vibrio anguillarum* as a cause of disease in winter flounder (*Pseudopleuronectes americanus*). *Canadian Journal of Microbiology* **18**, 1585–1892.

Lewin, C.S. and Hastings, T.S. (1990) *In vitro* activities of oxolinic acid, ciprofloxacin and norfloxacin against *Aeromonas salmonicida*. *Journal of Fish Diseases* **13**, 377–384.

Lewis, D.H., Grumble, L.C., McConnell, S. and Flowers, A.I. (1970) *Pasteurella*-like bacteria from a epizootic in menhaden and mullet in Galveston Bay. *Journal of Wildlife Diseases* **6**, 160–162.

Li, M.F. and Fleming, C. (1967) A proteolytic pseudomonad from skin lesions of rainbow trout (*Salmo gairdneri*). I. Characteristics of the pathogenic effects and the extracellular proteinase. *Canadian Journal of Microbiology* **13**, 405–416.

Li, M.F. and Jordan, C. (1968) A proteolytic pseudomonad from skin lesions of rainbow trout (*Salmo gairdneri*). II. Some properties of the proteinase. *Canadian Journal of Microbiology* **14**, 875–880.

Li, M.F. and Traxler, G.S. (1971) A proteolytic pseudomonad from skin lesions of rainbow trout (*Salmo gairdneri*). III. Morphological studies. *Journal of the Fisheries Research Board of Canada* **28**, 104–105.

Lim, C. and Lovell, R.T. (19 78) Pathology of vitamin C deficiency syndrome in channel catfish. *Journal of Nutrition* **108**, 1137–1141.

Lim, C., Sealey, W.M. and Klesius, P.H. (1996) Iron methionine and iron sulfate as sources of dietary iron for channel catfish *Ictalurus punctatus*. *Journal of the World Aquaculture Society* **27**, 290–296.

Lio-Po, G.D., Albright, L.J. and Leaño, E.M. (1996) Experiments on virulence dose and portals of entry for *Aeromonas hydrophila* in walking catfish. *Journal of Aquatic Animal Health* **8**, 340–343.

Lio-Po, G.D., Albright, L.J., Michel, C. and Leaño, E.M. (1998) Experimental induction of lesions in snakeheads (*Ophiocephalus striatus*) and catfish (*Clarias batrachus*) with *Aeromonas hydrophila, Aquaspirillum* sp., *Pseudomonas* sp. and *Streptococcus* sp. *Journal of Applied Ichthyology* **14**, 75–79.

Liu, P.V. (1961) Observations on the specificities of extra-cellular antigens of the genera *Aeromonas* and *Serratia*. *Journal of General Microbiology* **24**, 145–153.

Livesley, M.A., Smith, S.N., Armstrong, R.A. and Barker, G.A. (1997) Analysis of plasmid profile of *Aeromonas salmonicida* isolates by pulsed field gel electrophoresis. *FEMS Microbiology Letters* **146**, 297–301.

Llewellyn, L.C. (1980) A bacterium with similarities to the redmouth bacterium and *Serratia liquefaciens* (Grimes and Hennerty) causing mortalities in hatchery reared salmonids in Australia. *Journal of Fish Diseases* **3**, 29–39.

Llobrera, A.T. and Gacutan, R.Q. (1987) *Aeromonas hydrophila* associated with ulcerative disease epizootic in Laguna de Bay, Philippines. *Aquaculture* **67**, 273–278.

Lobb, C.J. and Rhoades, M. (1987) Rapid plasmid analysis for identification of *Edwardsiella ictaluri* from infected channel catfish (*Ictalurus punctatus*). *Applied and Environmental Microbiology* **53**, 1267–1272.

Lobb, C.J., Ghaffari, S.H., Hayman, J.R. and Thompson, D.T. (1992) Plasmid and serological differences between *Edwardsiella ictaluri* strains. *Applied and Environmental Microbiology* **59**, 2830–2836.

Locci, R., Baldacci, E. and Petrolini-Baldacci, B. (1969) The genus *Streptoverticillium*, a taxonomic study. *Gionale Microbiologica* **17**, 1–60.

Logan, N.A. (1989) Numerical taxonomy of violet-pigmented, Gram-negative bacteria and description of *Iodobacter fluviatile* gen. nov., comb. nov. *International Journal of Systematic Bacteriology* **39**, 450–456.

Loghothetis, P.N. and Austin, B. (1996) Antibody responses of rainbow trout *Oncorhynchus mykiss*, Walbaum) to live *Aeromonas hydrophila* as assessed by various antigen preparations. *Fish and Shellfish Immunology* **6**, 455–464.

Lönnström, L., Wiklund, T. and Bylund, G. (1994) *Pseudomonas anguilliseptica* isolated from Baltic herring *Clupea harengus membras* with eye lesions. *Diseases of Aquatic Organisms* **18**, 143–147.

Lopez, J.E. and Angulo, L. (1995) Survival of *Vibrio splendidus* biotype I in seawater. *Bulletin of the European Association of Fish Pathologists* **15**, 70–72.

Lorenzen, E. (1993) The importance of the brand of the beef extract in relation to the growth of *Flexibacter psychrophilus* in Anacker and Ordals medium. *Bulletin of the European Association of Fish Pathologists* **13**, 64–65.

Lorenzen, E., Dalsgaard, I., From, J., Hansen, E.M. Hørlyck, V. Korsholm, H., Mellergaard, S. and Olesen, N.J. (1991) Preliminary investigations of fry mortality syndrome in rainbow trout. *Bulletin of the European Association of Fish Pathologists* **11**, 77–79.

Lorenzen, E. and Karas, N. (1992) Detection of *Flexibacter psychrophilus* by immunofluorescence in fish suffering from fry mortality syndrome: a rapid diagnostic method. *Diseases of Aquatic Organisms* **13**, 231–234.

Lorenzen, E., Olesen, N.J., Korsholm, H., Heuer, O.E. and Evensen, Ø. (1996) First demonstration of *Renibacterium salmoninarum*/BKD in Denmark. *Bulletin of the European Association of Fish Pathologists* **17**, 140–144.

Lorenzen, E., Dalsgaard, I. and Bernadet, J.-F. (1997) Characterization of isolates of *Flavobacterium psychrophilum* associated with coldwater disease or rainbow trout fry syndrome. I. Phenotypic and genotypic studies. *Diseases of Aquatic Organisms* **31**, 197–208.

Lorenzen, E. and Olesen, N.J. (1997) Characterization of isolates of *Flavobacterium psychrophilum* associated with coldwater disease or rainbow trout fry syndrome. II. Serological studies. *Diseases of Aquatic Organisms* **31**, 209–220.

Love, M., Teebken-Fisher, D., Hose, J.E., Farmer III, J.J., Hickman, F.W. and Fanning, G.R. (1981) *Vibrio damsela*, a marine bacterium, causes skin ulcers on the damselfish *Chromis punctipinnis*. *Science (New York)* **214**, 1139–1140.

Lovely, J.E., Cabo, C., Griffiths, S.G. and Lynch, W.H. (1994) Detection of *Renibacterium salmoninarum* in asymptomatic Atlantic salmon. *Journal of Aquatic Animal Health* **6**, 126–132.

Lowe, G.H. (1962) The rapid detection of lactose fermentation in paracolon organisms by the demonstration of β-galactosidase. *Journal of Medical Laboratory Technology* **19**, 21–31.

Lund, M. (1967) A study of the biology of *Aeromonas salmonicida* (Lehmann and Neumann 1896) Griffin 1954. M.Sc. thesis, University of Newcastle upon Tyne, UK.

Lunder, T., Evensen, Ø., Holstad, G. and Håstein, T. (1995) 'Winter ulcer' in the Atlantic salmon *Salmo salar*. Pathological and bacteriological investigations and transmission experiments. *Diseases of Aquatic Organisms* **23**, 39–49.

Lupiani, B., Dopazo, C.P., Ledo, A., Fouz, B., Barja, J.L., Hetrick, F.M. and Toranzo, A.E. (1989) New syndrome of mixed bacterial and viral etiology in cultured turbot *Scophthalmus maximus*. *Journal of Aquatic Animal Health* **1**, 197–204.

Lutwyche, P., Exner, M.M., Hancock, R.E.W. and Trust, T.J. (1995) A conserved *Aeromonas salmonicida* porin provides protective immunity to rainbow trout. *Infection and Immunity* **63**, 3137–3142.

Lygren, B., Fausa Pettersen, E., Wergeland, H.I. and Endresen, C. (1998) An extracellular low molecular weight protease with activity against gelatin and casein produced earlier during growth of *Aeromonas salmonicida* ssp. *salmonicida* than the serine protease. *Bulletin of the European Association of Fish Pathologists* **18**, 96–101.

MacDonell, M.T. and Colwell, R.R. (1985) Phylogeny of the Vibrionaceae and recommendations for two new genera, *Listonella* and *Shewanella*. *Systematic and Applied Microbiology* **6**, 171–182.

MacDonell, M.T., Swartz, D.G., Ortiz-Conde, B.A., Last, G.A. and Colwell, R.R. (1986)

Ribosomal RNA phylogenies for the vibrio-enteric group of eubacteria. *Microbiological Sciences* **3**, 172–179.

MacFarlane, R.D., Bullock, G.L. and McLaughlin, J.J.A. (1986) Effects of five metals on susceptibility of striped bass to *Flexibacter columnaris*. *Transactions of the American Fisheries Society* **115**, 227–231.

Mackie, C. and Birkbeck, T.H. (1992) Siderophores produced by *Vibrio anguillarum in vitro* and in infected rainbow trout, *Oncorhynchus mykiss* (Walbaum). *Journal of Fish Diseases* **15**, 37–45.

Mackie, T.J., Arkwright, J.A., Pryce-Tannatt, T.E., Motram, J.C. and Johnstone, W.R. (1930, 1933 and 1935) *Interim, Second and Final Reports of the Furunculosis Committee*. Edinburgh, H.M.S.O.

Mackie, T.J. and Menzies, W.J.M. (1938) Investigations in Great Britain of furunculosis of the Salmonidae. *Journal of Comparative Pathology and Therapeutics* **51**, 25.

MacInnes, J.I., Trust, T.J. and Crosa, J.H. (1979) Deoxyribonucleic acid relationships among members of the genus *Aeromonas*. *Canadian Journal of Microbiology* **25**, 579–586.

MacLean, D.G. and Yoder, W.G. (1970) Kidney disease among Michigan salmon in 1967. *Progressive Fish Culturist* **32**, 26–30.

MacPhee, D.D., Ostland, V.E., Lumsden, J.S. and Ferguson, H.W. (1995) Development of an enzyme-linked immunosorbent assay (ELISA) to estimate the quantity of *Flavobacterium branchiophilum* on the gills of rainbow trout *Oncorhynchus mykiss*. *Diseases of Aquatic Organisms* **21**, 13–23.

Magariños, B., Romalde, J.L., Bandín, I., Fouz, B. and Toranzo, A.E. (1992) Phenotypic, antigenic, and molecular characterization of *Pasteurella piscicida* strains isolated from fish. *Applied and Environmental Microbiology* **58**, 3316–3322.

Magariños, B., Romalde, J.L., Barja, J.L. and Toranzo, A.E. (1994a) Evidence of a dormant but infective state of the fish pathogen *Pasteurella piscicida* in seawater and sediment. *Applied and Environmental Microbiology* **60**, 180–186.

Magariños, B., Romalde, J.L., Lemos, M.L., Barja, J.L. and Toranzo, A.E. (1994b) Iron uptake by *Pasteurella piscicida* and its role in pathogenicity for fish. *Applied and Environmental Microbiology* **60**, 2990–2998.

Magariños, B., Romalde, J., Santos, Y., Casal, J.F., Barja, J.L. and Toranzo, A.E. (1994c) Vaccination trials on gilthead sea bream (*Sparus aurata*) against *Pasteurella piscicida*. *Aquaculture* **120**, 201–208.

Magariños, B., Pazos, F., Santos, Y., Romalde, J.L. and Toranzo, A.E. (1995) Response of *Pasteurella piscicida* and *Flexibacter maritimus* to skin mucus of marine fish. *Diseases of Aquatic Organisms* **21**, 103–108.

Magariños, B., Romalde, J.L., Noya, M., Barja, J.L. and Toranzo, A.E. (1996a) Adherence and invasive capacities of the fish pathogen *Pasteurella piscicida*. *FEMS Microbiology Letters* **138**, 29–34.

Magariños, B., Bonet, R., Romalde, J.L., Martínez, M.L., Congregado, F. and Toranzo, A.E. (1996b) Influence of the capsular layer on the virulence of *Pasteurella piscicida* for fish. *Microbial Pathogenesis* **21**, 289–297.

Magariños, B., Romalde, J.L., Cid, A. and Toranzo, A.E. (1997a) Viability of starved

Pasteurella piscicida in seawater monitored by flow cytometry and the effects of antibiotics on its resuscitation. *Letters in Applied Microbiology* **24**, 122–126.

Magariños, B., Osorio, C.R., Toranzo, A.E. and Romalde, J.L. (1997b) Applicability of ribotyping for intraspecific classification and epidemiological studies of *Photobacterium damsela* subsp. *piscicida. Systematic and Applied Microbiology* **20**, 634–639.

Magneussen, H.B., Fridjónsson, Ó.H., Andrésson, Ó.S., Benediktsdóttir, E., Gudmundsóttir, S. and Andrésdottír, V. (1994) *Renibacterium salmoninarum*, the causal agent of bacterial kidney disease in salmonid fish, detected by nested reverse transcription-PCR of 16S rRNA sequences. *Applied and Environmental Microbiology* **60**, 4580–4583.

Maher, M., Palmer, R., Gannon, F. and Smith, T. (1995) Relationship of a novel bacterial fish pathogen to *Streptobacillus moniliformis* and the fusobacteria group, based on 16S ribosomal RNA analysis. *Systematic and Applied Microbiology* **18**, 79–84.

Maheshkumar, S., Goyal, S.M. and Economon, P.P. (1990) Concentration and detection of *Aeromonas salmonicida* from hatchery water. *Journal of Fish Diseases* **13**, 513–518.

Mahnken, C.V.W. (1975) Status report on commercial salmon culture in Puget Sound. *Commercial Fish Farmer, Aquaculture News* **2**, 8–11.

Mahoney, J.B., Midlige, F.H. and Deuel, D.G. (1973) A fin rot disease of marine and euryhaline fishes in the New York Bight. *Transactions of the American Fisheries Society* **102**, 597–605.

Mano, N., Inui, T., Arai, D., Hirose, H. and Deguchi, Y. (1996) Immune response in the skin of eel against *Cytophaga columnaris. Fish Pathology* **31**, 65–70.

Markovic, M., Radojicic, M., Cosic, S. and Levnaic, D. (1996) Massive death of silver carp (*Hypophthalmichthys molitrix* Val.) and big head (*Aristichthys nobilis* Rich.) caused by *Pseudomonas fluorescens* bacteria. *Veterinarski Glasnik* **50**, 761–765.

Markwardt, N.M., Gocha, Y.M. and Klontz, G.W. (1989) A new application for coomassie brilliant blue agar: detection of *Aeromonas salmonicida* in clinical samples. *Diseases of Aquatic Organisms* **6**, 231–233.

Marsden, M.J., Vaughan, L.M., Foster, T.J. and Secombes, C.J. (1996a) A live (Δ*aroA*) *Aeromonas salmonicida* vaccine for furunculosis preferentially stimulates T-cell responses relative to B-cell responses in rainbow trout (*Oncorhynchus mykiss*). *Infection and Immunity* **64**, 3863–3869.

Marsden, M.J., Collins, E. and Secombes, C.J. (1996b) Factors influencing the clearance of a genetically attenuated (Δ*aroA*) strain of *Aeromonas salmonicida* from rainbow trout *Oncrohynchus mykiss. Diseases of Aquatic Organisms* **27**, 89–94.

Marsden, M.J., Freeman, L.C., Cox, D and Secombes, C.J. (1996c) Non-specific immune response in families of Atlantic salmon, *Salmo salar*, exhibiting differential resistance to furunculosis. *Aquaculture* **146**, 1–16.

Marsh, M.C. (1902) *Bacterium truttae*, a new bacterium pathogenic to trout. *Science* **16**, 706.

Marshall, S., Heath, S., Henríquez, V. and Orrego, C. (1998) Minimally invasive detection of *Piscirickettsia salmonis* in cultivated salmonids via the PCR. *Applied and Environmental Microbiology* **64**, 3066–3069.

Martinez-Murcia, A.J., Esteve, C., Garay, E. and Collins, M.D. (1992) *Aeromonas allosaccharophila* sp. nov., a new mesophilic member of the genus *Aeromonas*. *FEMS Microbiology Letters* **91**, 199–206.

Martínez-Picardo, J., Blanch, A.R. and Jofre, J. (1994) Rapid detection and identification of *Vibrio anguillarum* by using a specific oligonucleotide probe complementary to 16S rRNA. *Applied and Environmental Microbiology* **60**, 732–737.

Martinsen, B., Myhr, E., Reed, E. and Håstein, T. (1991) In vitro antimicrobial activity of sarafloxacin against clinical isolates of bacteria pathogenic to fish. *Journal of Aquatic Animal Health* **3**, 235–241.

Massad, G., Arceneaux, J.E.L. and Byers, B.R. (1991) Acquisition of iron from host sources by mesophilic *Aeromonas* species. *Journal of General Microbiology* **137**, 237–241.

Masterman, A.T. and Arkwright, J.A. (1911) Report, Board of Agriculture and Fisheries, UK.

Masumura, K. and Wakabayashi, H. (1977) An outbreak of gliding bacterial disease in hatchery-borne red sea bream (*Pagnes major*) and gilthead (*Acanthopagrus schlegeli*) fry in Hiroshima. *Fish Pathology* **12**, 171–177.

Masumura, K., Yasunobu, H., Okada, N. and Muroga, K. (1989) Isolation of a *Vibrio* ssp., the causative bacterium of intestinal necrosis of Japanese flounder larvae. *Fish Pathology* **24**, 135–141.

Mateos, D. and Paniagua, C. (1995) Surface characteristics of *Aeromonas hydrophila* recovered from trout tissues. *Journal of General and Applied Microbiology* **41**, 249–254.

Matsuo, K. and Miyazono, I. (1993) The influence of long-term administration of peptidoglycan on disease resistance and growth of juvenile rainbow trout. *Nippon Suisan Gakkaishi* **59**, 1377–1379.

Matsuoka, S. and Kamada, S. (1995) Discharge of *Pasteurella piscicida* cells from experimentally infected yellowtail. *Fish Pathology* **30**, 221–225.

Matsuyama, H., Mangindaan, R.E.P. and Yano, Y. (1992) Protective effect of schizophyllan and scleroglucan against *Streptococcus* sp. infection in yellowtail (*Seriola quinqueradiata*). *Aquaculture* **101**, 197–203.

Mattheis, T. (1964) Das Vorkommen von *Vibrio anguillarum* in Ostseefischen. *Zentralblatt für Fischerei N.F.* **XII**, 259–263.

Mauel, M.J., Giovannoni, S.J. and Fryer, J.L. (1996) Development of polymerase chain reaction assays for detection, identification and differentiation of *Piscirickettsia salmonis*. *Diseases of Aquatic Organisms* **26**, 189–195.

Maugeri, T.L., Crisafi, E., Genovese, L. and Scoglio, M.E.R. (1983) Identification of *Vibrio anguillarum* with the API-20E system. *Microbiologica* **1**, 73–79.

Maulén, N.P., Morales, P.J., Aruti, D., Figueroa, J.E., Concha, M.I., Krauskopf, M. and Maulén, G. (1996) Identification of *Renibacterium salmoninarum* DNA fragment associated with bacterial internalization into CHSE-cultured cells. *FEMS Microbiology Letters* **135**, 37–43.

Mawdesley-Thomas, L.E. (1969) Furunculosis in the goldfish *Carassius auratus* (L.). *Journal of Fish Biology* **1**, 19–23.

McArdle, J.F. and Dooley-Martin, C. (1985) Isolation of *Yersinia ruckeri* type 1 (Hagerman strain) from goldfish *Carassius auratus* (L.). *Bulletin of the European Association of Fish Pathologists* **5**, 10–11.

McCarthy, D.H. (1975a) Detection of *Aeromonas salmonicida* antigen in diseased fish tissue. *Journal of General Microbiology* **88**, 185–187.

McCarthy, D.H. (1975b) Fish furunculosis. *Journal of the Institute of Fisheries Management* **6**, 13-18.

McCarthy, D.H. (1975c) Columnaris disease. *Journal of the Institute of Fisheries Management* **6**, 44–47.

McCarthy, D.H. (1976) Laboratory techniques for the diagnosis of fish-furunculosis and whirling disease. *Ministry of Agriculture, Fisheries and Foods, Fisheries Research Technical Report* no. 23, 5p.

McCarthy, D.H. (1977a) The identification and significance of atypical strains of *Aeromonas salmonicida. Bulletin de l'Office International des Epizooties* **87**, 459–463.

McCarthy, D.H. (1977b) Some ecological aspects of the bacterial fish pathogen, *Aeromonas salmonicida. Society of Applied Bacteriology Symposium Aquatic Microbiology* **6**, 299–324.

McCarthy, D.H. (1978) A study of the taxonomic status of some bacteria currently assigned to the genus *Aeromonas*. Ph.D. thesis, Council of National Academic Awards, UK.

McCarthy, D.H. (1980) Some ecological aspects of the bacterial fish pathogen— *Aeromonas salmonicida*. In: *Aquatic Microbiology*. Symposium of the Society of Applied Bacteriology No. 6, p.299–324.

McCarthy, D.H., Stevenson, J.P. and Salsbury, A.W. (1974) Therapeutic efficacy of a potentiated sulphonamide in experimental furunculosis. *Aquaculture* **4**, 407–410.

McCarthy, D.H. and Rawle, C.T. (1975) Rapid serological diagnosis of fish furunculosis caused by smooth and rough strains of *Aeromonas salmonicida. Journal of General Microbiology* **86**, 185–187.

McCarthy, D.H. and Whitehead, P. (1977) An immuno-india ink technique for rapid laboratory diagnosis of fish furunculosis. *Journal of Applied Bacteriology* **42**, 429–431.

McCarthy, D.H. and Roberts, R.J. (1980) Furunculosis of fish—the present state of our knowledge. In: Droop, M.A. and Jannasch, H.W. (eds), *Advances in Aquatic Microbiology*. London, Academic Press, p.293–341.

McCarthy, D.H., Amend, D.F., Johnson, K.A. and Bloom, J.V. (1983) *Aeromonas salmonicida*: determination of an antigen associated with protective immunity and evaluation of an experimental bacterin. *Journal of Fish Diseases* **6**, 155–174.

McCarthy, D.H., Cory, T.R. and Amend, D.F. (1984) Immunization of rainbow trout, *Salmo gairdneri* Richardson, against bacterial kidney disease: preliminary efficacy evaluation. *Journal of Fish Diseases* **7**, 65–71.

McCashion, R.N. and Lynch, W.H. (1987) Effects of polymyxin B nonapeptide on *Aeromonas salmonicida. Antimicrobial Agents and Chemotherapy* **31**, 1414–1419.

McCormick, J.I. and McLoughlin, M.F. (1993) The characterization and pathogenicity of the first isolate of *Yersinia ruckeri* from rainbow trout (*Oncorhynchus mykiss,* Walbaum) in Northern Ireland. *Bulletin of the European Association of Fish Pathologists* **13**, 138–140.

McCormick, J .I., Hughes, M.S. and McLoughlin, M.F. (1995) Identification of *Mycobacterium chelonae* in a chichlid oscar *Astronotus ocellatus* Cuvier, by direct cycle sequencing of polymerase chain reaction amplified 16S rRNA gene sequences. *Journal of Fish Diseases* **18**, 459–461.

McCoy, R.H. and Pilcher, K.S. (1974) peptone beef extract glycogen agar, selective and differential *Aeromonas* medium. *Journal of the Fisheries Research Board of Canada* **31**, 1553–1555.

McCracken, A. and Fidgeon, S. (1977) The effect of storage on drugs incorporated into pelleted fish food. *Journal of Applied Bacteriology* **42**, 289–290.

McCraw, B.M. (1952) Furunculosis of fish. *United States Fish and Wildlife Service, Special Scientific Report* 84, 87p.

McDaniel, D.W. (1972) *Hatchery Biologist, Quarterly report*, First Quarter 1972.

McIntosh, D. and Austin, B. (1988) Comparison of methods for the induction, propagation and recovery of L-phase variants of *Aeromonas* spp. *Journal of Diarrhoeal Disease Research* **6**, 131–136.

McIntosh, D. and Austin, B. (1990a). Recovery of cell wall deficient forms (L-forms) of the fish pathogens *Aeromonas salmonicida* and *Yersinia ruckeri. Systematic and Applied Microbiology* **13**, 378–381.

McIntosh, D. and Austin, B. (1990b) Recovery of an extremely proteolytic form of *Serratia liquefaciens* as a pathogen of Atlantic salmon, *Salmo salar*, in Scotland. *Journal of Fish Biology* **36**, 765–772.

McIntosh, D. and Austin, B. (1991a) Atypical characteristics of the salmonid pathogen *Aeromonas salmonicida. Journal of General Microbiology* **137**, 1341–1343.

McIntosh, D. and Austin, B. (1991b) The role of cell wall deficient bacteria (L-forms; sphaeroplasts) in fish diseases. *Journal of Applied Bacteriology* **70** 1S–7S.

McIntosh, D. and Austin, B. (1993) Potential use of vaccines based on cell-wall-defective or deficient (L-form) *Aeromonas salmonicida* for the control of furunculosis. *Journal of Aquatic Animal Health* **5**, 254–258.

McIntosh, D., Meaden, P.G. and Austin, B. (1996) A simplified PCR-based method for the detection of *Renibacterium salmoninarum* utilizing preparations of rainbow trout (*Oncorhynchus mykiss*, Walbaum) lymphocytes. *Applied and Environmental Microbiology* **62**, 3929–3932.

McIntosh, D., Flaño, E., Grayson, T.H., Gilpin, M.L. and Austin, B. (1997) Production of putative virulence factors by *Renibacterium salmoninarum* grown in cell culture. *Microbiology* **143**, 3349–3356.

McKell, J. and Jones, D. (1976) A numerical taxonomic study of *Proteus-Providencia* bacteria. *Journal of Applied Bacteriology* **41**, 143–161.

Mekuchi, T., Kiyokawa, T., Honda, K., Nakai, T. and Muroga, K. (1995a) Infection experiments with *Edwardsiella tarda* in the Japanese flounder. *Fish Pathology* **30**, 247–250.

Mekuchi, T., Kiyokawa, T., Honda, K., Nakai, T. and Muroga, K. (1995b) Vaccination trials in the Japanese flounder against edwardsiellosis. *Fish Pathology* **30**, 251–256.

Mellergaard, S. and Larsen, J.L. (1981) Haemagglutination activity of *Aeromonas salmonicida* and *Vibrio anguillarum* strains isolated from diseased rainbow trout (*Salmo gairdneri*). *Bulletin of the European Association of Fish Pathologists* **1**, 26–28.

Merino, S., Aguilar, A., Rubires, X., Simón-Pujol, D., Congregado, F. and Tomás, J.M. (1996) The role of the capsular polysaccharide of *Aeromonas salmonicida* in the adherence and invasion of fish cell lines. *FEMS Microbiology Letters* **142**, 185–189.

Merino, S., Aguilar, A., Rubires, X., Abitiu, N., Regué, M. and Tomás, J.M. (1997a) The role of the capsular polysaccharide of *Aeromonas hydrophila* serogroup O:34 in the adherence to and invasion of fish cell lines. *Research in Microbiology* **148**, 625–631.

Merino, S., Aguilar, A., Tomás, J.M., Bonet, R., Martinez, M.J., Simón-Pujol, D. and Congregado, F. (1997b) Complement resistance of capsulated strains of *Aeromonas salmonicida*. *Microbial Pathogenesis* **22**, 315–320.

Meshram, S.U., Joshi, S., Kamdi, R. and Peshwe, S. (1998) *In vitro* interaction of microbial biopesticides with fish pathogens prevailing in aquaculture food industry. *Journal of Food Science and Technology, Mysore* **35**, 177–178.

Meyer, F.P. (1964) Field treatments of *Aeromonas liquefaciens* infections in golden shiners. *Progressive Fish Culturist* **26**, 33–35.

Meyer, F.P. and Bullock, G.L. (1973) *Edwardsiella tarda*, a new pathogen of channel catfish (*Ictalurus punctatus*). *Applied Microbiology* **25**, 155–156.

Meyers, T.R., Short, S., Farrington, C., Lipson, K., Geiger, H.J. and Gates, R. (1993) Comparison of the enzyme-linked immunosorbent assay (ELISA) and the fluorescent antibody test (FAT) for measuring the prevalences and levels of *Renibacterium salmoninarum* in wild and hatchery stocks of salmonid fishes in Alaska, USA. *Diseases of Aquatic Organisms* **16**, 181–189.

Michel, C. (1979) Furunculosis of salmonids: vaccination attempts in rainbow trout (*Salmo gairdneri*) by formalin-killed germs. *Annales de Recherche Vétérinaire* **10**, 33–40.

Michel, C. (1980) A standardized model of experimental furunculosis in rainbow trout (*Salmo gairdneri*). *Canadian Journal of Fisheries and Aquatic Sciences* **37**, 746–750.

Michel, C. (1981) A bacterial disease of perch (*Perca fluviatilis* L.) in an Alpine lake: isolation and preliminary study of the causative organism. *Journal of Wildlife Diseases* **17**, 505–510.

Michel, C. and Dubois-Darnaudpeys, A. (1980) Persistence of the virulence of *Aeromonas salmonicida* kept in river sediments. *Annales de Recherche Vétérinaire* **11**, 375–386.

Michel, C., Gerard, J.-P., Fourbet, B., Collas, R. and Chevalier, R. (1980) Emploi de la flumequine contre la furunculose des salmonides; essais therapeutiques et perspectives pratiques. *Bulletin Français de Pisciculture* **52**, 154–162.

Michel, C., Faivre, B. and De Kinkelin, P. (1986a) A clinical case of enteric redmouth in minnows (*Pimephales promelas*) imported in Europe as bait-fish. *Bulletin of the European Association of Fish Pathologists* **6**, 97–99.

Michel, C., Faivre, B. and Kerouault, B. (1986b) Biochemical identification of *Lactobacillus piscicola* strains from France and Belgium. *Diseases of Aquatic Organisms* **2**, 27–30.

Michel, C. and Faivre, B. (1987) *In vitro* and *in vivo* study of an antimicrobial activity displayed by the redmouth disease agent, *Yersinia ruckeri. Annales Recherches Vétérinaire* **18**, 43–46.

Michel, C., Gonzalez, R., Bonjour, E. and Avrameas, S. (1990) A concurrent increasing of natural antibodies and enhancement of resistance to furunculosis in rainbow trout. *Annales de Recherche Vétérinaire* **21**, 211–218.

Michel, C., Nougayrède, P., Eldar, A., Sochon, E. and de Kinkelin, P. (1997) *Vagococcus salmoninarum*, a bacterium of pathological significance in rainbow trout *Oncorhynchus mykiss* farming. *Diseases of Aquatic Organisms* **30**, 199–208.

Midtlyng, P.J. (1996) A field study on intraperitoneal vaccination of Atlantic salmon (*salmo salar* L.) against furunculosis. *Fish and Shellfish Immunology* **6**, 553–565.

Midtlyng, P.J., Reitan, L.J., Lillehaug, A. and Ramstad, A. (1996) Protection, immune responses and side effects in Atlantic salmon (*Salmo salar* L.) vaccinated against furunculosis by different routes. *Fish and Shellfish Immunology* **6**, 599–600.

Millan, M. (1977) New diseases in salmon culture in Spain. *Bulletin de l'Office International des Epizooties* **87**, 515–516.

Milton, D.L., O'Toole, R., Hörstedt, P. and Wolf-Watz, H. (1996) Flagellin A is essential for the virulence of *Vibrio anguillarum. Journal of Bacteriology* **178**, 1310–1319.

Minagawa, T., Nakai, T. and Muroga, K. (1983) *Edwardsiella tarda* in eel culture environment. *Fish Pathology* **17**, 243–250.

Minami, T. (1979) *Streptococcus* sp., pathogenic to cultured yellowtail, isolated from fishes for diets. *Fish Pathology* **14**, 15–19.

Minami, T., Nakamura, M., Ikeda, Y. and Ozaki, H. (1979) A beta-hemolytic *Streptococcus* isolated from cultured yellowtail. *Fish Pathology* **14**, 33–38.

Mitchum, D.L. (1981) Concurrent infections: ERM and furunculosis found in emerald shiners. *Fish Health Section/American Fisheries Society Newsletter* **9**, 2.

Mitchum, D.L., Sherman, L.E. and Baxter, G.T. (1979) Bacterial kidney disease in feral populations of brook trout (*Salvelinus fontinalis*), brown trout (*Salmo trutta*), and rainbow trout (*Salmo gairdneri*). *Journal of the Fisheries Research Board of Canada* **36**, 1370–1376.

Mitchum, D.L. and Sherman, L.E. (1981) Transmission of bacterial kidney disease from wild to stocked hatchery trout. *Canadian Journal of Fisheries and Aquatic Sciences* **38**, 547–551.

Mitoma, Y., Aoki, T. and Crosa, J.H. (1984) Phylogenetic relationships among *Vibrio anguillarum* plasmids. *Plasmid* **12**, 143–148.

Mittal, K.R., Lalonde, G., Leblanc, D., Olivier, G. and Lallier, R. (1980) *Aeromonas hydrophila* in rainbow trout: relation between virulence and surface characteristics. *Canadian Journal of Microbiology* **26**, 1501–1503.

Miyata, M., Aoki, T., Inglis, V., Yoshida, T. and Endo, M. (1995) RAPD analysis of *Aeromonas salmonicida* and *Aeromonas hydrophila. Journal of Applied Bacteriology* **79**, 181–185.

Miyata, M., Inglis, V. and Aoki, T. (1996) Rapid identification of *Aeromonas salmonicida* subspecies *salmonicida* by the polymerase chain reaction. *Aquaculture* **141**, 13–24.

Miyazaki, T. and Egusa, S. (1976) Histopathological studies on *Edwardsiella tarda* infection of Japanese eel. Part 1. Natural infection—suppurative interstitial nephritis. *Fish Pathology* **11**, 33–43.

Miyazaki, T., Jo, Y., Kubota, S.S. and Egusa, S. (1977) Histopathological studies on vibriosis of the Japanese eel *Anguilla japonica*. Part 1. Natural infection. *Fish Pathology* **12**, 163–170.

Miyazaki, T. and Kaige, N. (1985) A histopathological study on motile aeromonad disease of Crucian carp. *Fish Pathology* **21**, 181–185.

Moki, S.T., Nomura, T. and Yoshimizu, M. (1995) Effects of incubation temperature for isolation on auto-agglutination of *Aeromonas salmonicida*. *Fish Pathology* **30**, 67–68.

Moles, A. (1997) Effect of bacterial kidney disease on saltwater adaptation of coho salmon smolts. *Journal of Aquatic Animal Health* **9**, 230–233.

Møller, V. (1955) Simplified tests for some amino acid decarboxylases and for the arginine dihydrolase system. *Acta Pathologie et Microbiologie Scandinavia* **36**, 58–172.

Mooney, J., Powell, E., Clabby, C. and Powell, R. (1995) Detection of *Aeromonas salmonicida* in wild Atlantic salmon using a specific DNA probe test. *Diseases of Aquatic Organisms* **21**, 131–135.

Moore, W.E.C. and Holdeman-Moore, L.V. (1986) Genus *Eubacterium* Prévot 1938, 294[AL]. In: Sneath, P.H.A., Mair, N.S., Sharpe, M.E. and Holt, J.G. (eds), *Bergey's Manual of Systematic Bacteriology*, Vol. 2. Baltimore, Williams & Wilkins, p.1353–1373.

Morgan, J.A.W., Cranwell, P.A. and Pickup, R.W. (1991) Survival of *Aeromonas salmonicida* in lake water. *Applied and Environmental Microbiology* **57**, 1777–1782.

Morgan, J.A.W., Rhodes, G. and Pickup, R.W. (1993) Survival of non-culturable *Aeromonas salmonicida* in lake water. *Applied and Environmental Microbiology* **59**, 874–880.

Morrison, C., Cornick, J., Shum, G. and Zwicker, B. (1981) Microbiology and histopathology of 'saddleback' disease of underyearling Atlantic salmon, *Salmo salar* L. *Journal of Fish Diseases* **4**, 243–258.

Morrison, E.E. and Plumb, J.A. (1994) Olfactory organ of channel catfish as a site of experimental *Edwardsiella ictaluri* infection. *Journal of Aquatic Animal Health* **6**, 101–109.

Morse, E.G., Greenwood, D.E., Meyers, E.P., Anderson, V.L. and Duncan, M.A. (1978) Experimental *Salmonella* infections in *Carassius auratus* goldfish. *Journal of Experimental Science and Health Part A. Environmental Science and Engineering* **13**, 325–335.

Moss, M.O. and Ryall, C. (1980) The genus *Chromobacterium*. In: Starr, M.P., Stolp, H., Trüper, H.G., Balows, A. and Schlegel, H.G. (eds), *The Prokaryotes: A Handbook on Habitats, Isolation and Identification of Bacteria*, Vol. 2, Berlin, Springer-Verlag, p.1355–1364.

Mudarris, M. and Austin, B. (1989) Systemic disease in turbot *Scophthalmus maximus* caused by a previously unrecognized *Cytophaga*-like bacterium. *Diseases of Aquatic Organisms* **6**, 161–166.

Mudarris, M. and Austin, B. (1992) Histopathology of a gill and systemic disease of turbot (*Scophthalmus maximus*) caused by a *Cytophaga*-like bacterium (CLB). *Bulletin of the European Association of Fish Pathologists* **12**, 120–123.

Mudarris, M., Austin, B., Segers, P., Vancanneyt, M., Hoste, B. and Bernardet, J.F. (1994) *Flavobacterium scophthalmum* sp. nov., a pathogen of turbot (*Scophthalmus maximus* L.). *International Journal of Systematic Bacteriology* **44**, 447–453.

Munn, C.B. (1978) Haemolysin production by *Vibrio anguillarum*. *FEMS Microbiology Letters* **3**, 265–268.

Munn, C.B. (1980) Production and properties of a haemolytic toxin by *Vibrio anguillarum*. In: Ahne, W. (ed.), *Fish Diseases, Third COPRAQ Session*. Berlin, Springer-Verlag, p. 69–74.

Munn, C.B. and Trust, T.J. (1984) Role of additional protein layer in virulence of *Aeromonas salmonicida*. In: Acuigrup (ed.), *Fish Diseases, Fourth COPRAQ Session*. Madrid, Editora ATP, p.69–73.

Munn, C.B., Ishiguro, E.E., Kay, W.W. and Trust, T.J. (1982) Role of surface components in serum resistance of virulent *Aeromonas salmonicida*. *Infection and Immunity* **36**, 1069–1075.

Munro, A.L.S. (1984) A furunculosis vaccine—illusion or achievable objective. In: de Kinkelin, P.K. (ed.), *Symposium on Fish Vaccination*. Paris, Office International des Epizooties, p.97–120.

Munro, A.L.S., Hastings, T.S., Ellis, A.E. and Liversidge, J. (1980) Studies on an ichthyotoxic material produced extracellularly by the furunculosis bacterium *Aeromonas salmonicida*. In: Ahne, W. (ed.), *Fish Diseases, Third COPRAQ Session*. Berlin, Springer-Verlag, p.98–106.

Muraoka, A., Ogawa, K., Hashimoto, S. and Kusuda, R. (1991) Protection of yellowtail against pseudotuberculosis by vaccination with a potassium thiocyanate extract of *Pasteurella piscicida* and co-operating protective effect of acid-treated, naked bacteria. *Nippon Suisan Gakkaishi* **57**, 249–253.

Muroga, K. (1975) Studies on *Vibrio anguillarum* and *V. anguillicida* infections. *Journal of the Faculty of Fisheries and Animal Husbandry, Hiroshima University* **14**, 101–205.

Muroga, K. (1990) Bacterial infections in cultured fishes in Japan. In: Hirano, R. and Hanyu, I. (eds), *The Second Asian Fisheries Forum*, Manila, Asian Fisheries Society, p.963–966.

Muroga, K., Jo, Y. and Sawada, T. (1975) Studies on red spot disease of pond cultured eels. Part II. Pathogenicity of the causative bacterium *Pseudomonas anguilliseptica*. *Fish Pathology* **9**, 107–114.

Muroga, K., Jo, Y. and Nishibuchi, M. (1976a) Pathogenic *Vibrio* isolated from cultured eels. I. Characteristics and taxonomic status. *Fish Pathology* **12**, 141–145.

Muroga, K., Nishibuchi, M. and Jo, Y. (1976b) Pathogenic *Vibrio* isolated from cultured eels. II. Physiological characteristics and pathogenicity. *Fish Pathology* **12**, 147–151.

Muroga, K., Sugiyama, T. and Ueki, N. (1977a) Pasteurellosis in cultured black sea bream (*Mylio macrocephalus*). *Journal of the Faculty of Fisheries and Animal Husbandry, Hiroshima University* **16**, 17–21.

Muroga, K., Nakai, T. and Sawada, T. (1977b) Studies on red spot disease of pond cultured eels. Part IV. Physiological characteristics of the causative bacterium *Pseudomonas anguilliseptica*. *Fish Pathology* **12**, 33–38.

Muroga, K., Takahashi, S. and Yamanoi, H. (1979) Non-cholera *Vibrio* isolated from diseased ayu. *Bulletin of the Japanese Society of Scientific Fisheries* **45**, 829–834.

Muroga, K., Lio-Po, G., Pitogo, C. and Imada, R. (1984a) *Vibrio* sp. isolated from milkfish (*Chanos chanos*) with opaque eyes. *Fish Pathology* **19**, 81–87.

Muroga, K., Yamanoi, H., Hironaka, Y., Yamamoto, S., Tatani, M., Jo, Y., Takahashi, S. and Hanada, H. (1984b) Detection of *Vibrio anguillarum* from wild fingerlings of ayu *Plecoglossus altivelis*. *Bulletin of the Japanese Society of Scientific Fisheries* **50**, 591–596.

Muroga, K., Iida, M., Matsumoto, H. and Nakai, T. (1986) Detection of *Vibrio anguillarum* from waters. *Bulletin of the Japanese Society of Scientific Fisheries* **52**, 641–647.

Muroga, K. and De La Cruz, M. (1987) Fate and location of *Vibrio anguillarum* in tissues of artificially infected ayu (*Plecoglossus altivelis*). *Fish Pathology* **22**, 99–103.

Muroga, K. and Yasunobu, H. (1987) Uptake of bacteria by rotifer. *Nippon Suisan Gakkaishi* **53**, 2091.

Muroga, K., Yasunobu, H., Okada, N. and Masumura, K. (1990) Bacterial enteritis of cultured flounder *Paralichthys olivaceus* larvae. *Diseases of Aquatic Organisms* **9**, 121–125.

Murray, C.B., Evelyn, T.P.T., Beacham, T.D., Barner, L.W., Ketcheson, J.E. and Prosperi-Porta, L. (1992) Experimental induction of bacterial kidney disease in chinook salmon by immersion and cohabitation challenges. *Diseases of Aquatic Organisms* **12**, 91–96.

Murray, R.G.E., Dooley, J.S.G., Whippey, P.W. and Trust, T.J. (1988) Structure of an S layer on a pathogenic strain of *Aeromonas hydrophila*. *Journal of Bacteriology* **170**, 2625–2630.

Mushiake, K., Muroga, K. and Nakai, T. (1984) Increased susceptibility of Japanese eel *Anguilla japonica* to *Edwardsiella tarda* and *Pseudomonas anguilliseptica* following exposure to copper. *Bulletin of the Japanese Society of Scientific Fisheries* **50**, 1797–1801.

Mushiake, K., Nakai, T. and Muroga, K. (1985) Lowered phagocytosis in the blood of eels exposed to copper. *Fish Pathology* **20**, 49–53.

Muthiara, L.W., Raymond, B.T., Dekievit, T.R. and Stevenson, R.H.W. (1993) Antibody specificities of polyclonal rabbit and rainbow trout antisera against *Vibrio ordalii* and serotype O:2 strains of *Vibrio anguillarum*. *Canadian Journal of Microbiology* **39**, 492–499.

Myhr, E., Larsen, J.L., Lillehaug, A., Gudding, R., Heum, M. and Håstein, T. (1991) Characterization of *Vibrio anguillarum* and closely related species isolated from farmed fish in Norway. *Applied and Environmental Microbiology* **57**, 2756–2757.

Nakai, T. and Muroga, K. (1979) Studies on red spot disease of pond cultured eels—V.

Immune response of the Japanese eel to the causative bacterium *Pseudomonas anguilliseptica*. *Bulletin of the Japanese Society of Scientific Fisheries* **45**, 817–821.

Nakai, T. and Muroga, K. (1982) *Pseudomonas anguilliseptica* isolated from European eels (*Anguilla anguilla*) in Scotland. *Fish Pathology* **17**, 147–150.

Nakai, T., Muroga, K. and Wakabayashi, H. (1981) Serological properties of *Pseudomonas anguilliseptica* in agglutination. *Bulletin of the Japanese Society of Scientific Fisheries* **47**, 699–703.

Nakai, T., Muroga, K. and Wakabayashi, H. (1982a) An immuno-electrophoretic analysis of *Pseudomonas anguilliseptica*. *Bulletin of the Japanese Society of Scientific Fisheries* **48**, 363–367.

Nakai, K., Muroga, K. and Ohnishi, K., Jo, Y. and Tanimoto, H. (1982b) Studies on the red spot disease of pond cultured eel—IV. A field vaccination trial. *Aquaculture* **30**, 131–135.

Nakai, T., Kanno, T., Cruz, E.R. and Muroga, K. (1987) The effects of iron compounds on the virulence of *Vibrio anguillarum* in Japanese eels and ayu. *Fish Pathology* **22**, 185–189.

Nakai, T., Miyakawa, M., Muroga, K. and Kamito, K. (1989a) The tissue distribution of atypical *Aeromonas salmonicida* in artificially infected Japanese eels, *Anguilla japonica*. *Fish Pathology* **24**, 23–28.

Nakai, T., Muroga, K. and Masumura, K. (1989b) Immersion vaccination of juvenile rockfish *Sebastes schlegeli* by *Vibrio ordalii* and *Vibrio anguillarum* antigens. *Suisanzoshoku* **37**, 129–132.

Nakajima, M. and Chikahata, H. (1979) Efficacy of oral and hyperosmotic vaccinations for vibriosis in ayu. *Fish Pathology* **14**, 9–13.

Nakajima, K., Muroga, K. and Hancock, R.E.W. (1983) Comparison of fatty acid, protein and serological properties distinguishing outer membrane of *Pseudomonas anguilliseptica* strains from those of fish pathogens and other pseudomonads. *International Journal of Systematic Bacteriology* **33**, 1–8.

Nakamura, Y. (1982) Doxycycline. *Fish Pathology* **17**, 67–76.

Nakamura, Y., Nakai, T. and Muroga, K. (1990) Induction of antibody producing cells in some fishes by immunization with *Vibrio anguillarum* O-antigen. *Fish Pathology* **25**, 225–230.

Nakatsugawa, T. (1983) *Edwardsiella tarda* isolated from cultured young flounder. *Fish Pathology* **18**, 99–101.

Nakatsugawa, T. (1994) Atypical *Aeromonas salmonicida* isolated from cultured shotted halibut. *Fish Pathology* **29**, 193–198.

Naudé, T.W. (1975) The occurrence and diagnosis of certain trout disease. *Fish Farmer* **9**, 15–20.

Navarre, O. and Halver, J.E. (1989) Disease resistance and humoral antibody production in rainbow trout fed high levels of vitamin C. *Aquaculture* **79**, 207–221.

Nelson, D.R., Sadlowski, Y., Eguchi, M. and Kjelleberg, S. (1997) The starvation-stress response of *Vibrio (Listonella) anguillarum*. *Microbiology* **143**, 2305–2312.

Nelson, J. S., Kawahara, E., Kawai, K. and Kusuda, R. (1989) Macrophage infiltration in pseudotuberculosis of yellowtail, *Seriola quinqueradiata. Bulletin of Marine Sciences and Fisheries, Kochi University* **11**, 17–22.

Nese, L. and Enger, Ø. (1993) Isolation of *Aeromonas salmonicida* from salmon lice *Lepeophtheirus salmonis* and marine plankton. *Diseases of Aquatic Organisms* **16**, 79–81.

Newman, S.G. and Majnarich, J.J. (1982) Direct immersion vaccination of juvenile rainbow trout, *Salmo gairdneri* Richardson, and juvenile coho salmon, *Oncorhynchus kisutch* (Walbaum), with a *Yersinia ruckeri* bacterin. *Journal of Fish Diseases* **5**, 339–341.

Newton, J.C., Wood, T.M. and Hartley, M.M. (1997) Isolation and partial characterization of extracellular proteases produced by isolates by *Flavobacterium columnare* derived from channel catfish. *Journal of Aquatic Animal Health* **9**, 75–85.

Nielsen, B. and Dalsgaard, I. (1991) Plasmids in *Vibrio salmonicida* isolates from the Faroe Islands. *Bulletin of the European Association of Fish Pathologists* **11**, 206–207.

Nielsen, B., Olsen, J.E. and Larsen, J.L. (1993) Plasmid profiling as an epidemiological marker within *Aeromonas salmonicida. Diseases of Aquatic Organisms* **15**, 129–135.

Nielsen, B., Olsen, J.E. and Larsen, J.L. (1994) Ribotyping of *Aeromonas salmonicida* subsp. *salmonicida. Diseases of Aquatic Organisms* **18**, 151–158.

Nieto, T.P., Corcobado, M.J.R., Toranzo, A.E. and Barja, J.L. (1985) Relation of water temperature to infection of *Salmo gairdneri* with motile *Aeromonas. Fish Pathology* **20**, 99–105.

Nieto, T.P. and Ellis, A.E. (1986) Characterization of extracellular metalo- and serine-proteases of *Aeromonas hydrophila* strain B_{51}. *Journal of General Microbiology* **132**, 1975–1979.

Nieto, T.P., López, L.R., Santos, Y., Núñez, S. and Toranzo, A.E. (1990) Isolation of *Serratia plymuthica* as an opportunistic pathogen in rainbow trout, *Salmo gairdneri* Richardson. *Journal of Fish Diseases* **13**, 175–177.

Nieto, T.P., Santos, Y., Rodriguez, L.A. and Ellis, A.E. (1991) An extracellular acetylcholinesterase produced by *Aeromonas hydrophila* is a major lethal toxin for fish. *Microbial Pathogenesis* **11**, 101–110.

Nigrelli, R.F. (1943) Causes of disease and death of fishes in captivity. *Zoologica* **28**, 208–216.

Nigrelli, R.F. and Hutner, S.H. (1945) The presence of a myxobacterium, *Chondrococcus columnaris* (Davis) Ordal and Rucker (1944) on *Fundulus heteroclitus* (Linn.). *Zoologica* **30**, 101–103.

Nigrelli, R.F. and Vogel, H. (1963) Spontaneous tuberculosis in fishes and in other cold-blooded vertebrates with special reference to *Mycobacterium fortuitum* Cruz from fish and human lesions. *Zoologica (New York)* **48**, 130–143.

Nikl, L., Albright, L.J. and Evelyn, T.P.T. (1991) Influence of seven immunostimulants on the immune response of coho salmon to *Aeromonas salmonicida. Diseases of Aquatic Organisms* **12**, 7–12.

Nikl, L., Evelyn, T.P.T. and Albright, L.J. (1993) Trials with an orally and immersion-administered β-1,3 glucan as an immunoprophylactic against *Aeromonas salmonicida* in juvenile chinook salmon *Oncorhynchus tshawytscha. Diseases of Aquatic Organisms* **17**, 191–196.

Ninomiya, M., Muraoka, A. and Kusuda, R. (1989) Effect of immersion vaccination of cultured yellowtail with a ribosomal vaccine prepared from *Pasteurella piscicida*. *Nippon Suissan Gakkaishi* **55**, 1773–1776.

Nishibuchi, M. and Muroga, K. (1977) Pathogenic *Vibrio* isolated from cultured eels. III. NaCl tolerance and flagellation. *Fish Pathology* **12**, 87–92.

Nishibuchi, M., Muroga, K., Seidler, R.J. and Fryer, J.L. (1979) Pathogenic *Vibrio* isolated from cultured eels. IV. Deoxyribonucleic acid studies. *Bulletin of the Japanese Society of Scientific Fisheries* **45**, 1469–1473.

Nishibuchi, M. and Muroga, K. (1980) Pathogenic *Vibrio* isolated from cultured eels. V. Serological studies. *Fish Pathology* **14**, 117–127.

Nishibuchi, M., Muroga, K. and Jo, Y. (1980) Pathogenic *Vibrio* isolated from cultured eels. VI. Diagnostic tests for the disease due to the present bacterium. *Fish Pathology* **14**, 125–131.

Nomura, S. and Saito, H. (1982) Production of the extracellular hemolytic toxin by an isolated strain of *Aeromonas salmonicida*. *Bulletin of the Japanese Society of Scientific Fisheries* **48**, 1589–1597.

Nomura, S., Fujino, M., Yamakawa, M. and Kawahara, E. (1988) Purification and characterization of salmolysin, an extracellular hemolytic toxin from *Aeromonas salmonicida*. *Journal of Bacteriology* **170**, 3694–3702.

Noonan, B. and Trust, T.J. (1995) The leucine zipper of *Aeromonas salmonicida* AbcA is required for the transcriptional activation of the P2 promoter of the surface-layer structural gene, *vapA*, in *Escherichia coli*. *Molecular Microbiology* **17**, 379–386.

Nordmo, R., Varma, K.J., Sutherland, I.H. and Brokken, E.S. (1994) Florfenicol in Atlantic salmon, *Salmo salar* L.: Field evaluation of efficacy against furunculosis in Norway. *Journal of Fish Diseases* **17**, 239–244.

Nordmo, R., Ramstad, A. and Holth Riseth, J.M. (1998) Induction of experimental furunculosis in heterogeneous test populations of Atlantic salmon (*Salmo salar* L.) by use of a cohabitation method. *Aquaculture* **162**, 11–21.

Norqvist, A., Hagström, Å. and Wolf-Watz, H. (1989) Protection of rainbow trout against vibriosis and furunculosis by the use of attenuated strains of *Vibrio salmonicida*. *Applied and Environmental Microbiology* **55**, 1400–1405.

Norquist, A., Norman, B. and Wolf-Watz, H. (1990) Identification and characterization of a zinc metalloprotease associated with invasion by the fish pathogen *Vibrio anguillarum*. *Infection and Immunity* **58**, 3731–3736.

Norquist, A., Bergman, A., Skogman, G. and Wolf-Watz, H. (1994) A field trial with the live attenuated fish vaccine strain *Vibrio anguillarum* VAN1000. *Bulletin of the European Association of Fish Pathologists* **14**, 156–158.

Novotny, A.J. (1978) Vibriosis and furunculosis in marine cultured salmon in Puget Sound. *Washington Marine Fisheries Review* **40**, 52–55.

Noya, M., Magariños, B. and Lamas, J. (1995) Interactions between peritoneal exudate cells (PECs) of gilthead bream (*Sparus aurata*) and *Pasteurella piscicida*. A morphological study. *Aquaculture* **131**, 11–21.

Nybelin, O. (1935) Untersuchungen über den bei Fischen Krankheitsegen den Spatpliz *Vibrio anguillarum*. *Meddelanden frau Statens Undersöknings—och Fasok—sanstalt für Sotvattenfisket Stockholm* **8**, 1–62.

416 **Bibliography**

Nzeako, B.C. (1991) Variation in *Aeromonas hydrophila* surface structures. *Bulletin of the European Association of Fish Pathologists* **11**, 176–179.

Oakey, H.J., Ellis, J.T. and Gibson, L.F. (1998) The development of random DNA probes specific for *Aeromonas salmonicida*. *Journal of Applied Microbiology* **84**, 37–46.

O'Brien, D., Mooney, J., Ryan, D., Powell, E., Hiney, M., Smith, P.R. and Powell, R. (1994) Detection of *Aeromonas salmonicida*, causal agent of furunculosis in salmonid fish, from the tank effluent of hatchery-reared Atlantic salmon smolts. *Applied and Environmental Microbiology* **60**, 3874–3877.

Ogara, W.O., Mbuthia, P.G., Kaburia, H.F.A., Sørum, H., Kagunya, D.K., Nduthu, D.I. and Colquhoun, D. (1998) Motile aeromonads associated with rainbow trout (*Oncorhynchus mykiss*) mortality in Kenya. *Bulletin of the European Association of Fish Pathologists* **18**, 7–9.

Ohnishi, K. and Muroga, K. (1976) *Vibrio* sp. as a cause of disease in rainbow trout cultured in Japan. 1. Biochemical characteristics. *Fish Pathology* **11**, 159–165.

Ohnishi and Shiro (1978) Streptococcal infections in cultured sweetfish and herring. Abstract presented at the Spring Meeting of the Japanese Marine Products Association, p.48.

Ohnishi, K., Watanabe, K. and Jo, Y. (1982) *Pasteurella* infection in young black seabream. *Fish Pathology* **16**, 207–210.

Ohtsuka, H., Nakai, T., Muroga, K. and Jo, Y. (1984) Atypical *Aeromonas salmonicida* isolated from diseased eels. *Fish Pathology* **19**, 101–107.

Oladosu, G.A., Ayinla, O.A. and Ajiboye, M.O. (1994) Isolation and pathogenicity of a *Bacillus* sp. associated with a septicaemic condition in some tropical freshwater fish species. *Journal of Applied Ichthyology* **10**, 69–72.

Olafsen, J.A., Christie, M. and Raa, J. (1981) Biochemical ecology and psychrotrophic strains of *Vibrio anguillarum* isolated from outbreaks of vibriosis at low temperature. *Zentralblatt für Bakteriologie and Hygiene, Erste Abteilung Originale* **C2**, 339–348.

Olea, I., Bruno, D.W. and Hastings, T.S. (1993) Detection of *Renibacterium salmoninarum* in naturally infected Atlantic salmon, *Salmo salar* L., and rainbow trout, *Oncorhynchus mykiss* (Walbaum) using an enzyme-linked immunosorbent assay. *Aquaculture* **116**, 99–110.

O'Leary, W.M., Panos, C. and Helz, G.E. (1956) Studies on the nutrition of *Bacterium salmonicida*. *Journal of Bacteriology* **72**, 673–676.

O'Leary, P.J. (1977) Enteric redmouth of salmonids: a biochemical and serological comparison of selected isolates. M.S. thesis, Oregon State University.

Olesen, J.E. and Larsen, J.L. (1993) Ribotypes and plasmid content of *Vibrio anguillarum* strains in relation to serovar. *Applied and Environmental Microbiology* **59**, 3863–3870.

Oliver, J.D., Warner, R.A. and Cleland, D.R. (1983) Distribution of *Vibrio vulnificus* and other lactose-fermenting vibrios in the marine environment. *Applied and Environmental Microbiology* **45**, 985–998.

Olivier, G. (1990) Virulence of *Aeromonas salmonicida*: lack of relationship with phenotypic characteristics. *Journal of Aquatic Animal Health* **2**, 119–127.

Olivier, G., Evelyn, T.P.T. and Lallier, R. (1985a) Immunogenicity of vaccines from a virulent and an avirulent strain of *Aeromonas salmonicida*. *Journal of Fish Diseases* **8**, 43–55.

Olivier, G., Evelyn, T.P.T. and Lallier, R. (1985b) Immunity to *Aeromonas salmonicida* in coho salmon (*Oncorhynchus kisutch*) induced by modified Freunds complete adjuvant: its non-specific nature and the probable role of macrophages in the phenomenon. *Developmental and Comparative Immunology* **9**, 419–432.

Olivier, G., Eaton, C.A. and Campbell, N. (1986) Interaction between *Aeromonas salmonicida* and peritoneal macrophages of brook trout (*Salvelinus fontinalis*). *Veterinary Immunology and Immunopathology* **12**, 223–234.

Olivier, G., Griffiths, S.G., Fildes, J. and Lynch, W.H. (1992) The use of Western blot and electroimmunotransfer blot assays to monitor bacterial kidney disease in experimentally challenged Atlantic salmon, *Salmo salar* L. *Journal of Fish Diseases* **15**, 229–241.

Olsen, A.B., Melby, H.P., Speilberg, L., Evensen, Ø. and Håstein, T. (1997) *Piscirickettsia salmonis* infection in Atlantic salmon *Salmo salar* in Norway—epidemiological, pathological and microbiological findings. *Diseases of Aquatic Organisms* **31**, 35–48.

Olsen, J.E. and Larsen, J.L. (1990) Restriction fragment length polymorphism of the *Vibrio anguillarum* serovar O1 virulence plasmid. *Applied and Environmental Microbiology* **56**, 3130–3132.

Olson, B.H. (1978) Enhanced accuracy of coliform testing in seawater by a modification of the Most Probable Number method. *Applied and Environmental Microbiology* **36**, 438–444.

Olsson, J.C., Jöborn, A., Westerdahl, A., Blomberg, L., Kjelleberg, S. and Conway, P.L. (1996) Is the turbot, *Scophthalmus maximus* (L.), intestine a portal of entry for the fish pathogen *Vibrio anguillarum*. *Journal of Fish Diseases* **19**, 225–234.

Olsson, J.C., Jöborn, A., Westerdahl, A., Blomberg, L., Kjelleberg, S. and Conway, P.L. (1998) Survival, persistence and proliferation of *Vibrio anguillarum* in juvenile turbot, *Scophthalmus maximus* (L.), intestine and faeces. *Journal of Fish Diseases* **21**, 1–9.

Onarheim, A.M. (1992) The glucan way to fish health. *Fish Farming International* **19**, 32–33.

Oppenheimer, C.H. (1962) Marine fish diseases. In: *Fish as Food*, Vol. 2. New York, Academic Press, p.541–572.

Ordal, E.J. and Rucker, R.R. (1944) Pathogenic myxobacteria. *Proceedings of the Society of Experimental Biology and Medicine* **56**, 15–18.

Ordal, E.J. and Earp, B.J. (1956) Cultivation and transmission of etiological agent of bacterial kidney disease in salmonid fishes. *Proceedings of the Society of Experimental Biology and Medicine* **92**, 85–88.

Ortega, C., Ruiz, I., De Blas, I., Muzquiz, J.L., Fernandez, A. and Alonso, J.L. (1996) Furunculosis control using a paraimmunization stimulant (Baypamun) in rainbow trout. *Veterinary Research (Paris)* **27**, 561–568.

Otis, V.S. and Behler, J.L. (1973) The occurrence of salmonellae and *Edwardsiella* in the turtles of New York zoological park. *Journal of Wildlife Diseases* **9**, 4–6.

O'Toole, R., Milton, D.L. and Wolf-Watz, H. (1996) Chemotactic motility is required for invasion of the host by the fish pathogen *Vibrio anguillarum*. *Molecular Microbiology* **19**, 625–637.

O'Toole, R., Milton, D.L., Hörstedt, P. and Wolf-Watz, H. (1997) RpoN of the fish pathogen *Vibrio (Listonella) anguillarum* is essential for flagellum production and virulence by the water-borne but not intraperitoneal route of inoculation. *Microbiology* **143**, 3849–3859.

Otte, E. (1963) Die heutigen Ansichten über die Atiologie der Infektiösen Bachwassersucht der Karpfen. *Wiener Tierärztliche Monatsschrift* **50**, 995–1005.

Owens, L., Austin, D.A. and Austin, B. (1996) Effect of strain origin on siderophore production in *Vibrio harveyi* isolates. *Diseases of Aquatic Organisms* **28**, 93–106.

Pacha, R.E. (1961) Columnaris disease in fishes in the Columbia River Basin. Ph.D. thesis, University of Washington, Seattle.

Pacha, R.E. (1968) Characteristics of *Cytophaga psychrophila* (Borg) isolated during outbreaks of bacterial cold-water disease. *Applied Microbiology* **16**, 97–101.

Pacha, R.E. and Ordal, E.J. (1963) Epidemiology of columnaris disease in salmon. *Bacteriological Proceedings* p.3.

Pacha, R.E. and Ordal, E.J. (1967) Histopathology of experimental columnaris disease in young salmon. *Journal of Comparative Pathology* **77**, 419–423.

Pacha, R.E. and Porter, S. (1968) Characteristics of myxobacteria isolated from the surface of freshwater fish. *Applied Microbiology* **16**, 1901–1906.

Pacha, R.E. and Kiehn, E.D. (1969) Characterization and relatedness of marine vibrios pathogenic to fish: physiology, serology and epidemiology. *Journal of Bacteriology* **100**, 1242–1247.

Pacha, R.E. and Ordal, E.J. (1970) Myxobacterial diseases of salmonids. In: Snieszko, S.F. (ed.), *A Symposium on Diseases of Fishes and Shellfishes*, Washington, D.C., American Fisheries Society, Special Publication No. 5, p.243–257.

Paclibare, J.O., Evelyn, T.P.T., Albright, L.J. and Prosperi-Porta, L. (1994) Clearing of the kidney disease bacterium *Renibacterium salmoninarum* from seawater by the blue mussel *Mytilus edulis*, and the status of the mussel as a reservoir of the bacterium. *Diseases of Aquatic Organisms* **18**, 129–133.

Palleroni, N.J. (1984) Genus 1 *Pseudomonas* Migual 1894, 237[AL] (Nom. cons. Opin. 5, Jud. Comm. 1952, 237). In: Krieg, N.R. and Holt, J.G. (eds), *Bergey's Manual of Systematic Bacteriology*, Vol. 1, Baltimore, Williams & Wilkins, p.141–199.

Palmer, R. and Smith, P.R. (1980) Studies on vaccination of Atlantic salmon against furunculosis. In: Ahne, W. (ed.), *Fish Diseases, Third COPRAQ Session*. Berlin, Springer-Verlag, p.107–112.

Palmer, R., Drinan, E. and Murphy, T. (1994) A previously unknown disease of farmed Atlantic salmon: pathology and establishment of bacterial aetiology. *Diseases of Aquatic Organisms* **19**, 7–14.

Palmer, R., Ruttledge, M., Callanan, K. and Drinan, E. (1997) A piscirickettsiosis-like disease in farmed Atlantic salmon in Ireland—isolation of the agent. *Bulletin of the European Association of Fish Pathologists* **17**, 68–72.

Paniagua, C., Rivero, O., Anguita, J. and Naharro, G. (1990) Pathogenicity factors and virulence for rainbow trout (*Salmo gairdneri*) of motile *Aeromonas* spp. isolated from a river. *Journal of Clinical Microbiology* **28**, 350–355.

Parisot, T.J. and Wood, E.M. (1960) A comparative study of the causative agent of a mycobacterial disease of salmonid fishes. II. A description of the histopathology of the disease in chinook salmon (*Oncorhynchus tshawytscha*) and a comparison of the strain characteristics of the fish disease with leprosy and human tuberculosis. *Annual Review of Respiratory Diseases* **82**, 212–222.

Park, K.H., Kato, H., Nakai, T. and Muroga, K. (1998) Phage typing of *Lactococcus garvieae* (formerly *Enterococcus seriolicida*) a pathogen of cultured yellowtail. *Fisheries Science (Tokyo)* **64**, 62–64.

Park, S.I. (1986) Distribution of *Edwardsiella tarda* in eel culture pond and its pathogenicity to eel. Doctoral thesis, Faculty of Agriculture, University of Tokyo.

Park, S.-I., Wakabayashi, H. and Watanabe, Y. (1983) Serotype and virulence of *Edwardsiella tarda* isolated from eel and their environment. *Fish Pathology* **18**, 85–89.

Parker, N.D. and Munn, C.B. (1985) Cell surface properties of virulent and attenuated strains of *Aeromonas salmonicida*. In: Ellis, A.E. (ed.), *Fish and Shellfish Pathology*. London, Academic Press, p.97–105.

Pascho, R.J., Elliott, D.G., Mallett, R.W. and Mulcahy, D. (1987) Comparison of five techniques for the detection of *Renibacterium salmoninarum* in adult coho salmon. *Transactions of the American Fisheries Society* **116**, 882–890.

Pascho, R.J., Landolt, M.L. and Ongerth, J.E. (1995) Inactivation of *Renibacterium salmoninarum* by free chlorine. *Aquaculture* **131**, 165–175.

Pascho, R.J., Chase, D. and McKibben, C.L. (1998) Comparison of the membrane-filtration fluorescent antibody test, the enzyme-linked immunosorbent assay, and the polymerase chain reaction to detect *Renibacterium salmoninarum* in salmonid ovarian fluid. *Journal of Veterinary Diagnostic Investigation* **10**, 60–66.

Paterson, W.D. (1974) Biochemical and serological differentiation of several pigment-producing aeromonads. *Journal of the Fisheries Research Board of Canada* **31**, 1259–1261.

Paterson, W.D. (1982) Furunculosis and other associated diseases caused by *Aeromonas salmonicida*. In: Anderson, D.P., Dorson, M. and Dubouget, P. (eds), *Antigens of Fish Pathogens: Development and Production for Vaccines and Serodiagnostics*. Lyon, Collection Foundation Marcel Merieux, p.119–137.

Paterson, W.D. and Fryer, J.L. (1974a) Immune response of juvenile coho salmon (*Oncorhynchus kisutch*) to *Aeromonas salmonicida* endotoxin. *Journal of the Fisheries Research Board of Canada* **31**, 1743–1749.

Paterson, W.D. and Fryer, J.L. (1974b) Immune response of juvenile coho salmon (*Oncorhynchus kisutch*) to *Aeromonas salmonicida* cells administered intraperitoneally in Freund's complete adjuvant. *Journal of the Fisheries Research Board of Canada* **31**, 1751–1755.

Paterson, W.D., Gallant, C., Desautels, D. and Marshall, L. (1979) Detection of bacterial kidney disease in wild salmonids in the Margaree River system and adjacent waters using an indirect fluorescent antibody technique. *Journal of the Fisheries Research Board of Canada* **36**, 1464–1468.

Paterson, W.D., Douey, D. and Desautels, D. (1980) Relationships between selected strains of typical and atypical *Aeromonas salmonicida, Aeromonas hydrophila,* and *Haemophilus piscium. Canadian Journal of Microbiology* **26**, 588–598.

Paterson, W.D., Lall, S.P. and Desautels, D. (1981) Studies on bacterial kidney disease in Atlantic salmon (*Salmo salar*) in Canada. *Fish Pathology* **15**, 283–292.

Pathiratne, A., Widanapathirana, G.S. and Chandrarkanthi, W.H.S. (1994) Association of *Aeromonas hydrophila* with epizootic ulcerative syndrome (EUS) in freshwater fish in Sri Lanka. *Journal of Applied Ichthyology* **10**, 204–208.

Paula, S.J., Duffey, P.S., Abbott, S.L., Kokka, R.P., Oshio, L.S., Janda, J.M., Shimada, T. and Sakazaki, R. (1988) Surface properties of autoagglutinating mesophilic aeromonads. *Infection and Immunity* **56**, 2658–2665.

Pazos, F., Santos, Y., Magariños, B., Bandín, I., Núñez, S. and Toranzo, A.E. (1993) Phenotypic characteristics and virulence of *Vibrio anguillarum*-related organisms. *Applied and Environmental Microbiology* **59**, 2969–2976.

Pedersen, K., Kofod, H., Dalsgaard, I. and Larsen, J.L. (1994) Isolation of oxidase-negative *Aeromonas salmonicida* from diseased turbot *Scophthalmus maximus. Diseases of Aquatic Organisms* **18**, 149–154.

Pedersen, K. and Larsen, J.L. (1995) Evidence for the existence of distinct populations of *Vibrio anguillarum* serogroup O1 based on plasmid contents and ribotypes. *Applied and Environmental Microbiology* **61**,2292–2296.

Pedersen, K., Dalsgaard, I. and Larsen, J.L. (1996a) Characterization of atypical *Aeromonas salmonicida* isolates by ribotyping and plasmid profiling. *Journal of Applied Bacteriology* **80**, 37–44.

Pedersen, K., Koblavi, S., Tiainen, T. and Larsen, J.L. (1996b) Restriction fragment length polymorphism of the pMJ101-like plasmid and ribotyping in the fish pathogen *Vibrio ordalii. Epidemiology and Infection* **117**, 386–391.

Pedersen, K., Tiainen, T. and Larsen, J.L. (1996c) Plasmid profiles, restriction length polymorphisms and O-serotypes among *Vibrio anguillarum* isolates. *Epidemiology and Infection* **117**, 471–478.

Pedersen, K., Dalsgaard, I. and Larsen, J.L. (1997a) *Vibrio damsela* associated with diseased fish in Denmark. *Applied and Environmental Microbiology* **63**, 3711–3715.

Pedersen, K., Gram, L., Austin, D.A. and Austin, B. (1997b) Pathogenicity of *Vibrio anguillarum* serogroup O1 compared to plasmids, outer membrane protein profile, and siderophore production. *Journal of Applied Microbiology* **82**, 365–371.

Pedersen, K., Verdonck, L., Austin, B., Austin, D.A., Blanch, A.R., Grimont, P.A.D., Jofre, J., Koblavi, S., Larsen, J.L., Tiainen, T., Vigneulle, M. and Swings, J. (1998) Taxonomic evidence that *Vibrio carchariae* Grimes *et al.* 1985 is a junior synonym of *Vibrio harveyi* (Johnson and Shunk 1936) Baumann *et al.* 1981. *International Journal of Systematic Bacteriology* **48**, 749–758.

Penner, J.L. (1984) Genus XII. *Providencia* Ewing 1962, 96[AL]. In: Krieg, N.R. and Holt, J.G. (eds), *Bergey's Manual of Systematic Bacteriology*, Vol. 1, Baltimore, Williams & Wilkins, p.494–496.

Perera, R.P., Johnson, S.K., Collins, M.D. and Lewis, D.H. (1994) *Streptococcus iniae* associated with mortality in *Tilapia nilotica* × *T. aurea* hybrids. *Journal of Aquatic Animal Health* **6**, 335–340.

Peréz, M.J., Rodríguez, L.A. and Nieto, T.P. (1998) The acetylcholinesterase ichthyotoxin is a common component of the extracellular products of Vibrionaceae strains. *Journal of Applied Microbiology* **84**, 47–52.

Petrie, J., Bruno, D.W. and Hastings, T.S. (1996) Isolation of *Yersinia ruckeri* from wild Atlantic salmon, *Salmo salar* L., in Scotland. *Bulletin of the European Association of Fish Pathologists* **16**, 83–84.

Phipps, B.M., Trust, T.J., Ishiguro, E.E. and Kay, W.W. (1983) Purification and characterization of the cell surface virulent A protein from *Aeromonas salmonicida*. *Biochemistry* **22**, 2934–2939.

Pickup, R.W., Rhodes, G., Cobban, R.J. and Clarke, K.J. (1996) The postponement of non-culturability in *Aeromonas salmonicida*. *Journal of Fish Diseases* **19**, 65–74.

Pier, G.B. and Madin, S.H. (1976) *Streptococcus iniae* sp. nov., a beta hemolytic stretococcus from an Amazon freshwater dolphon, *Inia geoffrensis*. *International Journal of Systematic Bacteriology* **26**, 545–553.

Pippy, J.H.C. (1969) Kidney disease in juvenile Atlantic salmon (*Salmo salar*) in the Margaree River. *Journal of the Fisheries Research Board of Canada* **26**, 2535–2437.

Pippy, J.H.C. and Hare, G.M. (1969) Relationship of river pollution to bacterial infection in salmon (*Salmo salar*) and suckers (*Catostomus commersoni*). *Transactions of the American Fisheries Society* **98**, 685–690.

Plehn, M. (1909) Die Furunculose—epidemie der Salmoniden in Suddeutschland. *Zentralblatt für Bakteriologie, Parasitenkunde, Infektionskrankheiten und Hygiene, Originale, Abteilung 1* **L11**, 468.

Plehn, M. (1911) Die Furunculose der Salmoniden. *Zentralblatt für Bakteriologie, Parasitenkunde, Infektionskrankheiten und Hygiene, Originale, Abteilung 1* **60**, 609–624.

Plumb, J.A. (1984) Immunization of five species of warm water fish against *Edwardsiella ictaluri*. *Journal of Fish Diseases* **6**, 261–266.

Plumb, J.A., Schachte, J.H., Gaines, J.L., Peltier, W. and Carrol, B. (1974) *Streptococcus* sp. from marine fishes along the Alabama and northwest Florida coast of the Gulf of Mexico. *Transactions of the American Fisheries Society* **103**, 358–361.

Plumb, J.A. and Sanchez, D.J. (1983) Susceptibility of five species of fish to *Edwardsiella ictaluri*. *Journal of Fish Diseases* **6**, 261–266.

Plumb, J.A. and Vinitnantharat, S. (1993) Vaccination of channel catfish, *Ictalurus punctatus* (Rafinesque), by immersion and oral booster against *Edwardsiella ictaluri*. *Journal of Fish Diseases* **16**, 65–71.

Popoff, M. (1969) Étude sur les *Aeromonas salmonicida*. I. Caractéres biochemiques et antigéniques. *Recherches Vétérinaire* **3**, 49–57.

Popoff, M. (1970) Bacterioses, generalities—*Aeromonas* et Aeromonose. Mimeograph, p.38. Institut national de la Recherche Agronomique.

Popoff, M. (1971a) Étude sur les *Aeromonas salmonicida*. II. Caracterisation des bactériophages actifs sur les '*Aeromonas salmonicida*' et lysoypie. *Annales de Recherches Vétérinaire* **2**, 33–45.

Popoff, M. (1971b) Interest diagnostique d'un bactériophage specifique des *Aeromonas salmonicida*. *Annales de Recherches Vétérinaire* **2**, 137–139.

Popoff, M. (1984) Genus III. *Aeromonas* Kluyver and van Niel 1936, 398[AL]. In: Krieg,

N.R. and Holt, J.G. (eds), *Bergey's Manual of Systematic Bacteriology*, Vol. 1. Baltimore, Williams & Wilkins, p.545–548.

Poppe, T.T., Håstein, T. and Salte, R. (1985) Hitra disease (haemorrhagic syndrome) in Norwegian salmon farming: present status. In: Ellis, A.E. (ed.), *Fish and Shellfish Pathology*. New York, Academic Press, p.223–229.

Pot, B., Devriese, L.A., Ursi, D., Vandamme, P., Haesebrouck, F. and Kersters, K. (1996) Phenotypic identification and differentiation of *Lactococcus* strains isolated from animals. *Systematic and Applied Microbiology* **19**, 213–222.

Powell, J.L. and Loutit, M.W. (1994a) Development of a DNA probe using differential hybridization to detect the fish pathogen *Vibrio anguillarum. Microbial Ecology* **28**, 365–373.

Powell, J.L. and Loutit, M.W. (1994b) The detection of the fish pathogen *Vibrio anguillarum* in water and fish using a species specific DNA probe combined with membrane filtration. *Microbial Ecology* **28**, 375–383.

Pratschner, G.A. (1978) The relative resistance of six transferrin phenotypes of coho salmon (*Oncorhynchus kisutch*) to cytophagosis, furunculosis, and vibriosis. M.Sc. thesis, University of Washington, Seattle.

Pybus, V., Loutit, M.W. and Tagg, J.R. (1994) Siderophore production by New Zealand strains of *Vibrio anguillarum. New Zealand Journal of Marine and Freshwater Research* **28**, 309–315.

Pychynski, T., Malanowska, T. and Kozlowski, M. (1981) Bacterial flora in branchionecrosis of carp (particularly *Bacillus cereus* and *Bacillus subtilis*). *Medycyna Weterynaryjna* **37**, 742–743.

Pyle, S.W. and Shotts, E.B. (1980) A new approach for differentiating flexibacteria isolated from cold-water and warm-water fish. *Canadian Journal of Fisheries and Aquatic Sciences* **37**, 1040–1042.

Pyle, S.W. and Shotts, E.B. (1981) DNA homology studies of selected flexibacteria associated with fish disease. *Canadian Journal of Fisheries and Aquatic Sciences* **38**, 146–151.

Raa, J., Rørstad, G., Engstad, R. and Robertsen, B. (1990) The use of immunostimulants to increase resistance of aquatic organisms to microbial infections. *Disease in Asian Aquaculture*. Bali, Indonesia.

Rabb, L., Cornick, J.W. and MacDermott, L.A. (1964) A macroscopic slide agglutination test for the presumptive diagnosis of furunculosis in fish. *Progressive Fish Culturist* **26**, 118–120.

Rahman, M.H., Kawai, K. and Kusuda, R. (1997) Virulence of starved *Aeromonas hydrophila* to cyprinid fish. *Fish Pathology* **32**, 163–168.

Rangdale, R.E. and Way, K. (1995) Rapid identification of *C. psychrophila* from infected spleen tissue using an enzyme-linked immunosorbent assay. *Bulletin of the European Association of Fish Pathologists* **15**, 213–216.

Rangdale, R.E., Richards, R.H. and Alderman, D.J. (1996) Isolation of *Cytophaga psychrophila*, causal agent of rainbow trout fry syndrome (RTFS) from reproductive fluids and egg surfaces of rainbow trout (*Oncorhynchus mykiss*). *Bulletin of the European Association of Fish Pathologists* **16**, 63–67.

Rangdale, R.E., Richards, R.H. and Alderman, D.J. (1997a) Minimum inhibitory concentrations of selected antimicrobial compound against *Flavobacterium psychrophilum* the causal agent of rainbow trout fry syndrome (RTFS). *Aquaculture* **158**, 193–201.

Rangdale, R.E., Richards, R.H. and Alderman, D.J. (1997b) Colonisation of eyed rainbow trout ova with *Flavobacterium psychrophilum* leads to rainbow trout fry syndrome in fry. *Bulletin of the European Association of Fish Pathologists* **17**, 108–111.

Ransom, D.P. (1978) Bacteriologic, immunologic and pathologic studies of *Vibrio* sp. pathogenic to salmonids. Ph.D. thesis, Oregon State University, Corvallis.

Ransom, D.P., Lannan, C.N., Rohovec, J.S. and Fryer, J.L. (1984) Comparison of histopathology caused by *Vibrio anguillarum* and *Vibrio ordalii* and three species of Pacific salmon. *Journal of Fish Diseases* **7**, 107–115.

Rashid, M.M., Honda, K., Nakai, T. and Muroga, K. (1994a) An ecological study of *Edwardsiella tarda* in flounder farms. *Fish Pathology* **29**, 221–227.

Rashid, M.M., Mekuchi, T., Nakai, T. and Muroga, K. (1994b) A serological study on *Edwardsiella tarda* strains isolated from diseased Japanese flounder (*Paralichthys olivaceus*). *Fish Pathology* **29**, 277.

Rasmussen, H.B. and Larsen, J.L. (1987) Further antigenic analyses of the fish pathogenic bacterium *Vibrio anguillarum*. *Current Microbiology* **16**, 145–148.

Real, F., Acosta, B., Déniz, S., Oros, J. and Rodriguez, E. (1994) *Aeromonas salmonicida* infection in *Sparus aurata* in the Canaries. *Bulletin of the European Association of Fish Pathologists* **14**, 153–155.

Reali, D., Pretti, C., Tavanti, L. and Cognetti-Varriale, A.M. (1997) *Pasteurella piscicida* (Janssen & Surgalla, 1964): a simple method of isolation and identification from rearing-ponds. *Bulletin of the European Association of Fish Pathologists* **17**, 51–53.

Reddacliff, G.L., Hornitsky, M., Carson, J., Petersen, R. and Zelski, R. (1993) Mortalities of goldfish, *Carassius auratus* (L.), associated with *Vibrio cholerae* (non-O1). *Journal of Fish Diseases* **16**, 517–520.

Rehnstam, A.-S., Norquist, A., Wolf-Watz, H. and Hagström, Å. (1989) Identification of *Vibrio anguillarum* in fish by using partial 16S rRNA sequences and a specific 16S rRNA oligonucleotide probe. *Applied and Environmental Microbiology* **55**, 1907–1910.

Reichenbach, H. (1989) Genus 1. *Cytophaga* Winogradsky 1929, 577[AL], emend. In: Staley, J.T., Bryant, M.P., Pfennig, N. and Holt, J.G. (eds), *Bergey's Manual of Systematic Bacteriology*, Vol. 3, Baltimore, Williams & Wilkins, p.2015–2020.

Report (1969) *Joint Committee on the Use of Antibiotics in Animal Husbandry and Veterinary Medicine (Swann Report)*, London, H.M.S.O.

Richard, C. (1984) Genus VI. *Enterobacter* Hormaeche and Edwards 1960, 7215[AL]; nom. Cons. Opin 28, Jud. Comm. 1963, 38. In: Krieg, N.R. and Holt, J.I. (eds), *Bergey's Manual of Systematic Bacteriology*, Vol. I., Baltimore, Williams & Wilkins, p.465–469.

Richards, R.H. and Roberts, R.J. (1978) Bacteriology of teleosts. In: Roberts R.J. (ed.) *Fish Pathology*, London, Baillière Tindall, p.183–204.

Rintamäki-Kinnunen, P., Bernadet, J.-F. and Bloigu, A. (1997) Yellow pigmented filamentous bacteria connected with farmed salmonid fish mortality. *Aquaculture* **149**, 1–14.

Roald, S.O. (1977) Effects of sublethal concentrations of lignosulphonates on growth, intestinal microflora and some digestive enzymes of rainbow trout (*Salmo gairdneri*). *Aquaculture* **12**, 327–335.

Roald, S.O. and Hastein, T. (1980) Infection with an acinetobacter-like bacterium in Atlantic salmon (*Salmo salar*) broodfish. In: Ahne, W. (ed.), *Fish Diseases, Third COPRAQ Session*. Berlin, Springer-Verlag, p.154–156.

Roberts, M.S. (1983) A report of an epizootic in hatchery rainbow trout, *Salmo gairdneri* Richardson, at an English trout farm, caused by *Yersinia ruckeri*. *Journal of Fish Diseases* **6**, 551–552.

Roberts, R.J. (1976) Bacterial diseases of farmed fishes. In: Skinner, F.A. and Carr, J.G. (eds), *Microbiology in Agriculture, Fisheries and Food*. London, Academic Press, p.55–61.

Roberts, R.J. (1978) *Fish Pathology*. London, Baillière Tindall.

Roberts, R.J. and Horne, M.T,. (1978) Bacterial meningitis in farmed rainbow trout, *Salmo gairdneri* Richardson, affected with chronic pancreatic necrosis. *Journal of Fish Diseases* **1**, 157–164.

Roberts, S.D. (1980) A method of reducing the carrier state of *Aeromonas salmonicida* in juvenile Pacific salmon. M.S. thesis, University of Idaho, USA.

Robertsen, B., Rørstad, G., Engstad, E. and Raa, J. (1990) Enhancement of non-specific disease resistance in Atlantic salmon, *Salmo salar* L., by a glucan from *Saccharomyces cerevisiae* cell walls. *Journal of Fish Diseases* **13**, 391–400.

Robertson, M.H., Clarke, I.R., Coghlan, J.D. and Gill, O.N. (1991) Leptospirosis in trout farmers. *Lancet* **ii**, 626–627.

Robinson, J.A. and Meyer, F.P. (1966) Streptococcal fish pathogen. *Journal of Bacteriology* **92**, 512.

Rockey, D.D., Fryer, J.L. and Rohovec, J.S. (1988) Separation and in vivo analysis of two extracellular proteases and the T-hemolysin from *Aeromonas salmonicida*. *Diseases of Aquatic Organisms* **5**, 197–204.

Rockey, D.D., Shook, L.A., Fryer, J.L. and Rohovec, J.S. (1989) Salmonid serum inhibits hemolytic activity of the secreted hemolysin of *Aeromonas salmonicida*. *Journal of Aquatic Animal Health* **1**, 263–268.

Rockey, D.D., Dungan, C.F., Lundar, T. and Rohovec, J.S. (1991) Monoclonal antibodies against *Aeromonas salmonicida* lipopolysacharide identify differences among strains. *Diseases of Aquatic Organisms* **10**, 115–120.

Rodger, H.D. and Drinan, E.M. (1993) Observation of a rickettsia-like organism in Atlantic salmon, *Salmo salar* L., in Ireland. *Journal of Fish Diseases* **16**, 361–369.

Rodgers, C.J. (1990) Immersion vaccination for control of fish furunculosis. *Diseases of Aquatic Organisms* **8**, 69–72.

Rodgers, C.J. (1992) Development of a selective-differential medium for the isolation of *Yersinia ruckeri* and its application in epidemiological studies. *Journal of Fish Diseases* **15**, 243–254.

Rodgers, C.J., Pringle, J.H., McCarthy, D.H. and Austin, B. (1981) Quantitative and qualitative studies of *Aeromonas salmonicida* bacteriophage. *Journal of General Microbiology* **125**, 335–345.

Rodgers, C.J. and Austin, B. (1982) Oxolinic acid for control of enteric redmouth disease in rainbow trout. *Veterinary Record* **112**, 83.

Rodgers, C.J. and Austin, B. (1985) Oral immunisation against furunculosis: an evaluation of two field trials. In: Manning, M.J. and Tatner, M.F. (eds), *Fish Immunology*. London, Academic Press, p.185–194.

Rodriguez, L.A., Fernandez, A.I.G., Santos, Y. and Nieto, T.P. (1990) Surface properties of *Serratia plymuthica* strains isolated from rainbow trout. In: Lésel, R. (ed.), *Microbiology in Poecilotherms*, Amsterdam, Elsevier, p.265–268.

Rodriguez, L.A., Ellis, A.E. and Nieto, T.P. (1993a) Effects of the acetylcholinesterase toxin of *Aeromonas hydrophila* on the central nervous system of fish. *Microbial Pathogenesis* **14**, 411–415.

Rodriguez, L.., Fernandez, A.I.G. and Nieto, T.P. (1993b) Production of the lethal acetylcholinesterase toxin by different *Aeromonas hydrophila* strains. *Journal of Fish Diseases* **16**, 73–78.

Rødsaether, M.C., Olafsen, J., Raa, J., Myhre, K. and Steen, J.B. (1977) Copper as an initiating factor of vibriosis (*Vibrio anguillarum*) in eel (*Anguilla anguilla*). *Journal of Fish Biology* **10**, 17–21.

Rogers, W.A. (1981) Serological detection of two species of *Edwardsiella* infecting catfish. *Developments in Biological Standardization* **49**, 169–172.

Rogosa, M., Franklin, J.G. and Perry, K.D. (1961) Correlation of the vitamin requirements with cultural and biochemical characters of *Lactobacillus* spp. *Journal of General Microbiology* **25**, 473–482.

Romalde, J.L., Conchas, R.F. and Toranzo, A.E. (1991) Evidence that *Yersinia ruckeri* possesses a high affinity iron uptake system. *FEMS Microbiology Letters* **80**, 121–126.

Romalde, J.L. and Toranzo, A.E. (1993) Pathological activities of *Yersinia ruckeri*, the enteric redmouth (ERM) bacterium. *FEMS Microbiology Letters* **112**, 291–300.

Romalde, J.L. Barja, J.L., Magariños, B. and Toranzo, A.E. (1994) Starvation-survival processes of the bacterial fish pathogen *Yersinia ruckeri*. *Systematic and Applied Microbiology* **17**, 161–168.

Romalde, J.L., Magariños, B., Fouz, B., Bandín, I., Núñez, S. and Toranzo, A.E. (1995) Evaluation of BIONOR Mono-kits for rapid detection of bacterial fish pathogens. *Diseases of Aquatic Organisms* **21**, 25–34.

Roode, M.C. (1977) *Streptococcus* sp. infection in rainbow trout. *Fish Farmer* **18**, 6–8.

Rose, A.S., Ellis, A.E. and Munro, A.L.S. (1989) The infectivity of different routes of exposure and shedding rates of *Aeromonas salmonicida* subsp. *salmonicida* in Atlantic salmon, *Salmo salar* L., held in sea water. *Journal of Fish Diseases* **12**, 573–578.

Rose, A.S., Ellis, A.E. and Munro, A.L.S. (1990a) Evidence against dormancy in the bacterial fish pathogen *Aeromonas salmonicida* subsp. *salmonicida*. *FEMS Microbiology Letters* **68**, 105–108.

Rose, A.S., Ellis, A.E. and Munro, A.L.S. (1990b) The survival of *Aeromonas salmonicida* subsp. *salmonicida* in seawater. *Journal of Fish Diseases* **13**, 205–214.

Ross, A.D. (1962) Isolation of a pigment-producing strain of *Aeromonas liquefaciens* from silver salmon (*Oncorhynchus kisutch*). *Journal of Bacteriology* **84**, 590–591.

Ross, A.D. (1966) Endotoxin studies. In: *Progress in Sport Fishery Research*. Resource Publication **39**, 77p.

Ross, A.J. and Johnson, H.E. (1962) Studies of transmission of mycobacterial infections of chinook salmon. *Progressive Fish Culturist* **24**, 147–149.

Ross, A.J., Rucker, R.R. and Ewing, W.H. (1966) Description of a bacterium associated with redmouth disease of rainbow trout (*Salmo gairdneri*). *Canadian Journal of Microbiology* **12**, 763–770.

Ross, A.J. and Brancato, F.P. (1959) *Mycobacterium fortuitum* Cruz from the tropical fish *Hyphessobrycon innesi*. *Journal of Bacteriology* **78**, 392–395.

Ross, A.J. and Klontz, G.W. (1965) Oral immunization of rainbow trout (*Salmo gairdneri*) against an etiological agent of 'Redmouth disease'. *Journal of the Fisheries Research Board of Canada* **22**, 713–719.

Ross, A.J. and Smith, C.A. (1972) Effect of two iodophors on bacterial and fungal fish pathogens. *Journal of the Fisheries Research Board of Canada* **29**,1359–1361.

Ross, A.J. and Toth, R.J. (1974) *Lactobacillus*—a new fish pathogen? *Progressive Fish Culturist* **36**, 191.

Rouhbakhsh-Khaleghdoust, A. (1975) The incidence of *Clostridium botulinum* type E in fish and bottom deposits in the Caspian Sea coastal water. *Pahlavi Medical Journal* **6**, 550–556.

Ruangpan, L., Kitao, T. and Yoshida, T. (1986) Protective efficacy of *Aeromonas hydrophila* vaccines in nile tilapia. *Veterinary Immunology and Immunopathology* **12**, 345–350.

Rucker, R.R. (1949) A streptomycete pathogenic to fish. *Journal of Bacteriology* **58**, 659–664.

Rucker, R.R. (1959) Vibrio infections among marine and fresh-water fish. *Progressive Fish Culturist* **21**, 22–25.

Rucker, R.R. (1963) Status of fish diseases and relation to production. *Report of the Second Governor's Conference on Pacific Salmon*, Seattle, January 1963, p.98–101.

Rucker, R.R. (1966) Redmouth disease of rainbow trout (*Salmo gairdneri*). *Bulletin de l'Office International des Epizooties* **65**, 825–830.

Rucker, R.R., Bernier, A.F., Whipple, W.J. and Burrows, R.E. (1951) Sulfadiazine for kidney disease. *Progressive Fish Culturist* **13**, 135–137.

Rucker, R.R., Earp, B.J. and Ordal, E.J. (1953) Infectious diseases of Pacific salmon. *Transactions of the American Fisheries Society* **83**, 297–312.

Runyon, E.H., Wayne, L.G. and Kubica, G.P. (1974) Genus 1. *Mycobacterium* Lehmann and Neumann 1896. In: Buchanan, R.E. and Gibbons, N.E. (eds), *Bergey's Manual of Determinative Bacteriology*, 8th edn. Baltimore, Williams & Wilkins, p.682–701.

Ryan, J.M. and Bryant, D.R. (1997) Fish tank granuloma—a frequently misdiagnosed infection of the upper limb. *Journal of Accident and Emergency Medicine* **14**, 398–399.

Saeed, M. (1983) Chemical characterization of the lipopolysaccharides of *Edwardsiella ictaluri* and the immune response of channel catfish to this function and to whole cell antigen with histopathological comparisons. Ph.D. dissertation, Auburn University, USA.

Saeed, M.O. (1995) Association of *Vibrio harveyi* with mortalities in cultured marine fish in Kuwait. *Aquaculture* **136**, 21–29.

Saeed, M.O., Alamoudi, M.M. and Al-Harbi, A.H. (1987) A *Pseudomonas* associated with disease in cultured rabbitfish *Siganus rivulatus* in the Red Sea. *Diseases in Aquatic Organisms* **3**, 177–180.

Saeed, M.O. and Plumb, J.A. (1987) Serological detection of *Edwardsiella ictaluri* Hawke lipopolysaccharide antibody in serum of channel catfish *Ictalurus punctatus* Rafinesque. *Journal of Fish Diseases* **10**, 205–209.

Sae-Oui, D., Muroga, K. and Nakai, T. (1984) A case of *Edwardsiella tarda* infection in cultured colored carp *Cyprinus carpio*. *Fish Pathology* **19**, 197–199.

Sakai, D.K. (1977) Causative factors of *Aeromonas salmonicida* in salmonid furunculosis: extracellular protease. *Scientific Reports of the Hokkaido Fish Hatchery* **32**, 61–89.

Sakai, D.K. (1978) Colliquative activity of purified protease for muscular tissue in *Aeromonas salmonicida* subsp. *salmonicida*. *Scientific Reports of the Hokkaido Fish Hatchery* **33**, 55–73.

Sakai, D.K. (1984) The non-specific activation of rainbow trout, *Salmo gairdneri* Richardson complement by *Aeromonas salmonicida* extracellular products and the correlation of complement activity with the inactivation of lethal toxicity products. *Journal of Fish Diseases* **7**, 329–338.

Sakai, D.K. (1985) Significance of extracellular protease for growth of a heterotrophic bacterium, *Aeromonas salmonicida*. *Applied and Environmental Microbiology* **50**, 1031–1037.

Sakai, D.K. (1986a) Kinetics of adhesion associated with net electrical charges in agglutinating *Aeromonas salmonicida* cells and their spheroplasts. *Bulletin of the Japanese Society of Scientific Fisheries* **52**, 31–36.

Sakai, D.K. (1986b) Electrostatic mechanism of survival of virulent *Aeromonas salmonicida* strains in river water. *Applied and Environmental Microbiology* **51**, 1343–1349.

Sakai, D.K. (1987) Adhesion of *Aeromonas salmonicida* strains associated with net electrostatic charges to host tissue cells. *Infection and Immunity* **55**, 704–710.

Sakai, M., Atsuta, S. and Kobayashi, M. (1986) A streptococcal disease of cultured Jacopever, *Sebastes schlegeli*. *Suisanzoshuku (Aquiculture)* **34**, 171–177.

Sakai, M., Kubota, R., Atsuta, S. and Kobayashi, M. (1987) Vaccination of rainbow trout *Salmo gairdneri* against β-haemolytic streptococcal disease. *Nippon Suisan Gakkaishi* **53**, 1373–1376.

Sakai, M., Atsuta, S. and Kobayashi, M. (1989a) *Pseudomonas fluorescens* isolated from the diseased rainbow trout, *Oncorhynchus mykiss*. *Kitasato Archives of Experimental Medicine* **62**, 157–162.

Sakai, M., Atsuta, S. and Kobayashi, M. (1989b) Comparison of methods used to detect *Renibacterium salmoninarum*, the causative agent of bacterial kidney disease. *Journal of Aquatic Animal Health* **1**, 21–24.

Sakai, M., Atsuta, S. and Kobayashi, M. (1989c) Bacterial kidney disease in Masu salmon, *Oncorhynchus masou*. *Physiology and Ecology, Japan, Special Volume* **1**, 577–586.

Sakai, M., Atsuta, S. and Kobayashi, M. (1989d) Protective immune response in rainbow trout *Oncorhynchus mykiss*, vaccinated with β-haemolytic streptococcal bacterin. *Fish Pathology* **24**, 169–173.

Sakai, M., Atsuta, S. and Kobayashi, M. (1989e) Attempted vaccination of rainbow trout *Oncorhynchus mykiss* against bacterial kidney disease. *Nippon Suisan Gakkaishi* **55**, 2105–2109.

Sakai, M., Sugawara, M., Atsuta, S. and Kobayashi, M. (1990) The confirmatory diagnosis of bacterial kidney disease (BKD) using dot and western blotting assay. *Bulletin of the European Association of Fish Pathologists* **10**, 77–80.

Sakai, M. and Kobayashi, M. (1992) Detection of *Renibacterium salmoninarum*, the causative agent of bacterial kidney disease in salmonid fish, from pen-cultured coho salmon. *Applied and Environmental Microbiology* **58**, 1061–1063.

Sakai, M., Otubo, T., Atsuta, S. and Kobayashi, M. (1993a) Enhancement of resistance to bacterial infections in rainbow trout, *Oncorhynchus mykiss* (Walbaum), by oral administration of bovine lactoferrin. *Journal of Fish Diseases* **16**, 239–247.

Sakai, M., Atsuta, S. and Kobayashi, M. (1993b) The immune response of rainbow trout (*Oncorhynchus mykiss*) injected with five *Renibacterium salmoninarum* bacterins. *Aquaculture* **113**, 11–18.

Sakai, M., Atsuta, S. and Kobayashi, M. (1994) Survival of fish pathogen *Edwardsiella tarda* in sea water and fresh water. *Bulletin of the European Association of Fish Pathologists* **14**, 188–190.

Sakai, M., Yoshida, T., Atsuta, S. and Kobayashi, M. (1995) Enhancement of resistance to vibriosis in rainbow trout, *Oncorhynchus mykiss* (Walbaum) by oral administration of *Clostridium butyricum* bacterin. *Journal of Fish Diseases* **18**, 187–190.

Sakata, T., Nakaji, M. and Kakimoto, D. (1978) Microflora in the digestive tract of marine fish—I. General characterization of the isolates from yellowtail. *Memoirs of the Faculty of Fisheries of Kagoshima University* **27**, 65–71.

Sakata, T., Sugita, H., Mitsuoka, T., Kakimoto, D. and Kadota, H. (1980) Isolation of obligate anaerobes from the intestinal tracts of freshwater fish. *Bulletin of the Japanese Society of Scientific Fisheries* **46**, 511.

Sakata, T., Matsuura, M. and Shimokawa, Y. (1989) Characteristics of *Vibrio damsela* isolated from diseased yellowtail *Seriola quinqueradiata*. *Nippon Suisan Gakkaishi* **55**, 135–141.

Sakazaki, R. (1984) Genus IV. *Citrobacter* Werkman and Gillen 1932, 17315[AL]. In: Krieg, N.R. and Holt, J.G. (eds), *Bergey's Manual of Systematic Bacteriology*, Vol. I., Baltimore, Williams & Wilkins, p.458–461.

Sakazaki, R. and Tamura, K. (1975) Priority of the specific epithet *anguillimortiferum* over the specific epithet *tarda* in the name of the organism presently known as *Edwardsiella tarda*. *International Journal of Systematic Bacteriology* **25**, 219–220.

Salati, F., Ikeda, Y. and Kusuda, R. (1987a) Effect of *Edwardsiella tarda* lipopolysaccharide immunization on phagocytosis in the eel. *Nippon Suisan Gakkaishi* **53**, 201–204.

Salati, F., Hamguchi, M. and Kusuda, R. (1987b) Immune response of red sea bream to *Edwardsiella tarda* antigens. *Fish Pathology* **22**, 93–98.

Salati, F., Watanabe, K., Kawai, K. and Kusuda, R. (1989a) Immune response of ayu against *Vibrio anguillarum* lipopolysaccharide. *Nippon Suisan Gakkaishi* **55**, 45–49.

Salati, F., Kawai, S. and Kusuda, R. (1989b) Characteristics of the lipopolysaccharide from *Pasteurella piscicida*. *Fish Pathology* **24**, 143–147.

Salinas, P.C. and Crosa, J.H. (1995) Regulation of *angR*, a gene with regulatory and biosynthetic functions in the pJM1 plasmid-mediated iron uptake system of *Vibrio anguillarum*. *Gene* **160**, 17–23.

Salte, R., Rørvik, K.-A., Reed, E. and Norberg, K. (1994) Winter ulcers of the skin in Atlantic salmon, *Salmo salar* L.: pathogenesis and possible aetiology. *Journal of Fish Diseases* **17**, 661–665.

Samuelsen, O.B., Hjeltnes, B. and Glette, J. (1998) Efficacy of orally administered florfenicol in the treatment of furunculosis in Atlantic salmon. *Journal of Aquatic Animal Health* **10**, 56–61.

Sanarelli, G. (1891) Über einen neuen Mikroorganisms des Wassers, welcher für Thiere mit veraenderlicher und konstanter Temperatur pathogen ist. *Zentralblatt für Bakteriologie, Parasitenkunde, Infektionskrankheiten and Hygiene* **9**, 193–228.

Sanders, J.E. and Fryer, J.L. (1978) *Corynebacterium salmoninus* sp. nov. the causative agent of bacterial kidney disease; selected biochemical properties and pathogenesis in salmonid fishes. *Proceedings of the Joint 3rd Biennial Fish Health Section/American Fisheries Society and 9th Annual Midwest Fish Disease Workshops, Kansas City,* p.28–33.

Sanders, J.E., Pilcher, K.S. and Fryer, J.L. (1978) Relation of water temperature to bacterial kidney disease in coho salmon (*Oncorhynchus kisutch*), sockeye salmon (*O. nerka*) and steelhead trout (*Salmo gairdneri*). *Journal of the Fisheries Research Board of Canada* **35**, 8–11.

Sanders, J.E. and Fryer, J.L. (1980) *Renibacterium salmoninarum* gen. nov., sp. nov., the causative agent of bacterial kidney disease in salmonid fishes. *International Journal of Systematic Bacteriology* **30**, 496–502.

Sandvick, O. and Hagan, O. (1968) Serological studies on proteinases produced by *Aeromonas salmonicida* and other aeromonads. *Acta Veterinaria Scandinavica* **9**, 1–9.

Sano, M., Nakano, H., Kimura, T. and Kusuda, R. (1994) Therapeutic effect of fosfomycin on experimentally induced pseudotuberculosis in yellowtail. *Fish Pathology* **29**, 187–192.

Santos, Y., Toranzo, A.E., Dopazo, C.P., Nieto, T.P. and Barja, J.L. (1987) Relationship among virulence for fish, enterotoxigenicity, and phenotypic characteristics of motile *Aeromonas*. *Aquaculture* **67**, 29–39.

Santos, Y., Toranzo, A.E., Barja, J.L., Nieto, T.P. and Villa, T.G. (1988) Virulence properties and enterotoxin production of *Aeromonas* strains isolated from fish. *Infection and Immunity* **56**, 3285–3293.

Santos, Y., Bandín, I., Núñez, S., Nieto, T.P. and Toranzo, A.E. (1991) Serotyping of motile *Aeromonas* species in relation to virulence phenotype. *Bulletin of the European Association of Fish Pathologists* **11**, 153–155.

Santos, Y., Romalde, J.L., Bandín, I., Magariños, B., Núñez, S., Barja, J.L. and Toranzo, A.E. (1993) Usefulness of the API-20E system for the identification of bacterial fish pathogens. *Aquaculture* **116**, 111–120.

Sanz, F. (1991) Rainbow trout mortalities associated with a mixed infection with

Citrobacter freundii and IPN virus. *Bulletin of the European Association of Fish Pathologists* **11**, 222.

Sarti, M., Giorgetti, G. and Manfrin, A. (1992) Method for the rapid diagnosis of visceral myxobacteriosis in reared trout in Italy. *Bulletin of the European Association of Fish Pathologists* **12**, 53.

Sasaki, T. and Aita, A. (1975) A study on fish diseases. Part 1. *Edwardsiella* isolated from the sea urchin. *Japanese Journal of Microbiology* **30**, 368.

Sato, N., Yamane, N. and Kawamura, T. (1982) Systemic *Citrobacter freundii* infection among sunfish *Mola mola* in Matsushima aquarium. *Bulletin of the Japanese Society of Scientific Fisheries* **48**, 1551–1557.

Sawyer, E.S. and Strout, R.G. (1977) Survival and growth in vaccinated, medicated and untreated coho salmon (*Oncorhynchus kisutch*) exposed to *Vibrio anguillarum*. *Aquaculture* **10**, 311–315.

Sawyer, E.S., Strout, R.G. and Coutermarsh, B.A. (1979) Comparative susceptibility of Atlantic salmon (*Salmo salar*) and coho (*Oncorhynchus kisutch*) salmon to three strains of *Vibrio anguillarum* from the Maine–New Hampshire coast. *Journal of the Fisheries Research Board of Canada* **36**, 280–282.

Scallan, A. and Smith, P.R. (1985) Control of asymptomatic carriers of *Aeromonas salmonicida* in Atlantic salmon smolts with flumequine. In: Ellis, A.E. (ed.), *Fish and Shellfish Pathology*. New York, Academic Press, p.119–127.

Schachte, J.T. (1978) Immunization of channel catfish, *Ictalurus punctatus*, against two bacterial diseases. *Marine Fisheries Review* **40**, 18–19.

Schachte, J.H. and Mora, E.C. (1973) Production of agglutinating antibodies in the channel catfish (*Ictalurus punctatus*) against *Chondrococcus columnaris*. *Journal of the Fisheries Research Board of Canada* **30**, 116–118.

Schäfer, J.-W., Alvarado, V., Enriquez, R. and Monrás, M. (1990) The 'coho salmon syndrome' (CSS): a new disease in Chilean salmon reared in sea water. *Bulletin of the European Association of Fish Pathologists* **10**, 130.

Schäperclaus, W. (1930) *Pseudomonas punctata* als Krankheitserreger bei Fischen. *Zeitung für Fischerei* **28**, 289–370.

Schäperclaus, W. (1959) Grossversuche mit Streptomycin zur Bekämpfung der infektiösen Bauchwassersucht des Karpfens. *Deutsches Fischeren Zeitung* **6**, 176–179.

Schäperclaus, W. (1979) *Fischkrakheiten*. Berlin, Akademie Verlag.

Schäperclaus, W. and Brauer, M. (1964) Bedeutung der Fluoreszenten für die Entstehung und Bekämpfung der Infektiösen Bauchwassersucht der Karpfen. *Zeitschrift für Fischerei* **12**, 75–76.

Schiewe, M.H. (1981) Taxonomic status of marine vibrios pathogenic for salmonid fish. *Developments in Biological Standardization* **49**, 149–158.

Schiewe, M.H., Crosa, J.H. and Ordal, E.J. (1977) Deoxyribonucleic acid relationships among marine vibrios pathogenic to fish. *Canadian Journal of Microbiology* **23**, 954–958.

Schiewe, M.H., Trust, T.J. and Crosa, J.H. (1981) *Vibrio ordalii* sp. nov.: a causative agent of vibriosis in fish. *Current Microbiology* **6**, 343–348.

Schmidtke, L.M. and Carson, J. (1994) Characteristics of *Vagococcus salmoninarum* isolated from diseased salmonid fish. *Journal of Applied Bacteriology* **77**, 229–236.

Schmidtke, L.M. and Carson, J. (1995) Characteristics of *Flexibacter psychrophilus* isolated from Atlantic salmon in Australia. *Diseases of Aquatic Organisms* **21**, 157–161.

Schneider, R. and Nicholson, B.L. (1980) Bacteria associated with fin rot disease in hatchery reared Atlantic salmon (*Salmo salar*). *Canadian Journal of Fisheries and Aquatic Sciences* **37**, 1505–1513.

Schrøder, M.B., Espelid, S. and Jørgense, T.Ø. (1992) Two serotypes of *Vibrio salmonicida* isolated from diseased cod (*Gadus morhua*); virulence, immunological studies and vaccination experiments. *Fish and Shellfish Immunology*.

Schubert, R.H.W. (1961) Über die biochemischen Merkmale von *Aeromonas salmonicida*. *Zentralblatt für Bakteriologie, Parasitenkunde, Infektionskrankheiten and Hygiene, 1. Originale* **183**, 485–494.

Schubert, R.H.W. (1967a) The taxonomy and nomenclature of the genus *Aeromonas* Kluyver and van Niel 1936. Part 1. Suggestions on the taxonomy and nomenclature of the aerogenic *Aeromonas* species. *International Journal of Systematic Bacteriology* **17**, 23–37.

Schubert, R.H.W. (1967b) The taxonomy and nomenclature of the genus *Aeromonas* Kluyver and van Niel 1936. Part II. Suggestions on the taxonomy and nomenclature of the anaerogenic *Aeromonas* species. *International Journal of Systematic Bacteriology* **17**, 237–279.

Schubert, R.H.W. (1974) Genus II. *Aeromonas* Kluyver and Van Niel 1936, 398. In: Buchanan, R.E. and Gibbons, N.E. (eds), *Bergey's Manual of Determinative Bacteriology*, 8th edn. Baltimore, Williams & Wilkins, p.345–348.

Scott, M. (1968) The pathogenicity of *Aeromonas salmonicida* (Griffin) in sea and brackish water. *Journal of General Microbiology* **50**, 864–868.

Sealey, W.M., Lim, C. and Klesius, P.H. (1997) Influence of the dietary level of iron from iron methionine and iron sulfate on immune response and resistance of channel catfish to *Edwardsiella ictaluri*. *Journal of the World Aquaculture Society* **28**, 142–149.

Sendra, R.M. Esteve, C. and Alcaide, E. (1997) Enzyme-linked immunosorbent assay for detection of *Aeromonas hydrophila* serogroup O:19. *FEMS Microbiology Letters* **157**, 123–129.

Shah, K.L. and Tyagi, B.C. (1986) An eye disease in silver carp, *Hypophthalmichthys molitrix*, held in tropical ponds, associated with the bacterium. *Staphylococcus aureus*. *Aquaculture* **55**, 1–4.

Sharma, V.K., Kaura, Y.K. and Singh, I.P. (1974) Frogs as carriers of *Salmonella* and *Edwardsiella*. *Antonie van Leeuwenhoek* **40**, 171–175.

Shaw, D.H. and Hodder, H.J. (1978) Lipopolysaccharides of the motile aeromonads; core aligosaccharide analysis as an aid to taxonomic classification. *Canadian Journal of Microbiology* **24**, 864–868.

Shayegani, M., Stone, W.B., DeForge, I., Root, T., Parsons, L.M. and Maupin, P. (1986)

Yersinia enterocolitica and related species isolated from wildlife in New York state. *Applied and Environmental Microbiology* **52**, 420–424.

Sheeran, B. and Smith, P.R. (1981) A second extracellular proteolytic activity associated with the fish pathogen *Aeromonas salmonicida*. *FEMS Microbiology Letters* **11**, 73–76.

Sheeran, B., Drinan, E. and Smith, P.R. (1984) Preliminary studies on the role of extracellular proteolytic enzymes in the pathogenesis of furunculosis. In: Acuigrup (ed.), *Fish Diseases, Fourth COPRAQ Session*. Madrid, Editora ATP, p.89–100.

Shewan, J.M. (1961) The microbiology of sea-water fish. In: Bergstöm, G. (ed.), *Fish as Food*, Vol. 1. London, Academic Press, p.487–560.

Shewan, J.M., Hobbs, G. and Hodgkiss, W. (1960) The *Pseudomonas* and *Achromobacter* groups of bacteria in the spoilage of marine white fish. *Journal of Applied Bacteriology* **23**, 463–468.

Shieh, H.S. (1985) Vaccination of Atlantic salmon, *Salmo salar* L., against furunculosis with protease from an avirulent strain of *Aeromonas salmonicida*. *Journal of Fish Biology* **27**, 97–101.

Shieh, H.S. (1989a) Blood-free media for the cultivation of the fish kidney disease bacterium, *Renibacterium salmoninarum*. *Microbios Letters*.

Shieh, H.S. (1989b) An extracellular toxin produced by the fish kidney disease bacterium, *Renibacterium salmoninarum*. *Microbios Letters*.

Shieh, H.S. and MacLean, J.R. (1975) Purification and properties of an extra-cellular protease of *Aeromonas salmonicida*, the causative agent of furunculosis. *International Journal of Biochemistry* **6**, 653–656.

Shinobu, R. (1965) Taxonomy of the whorl-forming *Streptomycetaceae*. *Memoirs of Osaka University Liberal Arts, Education and Natural Sciences* **14**, 72–201.

Shiomitsu, K., Kusuda, R., Osuga, H. and Munekiyo, M. (1980) Studies on chemotherapy of fish disease with erythromycin—II. Its clinical studies against streptococcal infection in cultured yellowtails. *Fish Pathology* **15**, 17–23.

Shiraki, K., Miyamoto, F., Sato, T., Sonezaki, I. and Sano, K. (1970) Studies on a new chemotherapeutic agent nifurprazine (HB-115) against infectious diseases. Part 1. *Fish Pathology* **4**, 130–137.

Shotts, E.B. and Rimler, R. (1973) Medium for the isolation of *Aeromonas hydrophila*. *Applied Microbiology* **26**, 550–553.

Shotts, E.B., Hsu, T.C. and Waltman, W.D. (1984) Extracellular proteolytic activity of *Aeromonas hydrophila* complex. *Fish Pathology* **20**, 37–44.

Shotts, E.B., Hsu, T.C. and Waltman, W.D. (1985) Extracellular proteolytic activity of *Aeromonas hydrophila* complex. *Fish Pathology* **20**, 37–44.

Shotts, E.B., Blazer, V.S. and Waltman, W.D. (1986) Pathogenesis of experimental *Edwardsiella ictaluri* infections in channel catfish (*Ictalurus punctatus*). *Canadian Journal of Fisheries and Aquatic Sciences* **43**, 36–42.

Shotts, E.B. and Waltman, W.D. (1990) A medium for the selective isolation of *Edwardsiella ictaluri*. *Journal of Wildlife Diseases* **26**, 214–218.

Shotts, Jr., E.B., Talkington, F.D., Elliot, D.G. and McCarthy, D.H. (1980) Aetiology of an ulcerative disease in goldfish, *Carassius auratus* (L.): characterization of the causative agent. *Journal of Fish Diseases* **3**, 181–186.

Siddall, R., Pike, A.W. and McVicar, A.H. (1994) Parasites of flatfish in relation to sewage dumping. *Journal of Fish Biology* **45**, 193–209.

Siegel, D.C. and Congleton, J.L. (1997) Bactericidal activity of juvenile chinook salmon macrophages against *Aeromonas salmonicida* after exposure to live or heat killed *Renibacterium salmoninarum* or to soluble proteins produced by *Renibacterium salmoninarum. Journal of Aquatic Animal Health* **9**, 180–189.

Simidu, U. and Egusa, S. (1972) A re-examination of the fish-pathogenic bacterium that has been reported as a *Pasteurella* species. *Bulletin of the Japanese Society of Scientific Fisheries* **38**, 803–812.

Simón, M., Mathes, A., Blanch, A. and Engelhardt, H. (1996) Characterization of a porin from the outer membrane of *Vibrio anguillarum. Journal of Bacteriology* **178**, 4182–4188.

Simón, M., Jofre, J. and Blanch, A.R. (1998) Evaluation of the immunospecificity of the porin Oml of *Vibrio anguillarum* serotype O1. *Journal of Applied Microbiology* **84**, 709–714.

Singer, J.T., Schmidt, K.A. and Reno, P.W. (1991) Polypeptides p40, pOM2, and pAngR are required for iron uptake and for virulence of the marine fish pathogen *Vibrio anguillarum 775. Journal of Bacteriology* **173**, 1347–1352.

Singleton, F.L., Attwell, R.W., Jangi, M.S. and Colwell, R.R. (1982a) Influence of salinity and organic nutrient concentration on survival and growth of *Vibrio cholerae* in aquatic microcosms. *Applied and Environmental Microbiology* **43**, 1080–1085.

Singleton, F.L., Attwell, R., Jangi, S. and Colwell, R.R. (1982b) Effects of temperature and salinity on *Vibrio cholerae* growth. *Applied and Environmental Microbiology* **44**, 1047–1058.

Siwicki, A.K., Klein, P., Morand, M., Kiczka, W. and Studnicka, M. (1998) Immunostimulatory effects of dimerized lysozyme (KLP-602) on the nonspecific defense mechanisms and protection against furunculosis in salmonids. *Veterinary Immunology and Immunopathology* **61**, 369–378.

Skarmeta, A.M., Bandín, I., Santos, Y. and Toranzo, A.E. (1995) *In vitro* killing of *Pasteurella piscicida* by fish macrophages. *Diseases of Aquatic Organisms* **23**, 51–57.

Skerman, V.B.D., McGowan, V. and Sneath, P.H.A. (1980) Approved Lists of Bacterial Names. *International Journal of Systematic Bacteriology* **30**, 225–420.

Skov, M.N., Pedersen, K. and Larsen, J.L. (1995) Comparison of pulsed-field gel electrophoresis, ribotyping, and plasmid profiling for typing of *Vibrio anguillarum* serovar O1. *Applied and Environmental Microbiology* **61**, 1540–1545.

Smibert, R.M. and Krieg, W.R. (1981) 20. General characterization. In: Gerhardt, P. (ed.), *Manual of Methods for General Bacteriology.* Washington, D.C., American Society of Microbiology, p.409–443.

Smith (1939) Cited by McFaddin, T.W. (1969).

Smith, H.L. and Goodner, I.K. (1958) Detection of bacterial gelatinases by gelatin-agar plate method. *Journal of Bacteriology* **76**, 662–665.

Smith, I.W. (1961) A disease of finnock due to *Vibrio anguillarum. Journal of General Microbiology* **74**, 247–252.

Smith, I.W. (1962) Furunculosis in kelts. Department of Agriculture and Fisheries for Scotland. *Freshwater and Salmon Fisheries Research* **27**, 1–5.

Smith, I.W. (1963) The classification of *Bacterium salmonicida*. *Journal of General Microbiology* **33**, 263–274.

Smith, I.W. (1964) The occurrence and pathology of Dee disease. *Department of Agriculture and Fisheries for Scotland, Freshwater Salmon Fisheries Research* **34**, 1–12.

Smith, P. and Davey, S. (1993) Evidence for the competitive exclusion of *Aeromonas salmonicida* from fish with stress-inducible furunculosis by a fluorescent pseudomonad. *Journal of Fish Diseases* **16**, 521–524.

Smith, P.A., Lannan, C.N., Garces, L.H., Jarpa, M., Larenas, J., Caswell-Reno, P., Whipple, M. and Fryer, J.L. (1995) Piscirickettsiosis: a bacterin field trial in coho salmon (*Oncorhynchus kisutch*). *Bulletin of the European Association of Fish Pathologists* **14**, 137–141.

Smith, P.D., McCarthy, D.H. and Paterson, W.D. (1980) Further studies on furunculosis vaccination. In: Ahne, W. (ed.), *Fish Diseases, Third COPRAQ Session*. Berlin, Springer-Verlag, p.113–119.

Smith, P.R. (1991) Stress test picks up furunculosis carriers. *Fish Farmer* **14**, 42.

Smith, P.R., Brazil, G.M., Drinan, E.M., O'Kelly, J., Palmer, R. and Scallan, A. (1982) Lateral transmission of furunculosis in sea water. *Bulletin of the European Association of Fish Pathologists* **3**, 41–42.

Smith, S.K., Sutton, D.C., Fuerst, J.A. and Reichelt, J.L. (1991) Evaluation of the genus *Listonella* and reassignment of *Listonella damsela* (Love et al) MacDonell and Colwell to the genus *Photobacterium* as *Photobacterium damsela* comb. nov. with an emended description. *International Journal of Systematic Bacteriology* **41**, 529–534.

Smith, W.W. (1942) Action of alkaline acriflavine solution on *Bacterium salmonicida* and trout eggs. *Proceedings of the Society of Experimental Biology and Medicine* **51**, 324–326.

Sneath, P.H.A. (1984) Genus *Chromobacterium* De Ley, Segers and Gillis 1978, 164[AL]. In: Krieg, N.R. and Holt, J.G. (eds), *Bergey's Manual of Systematic Bacteriology*, Vol. 1, Baltimore, Williams & Wilkins, p.376–377.

Snieszko, S.F. (1954) Therapy of bacterial fish diseases. *Transactions of the American Fisheries Society* **83**, 313–330.

Snieszko, S.F. (1957) Genus IV. *Aeromonas* Kluyver and van Niel 1936. In: Breed, R.S., Murray, E.G.D. and Smith, N.R. (eds), *Bergey' Manual of Determinative Bacteriology*, 7th edn. Baltimore, Williams & Wilkins, p.189–193.

Snieszko, S.F. (1958a) Fish furunculosis. *Fishery Leaflet* 467. United States Fish and Wildlife Service, Washington, D.C., 4p.

Snieszko, S.F. (1958b) Columnaris disease of fishes. *United States Department of the Interior, Fish and Wildlife Service, Fisheries Leaflet* 46, 1–3.

Snieszko, S.F. (1964) Remarks on some facts of epizootiology of bacterial fish diseases. *Developments in Industrial Microbiology* **5**, 97–100.

Snieszko, S.F. (1978) Control of fish diseases. *Marine Fisheries Review* **40**, 65–68.

Snieszko, S.F. and Friddle, S.B. (1949) Prophylaxis of furunculosis in brook trout (*Salvelinus fontinalis*) by oral immunization and sulphamerazine. *Progressive Fish Culturist* **1**, 161–168.

Snieszko, S.F., Griffin, P. and Friddle, S.B. (1950) A new bacterium (*Haemophilus piscium* n.sp.) from ulcer disease of trout. *Journal of Bacteriology* **59**, 699–710.

Snieszko, S.F. and Griffin, P.J. (1951) Successful treatment of ulcer disease in brook trout with terramycin. *Science* **112**, 717–718.

Snieszko, S.F. and Griffin, P.J. (1955) Kidney disease in brook trout and its treatment. *Progressive Fish Culturist* **17**, 3–13.

Snieszko, S.F. and Bullock, G.L. (1957) Treatment of sulfonamide-resistant furunculosis in trout and determination of drug sensitivity. *Fishery Bulletin* 125. United States Fish and Wildlife Service, Washington, D.C. p.555–564.

Snieszko, S.F., Dunbar, C.E. and Bullock, G.L. (1959) Resistance to ulcer disease and furunculosis in eastern brook trout, *Salvelinus fontinalis*. *Progressive Fish Culturist* **21**, 111–116.

Snieszko, S.F. and Hoffman, G.L. (1963) Control of fish diseases. *Laboratory Animal Care* **13**, 197–206.

Snieszko, S.F., Bullock, G.L., Hollis, E. and Boone, J.G. (1964a) *Pasteurella* sp. from an epizootic of white perch (*Roccus americanus*) in Chesapeake Bay tidewater areas. *Journal of Bacteriology* **88**, 1814–1815.

Snieszko, S.F., Bullock, G.L., Dunbar, C.E. and Pettijohn, L.L. (1964b) Nocardial infection in hatchery-reared fingerling rainbow trout (*Salmo gairdneri*). *Journal of Bacteriology* **88**, 1809–1810.

Soltani, M., Munday, B. and Carson, J. (1994) Susceptibility of some freshwater species of fish to infection by *Cytophaga johnsonae*. *Bulletin of the European Association of Fish Pathologists* **14**, 133–135.

Soltani, M., Shanker, S. and Munday, B.L. (1995) Chemotherapy of *Cytophaga/Flexibacter*-like bacteria (CFLB) infections in fish: studies validating clinical efficacies of selected antimicrobials. *Journal of Fish Diseases* **18**, 555–565.

Song, Y.L. and Kou, G.H. (1981) Immune response of eels (*Anguilla japonica*) against *Aeromonas hydrophila* and *Edwardsiella anguillimortiferum* (*E. tarda*) infection. *Proceedings of the Republic of China/U.S. Cooperative Science Seminar on Fish Diseases, National Science Council Series*, Vol. 3, p.107–115.

Song, Y.L., Kou, G.H. and Chen, K.Y. (1982) Vaccination conditions for the eel (*anguilla japonica*) with *Edwardsiella anguillimortiferum* bacterin. *CAPD (Taiwan) Fisheries Series No. 8, Fish Disease Research* **4**, 18–25.

Song, Y.L., Fryer, J.L. and Rohovec, J.S. (1988) Comparison of gliding bacteria isolated from fish in North America and other areas of the Pacific rim. *Fish Pathology* **23**, 197–202.

Sørensen, U.B.S. and Larsen, J.L. (1986) Serotyping of *Vibrio anguillarum*. *Applied and Environmental Microbiology* **51**, 593–597.

Sorimachi, M., Maeno, Y., Nakajima, K., Inouye, K. and Inui, Y. (193) Causative agent of jaundice of yellowtail, *Seriola quinqueradiata*. *Fish Pathology* **28**, 119–124.

Sørum, H., Hvaal, A.B., Heum, M., Daae, F.L. and Wiik, R. (1990) Plasmid profiling of *Vibrio salmonicida* for epidemiological studies of cold-water vibriosis in Atlantic salmon (*Salmo salar*) and cod (*Gadus morhua*). *Applied and Environmental Microbiology* **56**, 1033-1037.

Sørum, H., Myhr, E., Zwicker, B.M. and Lillehaug, A. (1993a) Comparison by plasmid profiling of *Vibrio salmonicida* strains isolated from diseased fish from different North European and Canadian coastal areas of the Atlantic Ocean. *Canadian Journal of Fisheries and Aquatic Sciences* **50**, 247-250.

Sørum, H., Kvello, J.H. and Håstein, T. (1993b) Occurrence and stability of plasmids in *Aeromonas salmonicida* ss. *salmonicida* isolated from salmonids with furunculosis. *Diseases of Aquatic Organisms* **16**, 199–206.

Sövényi, J.F., Yamamoto, H., Fujimoto, S. and Kusuda, R. (1990) Lymphomyeloid cells, susceptibility to erythrodermatitis of carp and bacterial antigens. *Developmental and Comparative Immunology* **14**, 185–200.

Speare, D.J. (1997) Differences in patterns of meningoencephalitis due to bacterial kidney disease in farmed Atlantic and chinook salmon. *Research in Veterinary Science* **62**, 79–80.

Speare, D.J., Brocklebank, J., Macnair, N. and Bernard, K.A. (1995) Experimental transmission of a salmonid *Rhodococcus* sp. isolate to juvenile Atlantic salmon, *Salmo salar* L. *Journal of Fish Diseases* **18**, 587–597.

Spence, K.D., Fryer, J.L. and Pilcher, K.S. (1965) Active and passive immunization of certain salmonid fishes against *Aeromonas salmonicida*. *Canadian Journal of Microbiology* **11**, 397–405.

Speyerer, P.D. and Boyle, J.A. (1987) The plasmid profile of *Edwardsiella ictaluri*. *Journal of Fish Diseases* **10**, 461–469.

Stackebrandt, E., Wehmeyer, U., Nader, H. and Fiedler, F. (1988) Phylogenetic relationship of the fish pathogenic *Renibacterium salmoninarum* to *Arthrobacter, Micrococcus* and related taxa. *FEMS Microbiology Letters* **50**, 117–120.

Stamm, J.M., Hanson, C.W., Chu, D.T.W., Bailer, R., Vojtko, C. and Fernandes, P.B. (1986) *In vitro* evaluation of A-56619 (Difloxacin) and A-56620: new arylfluoroquinolones. *Antimicrobial Agents and Chemotherapy* **29**, 193–200.

Stanier, R.Y. (1942) The *Cytophaga* group: a contribution to the biology of myxobacteria. *Bacteriological Reviews* **6**, 143–196.

Stanier, R.Y., Palleroni, M.J. and Doudoroff, M. (1966) The aerobic pseudomonads: a taxonomic study. *Journal of General Microbiology* **43**, 159–271.

Stanley, L.A., Hudson , J.S., Schwedler, T.E. and Hayasaki, S.S. (1994) Extracellular products associated with virulent and avirulent strains of *Edwardsiella ictaluri* from channel catfish. *Journal of Aquatic Animal Health* **6**, 36–43.

Starliper, C.E. (1996) Genetic diversity of North American isolates of *Renibacterium salmoninarum*. *Diseases of Aquatic Organisms* **27**, 207–213.

Starliper, C.E., Schill, W.B., Shotts, E.B. and Waltman, W.D. (1988) Isozyme analysis of *Edwardsiella ictaluri*. *Microbios Letters* **37**, 81–87.

Starliper, C.E., Shotts, E.B. and Brown, J. (1992) Isolation of *Carnobacterium piscicola* and an unidentified Gram-positive bacillus from sexually mature and post-spawning rainbow trout (*Oncorhynchus mykiss*). *Diseases of Aquatic Organisms* **13**, 181 187.

Starr, M.P. and Chatterjee, A.V. (1972) The genus *Erwinia*: enterobacteria pathogenic to plants and animals. *Annual Review of Microbiology* **26**, 289–426.

Steigerwalt, A.G., Fanning, G.R., Fife-Asbury, M.A. and Brenner, D.J. (1976) DNA relatedness among species of *Enterobacter* and *Serratia*. *Canadian Journal of Microbiology* **22**, 121–137.

Stevenson, L.H. (1978) A case for bacterial dormancy in aquatic systems. *Microbial Ecology* **4**, 127–133.

Stevenson, R.M.W. and Allan, B.J. (1981) Extracellular virulence factors in *Aeromonas hydrophila* disease processes in salmonids. *Developments in Biological Standardization* **49**, 173–180.

Stevenson, R.M.W. and Daly, J.G. (1982) Biochemical and serological characteristics of Ontario isolates of *Yersinia ruckeri*. *Canadian Journal of Fisheries and Aquatic Science* **39**, 870–876.

Stevenson, R.M.W. and Airdrie, D.E. (1984a) Serological variation among *Yersinia ruckeri* strains. *Journal of Fish Diseases* **7**, 247–254.

Stevenson, R.M.W. and Airdrie, D.E. (1984b) Isolation of *Yersinia ruckeri* bacteriophages. *Applied and Environmental Microbiology* **47**, 1201–1205.

Stewart, D.J., Waldemariam, K., Dear, G. and Mochaba, F.M. (1983) An outbreak of 'Sekiten-byo' among cultured European eels, *Anguilla anguilla* L., in Scotland. *Journal of Fish Diseases* **6**, 75–76.

Stoffregen, D.A., Chako, A.J., Backman, S. and Babish, J.G. (1993) Successful therapy of furunculosis in Atlantic salmon, *Salmo salar* L., using the fluoroquinolone antimicrobial agent enrofloxacin. *Journal of Fish Diseases* **16**, 219–228.

Stoffregen, D.A., Backman, S.C., Perham, R.E., Bowser, P.R. and Babish, J.G. (1996) Initial disease report of *Streptococcus iniae* infection in hybrid striped (sunshine) bass and successful therapeutic intervention with the fluoroquinolone antibacterial enrofloxacin. *Journal of the World Aquaculture Society* **27**, 420–434.

Strohl, W.R. and Tait, L.R. (1978) *Cytophaga aquatilis* sp. nov., a facultative anaerobe isolated from the gills of freshwater fish. *International Journal of Systematic Bacteriology* **28**, 293–303.

Stuart, C.A., van Stratum, E. and Rustigian, R. (1945) Further studies on urease production by *Proteus* and related organisms. *Journal of Bacteriology* **49**, 437.

Sugita, A. (1996) A case of streptococcicosis in dusky spinefoot. *Fish Pathology* **31**, 47–48.

Sugita, H., Matsuo, N., Hirose, Y., Iwato, M. and Deguchi, Y. (1997) *Vibrio* sp. strain NM10, isolated from the intestine of a Japanese coastal fish, has an inhibitory effect against *Pasteurella piscicida*. *Applied and Environmental Microbiology* **63**, 4986–4989.

Sugiyama, A. and Kusuda, R. (1981a) Studies on the characters of *Staphylococcus epidermidis* isolated from diseased fishes, Part II. Serological properties of the isolates. *Fish Pathology* **16**, 25–33.

Sugiyama, A. and Kusuda, R. (1981b) Studies on the characters of *Staphylococcus epidermidis* isolated from diseased fishes, Part II. A comparative study on the serological properties between the isolates and the strains of human origin. *Fish Pathology* **16**, 35–41.

Suprapto, H., Hara, T., Nakai, T. and Muroga, K. (1996) Purification of a lethal toxin of *Edwardsiella tarda*. *Fish Pathology* **31**, 203–207.

Sussman, A.S. and Halvorson, H.O. (1966) *Spores, their Dormancy and Germination.* New York, Harper & Row, 356 p.

Suzumoto, B.K., Schreck, C.B. and McIntyre, J.D. (1977) Relative resistance of three transferrin genotypes of coho salmon (*Oncorhynchus kisutch*) and their hematological responses to bacterial kidney disease. *Journal of the Fisheries Research Board of Canada* **34**, 1–8.

Sykes, J.B. (1976) *The Concise Oxford Dictionary of Current English*, 6th edn. Oxford, Oxford University Press.

Tajima, K., Yoshimizu, M., Ezura, Y. and Kimura, T. (1981) Studies on the causative organisms of vibriosis among the pen cultured coho salmon (*Oncorhynchus kisutch*) in Japan. *Bulletin of the Japanese Society of Scientific Fisheries* **47**, 35–42.

Tajima, K., Takahashi, T., Ezura, Y. and Kimura, T. (1984) Emzymatic properties of the purified extracellular protease of *Aeromonas salmonicida*, Ar-4 (EFDL). *Bulletin of the Japanese Society of Scientific Fisheries* **50**, 145–150.

Tajima, K., Ezura, Y. and Kimura, T. (1987a) The possible use of a thermolabile antigen in detection of *Vibrio anguillarum*. *Fish Pathology* **22**, 237–242.

Tajima, K., Ezura, Y. and Kimura, T. (1987b) Pathogenicity of a non-protease secreting strain of *Aeromonas salmonicida*. *Bulletin of the Faculty of Fisheries, Hokkaido University* **38**, 139–150.

Takemaru, I. and Kusuda, R. (1988) Chemotherapeutic effect of josamycin against natural streptococcal infection in cultured yellowtail. *Bulletin of the Japanese Society of Scientific Fisheries* **54**.

Tan, E., Low, K.W., Wong, W.S.F. and Leung, K.Y. (1998) Internalization of *Aeromonas hydrophila* by fish epidermal cells can be inhibited with a tyrosine kinase inhibitor. *Microbiology* **144**, 299–307.

Taniguchi, M. (1982a) Experiment on peroral inoculation via food to induce yellowtail streptococcicosis. *Bulletin of the Japanese Society of Scientific Fisheries* **48**, 1717–1720.

Taniguchi, M. (1982b) Influence of food condition on artificial peroral infection of yellowtail streptococcicosis. *Bulletin of the Japanese Society of Scientific Fisheries* **48**, 1721–1723.

Tatani, M., Muroga, K., Sugiyama, T. and Hiramoto, Y. (1985) Detection of *Vibrio anguillarum* from reared fry and fingerlings of ayu. *Aquaculture* **33**, 59–66.

Tatner, M.F. (1987) The quantitative relationship between vaccine dilution, length of immersion time and antigen uptake, using a radiolabelled *Aeromonas salmonicida* bath in direct immersion experiments with rainbow trout, *Salmo gairdneri*. *Aquaculture* **62**, 173–185.

Tatner, M.F. (1989) The antibody response of intact and short term thymectomised rainbow trout (*Salmo gairdneri*) to *Aeromonas salmonicida*. *Developmental and Comparative Immunology* **13**, 387.

Tatner, M.F., Johnson, C.M. and Horne, M.T. (1984) The tissue localization of *Aeromonas salmonicida* in rainbow trout, *Salmo gairdneri* Richardson, following three methods of administration. *Journal of Fish Biology* **25**, 95–108.

Tebbit, G.L., Erickson, J.D. and vande Water, R.B. (1981) Development and use of

Yersinia ruckeri bacterins to control enteric redmouth disease. *Developments in Biological Standardization* **49**, 395–401.

Teixeira, L.M., Merquior, V.L.C., Vianni, M. da C.E., Carvalho, M. da G.S., Fracalanzza, S.E.L., Steigerwalt, A.G., Brenner, D.J. and Facklam, R.R. (1996) Phenotypic and genotypic characterization of atypical *Lactococcus garvieae* strains isolated from water buffalos with subclinical mastitis and confirmation of *L. garvieae* as the senior subjective synonym of *Enterococcus seriolicida*. *International Journal of Systematic Bacteriology* **46**, 664–668.

Teshima, C., Kudo, S., Ohtani, Y. and Saito, A. (1992) Kidney pathology from the bacterium *Hafnia alvei*: experimental evidence. *Transactions of the American Fisheries Society* **121**, 599–607.

Teska, J.H., Shotts, E.B. and Hsu, T.C. (1989) Automated biochemical identification of bacterial fish pathogens using the Abbott Quantum II. *Journal of Wildlife Diseases* **25**, 103–107.

Teska, J.H., Twerdok, L.E., Beaman, J., Curry, M. and Finch, R.A. (1997) Isolation of *Mycobacterium abscessus* from Japanese Meduka. *Journal of Aquatic Animal Health* **9**, 234–238.

Teskeredzic, Grahek, D., Malnar, L., Teskeredzic, Z. and Hacmanjek, M. (1993) Bakterijska bolest americkog somica (*Amiurus nebulosus* L.). *Ribarstvo* **48**, 5–11.

Thornton, J.C., Garduño, R.A., Newman, S.G. and Kay, W.W. (1991) Surface-disorganized, attenuated mutants of *Aeromonas salmonicida* as furunculosis live vaccines. *Microbial Pathogenesis* **11**, 85–99.

Thornton, J.C., Garduño, R.A., Carlos, S.J. and Kay, W.W. (1993) Novel antigens expressed by *Aeromonas salmonicida* grown *in vivo*. *Infection and Immunity* **61**, 4582–4589.

Thornton, J.C., Garduño, R.A. and Kay, W.W. (1994) The development of live vaccines for furunculosis lacking the A-layer and O-antigen of *Aeromonas salmonicida*. *Journal of Fish Diseases* **17**, 195–204.

Thorsen, B.K., Enger, Ø., Norland, S. and Hoff, K.J. (1992) Long-term starvation survival of *Yersinia ruckeri* at different salinities studied by microscopical and flow cytometric methods. *Applied and Environmental Microbiology* **58**, 1624–1628.

Thune, R.L., Graham, T.E., Riddle, L.M. and Amborski, R.L. (1982a) Extracellular products and endotoxin from *Aeromonas hydrophila*: effect on age-0 channel catfish. *Transactions of the American Fisheries Society* **111**, 404–408.

Thune, R.L., Graham, T.E., Riddle, L.M. and Amborski, R.L. (1982b) Extracellular proteases from *Aeromonas hydrophila*: partial purification and effects on age-0 channel catfish. *Transactions of the American Fisheries Society* **111**, 749–754.

Thune, R.L., Johnson, M.C., Graham, T.E. and Amborski, R.L. (1986) *Aeromonas hydrophila* β-haemolysin: purification and examination of its role in virulence in O-group channel catfish, *Ictalurus punctatus* (Rafinesque). *Journal of Fish Diseases* **9**, 55–61.

Thune, R.L., Collins, R.A. and Peña, M.P. (1997) A comparison of immersion, immersion/oral combination and injection methods for the vaccination of channel

440 **Bibliography**

catfish *Ictalurus punctatus* against *Edwardsiella ictaluri. Journal of the World Aquaculture Society* **28**, 193–201.

Tiainen, T., Pedersen, K. and Larsen, J.L. (1995) Ribotyping and plasmid profiling of *Vibrio anguillarum* serovar O2 and *Vibrio ordalii. Journal of Applied Bacteriology* **73**, 384–392.

Tiainen, T., Pedersen, K. and Larsen, J.L. (1997) Immunological reactivity of *Vibrio anguillarum* sero-subgroups O2a and O2b, and comparison of their lipopolysaccharide profiles. *Current Microbiology* **34**, 38–42.

Tison, D.L., Nishibuchi, M., Greenwood, J.D. and Seidler, R.J. (1982) *Vibrio vulnificus* biogroup 2: a new biogroup pathogenic for eels. *Applied and Environmental Microbiology* **44**, 640–646.

Titball, R.W. and Munn, C.B. (1981) Evidence for two haemolytic activities from *Aeromonas salmonicida. FEMS Microbiology Letters* **12**, 27–30.

Titball, R.W. and Munn, C.B. (1983) Partial purification and properties of a haemolytic activity (T-lysin) from *Aeromonas salmonicida. FEMS Microbiology Letters* **20**, 207–210.

Titball, R.W. and Munn, C.B. (1985a) The purification and some properties of H-lysin from *Aeromonas salmonicida. Journal of General Microbiology* **131**, 1603–1609.

Titball, R.W. and Munn, C.B. (1985b) Interactions of extracellular products from *Aeromonas salmonicida.* In: Ellis, A.E. (ed.), *Fish and Shellfish Pathology.* London, Academic Press, p.61–68.

Todd, C. (1933) The presence of a bacteriophage for *B. salmonicida* in river waters. *Nature (London)* **131**, 360.

Tolmasky, M.E. and Crosa, J.H. (1984) Molecular cloning and expression of genetic determinants for the iron uptake system mediated by the *Vibrio anguillarum* plasmid pJM1. *Journal of Bacteriology* **160**, 860–866.

Tolmasky, M.E. and Crosa, J.H. (1995) Iron transport genes of the pJM1-mediated iron uptake system of *Vibrio anguillarum* are included in a transposon like structure. *Plasmid* **37**, 180–190.

Toranzo, A.E., Barja, J.L. and Hetrick, F.H. (1982) Survival of *Vibrio anguillarum* and *Pasteurella piscicida* in estuarine and freshwaters. *Bulletin of the European Association of Fish Pathologists* **3**, 43–45.

Toranzo, A.E., Barja, J.L., Colwell, R.R. and Hetrick, F.M. (1983a) Characterization of plasmids in bacterial fish pathogens. *Infection and Immunity* **39**, 184–192.

Toranzo, A.E., Barja, J.L., Potter, S.A., Colwell, R.R., Hetrick, F.M. and Crosa, J.H. (1983b) Molecular factors associated with virulence of marine vibrios isolated from striped bass in Chesapeake Bay. *Infection and Immunity* **39**, 1220–1227.

Toranzo, A.E., Santos, Y., Nieto, T.P. and Barja, J.L. (1986) Evaluation of different assay systems for identification of environmental *Aeromonas* strains. *Applied and Environmental Microbiology* **51**, 652–656.

Toranzo, A.E., Baya, A.M., Roberson, B.S., Barja, J.L., Grimes, D.J. and Hetrick, F.M. (1987) Specificity of slide agglutination test for detecting bacterial fish pathogens. *Aquaculture* **61**, 81–97.

Toranzo, A.E., Baya, A.M., Romalde, J.J. and Hetrick, F.M. (1989) Association of

Aeromonas sobria with mortalities of adult gizzard shad, *Dorosoma cepedianum* Lesueur. *Journal of Fish Diseases* **12**, 439–448.

Toranzo, A.E., Romalde, J.L., Núñez, S., Figueras, A. and Barja, J.L. (1993) An epizootic in farmed, market-size rainbow trout in Spain caused by a strain of *Carnobacterium piscicola* of unusual virulence. *Diseases of Aquatic Organisms* **17**, 87–99.

Toranzo, A.E., Devesa, S., Romalde, J.L., Lamas, J., Riaza, A., Leiro, J. and Barja, J.L. (1995) Efficacy of intraperitoneal vaccination and immersion vaccination against *Enterococcus* sp. infections in turbot. *Aquaculture* **134**, 17–27.

Touraki, M., Mourelatos, S., Karamanlidou, G., Kalaitzopoulou, S. and Kastritsis, C. (1996) Bioencapsulation of chemotherapeutics in *Artemia* as a means of prevention and treatment of infectious disease in marine fish fry. *Aquacultural Engineering* **15**, 133–147.

Toyama, T., Kita-Tsukamoto, K. and Wakabayashi, H. (1994) Identification of *Cytophaga psychrophila* by PCR targeted 16S ribosomal RNA. *Fish Pathology* **29**, 271–275.

Toyama, T., Tsukamoto, K.K. and Wakabayashi, H. (1996) Identification of *Flexibacter maritimus*, *Flavobacterium branchiophilum* and *Cytophaga columnaris* by PCR targeted 16 ribosomal DNA. *Fish Pathology* **31**, 25–33.

Traxler, G.S. and Li, M.F. (1972) *Vibrio anguillarum* isolated from a nasal abscess of the cod fish (*Gadus morhua*). *Journal of Wildlife Diseases* **8**, 207–214.

Treasurer, J. and Cox, D. (1991) The occurrence of *Aeromonas salmonicida* in wrasse (Labridae) and implications for Atlantic salmon farming. *Bulletin of the European Association of Fish Pathologists* **11**, 208–210.

Trüper, H.G. and de-Clari, L. (1997) Taxonomic note: Necessary correction of specific epithets formed as substantives (nouns) 'in apposition'. *International Journal of Systematic Bacteriology* **47**, 908–909.

Trust, T.J. (1975) Bacteria associated with the gills of salmonid fishes in freshwater. *Journal of Applied Bacteriology* **38**, 225–233.

Trust, T.J. and Sparrow, R.A.H. (1974) The bacterial flora in the alimentary tract of freshwater salmonid fishes. *Canadian Journal of Microbiology* **20**, 1219–1228.

Trust, T.J., Bull, L.M., Currie, B.R. and Buckley, J.T. (1979) Obligate anaerobic bacteria in the gastro-intestinal microflora of the grass carp, *Ctenopharyngodon idella*, goldfish *Carassius auratus* and rainbow trout *Salmo gairdneri*. *Journal of the Fisheries Research Board of Canada* **36**, 1174–1179.

Trust, T.J., Ishiguro, E.E. and Atkinson, H.M. (1980a) Relationship between *Haemophilus piscium* and *Aeromonas salmonicida* revealed by *Aeromonas hydrophila* bacteriophage. *FEMS Microbiology Letters* **9**, 199–201.

Trust, T.J., Khouri, A.G., Austen, R.A. and Ashburner, L.D. (1980b) First isolation in Australia of atypical *Aeromonas salmonicida*. *FEMS Microbiology Letters* **9**, 39–42.

Trust, T.J., Howard, P.S., Chamberlain, J.B., Ishiguro, E.E. and Buckley, J.T. (1980c) Additional surface protein in autoaggregating strains of atypical *Aeromonas salmonicida*. *FEMS Microbiology Letters* **9**, 35–38.

Trust, T.J., Courtice, I.D., Khouri, A.G., Crosa, J.H. and Schiewe, M.H. (1981) Serum resistance and haemagglutination ability of marine vibrios pathogenic for fish. *Infection and Immunity* **34**, 702–707.

Trust, T.J., Kay, W.W. and Ishiguro, E.E. (1983) Cell surface hydrophobicity and macrophage association of *Aeromonas salmonicida. Current Microbiology* **9**, 315–318.

Tsoumas, A., Alderman, D.J. and Rodgers, C.J. (1989) *Aeromonas salmonicida*: development of resistance to 4-quinolone antimicrobials. *Journal of Fish Diseases* **12**, 493–507.

Uddin, N., Chowdhury, B.R. and Wakabayashi, H. (1997) Optimum temperatures for the growth and protease production of *Aeromonas hydrophila. Fish Pathology* **32**, 117–120.

Udey, L.R. (1977) Pathogenic, antigenic and immunogenic properties of *Aeromonas salmonicida* studied in juvenile coho salmon (*Oncorhynchus kisutch*). Ph.D. thesis, Oregon State University, Corvallis.

Udey, L.R. (1982) A differential medium for distinguishing A/r^+ from A/r^- phenotypes in *Aeromonas salmonicida*. In: *Proceedings of the 13th Annual Conference and Workshop and 7th Eastern Fish Health Workshop.* Baltimore, International Association for Aquatic Animal Medicine, p.41.

Udey, L.R., Young, E. and Sallman, B. (1976) *Eubacterium* sp. ATCC 29255: an anaerobic bacterial pathogen of marine fish. *Fish Health News* **5**, 3–4.

Udey, L.R., Young, E. and Sallman, B. (1977) Isolation and characterization of an anaerobic bacterium, *Eubacterium tarantellus* sp. nov., associated with striped mullet (*Mugil cephalus*) mortality in Biscayne Bay, Florida. *Journal of the Fisheries Research Board of Canada* **34**, 402–409.

Udey, L.R. and Fryer, J.L. (1978) Immunization of fish with bacterins of *Aeromonas salmonicida. Marine Fisheries Review* **40**, 12–17.

Ugajin, M. (1979) Studies on the taxonomy of major microflora on the intestinal contents of salmonids. *Bulletin of the Japanese Society of Scientific Fisheries* **45**, 721–731.

Ugajin, M. (1981) Studies on *Streptococcus* sp. as a causal agent of an epizootic among the cultured ayu (*Plecoglossus altivelis*) in Tochigi Prefecture, Japan, 1980. *Fish Pathology* **16**, 119–127.

Ullah, M.A. and Arai, T. (1983a) Pathological activities of the naturally occurring strains of *Edwardsiella tarda. Fish Pathology* **18**, 65–70.

Ullah, M.A. and Arai, T. (1983b) Exotic substances produced by *Edwardsiella tarda. Fish Pathology* **18**, 71–75.

Umbreit, W.W. and Ordal, E.J. (1972) Infection of goldfish with *Vibrio anguillarum. American Society of Microbiology News* **32**, 93–96.

Umbreit, T.H. and Tripp, M.R. (1975) Characterization of the factors responsible for death of fish infected with *Vibrio anguillarum. Canadian Journal of Microbiology* **21**, 1272–1274.

Umelo, E. and Trust, T.J. (1997) Identification and molecular characterization of two randomly located flagellin genes from *Aeromonas salmonicida* A449. *Journal of Bacteriology* **179**, 5292–5299.

Umelo, E. and Trust, T.J. (1998) Physical map of the chromosome of *Aeromonas salmonicida* and genomic comparisons between *Aeromonas* strains. *Microbiology* **144**, 2141–2149.

Urakawa, H., Kita-Tsukamoto, K., Stevens, S.E., Ohwada, K. and Colwell, R.R. (1998) A proposal to transfer *Vibrio marinus* (Russell 1891) to a new genus *Moritella* gen. nov. as *Moritella marina* comb. nov. *FEMS Microbiology Letters* **165**, 373–378.

Uribe, J.C., Vial, M.V., Carvajal, J., Onate, R. and Teuber, C. (1995) Increasing prevalence of BD: an inadequate management during salmon transfer to seawater. *Medio Ambiente* **12**, 60–66.

Valdez, I.E. and Conroy, D.A. (1963) The study of a tuberculosis-like condition in neon tetras (*Hyphessobrycon innesi*) II. Characteristics of the bacterium isolated. *Microbiología española* **16**, 249–253.

Valla, S., Frydenlund, K., Coucheron, D.H., Haugan, K., Johansen, B., Jørgensen, T., Knudsen, G. and Strøm, A. (1992) Development of a gene transfer system for curing plasmids in the marine fish pathogen *Vibrio salmonicida*. *Applied and Environmental Microbiology* **58**, 1980–1985.

Van Alstine, J.M., Trust, T.J. and Brooks, D.E. (1986) Differential partition of virulent *Aeromonas salmonicida* and attenuated derivatives possessing specific cell surface alterations in polymer aqueous-phase systems. *Applied and Environmental Microbiology* **51**, 1309–1313.

Vandamme, P., Bernardet, J.-F., Segers, P., Kersters, K. and Holmes, B. (1994) New perspectives in the classification of the flavobacteria: description of *Chryseobacterium* gen. nov., *Bergeyella* gen. nov., and *Empedobacter* nom. rev. *International Journal of Systematic Bacteriology* **44**, 827–831.

Vandamme, P., Devriese, L.A., Kersters, B.P.K. and Melin, P. (1997) *Streptococcus difficile* is a nonhemolytic group B, type Ib streptococcus. *International Journal of Systematic Bacteriology* **47**, 81–85.

Vandepitte, J., VanDamme, L., Fofana, Y. and Desmyter, J. (1980) *Edwardsiella tarda* et *Plesiomonas shigelloides*. Leur rôle comme agents de diarrhées et leur épidemiologie. *Bulletin de Societé Pathologie Exotique* **73**, 139–149.

Vandepitte, J., Lemmens, P. and De Swert, L. (1983) Human edwardsiellosis tracted to ornamental fish. *Journal of Clinical Microbiology* **17**, 165–167.

Van Duijn, C. (1981) Tuberculosis in fishes. *Journal of Small Animal Practice* **22**, 391–411.

Vaughan, L.M., Smith, P.R. and Foster, T.J. (1993) An aromatic-dependent mutant of the fish pathogen *Aeromonas salmonicida* is attenuated in fish and is effective as a live vaccine against the salmonid disease furunculosis. *Infection and Immunity* **6**, 2172–2181.

Vedros, N.A. (1984) Genus I *Neisseria* Trevisan 1885, 105[AL]. In: Krieg, N.R. and Holt, J.G. (eds), *Bergey's Manual of Systematic Bacteriology*, Vol. 1, Baltimore, Williams & Wilkins, p.290–296.

Veenstra, J., Rietra, P.J.G.M., Stoutenbeek, C.P., Coster, J.M., de Gier, H.H.W. and DirksGo, S. (1992) Infection by an indole-negative variant of *Vibrio vulnificus* transmitted by eels. *Journal of Infectious Diseases* **166**, 209–210.

Velji, M.I., Albright, L.J. and Evelyn, T.P.T. (1990) Protective immunity in juvenile coho salmon *Oncorhynchus kisutch* following immunization with *Vibrio ordalii* lipolysaccharide or from exposure to live *V. ordalii* cells. *Diseases of Aquatic Organisms* **9**, 25–29.

Velji, M.I., Albright, L.J. and Evelyn, T.P.T. (1991) Nature of the immune response in coho salmon *Oncorhynchus kisutch* following vaccination with *Vibrio ordalii* lipolysaccharides by two different routes. *Diseases of Aquatic Organisms* **11**, 79–84.

Velji, M.I., Albright, L.J. and Evelyn, T.P.T. (1992) Immunogenicity of various *Vibrio ordalii* lipolysaccharide fractions in coho salmon *Oncorhynchus kisutch*. *Diseases of Aquatic Organisms* **12**, 97–101.

Vethaak, A.D. (1992) Diseases of flounder (*Platichthys flesus* L.) in the Dutch Wadden Sea, and their relation to stress factors. *Netherlands Journal of Sea Research* **29**, 257–272.

Vethaak, A.D. and ap Rheinallt, T. (1992) Fish disease as a monitor for marine pollution: the case of the North Sea. *Reviews in Fish Biology and Fisheries* **2**, 1–32.

Vethaak, A.D., Bucke, D., Lang, T., Wester, P., Johl, J. and Carr, M. (1992) Fish disease monitoring along a pollution transect: a case study using dab *Limanda limanda* in the German Bight, North Sea. *Marine Ecology Progress Series* **91**, 173–192.

Vethaak, A.D. and Jol, J.G. (1996) Diseases of flounder *Platichthys flesus* in Dutch coastal and estuarine waters, with particular reference to environmental stress factors. 1. Epizootiology of gross lesions. *Diseases of Aquatic Organisms* **26**, 81–97.

Vethaak, A.D., Jol, J.G., Meijboom, A., Eggens, M.L., Ap Rheinallt, T., Wester, P.W., Van De Zande, T., Bergman, A., Dankers, N., Ariese, F., Baan, R.A., Everts, J.M., Opperhuizen, A. and Marquenie, J.M. (1996) Skin and liver diseases induced in flounder (*Platichthys flesus*) after long-term exposure to contaminated sediments in large-scale mesocosms. *Environmental Health Perspectives* **104**, 1218–1229.

Vigneulle, M. (1990) Yersinose des salmonides: etude comparée de différents modes de vaccination. *Ichtyophysiologica Acta* **13**, 43–58.

Vigneulle, M. and Gérard, J.P. (1986) Incidence d'un apport polyvitaminique sur la yersiniose expérimentale de la truite arc-en-ciel (*Salmo gairdneri*). *Bulletin de l'Academie Vétérinaire de France* **59**, 77–86.

Vigneulle, M. and Baudin-Laurençin, F. (1995) *Serratia liquefaciens*: a case report in turbot (*Scophthalmus maximus*) cultured in floating cages in France. *Aquaculture* **132**, 121–124.

Vinitnantharat, S. and Plumb, J.A. (1993) Protection of channel catfish *Ictalurus punctatus* following natural exposure to *Edwardsiella ictaluri* and effects of feeding antigen on antibody titer. *Diseases of Aquatic Organisms* **15**, 31–34.

Vipond, R., Bricknell, I.R., Durant, E., Bowden, T.J., Ellis, A.E., Smith, M. and MacIntyre, S. (1998) Defined deletion mutants demonstrate that the major secreted toxins are not essential for the virulence of *Aeromonas salmonicida*. *Infection and Immunity* **66**, 1990–1998.

Vladik, P., Vitovac, J. and Carvinka, S. (1974) The taxonomy of gram-positive immobile diplobacilli isolated from necrotizing nephroses in American char and rainbow trout. *Veterinami Medicina (Praha)* **19**, 233–238.

Voigt, H.-R. (1994) Fish surveys in the Vaike Vain Strait between the islands of Saaremaa and Muhu, western Estonia. *Proceedings of the Estonian Academy of Science and Ecology* **4**, 128–135.

Von Betegh, L. (1910) Weitere Beiträge zur experimentellen Tuberkulose der

Meeresfischen. *Zentralblatt für Bakteriologie, Parasitenkunde, Infektionskrankheiten und Hygiene, Abteilung 1* **53**, 54.

Von Graevenitz, A. (1990) Revised nomenclature of *Campylobacter laridis, Enterobacter intermedium,* and '*Flavobacterium branchiophila*'. *International Journal of Systematic Bacteriology* **40**, 211.

Wakabayashi, H. and Egusa, S. (1966) Characteristics of a myxobacterium, *Chondrococcus columnaris,* isolated from diseased loaches. *Bulletin of the Japanese Society of Scientific Fisheries* **32**, 1015–1022.

Wakabayashi, H. and Egusa, S. (1972) Characteristics of a *Pseudomonas* sp. from an epizootic of pond-cultured eels (*Anguilla japonica*). *Bulletin of the Japanese Society of Scientific Fisheries* **38**, 577–587.

Wakabayashi, H. and Egusa, S. (1973) *Edwardsiella tarda* (*Paracolobactrum anguillimortiferum*) associated with pond-cultured eel diseases. *Bulletin of the Japanese Society of Scientific Fisheries* **39**, 931–939.

Wakabayashi, H., Egusa, S. and Fryer, J.L. (1980) Characteristics of filamentous bacteria isolated from the gills of salmonids. *Canadian Journal of Fisheries and Aquatic Sciences* **37**, 1499–1504.

Wakabayashi, H., Hikida, M. and Masumura, K. (1984) *Flexibacter* infections in cultured marine fish in Japan. *Helgoländer Meeresuntersuchungen* **37**, 587–593.

Wakabayashi, H., Hikida, M. and Masumura, K. (1986) *Flexibacter maritimus* sp. nov., a pathogen of marine fishes. *International Journal of Systematic Bacteriology* **36**, 396–398.

Wakabayashi, H., Huh, G.J. and Kimura, N. (1989) *Flavobacterium branchiophila* sp. nov., a causative agent of bacterial gill disease of freshwater fishes. *International Journal of Systematic Bacteriology* **39**, 213–216.

Wakabayashi, H., Toyama, T. and Iida, T. (1994) A study on serotyping of *Cytophaga psychrophila* isolated from fishes in Japan. *Fish Pathology* **29**, 101–104.

Wakabayashi, H., Sawada, K., Ninomiya, K. and Nishimori, E. (1996) Bacterial hemorrhagic ascites of ayu caused by *Pseudomonas* sp. *Fish Pathology* **31**, 239–240.

Wallbanks, S., Martinez-Murcia, A.J., Fryer, J.L., Phillips, B.A. and Collins, M.D. (1990) 16S rRNA sequence determination for members of the genus *Carnobacterium* and related lactic acid bacteria and description of *Vagococcus salmoninarum* sp. nov. *International Journal of Systematic Bacteriology* **40**, 224–230.

Walters, G.R. and Plumb, J.A. (1978) Modified oxidation/fermentation medium for use in the identification of bacterial fish pathogens. *Journal of the Fisheries Research Board of Canada* **35**, 1629–1630.

Waltman, W.D. and Shotts, E.B. (1984) A medium for the isolation and differentiation of *Yersinia ruckeri. Canadian Journal of Fisheries and Aquatic Sciences* **41**, 804–806.

Waltman, W.D., Shotts, E.B. and Blazer, V.S. (1985) Recovery of *Edwardsiella ictaluri* from Danio (*Danio devario*). *Aquaculture* **46**, 63–66.

Waltman, W.D. and Shotts, E.B. (1986a) Antimicrobial susceptibility of *Edwardsiella ictaluri. Journal of Wildlife Diseases* **22**, 173–177.

Waltman, W.D. and Shotts, E.B. (1986b) Antimicrobial susceptibility of *Edwardsiella tarda* from the United States and Taiwan. *Veterinary Microbiology* **12**, 277–282.

Waltman, W.D., Shotts, E.B. and Wooley, R.E. (1989) Development and transfer of plasmid-mediated antimicrobial resistance in *Edwardsiella ictaluri*. *Canadian Journal of Fisheries and Aquatic Sciences* **46**, 1114–1117.

Wang, W.-S., Chang, Y.-C., Shieh, M.-T. and Lin, C.-C. (1996) *Staphylococcus epidermidis* and cestode infection of cultured grass carp (*Ctenopharyngodon idella*) in Taiwan. *Reports on Fish Disease Research* **17**, 57–63.

Ward, P.D., Waters, C.A. and Sweeney, K.J. (1985) Autoagglutination of virulent *Aeromonas salmonicida* strains lacking additional surface layer. In: Ellis, A.E. (ed.), *Fish and Shellfish Pathology*. London, Academic Press, p.107–117.

Warren, J.W. (1963) Kidney disease of salmonid fishes and the analysis of hatchery waters. *Progressive Fish Culturist* **25**, 121–131.

Waterstrat, P., Ainsworth, J. and Capley, G. (1989) Use of an indirect enzyme-linked immunosorbent assay (ELISA) in the detection of channel catfish, *Ictalurus punctatus* (Rafinesque), antibodies to *Edwardsiella ictaluri*. *Journal of Fish Diseases* **12**, 87–94.

Weber, J.M. and Zwicker, B.M. (1979) *Aeromonas salmonicida* in Atlantic salmon (*Salmo salar*): occurrence of specific agglutinins to three bacterial pathogens. *Journal of the Fisheries Research Board of Canada* **36**, 1102–1107.

Wedemeyer, G.A. and Ross, A.J. (1973) Nutritional factors in the biochemical pathology of corynebacterial kidney disease in the coho salmon (*Oncorhynchus kisutch*). *Journal of the Fisheries Research Board of Canada* **30**, 296–298.

Weeks-Perkins, B.A. and Ellis, A.E. (1995) Chemotactic responses of Atlantic salmon (*Salmo salar*) macrophages to virulent and attenuated strains of *Aeromonas salmonicida*. *Fish and Shellfish Immunology* **5**, 313–323.

Weinstein, M.R., Litt, M., Kertesz, D.A., Wyper, P., Rose, D., Coulter, M., McGeer, A., Facklam, R., Ostach, C., Willey, B.M., Borczyk, A. and Low, D.E. (1997) Invasive infections due to a fish pathogen, *Streptococcus iniae*. *New England Journal of Medicine* **337**, 589–594.

Wells, N.A. and ZoBell, C.E. (1934) *Achromobacter ichthyodermis*, n. sp., the etiological agent of an infectious dermatitis of certain marine fishes. *Proceedings of the National Academy of Science, U.S.A.* **20**, 123–126.

Wenzel, S. Bach, R. and Müller-Prasuhn, G. (1971) Farmed trout as carriers of *Clostridium botulinum* and the cause of botulism—IV. Sources of contamination and contamination paths in fish farming and processing stations; ways to improve hygiene. *Archives für Lebensmittelhygiene* **22**, 131.

West, P.A. and Lee, J.V. (1982) Ecology of *Vibrio* species, including *Vibrio cholerae*, in natural waters of Kent, England. *Journal of Applied Bacteriology* **52**, 435–448.

West, P.A., Lee, J.V. and Bryant, T.N. (1983) A numerical taxonomic study of species of *Vibrio* isolated from the aquatic environment and birds in Kent, England. *Journal of Applied Bacteriology* **55**, 263–282.

Westfall, B.A. (1945) Coagulation film anoxia in fishes. *Ecology* **26**, 283–287.

Wiens, G.D. and Kaattari, S.L. (1991) Monoclonal antibody characterization of a leukoagglutinin produced by *Renibacterium salmoninarum*. *Infection and Immunity* **59**, 631–637.

Wiik, R. and Egidius, E. (1986) Genetic relationships of *Vibrio salmonicida* sp. nov. to other fish-pathogenic vibrios. *International Journal of Systematic Bacteriology* **36**, 521–523.

Wiik, R., Andersen, K., Daae, F.L. and Hoff, K.A. (1989) Virulence studies based on plasmid profiles of the fish pathogen *Vibrio salmonicida*. *Applied and Environmental Microbiology* **55**, 819–825.

Wiklund, T. (1995a) Survival of 'atypical' *Aeromonas salmonicida* in water and sediment microcosms of different salinities and temperatures. *Diseases of Aquatic Organisms* **21**, 137–143.

Wiklund, T. (1995b) Virulence of 'atypical' *Aeromonas salmonicida* isolated from ulcerated flounder *Platichthys flesus*. *Diseases of Aquatic Organisms* **21**, 145–150.

Wiklund, T., Sazonov, A., L'Iniova, G.P., Pugaewwa, V.P., Zoobaha, S.V. and Bylund, G. (1992) Characteristics of *Aeromonas salmonicida* subsp. *salmonicida* isolated from wild Pacific salmonids in Kamchatka, Russia. *Bulletin of the European Association of Fish Pathologists* **12**, 76–79.

Wiklund, T. and Bylund, G. (1993) Skin ulcer disease of flounder *Platichthys flesus* in the northern Baltic Sea. *Diseases of Aquatic Organisms* **17**, 165–174.

Wiklund, T., Lönnström, L. and Niiranen, H. (1993) *Aeromonas salmonicida* ssp. *salmonicida* lacking pigment production, isolated from farmed salmonids in Finland. *Diseases of Aquatic Organisms* **15**, 219–223.

Wiklund, T. and Lönnström, L. (1994) Occurrence of *Pseudomonas anguilliseptica* in Finnish fish farms during 1986–1991. *Aquaculture* **126**, 211–217.

Wiklund, T., Dalsgaard, I., Eerola, E. and Olivier, G. (1994) Characteristics of 'atypical' cytochrome-oxidase negative *Aeromonas salmonicida* isolated from ulcerated flounder *Platichthys flesus* (L.). *Journal of Applied Bacteriology* **76**, 511–520.

Wiklund, T. and Dalsgaard, I. (1995) Atypical *Aeromonas salmonicida* associated with ulcerated flatfish species in the Baltic sea and the North Sea. *Journal of Aquatic Animal Health* **7**, 218–224.

Wiklund, T. and Dalsgaard, I. (1998) Occurrence and significance of atypical *Aeromonas salmonicida* in non-salmonid and salmonid fish hosts. *Diseases of Aquatic Organisms* **32**, 49–69.

Williams, A.M., Fryer, J.L. and Collins, M.D. (1990) *Lactococcus piscium* sp. nov. a new *Lactococcus* species from salmonid fish. *FEMS Microbiology Letters* **68**, 109–114.

Williams, M.A. and Briggs, G.M. (1963) A new mineral mixture for experimental rat diets and evaluation of other mineral mixtures. *Federal Proceedings* **22**, 601–608.

Williams, P.J., Courtenay, S.C. and Vardy, C. (1997) Use of enrofloxacin to control atypical *Aeromonas salmonicida* in Atlantic tom cod. *Journal of Aquatic Animal Health* **9**, 216–222.

Williams, S.T., Locci, R., Vickers, J., Schofield, G.M., Sneath, P.H.A. and Mortimer, A.M. (1985) Probabilistic identification of *Streptoverticillium* species. *Journal of General Microbiology* **131**, 1681–1689.

Williamson, I.J.F. (1928) Furunculosis of the salmonidae. *Fisheries Board of Scotland, Salmon Fisheries* **5**, 1–17.

Williamson, I.J.F. (1929) A study of bacterial infection in fish and certain lower invertebrates. *Fisheries Board of Scotland, Salmon Fisheries* **11**, 3.

Winter, G.W., Schreck, C.B. and McIntyre, J.D. (1979) Resistance of different stocks and transferring genotypes of coho salmon, *Oncorhynchus kisutch*, and steelhead trout, *Salmo gairdneri*, to bacterial kidney disease and vibriosis. *Fishery Bulletin* **77**, 795–802.

Winton, J.R., Rohovec, J.S. and Fryer, J.L. (1983) Bacterial and viral diseases of cultured salmonids in the Pacific Northwest. In: Crosa, J.H. (ed.), *Bacterial and Viral Diseases of Fish: Molecular Studies, USA*, p.1–20.

Wise, D.J., Tomasso, J.R., Schwedler, T.E., Blazer, V.S. and Gatlin III, D.M. (1993) Effect of vitamin E on the immune response of channel catfish to *Edwardsiella ictaluri. Journal of Aquatic Animal Health* **5**, 183–188.

Wise, D.J., Schwedler, T.E. and Terhune, J.S. (1997) Uptake and clearance of *Edwardsiella ictaluri* in the peripheral blood of channel catfish *Ictalurus punctatus* fingerlings during immersion challenge. *Journal of the World Aquaculture Society* **28**, 45–51.

Withler, R.E. and Evelyn, T.P.T. (1990) Genetic variation in resistance to bacterial kidney disease within and between two strains of coho salmon from British Columbia. *Transactions of the American Fisheries Society* **119**, 1003–1009.

Witt, D. and Stackebrandt, E. (1990) Unification of the genera *Streptoverticillium* and *Streptomyces*, and amendation of *Streptomyces* Waksman and Henrici 1943, 339[AL]. *Systematic and Applied Microbiology* **13**, 361–371.

Wobeser, G. and Atton, F.M. (1973) An outbreak of columnaris disease in white suckers (*Catostomus commersoni*) in Saskatchewan. *Journal of the Fisheries Research Board of Canada* **30**, 681–683.

Wold, A. (1950) A promising drug … chloromycetin. *Aquarium*, September 1950.

Wolf, K. (1966) Bacterial kidney disease of salmonid fishes. U.S.A. Department of the Interior, Bureau of Sport Fisheries and Wildlife, Division of Fisheries Research, Washington, D.C. *Fish Disease Leaflet* No. 8.

Wolf, K. (1981) Chlamydia and rickettsia of fish. *Fish Health News* **10**, 1–5.

Wolf, K. and Dunbar, C.E. (1959) Test of 34 therapeutic agents for control of kidney disease in trout. *Transactions of the American Fisheries Society* **88**, 117–124.

Wolf, K. and Snieszko, S.F. (1963) Uses of antibiotics and other antimicrobials in therapy of diseases of fishes. *Antimicrobial Agents and Chemotherapy*, p.597–603.

Wolf, L.E. (1954) Development of disease resistant strains of fish. *Transactions of the American Fisheries Society* **8**, 342–349.

Wolf, M.K. and Crosa, J.H. (1986) Evidence for the role of a siderophore in promoting *Vibrio anguillarum* infections. *Journal of General Microbiology* **132**, 2949–2952.

Wolke, R.E. (1975) Pathology of bacterial and fungal diseases affecting fish. In: Ribelin, W.E. and Migaki, G. (eds), *The Pathology of Fishes*. Wisconsin, University of Wisconsin Press, p.76–78.

Wong, K.R., Green, M.J. and Buckley, J.T. (1989) Extracellular secretion of cloned aerolysin and phospholipase by *Aeromonas salmonicida. Journal of Bacteriology* **171**, 2523–2527.

Woo, N.Y.S., Ling, J.L.M. and Lo, K.M. (1995) Pathogenic *Vibrio* spp. in the sea bream, *Sparus sarba. Journal of Sun Yetsen University* **Supplement 3**, 192–193.

Wood, E.M. and Yasutake, W.T. (1956) Histopathology of kidney disease in fish. *American Journal of Pathology* **32**, 845–857.

Wood, J.W. (1968) *Diseases of Pacific Salmon: Their Prevention and Treatment.* State of Washington Department of Fisheries, Hatchery Division.

Wood, J.W. (1974) *Diseases of Pacific Salmon, their prevention and treatment.* Washington, Department of Fisheries.

Wood, J.W. and Ordal, E.J. (1958) Tuberculosis in Pacific salmon and steelhead trout. *Fish Commission of Oregon Control* **25**, 1–38.

Wood, J.W. and Wallis, J. (1955) Kidney disease in adult chinook salmon and its transmission by feeding to young chinook salmon. *Fisheries Commission of Oregon, Research Briefs* **6**, 32–40.

Wood, P.A., Wiens, G.D., Rohovec, J.S. and Rockey, D.D. (1995) Identification of an immunologically cross reactive 60-kilodalton *Renibacterium salmoninarum* protein distinct from p57: implications for immunodiagnostics. *Journal of Aquatic Animal Health*, **7**, 95–103.

Wood, P.A. and Kaattari, S.L. (1996) Enhanced immunogenicity of *Renibacterium salmoninarum* in chinook salmon after removal of the bacterial cell surface associated 57 kDa protein. *Diseases of Aquatic Organisms* **25**, 71–97.

Wood, S.C., McCashion, R.N. and Lynch, W.H. (1986) Multiple low-level antibiotic resistance in *Aeromonas salmonicida*. *Antimicrobial Agents and Chemotherapy* **29**, 992–996.

Woodall, A.N. and Laroche, G. (1964) Nutrition of salmonid fishes. XI. Iodide requirements of chinook salmon. *Journal of Nutrition* **824**, 475–482.

Wooster, G.A. and Bowser, P.R. (1996) The aerobiological pathway of a fish pathogen: survival and dissemination of *Aeromonas salmonicida* in aerosols and its implications in fish health management. *Journal of the World Aquaculture Society* **27**, 7–14.

Wyatt, L.E., Nickelson, R. and Van Derzant, C. (1979) *Edwardsiella tarda* in freshwater catfish and their environment. *Applied and Environmental Microbiology* **38**, 710–714.

Xu (1975) Studies on the gill diseases of the grass carp (*Ctenopharynogodon idelluls*). 1. Isolation of a myxobacterial pathogen. *Acta Hydrobiologica Sinica* **5**, 315–329.

Xu, H.-S., Roberts, N., Singleton, F.L., Attwell, R.W., Grimes, D.J. and Colwell, R.R. (1982) Survival and viability of non-culturable *Escherichia coli* and *Vibrio cholerae* in the estuarine and marine environment. *Microbial Ecology* **8**, 313–323.

Yamanoi, H., Muroga, K. and Takahashi, S. (1980) Physiological characteristics and pathogenicity of NAG vibrio isolated from diseased ayu. *Fish Pathology* **15**, 69–73.

Yasunaga, N., Hatai, K. and Tsukahara, J. (1983) *Pasteurella piscicida* from an epizootic of cultured red seabream. *Fish Pathology* **18**, 107–110.

Yasunobu, H., Muroga, K. and Maruyama, K. (1988) A bacteriological investigation on the mass mortalities of red seabream *Pagrus major* larvae with intestinal swelling. *Suisanzoshoku* **1**, 11–20.

Ye, J., Foo, R.W.T., Lo, K.M., Zeng, J.S., Ling, J.M.L., Woo, N.Y.S. and Xu, H.S. (1997) Studies on the pathogens of vibriosis in cultured sea bream (*Sparus sarba*) in Hong Kong. *Journal of Marine Science* (in press).

Yii, K.-C., Yang, T.I. and Lee, K.-K. (1997) Isolation and characterization of *Vibrio carchariae*, a causative agent of gastroenteritis in the groupers, *Epinephelus coioides. Current Microbiology* **35**, 109–115.

Yoshida, T., Kruger, R. and Inglis, V. (1995) Augmentation of non-specific protection in African catfish, *Clarias gariepinus* (Burchell), by the long-term oral administration of immunostimulants. *Journal of Fish Diseases* **18**, 195–198.

Yoshida, T., Endo, M., Sakai, M. and Inglis, V. (1997) A cell capsule with possible involvement in resistance to opsonophagocytosis in *Enterococcus seriolicida* isolated from yellowtail *Seriola quinqueradiata. Diseases of Aquatic Organisms* **29**, 233–235.

Yoshimizu, M., Ji, R., Nomura, T. and Kimura, T. (1987) A false positive reaction in the indirect fluorescent antibody test for *Renibacterium salmoninarum* ATCC 33209 caused by a *Pseudomonas* sp. *Scientific Report of the Hokkaido Salmon Hatchery* **41**, 121–127.

Young, C.L. and Chapman, G.B. (1978) Ultrastructural aspects of the causative agent and histopathology of bacterial kidney disease in brook trout (*Salvelinus fontinalis*). *Journal of the Fisheries Research Board of Canada* **35**, 1234–1248.

Zamora, J. and Enriquez, R. (1987) *Yersinia enterocolitica, Yersinia fredericksenii* and *Yersinia intermedia* in *Cyprinus carpio* (L.). *Journal of Veterinary Medicine B* **34**, 155–157.

Zhao, J. and Aoki, T. (1989) A specific DNA hybridization probe for detection of *Pasteurella piscicida. Diseases of Aquatic Organisms* **7**, 203–210.

Zimmermann, O.E.R. (1890) Die Bakterien unserer Trink- und Nutzwässer insbesondere der Wassers der Chemnitzer Wasseleitung. 1. Reihe. *Elfter Bericht der Naturwissenschaftlichen Gesellschaft zu Chemnitz* **1**, 38–39.

Zlotkin, A., Eldar, A., Ghittino, C. and Bercovier, H. (1998) Identification of *Lactococcus garvieae* by PCR. *Journal of Clinical Microbiology* **36**, 983–985.

Index of bacterial taxa

General index